Lecture Notes
Mathematics

Edited by A. Dold and B. Eckmann

1039

Analytic Functions Błażejewko 1982

Proceedings of a Conference
held in Błażejewko, Poland, August 19–27, 1982

Edited by J. Ławrynowicz

Springer-Verlag
Berlin Heidelberg New York Tokyo 1983

Editor

Julian Ławrynowicz
Institute of Mathematics of the Polish Academy of Sciences
Łódź Branch, Narutowicza 56, 90–136 Łódź, Poland

AMS Subject Classifications (1980): 20Hxx, 30-xx, 31-xx, 32-xx, 35-xx, 41-xx, 46-xx, 49-xx, 58-xx

ISBN 3-540-12712-7 Springer-Verlag Berlin Heidelberg New York Tokyo
ISBN 0-387-12712-7 Springer-Verlag New York Heidelberg Berlin Tokyo

Library of Congress Cataloging in Publication Data. Main entry under title: Analytic functions, Błażejewko 1982. (Lecture notes in mathematics; 1039) English and French. Selected papers from the 8th Conference on Analytic Functions organized by the Institute of Mathematics of the Polish Academy of Sciences and the Institute of Mathematics of Łódź University. 1. Analytic functions–Congresses. I. Ławrynowicz, Julian, 1939-. II. Conference on Analytic Functions (8th: 1982 : Błażejewko, Poland) III. Instytut Matematyczny (Polska Akademia Nauk) IV. Uniwersytet Łódzki. Instytut Matematyki. V. Series: Lecture notes in mathematics (Springer-Verlag); 1039.
QA3L28 no. 1039 [QA331] 510s [515.7] 83-20265
ISBN 0-387-12712-7 (U.S.)

Printing and binding: Beltz Offsetdruck, Hemsbach/Bergstr.
2146/3140-543210

Math

Sep

FOREWORD

These Proceedings contain selected papers from those submitted by a part of mathematicians lecturing at the 8th Conference on Analytic Functions held in Poland at Błażejewko (Lake District, Province of Poznań) during the eight days from August 19 to 27, 1982. These papers form the extended versions of their lectures.

According to the tradition of the preceding seven conferences (held in Łódź 1954, Lublin 1958, Kraków 1962, Łódź 1966, Lublin 1970, Kraków 1974, and Kozubnik 1979) the topics chosen are rather homogeneous. A considerable part of the papers is concerned with extremal methods and their applications to various branches of complex analysis: one and several complex variables, quasiconformal mappings and complex manifolds. This is however not a rule and the organizers decided to accept also papers on other subjects in complex analysis if they were of good quality.

The Organizing Committee consisted of: C. Andreian-Cazacu (Bucharest), Z. Charzyński (Łódź), P. Dolbeault (Paris), J. Eells (Coventry), A.A. Gončar (Moscow), J. Górski (Katowice), H. Grauert (Göttingen), L. Iliev (Sofia), S. Kobayashi (Berkeley), J. Krzyż (Lublin), O. Lehto (Helsinki), P. Lelong (Paris), J. Ławrynowicz (Łódź) - Chairman, S.N. Mergeljan (Erevan), J. Siciak (Kraków), W. Tutschke (Halle/Saale), and A. Marciniak (Łódź) - Secretary. The Conference was attended by 108 participants (70 from Poland) representing 14 countries.

The Conference was sponsored and organized by the Institute of Mathematics of the Polish Academy of Sciences in collaboration with the Institute of Mathematics of the Łódź University.

The Organizing Committee of the Conference expresses its gratitude to the Springer-Verlag for kind consent of publishing for the second time the Proceedings in the series "Lecture Notes in Mathematics".

Łódź, May 1983 Julian Ławrynowicz

C O N T E N T S

LIST OF SEMINARS HELD DURING THE CONFERENCE

O. TAMMI (Helsinki) [Chairman]: Seminar on extremal problems for analytic functions of one variable

C.O. KISELMAN (Uppsala) [Chairman]: Seminar on functions of several complex variables (including the theory of analytic functions in topological vector spaces)

M. OHTSUKA (Tokyo) [Chairman]: Seminar on quasiconformal mappings

P. DOLBEAULT (Paris) [Chairman]: Seminar on analysis on complex manifolds

During the seminars new problems were posed and discussed (see pp. 464-494.

LECTURES NOT INCLUDED IN THIS VOLUME
(* = one hour lecture)

L.A. AĬZENBERG (Krasnojarsk)*: Замечание к многомерному принципу Руше

V.V. ANDREEV (Sofia): Estimates of the divided difference of analytic functions

Cabiria ANDREIAN-CAZACU (Bucureşti)*: On interior mappings in the sense of Stoilov between Klein surfaces

B.N. APANASOV (Novosibirsk)*: On isomorphisms of Kleinian groups and supports of deformations

A. BAYOUMI (Uppsala): Weakly bounding subsets of some metric vector spaces

A. BOIVIN (Montréal): Meromorphic approximation on closed sets

Š.A. DAUTOV (Krasnojarsk): Весовые равномерные и интегральные оценки решения $\bar{\partial}$-задачи в строго псевдовыпуклых областях

J.T. DAVIDOV (Sofia): A note on the compactness principle

I.H. DIMOVSKI (Sofia): On two spectral problems in analytic function theory

L. DRUŻKOWSKI (Kraków): On Keller's Jacobian conjecture

R. DWILEWICZ (Warszawa)*: Some problems about Cauchy-Riemann functions

J. EELLS (Coventry)*: Stochastic differential equations on complex manifolds

J. FUKA (Praha): On the continuity of Faber's mapping

B. GAVEAU (Paris), J. ŁAWRYNOWICZ and L. WOJTCZAK (Łódź)*: On certain transformations of motion equations and of the corresponding manifolds

V.M. GOLDŠTEIN (Novosibirsk): Continuation of differentiable functions, capacity and quasiconformal mappings

N.A. GUSEVSKIĬ (Novosibirsk): On completions of the fundamental group of a compact negatively curved manifold

V.Ja. GUTLANSKIĬ (Donetsk): О параметрическом методе Лёвнера-Куфарева и экстремальных задачах для однолистных аналитических функций

F. HASLINGER (Wien): Bases in spaces of holomorphic functions

G. HUDAĬBERGANOV (Krasnojarsk): Некоторые замечания о полиномиальной выпуклости в $\mathbb{C}^n$

L.G. ILIEV (Sofia)*: Spline mit Laguerschen ganzen Funktionen

T. IWANIEC (Warszawa)*: Cauchy-Riemann operators in several real variables

P. JAKÓBCZAK (Kraków): Extension and the composition operators in products of strictly pseudoconvex sets

E. JANIEC (Łódź): A uniqueness theorem concerning bounded maps in the unit ball

M. JARNICKI (Kraków): Multiplicative linear functionals on some algebras of holomorphic functions with restricted growth

Burglind JÖRICKE (Berlin, GDR)*: The comparison of the modulus of continuity of an analytic function along the Šilov boundary and in the interior of a domain in $\mathbb{C}^n$

A.P. JUŽAKOV and M.A. MKRTGJAN (Krasnojarsk): On Laurent series of rational function in n-variables

A.P. JUŽAKOV and A.K. TSIH (Krasnojarsk): The properties of global residue with respect to polynomial mapping

Elena V. KARUPU (Kiev): О конечно-разностных локальных гладкостях конформных отображений

C.O. KISELMAN (Uppsala)*: On the growth of restrictions of plurisubharmonic functions

T. KRASIŃSKI (Łódź): Semi-norms on homology groups of complex manifolds

R. KÜHNAU (Halle an der Saale): Fredholmsche Eigenwerte und quasikonform fortsetzbare Abbildungen

A.M. KYTMANOV and M.A. MKRTGJAN (Krasnojarsk): On polynomial mappings with common zeros

E. LANCKAU (Karl-Marx-Stadt): Bergman operators for non-stationary processes in the plane

J. LEITERER (Berlin, GDR)*: The Penrose transform for bundles non-trivial on the general line

L. LEMPERT (Budapest): Imbedding pseudoconvex domains into a ball

P. LICZBERSKI (Łódź): Zu Differentialungleichungen für holomorphe Funktionen mehrerer komplexen Variablen

Agnieszka MACIEJKOWSKA (Lublin): On holomorphic continuation of certain univalent functions into the space $\mathbb{C}^2$

Andreana S. MADGUEROVA (Sofia): On the isomorphisms of some algebras of complex-valued functions on the torus

L.S. MAERGOJZ and E.I. JAKOVLEV (Krasnojarsk): О росте выпуклых и голоморфных функций многих переменных

R.P. MANANDHAR (Kathmandu): Post-Widder inversion operator of generalized functions

M. MATELJEVIĆ and M. PAVLOVIĆ (Beograd)*: On spaces of analytic functions with mixed norms

A.D. MEDNYH (Omsk): Branched coverings over a compact Riemann surfaces

I.P. MELNIČENKO (Kiev): Представление дифференцируемыми функциями потенциалов с осевой симметрией

O.K. MUŠKAROV (Sofia): Existence of almost holomorphic functions

M. NACINOVICH (Pisa)*: On the envelope of regularity for solutions of partial differential equations

NGUYEN Thanh Van (Toulouse)*: Caractérisation des mesures $\mu > 0$ sur K telles que (K,μ) vérifie (L^*) (K compact régulier de $\mathbb{C}^n$)

M. OHTSUKA (Tokyo)*: A theorem on extremal length

K. PETROV (Sofia): Bornological proof of the Oka-Weil theorem for holomorphic functions with values in Fréchet spaces

A. PIERZCHALSKI (Łódź)*: Transformations and deformations conformal on some distributions

W. PLEŚNIAK (Kraków)*: L -regularity condition in $\mathbb{C}^N$

A. PŁOSKI (Kielce): Sur les valeurs critiques des applications analytiques dans le plan

I.P. RAMADANOV (Sofia): On some extremal problems of analytic functions of several variables

J. RIIHENTAUS (Oulu): On the extension of holomorphic and meromorphic functions

Aleksandra ROST, Janina ŚLADKOWSKA-ZAHORSKA et R. TARGOSZ (Gliwice): Les inegalités du type de Grunsky pour les pairs d'Aharonov et de Guelfer

K. RUSEK (Kraków): Remarks on Keller's problem

Irena RUSZCZYK (Kielce): О некоторых классах регулярных функций двух перемменных

M. SAKAI (Tokyo): Applications of variational inequalities to the existence theorem on quadrature domains

J. SICIAK (Kraków)*: Pluripolar sets and capacities in $\mathbb{C}^N$

Maria SKOWIERŻAK (Kielce): Экстремальные задачи для некоторых классов функций двух комплексных переменных

Z. SŁODKOWSKI (Warszawa): On analytic set-valued functions

SUNG Chen-han (Notre Dame, IM)*: A refined defect relation for holomorphic mappings

Anna SZYNAL and J. SZYNAL (Lublin): The extension of Jenkins inequality

P.M. TAMRAZOV (Kiev): Holomorphic functions of one and of several complex variables: contour-and-solid properties, finite-difference smoothnesses and approximation

N.N. TARHANOV (Krasnojarsk): Grothendieck's duality theorem for elliptic complexes

T.V. TONEV (Sofia): Generalized-analytic coverings in the spectrum of a uniform algebra

S. TOPPILA (Helsinki): On the spherical derivative of a meromorphic function

Ju.Ju. TROHIMČUK (Kiev): Одна теорема об отображениях с постоянным растяжением

G. TSAGAS (Thessaloniki): The geometry of a homogeneous bounded domain

A.K. TSIH (Krasnojarsk): Локальные вычеты в $\mathbb{C}^n$ и теорема Нётера

W. TUTSCHKE (Halle an der Saale)*: Solution of initial-value problems in classes of generalized analytic functions

A. VAZ FERREIRA (Bologna): Characterizing holomorphic function algebras in the C^∞-class

J-L. VERDIER (Paris)*: Théorie de Yang-Mills en dimension 2

E. VESENTINI (Pisa)*: Idempotents and fixed points

M. VUORINEN (Helsinki): On Dirichlet finite functions

CONDITION OF CONFORMAL RIGIDITY OF HYPERBOLIC MANIFOLDS WITH BOUNDARIES

Boris Nikolaevič Apanasov (Novosibirsk)

Contents

1. Introduction

Let M be an n-dimensional manifold with boundary ∂M, such that int M is a hyperbolic manifold (of infinite volume), and let there be assigned a quasiconformal mapping f on a similar manifold M' whose contraction upon the boundary ∂M is conformal. Will the manifolds M and M' be isometric ? This problem generalizing the rigidity problem of hyperbolic manifolds without boundary (see [1-5]), was set up in the framework of the theory of deformations of Kleinian groups by Bers [6] and Kruškal [7-9]: will Kleinian groups on n-dimensional sphere S^n conjugate in the Möbius group M_n if they are conjugated by a quasiconformal homeomorphism f of the sphere, f being conformal on a discontinuity set ? The author has shown [10,11] that without imposing additional constrains this problem has a negative solution; besides, from the proposed proof it is clear that one cannot remove the quasiconformality condition of the cojugating mapping f (or, rather, the condition of keeping the measure on S^n by the mapping f).

For a Riemannian manifold X and a fixed point $p \in X$ we denote by $X(r)$ its submanifold consisting of points removed from p for the distance $\leq r$, and by $V(r,n)$ denote the volume of a ball of radius r in n-dimensional hyperbolic space. Then the main result of the paper can be formulated as (cf. [12,15]):

THEOREM A. *Let* M *and* M' *be n-dimensional manifolds with boundaries, whose interiors are hyperbolic manifolds;* $f: M \to M'$ *is a quasi-conformal mapping, conformal on the boundary* ∂M; $X_s \subset M$ *is the s--neighbourhood of the minimal convex retract* X *of the manifold* M. *If for some* $s > 0$

$$(1.1) \quad \lim_{r\to\infty}[\mathrm{Vol}\, X_s(r)/V(r,n)] = 0$$

then manifolds M *and* M' *are isometric.*

The main points of the proof of this theorem are the description of ergodic properties of discrete Möbius groups by Sullivan [5] (see Section 3) and the ideas close to [13] and Ch. 6 of [15].

2. Convex retracts in manifolds and the geometry of fundamental polyhedra

If on some M^n a hyperbolic structure is introduced, then we denote by G the image of the fundamental group $\pi_1(M^n)$ when mapping the holonomy H

$$(2.1) \quad H : \pi_1(M^n) \to G \subset \mathrm{Isom}\, H^n = M_{n-1}.$$

The group G is a discrete group of hyperbolic isometries. But if M^n is also the interior of some manifold with boundary, then G acts discontinuously on the sphere $S^n = \partial H^n$, i.e. it is the Kleinian group on S^n. Then manifold M^n is restored by factorizing the space H^n by the group G.

Definition 2.1. A convex (in hyperbolic geometry) domain of the limit set $L(G)$ of the group G, i.e. the set

$$(2.2) \quad H_G = \cap(Q \subset \overline{H^n} : L(G) \subset Q \text{ and } Q \text{ is convex})$$

is called the *convex Nielsen domain* of the discrete group $G \subset \mathrm{Isom}\, H^n$.

Except the groups which are the continuation of Fuchsian groups from R^{n-1} and have, as a Nielsen domain, a subset on some $(n-1)$-dimensional hyperbolic plane, the Nielsen domain ∂H_G of a discrete group G in H^n has dimension n. Its boundary ∂H_G consists of geodesics whose infinitely removed ends are the ends of Euclidean intervals in R^{n-1} lying in the discontinuity set $O(G)$. Therefore, ∂H_G is developed in a hyperbolic plane of dimension $\leq n-1$. In terms of Riemannian geometry the outer curvature of ∂H_G equals 0, and the inner curvature (sectional curvature of the space H^n) equals -1.

The group G leaves the domain H_G invariant and on its subset $H_G - L(G)$ acts discontinuously. The space of orbits of the group G on this set is the convex hyperbolic manifold denoted by M_G:

$$(2.3) \quad M_G = [H_G - L(G)]/G.$$

If the group G acts on $\partial H^n = S^n$ discontinuously, then the manifold M_G has boundary on which the hyperbolic structure is introduced; this structure is induced by the hyperbolic metric of the space H^n. The manifold M_G is a natural retract of the manifold $M(G) = (\overline{H^n} - L(G))/G$ whose interior is a hyperbolic manifold. This retraction $r: M(G) \to M_G$ is induced by the retraction

$$(2.4) \quad \bar{r}: \overline{H^n} \to H_G$$

defined as follows:

$$(2.5) \quad \bar{r}(x) = x \quad \text{for} \quad x \in H_G, \quad \bar{r}(x) = x_0 \quad \text{for} \quad x \in \overline{H^n} - H_G.$$

Here x_0 is the nearest point to x from the domain H_G in the case where $x \in H^n - H_G$; but if the point x is taken from $\partial H^n - L(G)$, then as the nearest point x_0 one takes the first point of contact with H_G of the horosphere with the centre in the point x. This definition is correct due to strict convexity in the hyperbolic geometry of a ball and a horoball.

Note that the limit set is the minimal closed set on sphere S^{n-1}, which is invariant with respect to the action of the group G. Hence, the manifold M_G is the minimal convex retract of the manifold $M(G)$.

By a <u>convex</u> <u>fundamental</u> <u>polyhedron</u> P of the discrete group $G \subset \mathrm{Isom}\, H^n$ we mean a polyhedron with the following properties:

1. P is an open domain in H^n; it is the intersection of no more than a countable family of hyperbolic half-spaces Q_i with boundary planes S_i; the intersection $\bar{P} \cap S_i$ is said to be a <u>side</u> of P.

2. Every compact set in H^n only intersects a finite number of sides of P (the boundary P in H^n only consists of sides).

3. P does not contain G-equivalent points and the images of its closure cover the whole discontinuity set in $\overline{H^n}$.

4. Sides of P are identified pairwise by elements of the group.

5. Every point in H^n has a neigbourhood which only intersects a finite number of images $g(P)$, $g \in G$.

The latter condition (the property of the local finiteness) in dimension $n \geq 3$ does not follow from the former ones. This was shown by Tetenov [14]. He also obtained sufficient conditions of local finiteness. In particular, the Dirichlet polyhedron is of such a kind.

Also the aforesaid about the minimal convex retract of the manifold M can be formulated as follows:

LEMMA 2.2. The minimal convex retract of the manifold M, whose interior is a hyperbolic manifold, is obtained by identifying G-equivalent sides of the polyhedron

$$(2.6) \quad P_H = H_G \cap [\bar{P} - L(G)],$$

where $P \subset H^n$ is a convex fundamental polyhedron of the group $G = H[\pi_1(M)]$ of hyperbolic isometries.

3. Ergodic properties of discrete Möbius groups

In this section we describe some results of Sullivan [5] we need further.

The action of the discrete group $G \subset \mathrm{Isom}\, H^n$ in the sphere $S^{n-1} = \partial H^n$ to which the set of zero measure is divided into two parts - dissipative and conservative. The dissipative part is the union of pairwise intersecting measurable sets represented by the elements of G (the analogy of the action in H^n). The conservative part K is characterized by the fact that for any subset $Y \subset K$, $m_{n-1}(Y) > 0$, there exists a sequence of distinct elements $g_i \in G$ such that for all the numbers i

$$m_{n-1}[Y \cap g_i(Y)] > 0.$$

D e f i n i t i o n 3.1. The point $s \in S^{n-1}$ is called a horospherical limit point of the group $G \subset \mathrm{Isom}\, H^n$, if the orbit $G(p)$ of some fixed point $p \in H^n$ enters any horosphere with the centre at the point s. We denote the set of horospherical limit points of group G by $L_h = L_h(G)$.

For the points $s \in S^{n-1} - L_h$ we increase the radius of the horosphere with the centre at s till we come across some point x from the orbit $G(p)$. If such a point x of the orbit is also unique we call it the nearest to s point of the orbit.

THEOREM 3.2. For any discrete group $G \subset \text{Isom } H^n$ and for any choice of the orbit $G(p) \subset H^n$, the union $L_h(G)$ with a set of points on the sphere ∂H^n having the nearest points of the orbit is a full measure set. Moreover, this division of the sphere is the division of the action of G on the conservative $(= L_h(G))$ and dissipative parts.

COROLLARY 3.3. Let $P \subset H^n$ be a convex fundamental polyhedron of the discrete group $G \subset \text{Isom } H^n$. Then the dissipative part of the action of G upon ∂H^n is the set

$$\bigcup_{g \in G} g[\partial H^n \cap \bar{P}]$$

From Corollary 3.3 and Theorem 3.2 itself, there follows directly the discription of the action of G on its limit set (cf. [15]).

COROLLARY 3.4. The group G acts conservatively on the set $L(G)$ $(m_{n-1}[L(G)] > 0)$ iff

$$(3.1) \quad m_{n-1}[L(G) \cap \bar{P}] = 0.$$

4. Proof of Theorem A

Let $P \subset H^n$ be a convex fundamental polyhedron of the group $G = H[\pi_1(M)] \subset \text{Isom } H^n$, and $\bar{P}$ be its closure in $\overline{R^n}$ (we assume that H^n is the Poincaré model in the half-space). Firstly, let us prove that the condition (1.1) of the theorem is equaivalent to the condition (3.1). Let $P^* = L(G) \cap \bar{P}$. Suppose that $m_{n-1}(P^*) > 0$. Then almost all points $x \in P^*$ are the density points for P^*, i.e. they are characterized by the fact that

$$(4.1) \quad \lim_{r \to 0} \{m_{n-1}[B^{n-1}(x,r) \cap P^*]/m_{n-1}[B^{n-1}(x,r)]\} = 1.$$

If now we fix in the polyhedron P_H from Lemma 2.2 the point x_0 corresponding to the initial point p in the minimal convex retract $(= M_G)$, then the spherical measure of the set P^* is the solid angle at which this set is seen from the point x_0. Hence, by (4.1), the limit in (1.1) tends to this measure which we assume to be positive.

Conversely, if $m_{n-1}(P^*) = 0$, then consider the sphere S_r of the radius $r > 0$ with the centre at x_0 from the polyhedron P_H. Denote by $w(r)$ the solid angle of the part of S_r which lies in the s-neighbourhood of the polyhedron P_H and intersects the polyhedron P. From the convexity of P_H, since the measure of its limit vertices is zero, it follows that with increasing the radius r the value of $w(r)$ decreases to zero. If $a(r)$ is the volume of $(n-1)$-dimensional sphere with radius r, then

$$(4.2) \qquad \frac{\operatorname{Vol} X_s(r)}{V(r,n)} = \frac{\operatorname{Vol}(M_G)_s(r)}{V(r,n)} = \int_0^r a(r)w(r)\,dr \Big/ \int_0^r a(r)\,dr.$$

Observe that for any $t > 0$ there exists an r_0 such that the angle $w(r)$ is less than t for $r > r_0$. Summing up, we obtain

$$(4.3) \qquad \frac{\operatorname{Vol} X_s(r)}{V(r,n)} = \frac{\int_0^{r_0} a(r)w(r)\,dr}{\int_0^r a(r)\,dr} + t\,\frac{\int_{r_0}^r a(r)\,dr}{\int_0^r a(r)\,dr}.$$

The first term on the right-hand side of inequality (4.3) with the increase of r tends to zero, and another one is not greater than any arbitrarily chosen $t > 0$. This proves that relation (1.1) holds.

Using Corollary 3.3 we conclude that the condition (1.1) is equivalent to the conservativeness of the action of G upon its limit set.

Yet, on the conservative part of the action of G on the sphere ∂H^n there does not exist a measurable tangent field of k-dimensional planes, $1 \le k \le n-1$, G-invariant almost everywhere. This fact, proved by Sullivan for the planar case [5], takes place for any n. Its complete proof can be found in [15]. Our version of the proof gives some improvement since it applies to any dimension. A different proof can be found in [16].

Hence, it follows that if W is a measurable conformal structure on the sphere ∂H^n (in tangent space), invariant a.e. relative to the group G, then W a.e. coincides with the standard conformal structure on the conservative part of the action of G on the sphere ∂H^n, i.e. on the set $L(G)$. This follows from the fact that when comparing the structure W on ∂H^n to the standard conformal structure, there arises a.e. a field of ellipsoids determined up to the dilatation. Thus, it is proved that the mapping f conjugating the groups G and G' on the limit set a.e. has a distortion coefficeint equal to 1 as well, i.e. it is conformal (Möbius), and the groups G and G' are conjugated in $\operatorname{Isom} H^n$. Hence, it follows that the manifolds M and M' corresponding to the groups are isometric.

R e m a r k 4.1. The condition of Theorem A is sufficient but not necessary. This is shown by the example of the functional group on the sphere S^n, $n \geq 3$, having the limit set of the positive measure (it is constructed similarly to the example of Abikoff [17] who uses the Peano curve). In this group a set of limit vertices of a convex fundamental polyhedron in $H^{n+1} = \text{int } S^n$ has the full measure m_n of "the Peano surface", which is positive by construction. At the same time, as it follows from the result of Kruškal [8] (Theorem 4, actually proved for functional groups only), this group is rigid in the above sense.

R e f e r e n c e s

[1] МАРГУЛИС, Г.А.: Изометричность замкнутых многообразий постоянной отрицательной кривизны с одинаковой фундаментальной группой. - Доклады АН СССР, 192 /1970/, 736-737.

[2] MOSTOW, G.D.: Strong rigidity of locally symmetric spaces /Ann. of Math. St. 78/, Princeton Univ. Press, Princeton 1973.

[3] АПАНАСОВ, Б.Н.: К теореме жесткости Мостова, Доклады АН СССР 243 /1978/, 829-832.

[4] —————— : Nontriviality of Teichmüller space for Kleinian group in space, Riemann Surfaces and Related Topics: Proceedings of the 1978 Stony Brook Conference /Ann. of Math.St.,97/, Princeton Univ. Press, Princeton 1981, pp.21-31.

[5] SULLIVAN, D.: On the ergodic theory at infinity of an arbitrary discrete group of hyperbolic motions, ibid, pp. 465-496.

[6] БЕРС, Л.: Униформизация. Модули и клейновы группы, Успехи матем. наук 28 /1973/, 153-198.

[7] КРУШКАЛЬ, С.Л.: О носителях дифференциалов Бельтрами для клейновых групп, Доклады АН СССР 231 /1976/, 799-801.

[8] ——————: Некоторые теоремы жесткости для разрывных групп.- Математический анализ и смежные вопросы, "Наука", Новосибирск 1978, pp.69-82.

[9] ——————: To the problem of the supports of Beltrami differentials for Kleinian groups, In: Romanian-Finnish Seminar on Complex Analysis, Proceedings, Bucharest 1976 /Lecture Notes in Mathematics 743/, Springer-Verlag, Berlin-Heidelberg-New York 1979, pp. 132-134.

[10] АПАНАСОВ, Б.Н.: Клейновы группы, пространство Тейхмюллера и теорема жесткости Мостова, Сиб. матем. ж. 21 /1980/, 3-15.

[11] КРУШКАЛЬ, С.Л., Б.Н. Апанасов, Н.А. Гусевский: Клейновы группы и униформизация в примерах и задачах,"Наука", Новосибирск 1981.

[12] APANASOV, B.N.: On isomorphisms of Kleinian groups and supports of deformations, Conference on Analytic Functions, Błażejewko, August 19-27, 1982, Abstracts, Univ. of Łódź, Łódź 1982, p. 2.

[13] ——————: Finiteness Theorems for Kleinian groups in Space, Abstracts for the International Congress Math. at Warsaw.

[14] ТЕТЕНОВ, А.В.: Локально конечные фундаментальные области дискретных групп в пространстве, Сиб. матем.ж. 23 /1982/, 102-107.

[15] АПАНАСОВ, Б.Н.: Дискретные группы преобразований и структуры многообразий, Изд. "Наука", Новосибирск 1983.

[16] AHLFORS, L.V.: Ergodic properties of groups of Möbius transformations, Analytic Functions, Kozubnik 1979, Proceedings, ed. by J. Ławrynowicz /Lecture Notes in Mathematics 798/, Springer--Verlag, Berlin-Heidelberg-New York 1980, pp. 1-9.

[17] ABIKOFF, W.: Some remarks on Kleinian groups, Advances in the Theory of Riemann Surfaces, ed. L. Ahlfors et al. /Ann. of Math. St. 66/, Princeton Univ. Press, Princeton 1971, pp. 1-5.

Institute of Mathematics
Siberian Branch of the
USSR Academy of Sciences
SU-630090 Novosibirsk 90, USSR

ON CARLEMAN APPROXIMATION BY MEROMORPHIC FUNCTIONS

André Boivin (Montréal)

I. CONDITIONS K AND G

Let E be a relatively closed subset of a domain D in the complex plane (closed in the D-topology). We denote by $H(D)$ the functions holomorphic in D and by $A(E)$ the functions continuous on E and holomorphic in the interior E^o of E. If for every pair of functions $\{f(z),\varepsilon(z)\}$, $f \in A(E)$ and $\varepsilon(z)$ positive and continuous on E, there exists a function $g \in H(D)$ such that

$$|f(z)-g(z)| < \varepsilon(z) \ , \ z \in E$$

then E is called a Carleman set in D. If we restrict ourselves to closed sets E with empty interior $(E^o = \phi)$, then this definition amounts to the definition given by P.M. Gauthier (in this volume). In 1927, T. Carleman [2] proved that the real line $\mathbb{R}$ is a Carleman set in $\mathbb{C}$. See also Kaplan [9], Sinclair [17] and Hoischen [7].

Let $D^* = D \cup \{*\}$ be the one-point compactification of D. We will say that E satisfies condition K if $D^* \setminus E$ is connected, and locally connected at infinity (i.e. locally connected at the point "$*$"). M. V. Keldysh [10] and A. Roth [15] introduced this condition in connection with problems in approximation. In 1968, N.U. Arakeljan [1] showed that this condition was equivalent to the possibility of uniform approximation of every function continuous on E and holomorphic in the interior E^o by functions holomorphic in all of D, i.e.

$$A(E) = \overline{H(D)}^E \iff \text{condition } K,$$

where $\overline{H(D)}^E$ denotes the uniform closure (i.e. closure in the sup norm) on E of the space of functions $H(D)$. Thus, indeed, this condition must be satisfied by all Carleman sets.

An other condition related to the characterization of Carleman sets was introduced in 1969 by P.M. Gauthier [5] and shown to be necessary. E is said to satisfy condition G, if for every compact $K \subset D$, there exists a compact Q

(depending on K) such that no components of E^o meets both K and $D \setminus Q$. ($D \setminus Q$ can be thought of as a neighborhood of the point "at infinity" $*$).

II. EXAMPLES

Actually these conditions characterize Carleman sets (see section III), so let us give a few examples. In these examples D will always be the whole complex plane $\mathbb{C}$.

1) long fingers

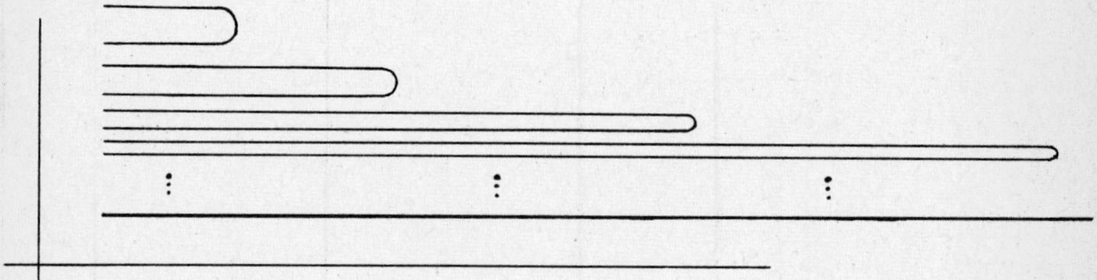

not K , G

2) long islands

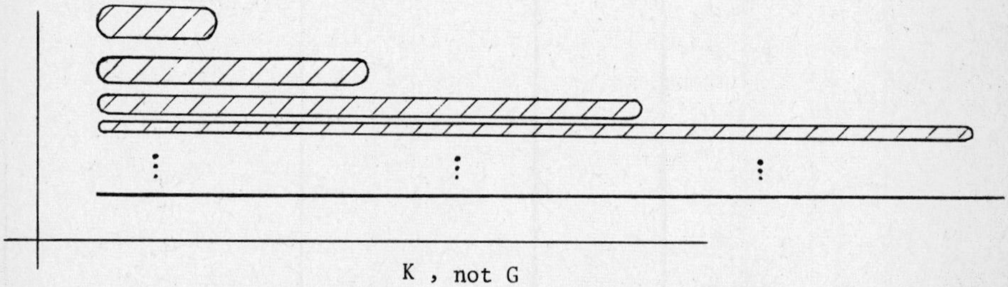

K , not G

3) tangent discs

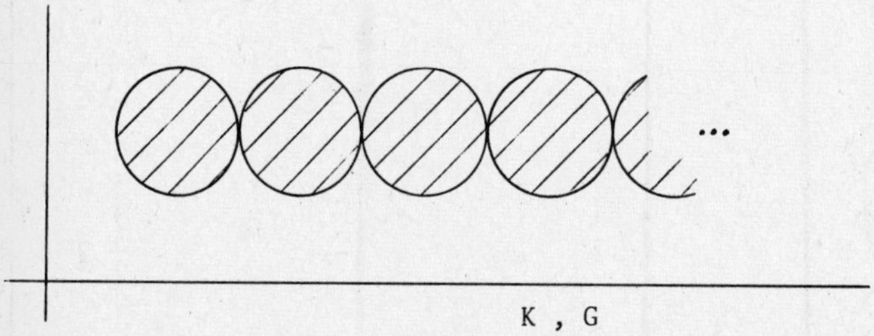

K , G

III. GOAL

As previously announced, we have

Theorem. (Nersesjan, 1971) [12]. If D is a domain in $\mathbb{C}$ and if E is a relatively closed subset of D, then E is a Carleman set in D if and only if E satisfies conditions K and G.

Now denote by $M_E(D)$, the space of functions meromorphic on D with no singularities on E. In the definition of a Carleman set we required that the function $f \in A(E)$ be approximated by a function $g \in H(D)$, that is

$$|f(z)-g(z)| < \varepsilon(z) \ , \ z \in E$$

with g holomorphic in D. Suppose now we allow g to be meromorphic on D. Of course g should have no singularities on E, so $g \in M_E(D)$. Can we, then, characterize those sets E of meromorphic Carleman approximation?

In this new setting, condition G remains necessary. And we will naturally replace condition K by the (necessary) condition $\hat{K}$ that E be a set of meromorphic uniform approximation. i.e.

$$A(E) = \overline{M_E(D)}^E$$

where $\overline{M_E(D)}^E$ is the uniform closure of the space $M_E(D)$ on E.

The purpose of this article is to show that, unlike the holomorphic case, these two conditions are not sufficient to characterize meromorphic Carleman sets.

IV. PRELIMINARIES

To construct a set which satisfies conditions G and $\hat{K}$, but fails to be a (meromorphic) Carleman set, we shall make use of the following results.

Let D be a domain in $\mathbb{C}$. A set $E \subsetneq D$ is a set of uniqueness if there exists a positive continuous function $\varepsilon(z)$, $z \in E$, such that for any f meromorphic in D, if f satisfies

$$|f(z)| < \varepsilon(z) \ , \ z \in E$$

then $f(z) \equiv 0$.

Theorem. (Gauthier, 1969) [5]. If E is a set of uniqueness, then E is not a Carleman set.

Let G be a domain in $\mathbb{C}$ bounded by finitely many Jordan arcs and $\alpha \subset \partial G$ certain given boundary arcs. We denote by

$$\omega(z,\alpha,G)$$

the harmonic measure of α at z with respect to G. That is $\omega(z,\alpha,G)$ is harmonic and bounded on G; on α, ω assumes the value 1; on the complementary arcs β, the value 0.

Two-constants Theorem. [14] Let $f \in H(G)$. Suppose that $|f(z)| \le M$, $\forall z \in G$ while at the points of α, $\overline{\lim}\, |f(z)| \le m < M$. Then at every point of the region

$$0 < \lambda < \omega(z,\alpha,G) < 1 ,$$

$$\log|f(z)| \le \lambda \log m + (1-\lambda)\log M .$$

V. A SET OF UNIQUENESS

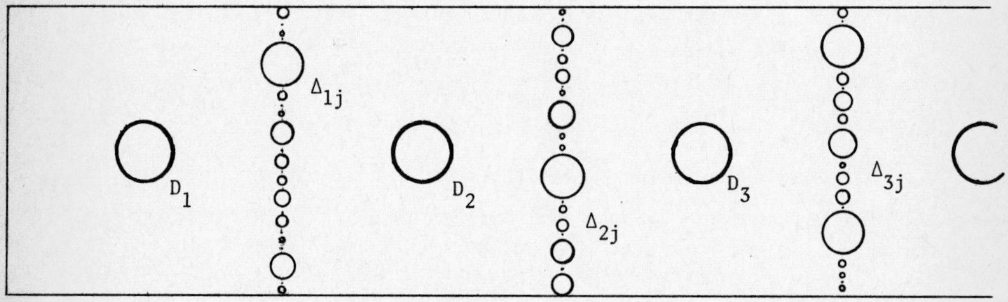

E will be a subset of the complex plane $\mathbb{C}$, constructed as follows. Set

$$E_o = \{\text{Re } z \ge 0\} \cap \{0 \le \text{Im } z \le 1\}$$

$$E_i = E_o \setminus \{ \bigcup_{k=1}^{i} \{ \bigcup_{j=1}^{\infty} \Delta_{kj} \}\} \; ; \; i = 1,2,\ldots$$

Δ_{kj} = small disjoint discs in E_o (radius $\le 1/8$) which cover a dense subset of $\{\text{Re } z = k\} \cap \{0 < \text{Im } z < 1\}$, $j = 1,2,\ldots$; $k = 1,2,\ldots$

$$E = \cap E_i$$

To choose Δ_{kj}, define

D_i = small discs such that $D_i \subset [E^o \cap \{\frac{4i-3}{4} < \text{Re } z < \frac{4i-1}{4}\}]$; $i = 1,2,\ldots$

$$G_i = [E^o_{i-1} \cap \{\frac{4i-3}{4} < \text{Re } z < \frac{4i+3}{4}\}] \setminus \bar{D}_{i+1} \; ; \; i = 1,2,\ldots$$

$$\gamma_i = \min_{z \in \bar{D}_i} \{\omega(z, \partial D_{i+1}, G_i)\} \; ; \; i = 1,2,\ldots$$

For a fixed i, take Δ_{i1} so small that

$$\min_{z\in\bar{D}_i} \{\omega(z,\partial D_{i+1}, G_i \setminus \Delta_{i1})\} = \gamma_{i1} > \frac{\gamma_i}{2} .$$

This is always possible because $\omega(z,\partial D_{i+1},G_i\setminus\Delta_{i1}) \nearrow \omega(z,\partial D_{i+1},G_i)$ uniformly on compact subsets of $G_i \setminus \{\text{centre of } \Delta_{i1}\}$ as radius $\Delta_{i1} \searrow 0$ (Harnack's principle). And by induction, choose Δ_{ij} so small that

$$\min_{z\in\bar{D}_i} \{\omega(z,\partial D_{i+1}, G_i \setminus \bigcup_{k=1}^{j} \Delta_{ik})\} = \gamma_{ij} > \frac{\gamma_i}{2} .$$

This can be done because $\gamma_{i(j-1)}$ is strictly bigger than $\frac{\gamma_i}{2}$.

Now suppose f is meromorphic on $\mathbb{C}$ without poles on E. Eventually f is without poles on $G_i \setminus \bigcup_{j=1}^{n_i} \Delta_{ij}$. Moreover suppose that $|f(z)| < \varepsilon(z)$, $z \in E$, where $\varepsilon(z)$ is a positive continuous function to be determined but strictly less than 1. Then by the two-constants theorem, if $0 < \lambda \le \omega(z,\partial D_{i+1},G_i\setminus \bigcup_{j=1}^{n_i} \Delta_{ij})$, we have,

$$\log|f(z)| \le \lambda \log \varepsilon_{i+1} + (1-\lambda)\log 1 = \lambda \log \varepsilon_{i+1}$$

where $\varepsilon_i = \max_{z\in\bar{D}_i} |f(z)| = |f(z)|_{D_i}$. By construction of E, $z \in D_i$ implies $\omega(z,\partial D_{i+1},G_i\setminus \bigcup_{j=1}^{n_i} \Delta_{ij}) > \frac{\gamma_i}{2}$, $i = 1,2,\ldots$ Thus

$$\log|f(z)| \le (\frac{\gamma_i}{2})\log \varepsilon_{i+1} \; , \; \forall z \in D_i \; ; \; i = 1,2,\ldots$$

$$|f(z)| \le \varepsilon_{i+1}^{(\frac{\gamma_i}{2})} \; , \; z \in D_i \; ; \; i = 1,2,\ldots$$

$$\varepsilon_i \le \varepsilon_{i+1}^{(\frac{\gamma_i}{2})}$$

$$\log|f(z)| \le (\frac{\gamma_{i-1}}{2})\log \varepsilon_i \le (\frac{\gamma_{i-1}}{2})(\frac{\gamma_i}{2})\log \varepsilon_{i+1} \; , \; z \in D_{i-1}$$

and consequently

$$|f(z)|_{D_1} \le (\varepsilon_{i+1})^{(\frac{\gamma_i}{2})(\frac{\gamma_{i-1}}{2})\dots(\frac{\gamma_1}{2})} = \delta_i$$

Now choose $\varepsilon(z)$ such that ε_{i+1} is so small that $\delta_i < \frac{1}{i}$. Then

$$|f(z)|_{D_1} \le \frac{1}{n} \; ; \; n = 1,2,\dots$$

Thus $f \equiv 0$ on $\mathbb{C}$.

VI. APPENDIX

The set E, being a set of uniqueness, is not a Carleman set for meromorphic approximation. Note that E satisfies condition G. That E is a set of uniform meromorphic approximation, i.e. that $A(E) = \overline{M_E(D)}$, results from the following theorems. As usual, if K is a compact subset of $\mathbb{C}$, denote the closure in the uniform norm (sup norm) of the rational functions with poles out of K by $R(K)$.

Theorem. (Nersesjan-1972, Roth-1976) [13,16]. Let E be a relatively closed subset of a domain $D \subset \mathbb{C}$. A necessary and sufficient condition in order that every function in $A(E)$ can be approximated uniformly by functions in $M_E(D)$ is that

$$A(E \cap \bar{\Delta}) = R(E \cap \bar{\Delta})$$

for every closed disc $\bar{\Delta}$, $\bar{\Delta} \subset D$.

The next theorem was first proved by M.S. Melnikov [11] and then generalized by A.G. Vitushkin [18]. It can be found, for example, in [19].

Theorem. (Melnikov-Vitushkin, 1966). Let K be a compact subset of $\mathbb{C}$. If the inner boundary Γ_I of K lies on the union of finitely many analytic curves, then $A(K) = R(K)$.

Recall that $\Gamma_I = \partial K \setminus \bigcup_{i=1}^{\infty} \partial V_i$, where the V_i are the connected components of the complement of K $(\mathbb{C} \setminus K = \bigcup_{i=1}^{\infty} V_i)$.

The reader interested by Carleman sets or approximation by meromorphic functions may consult the bibliography for a more complete (but non-exhaustive) list of articles or books treating these subjects.

There remains a most pleasing duty. This article is based on an original idea of Professor A.H. Nersesjan. Professor A.A. Goncar kindly explained to me the process involving harmonic measure used in the construction of E . My contact with these people was made possible by a friend, who happens to be my thesis supervisor, Professor P.M. Gauthier. Thus most of the credit lies with those mentionned and I wish to express all my gratitude to them.

Université de Montréal
Mathématiques
Montréal H3C 3J7, Canada

BIBLIOGRAPHY

[1] ARAKELJAN, N.U.: Uniform and tangential approximations by analytic functions, Izv. Akad. Nauk Armjan. SSR Ser. Mat. 3 (1968), 273-286 (Russian).

[2] CARLEMAN, T.: Sur un théorème de Weierstrass, Ark. Mat. Astronom. Fys. 20B 4 (1927), 1-5.

[3] GAIER, D.: Vorlesungen über Approximation im Komplexen, Birkhäuser-Verlag, Basel-Boston-Stuttgart (1980).

[4] GAMELIN, T.W.: Uniform Algebras, Prentice-Hall, Englewood Cliffs, N.J. (1969).

[5] GAUTHIER, P.M.: Tangential approximation by entire functions and functions holomorphic in a disc, Izv. Akad. Nauk Armjan. SSR Ser. Mat 4 (1969), 319-326.

[6] HADJIISKI, V.H.: Vitushkin's type theorems for meromorphic approximation on unbounded sets, Proc. of the Conf. on Complex Anal. and Appl., Varna 1981, to appear.

[7] HOISCHEN, L.: A note on the approximation of continuous functions by integral functions, J. London Math. Soc. 42 (1967), 351-354.

[8] HOISCHEN, L.: Eine Verschärfung eines Approximationsatzes von Carleman, J. Approx. Theory 9 (1973), 272-277.

[9] KAPLAN, W.: Approximation by entire functions, Michigan Math J. 3 (1955-1956) 43-52.

[10] KELDYCH, M. and LAVRENTIEFF, M.: Sur un problème de M. Carleman, C.R. (Doklady) Acad. Sci. URSS 23 (1939), 746-748.

[11] MELNIKOV, M.S.: A bound for the Cauchy integral along an analytic curve, Mat. Sb. (N.S.) 71 (113) (1966), 503-514 (Russian).

[12] NERSESJAN, A.H.: On Carleman sets, Izv. Akad. Nauk Armjan. SSR Ser. Mat. 6 (1971), 465-471 (Russian)

[13] NERSESJAN, A.H.: On uniform and tangential approximation by meromorphic functions, Izv. Akad. Nauk Armjan. SSR Ser. Mat. 7 (1972), 405-412 (Russian).

[14] NEVANLINNA, R.: Analytic Functions, Springer-Verlag, New York-Heidelberg-Berlin (1970).

[15] ROTH, A.: Approximationseigenshaften und Strahlengrenzwerte meromorpher und ganzen Funktionen, Comment. Math. Helv. 11 (1938), 77-125.

[16] ROTH, A.: Uniform and tangential approximation by meromorphic functions on closed sets, Can. J. Math. 28 (1976), 104-111.

[17] SINCLAIR, A.: A general solution for a class of approximation problems, Pacific J. Math. 8 (1958), 857-866.

[18] VITUSHKIN, A.G.: A bound for the Cauchy integral, Mat. Sb. (N.S.) 71 (113) (1966), 515-535 (Russian).

[19] ZALCMAN, L.: Analytic capacity and rational approximation, Lecture Notes No 50, Springer-Verlag, Berlin-Heidelberg-New York (1968).

POSITIVE DEFINITENESS AND HOLOMORPHY

Jacob Burbea (Pittsburgh, PA)

Contents page

Summary

Conditions for holomorphic extensions of operator-valued functions in domains D (or complex manifolds) of $\mathbb{C}^n$ are formulated in terms of positive-definiteness of order 3 of certain kernels on $D \times D$.

1. Introduction

The main concern of this paper is in establishing various generalizations of the following remarkable result, first proved by Hindmarsh [6] (see also Burbea [2,3] and FitzGerald and Horn [5]): Let $k_\Delta(z,\zeta)=(1-z\bar{\zeta})^{-1}$ be the Szegö kernel of the unit disk $\Delta=\{z\varepsilon\mathbb{C} : |z|<1\}$ and let Δ_o be a dense subset of Δ. Assume that $f_o:\Delta_o\to\mathbb{C}$ is a given function such that

$$\sum_{i,j=1}^{3} k_\Delta(z_i,z_j)[1-f_o(z_i)\overline{f_o(z_j)}]\alpha_i\overline{\alpha_j} \geq 0$$

for all $z_1,z_2,z_3\ \varepsilon\Delta_o$ and for all $\alpha_1,\alpha_2,\alpha_3\ \varepsilon\mathbb{C}$. Then there exists a unique holomorphic function f on Δ such that $f(z)=f_o(z)$ for all $z\varepsilon\Delta_o$. This result permits a strenghtening of the Pick-Nevanlinna theorem which characterizes the holomorphic mappings of the unit disk into itself.

This result has been extened by FitzGerald and Horn [5] by allowing the scalars $\alpha_1,\alpha_2,\alpha_3\ \varepsilon\mathbb{C}$ to be subject to the constraint $\alpha_1+\alpha_2+\alpha_3=0$. The same result has been recently generalized in several directions by Burbea [3].

In particular, the unit disk Δ has been replaced by a domain D (or a complex manifold) in $\mathbb{C}^n$, the Szegö kernel $k_\Delta(\cdot,\cdot)$ has been replaced by a positive-definite and sesqui-holomorphic kernel $k(\cdot,\cdot)$ on $D\times D$ and f_o has been replaced by an operator-valued function $T_o(\cdot)$ on a dense subset D_o of D having values in $\mathcal{B}(U:W)$, the Banach space of bounded linear operators from the Hilbert space U to the Hilbert space W. Under some additional smoothness assumptions on $k(\cdot,\cdot)$ it was proved in Burbea [3] that there exists a unique contraction operator $T(\cdot)$ in $\mathcal{B}(U:W)$, holomorphic in D and extending $T_o(\cdot)$ to D.

To some extent, the additional smoothness assumptions on $k(\cdot,\cdot)$ in [3] were somewhat unatural; they were needed only as technicalities based on standard mollification arguments. On the other hand, in the work of FitzGerald and Horn [5], mollifiers were avoided by using instead a simple and yet ingenious device based on Cauchy's representation formula. In this work, following the findings in [3] and [5], we also adopt the device of Cauchy's formula and thus avoid the use of mollifiers. This will enable us not only to extend our previous work [3] but also to generalize the results of FitzGerald and Horn [5] in various directions, including generalized vectorial versions of the basic result of Hindmarsh [6].

In section 2, we give the basic definitions concerning positive definitness and holomorphy. In section 3 we give our basic theorem (Theorem 1), stated in terms of operator-valued functions on arbitrary domains in $\mathbb{C}^n$ and having values in $\mathcal{B}(U:W)$. The proof of this theorem is motivated in part by the work of FitzGerald and Horn [5]. In section 4, we state and prove Theorems 2, 3 and 4; their proofs rely on Theorem 1 and they constitute the various generalizations of the results of Burbea [3], FitzGerald and Horn [5] and Hindmarsh [6].

2. Preliminaries and Definitions

In this paper all vector spaces are assumed to be over the complex field $\mathbb{C}$. The following notation will be used: $\mathcal{B}(U:W)$ stands for the Banach space of bounded linear operators from the Hilbert space U to the Hilbert space W. The subset of $\mathcal{B}(U:W)$ consisting of all contractive operators is denoted by $C(U:W)$. Thus $T\varepsilon C(U:W)$ means $||T_u||_W \leq ||u||_U$ for every $u\varepsilon U$ or, equivalently, $||T||\leq 1$. An operator $S\varepsilon\mathcal{B}(U:U)$ is said to be accretive if $\mathrm{Re}(Su,u)_U\geq 0$ for every $u\varepsilon U$ and the family of all such accretive operators is denoted by $A(U:U)$. The adjoint of

$T\varepsilon\mathcal{B}(U:W)$ is denoted by T^*; thus $T^*\varepsilon\mathcal{B}(W:U)$ and $T\varepsilon\mathcal{C}(U:W)$ or $S\varepsilon\mathcal{A}(U:U)$ if and only if $T^*\varepsilon\mathcal{C}(W:U)$ or $S^*\varepsilon\mathcal{A}(U:U)$. Let Ω be a subset of $\mathbb{C}$ and denote by $\hat{\Omega}$ its closure. An operator $S\varepsilon\mathcal{B}(U:U)$ is said to belong to $\Omega(U:U)$ if $(Su,u)_U\varepsilon\Omega$ for every unit vector $u\varepsilon U$. We write $\Delta=\{z\varepsilon\mathbb{C}:|z|<1\}$ for the unit disk and $\mathcal{R}=\{z\varepsilon\mathbb{C}:\mathrm{Re}z>0\}$ for the right-half plane. It follows that for $T\varepsilon\mathcal{B}(U:W)$, $T\varepsilon\mathcal{C}(U:W)$ if and only if $T^*T\varepsilon\hat{\Delta}(U:U)$. Moreover, $\mathcal{A}(U:U)=\hat{\mathcal{R}}(U:U)$. The identity-operator of U is denoted by I_U.

For an arbitrary non-empty set Λ, $(\Lambda:\mathcal{B}(U:W))$ designates the class of all operator-valued functions $T(\cdot)$ on Λ to $\mathcal{B}(U:W)$. Similarly, $(\Lambda\times\Lambda;\mathcal{B}(U:W))$ designates the class of all operator-valued kernel $K(\cdot,\cdot)$ on $\Lambda\times\Lambda$ to $\mathcal{B}(U:W)$. The classes $(\Lambda;\mathcal{C}(U:W))$, $(\Lambda;\mathcal{A}(U:U))$, $(\Lambda;\Omega(U:U))$, $(\Lambda\times\Lambda;\mathcal{C}(U:W))$ and so on are defined in a similar manner. A scalar-valued kernel $k(\cdot,\cdot)$ on $\Lambda\times\Lambda$ is said to be <u>positive-definite of order N</u>, in short $k\varepsilon\mathcal{P}_N(\Lambda)$, if

$$\sum_{i,j=1}^{N} k(z_i,z_j)\alpha_i\overline{\alpha_j} \geq 0$$

for every N points $z_1,\ldots,z_N\varepsilon\Lambda$ and equally many corresponding scalars $\alpha_1,\ldots,\alpha_N\varepsilon\mathbb{C}$. It is said to be <u>positive-definite</u>, in short $k\varepsilon\mathcal{P}(\Lambda)$, if $k\varepsilon\mathcal{P}_N(\Lambda)$ for every $N=1,2,\ldots$. An hermitian kernel $k(\cdot,\cdot)$ on $\Lambda\times\Lambda$, is said to be <u>almost positive-definite of order N</u>, in short $k\varepsilon A\mathcal{P}_N(\Lambda)$, if in the previous definition of $\mathcal{P}_N(\Lambda)$ the scalars $\alpha_1,\ldots,\alpha_N$ are subject to the constraint

$$\sum_{i=1}^{N} \alpha_i = 0.$$

It is said to be <u>almost positive-definite</u>, in short $k\varepsilon A\mathcal{P}(\Lambda)$, if $k\varepsilon A\mathcal{P}_N(\Lambda)$ for every $N=1,2,\ldots$. Of course, any hermitian $k(\cdot,\cdot)$ belongs trivially to $A\mathcal{P}_1(\Lambda)$ and, clearly, $\mathcal{P}_N(\Lambda)\subset A\mathcal{P}_N(\Lambda)$, $\mathcal{P}(\Lambda)\subset A\mathcal{P}(\Lambda)$. It is also well-known (cf. Donoghue [4, p. 135]) that $k\varepsilon A\mathcal{P}_N(\Lambda)$, $N\geq 2$, if and only if the scalar-valued kernel $a_\tau(\cdot,\cdot)$, defined by

$$a_\tau(z,\zeta) = k(z,\zeta)-k(\tau,\zeta)-k(z,\tau)+k(\tau,\tau) \quad ; \quad z,\zeta\varepsilon\Delta,$$

belongs to $\mathcal{P}_N(\Lambda)$ for any choice of $\tau\varepsilon\Lambda$.

An operator-valued kernel $K(\cdot,\cdot)$ of $(\Lambda\times\Lambda;\mathcal{B}(U:U))$ is said to be <u>hermitian</u> if $K(z,\zeta)^*=K(\zeta,z)$ for every $z,\zeta\varepsilon\Lambda$. It is said to be <u>(weakly) positive-definite of</u>

order N, in short $K\epsilon\mathcal{P}_N(\Lambda:U)$, if the scalar-valued kernel $(K(\cdot,\cdot)u,u)_U$ is in $\mathcal{P}_N(\Lambda)$ for every unit vector $u\epsilon U$. It is (weakly) positive-definite, in short $K\epsilon\mathcal{P}(\Lambda:U)$, if $K\epsilon\mathcal{P}_N(\Lambda:V)$ for every $N=1,2,\ldots$. The operator-valued kernel $K(\cdot,\cdot)$ is said to be (weakly) almost positive-definite of order N, in short $K\epsilon A\mathcal{P}_N(\Lambda:U)$, if $(K(\cdot,\cdot)u,u)_U$ is in $A\mathcal{P}_N(\Lambda)$ for every unit vector $u\epsilon U$. It is said to be (weakly) almost positive-definite, in short $K\epsilon A\mathcal{P}(\Lambda:U)$, if $K\epsilon A\mathcal{P}_N(\Lambda:U)$ for every $N=1,2,\ldots$.

The following proposition is almost immediate:

Proposition 1. *Let* $K(\cdot,\cdot)\epsilon(\Lambda\times\Lambda;\mathcal{B}(U:U))$ *belong to* $\mathcal{P}_N(\Lambda:U)$, $N\geq 2$. *Then* $K(\cdot,\cdot)$ *is hermitian.*

Proof. It follows that

$$(K(z,z)u,u)_U|\alpha|^2 + (K(z,\zeta)u,u)_U\alpha\bar{\beta} + (K(\zeta,z)u,u)_U\bar{\alpha}\beta + (K(\zeta,\zeta)u,u)_U|\beta|^2 \geq 0$$

for any vector $u\epsilon U$, any points $z,\zeta\epsilon\Lambda$ and any scalars $\alpha,\beta\epsilon\mathbb{C}$. In particular,

$$(K(z,\zeta)u,u)_U\alpha\bar{\beta} + (K(\zeta,z)u,u)_U\bar{\alpha}\beta \;\epsilon\; \mathbb{R}$$

which means that

$$((K(z,\zeta) - K(\zeta,z)^*)u,u)_U\alpha\bar{\beta} + ((K(\zeta,z) - K(z,\zeta)^*)u,u)_U\bar{\alpha}\beta = 0.$$

Upon putting successively $(\alpha,\beta)=(1,1)$ and $(\alpha,\beta)=(i,1)$ in this equation we obtain

$$((K(z,\zeta) - K(\zeta,z)^*)u,u)_U = 0$$

for any $u\epsilon U$ and any $z,\zeta\epsilon\Lambda$. In the last equation we replace u by u_1+u_2 and then u_1+iu_2, where u_1,u_2 are any vectors in U. This results in

$$((K(z,\zeta) - K(\zeta,z)^*)u_1,u_2)_U = 0$$

which concludes the proof.

Any positive-definite scalar-valued kernel $k\epsilon\mathcal{P}(\Lambda)$ induces a uniquely determined Hilbert space $H_k(\Lambda)$ of functions on Λ with $k(\cdot,\cdot)$ as the reproducing kernel (see Aronszajn [1]) of $H_k(\Lambda)$. It also follows that for any orthonormal basis $\{\psi_\alpha\}$ of $H_k(\Lambda)$,

$$k(z,\zeta) = \sum_\alpha \psi_\alpha(z)\overline{\psi_\alpha(\zeta)} \quad ; \quad z,\zeta\varepsilon\Lambda.$$

By $P^+(\Lambda)$ we denote all $k\varepsilon P(\Lambda)$ such that $k(z,z)>0$ for all $z\varepsilon\Lambda$.

Let D be a domain (or a complex manifold) in $\mathbb{C}^n$. By $H(D)$ and $\bar{H}(D)$ we denote the classes of all holomorphic and antiholomorphic functions (or forms) in D, respectively. A scalar-valued kernel $k(\cdot,\cdot)$ is said to be sesqui-holomorphic on $D\times D$ if $k(\cdot,z)\ \varepsilon\ H(D)$ and $k(z,\cdot)\ \varepsilon\ \bar{H}(D)$ for every $z\varepsilon D$. A theorem of Hartogs implies that any sesqui-holomorphic kernel on $D\times D$ belongs to $H(D\times\bar{D})$, where $\bar{D}$ is the complex-conjugate manifold of D. Evidently, any hermitian scalar-valued kernel $k(\cdot,\cdot)$ on $D\times D$ with the property that $k(\cdot,z)\varepsilon H(D)$ for any $z\varepsilon D$ is itself sesqui-holomorphic on $D\times D$. In particular, by Proposition 1, any $k\varepsilon P_N(D)$, $N\geq 2$, with the above mentioned property is also sesqui-holomorphic on $D\times D$. Moreover, the reproducing space $H_k(D)$ of a sesqui-holomorphic positive-definite kernel $k\varepsilon P(D)$ is a subset of $H(D)$ (see also Donoghue [4, p. 92]). In this case, convergence in the norm of $H_k(D)$ implies uniform convergence on compacta of D. In particular, for any orthonormal basis $\{\psi_m\}$ of $H_k(D)$,

$$k(z,\zeta) = \sum_{m=1}^{\infty} \psi_m(z)\overline{\psi_m(\zeta)} \quad ; \quad z,\zeta\varepsilon D, \tag{2.1}$$

and the bilinear sum converges absolutely and uniformly on compacta of $D\times D$.

A simple example of a sesqui-holomorphic and positive-definite kernel is the Szegö kernel $k_\Delta(z,\zeta)=(1-z\bar{\zeta})^{-1}$ of the unit disk. More generally, let Ω be an hyperbolic simply connected domain in $\mathbb{C}$ and let $\phi:\Omega\to\Delta$ be a Riemann mapping. Then

$$k_\Omega(z,\zeta) = \frac{\{\phi'(z)\overline{\phi'(\zeta)}\}^{1/2}}{1-\phi(z)\overline{\phi(\zeta)}} \quad ; \quad z,\zeta\varepsilon\Omega,$$

is the Szegö kernel of Ω. Clearly, $k_\Omega\varepsilon P^+(\Omega)$ and $k_\Omega(\cdot,\cdot)$ is sesqui-holomorphic on $\Omega\times\Omega$. Moreover, as is well-known, the reproducing space of $k_\Omega(\cdot,\cdot)$ is the Hardy-Szegö space $H_2(\Omega)$ (see also [2]), and hence $k_\Omega(\cdot,\cdot)$ is independent of the particular choice of the Riemann mapping ϕ. One may also generalize the above concept to any hyperbolic domain (or a Riemann surface) Ω by using a holomorphic cover map $\pi:\Delta\to\Omega$

instead of a Riemann mapping. We omit the details.

An operator-valued function $T(\cdot)\varepsilon(D;\mathcal{B}(U:W))$ is said to be holomorphic in D if $(T(\cdot)u,w)_W\varepsilon H(D)$ for every $(u,w)\varepsilon U\times W$. When $U=W$, this definition of holomorphy reduces in only requiring that $(T(\cdot)u,u)_U\varepsilon H(D)$ for every $u\varepsilon U$. The class of all such operator-valued functions is denoted by $H(D;\mathcal{B}(U:W))$. The corresponding class of operator-valued antiholomorphic functions $\bar{H}(D;\mathcal{B}(U:W))$ is defined analogously. Evidently, $T(\cdot)\varepsilon H(D;\mathcal{B}(U:W))$ if and only if $T(\cdot)^*\varepsilon\bar{H}(D;\mathcal{B}(W:U))$. The classes $H(D;\mathcal{C}(U:W))$, $H(D;\mathcal{A}(U:U))$, $H(D;\Omega(U:U))$, $\bar{H}(D;\mathcal{C}(U:W))$ and so on are defined in a similar manner. An operator-valued kernel $K(\cdot,\cdot)\varepsilon(D\times D;\mathcal{B}(U:W))$ is said to be sesqui-holomorphic on $D\times D$ if for any $(u,w)\varepsilon U\times W$, $(K(\cdot,\cdot)u,w)_W$ is sesqui-holomorphic on $D\times D$. In this case, $K(\cdot,\cdot)\varepsilon H(D\times\bar{D};\mathcal{B}(U:W))$

3. The Basic Theorem

Throughout this section and the remaining parts of the paper, D stands for a fixed domain (or a complex manifold) in $\mathbb{C}^n$. We use multinomial notation. Explicitly, expressing a point $z\varepsilon\mathbb{C}^n$ with superscripted components $z=(z^1,\ldots,z^n)$, the following conventions will be employed:

$$||z|| = \{\sum_{i=1}^{n}|z^i|^2\}^{1/2} \tag{3.1}$$

and

$$z_m^{(i)} = (z^1,\ldots,z_m^i,\ldots,z^n) \quad ; \quad 1\leq i\leq n,\quad m=1,2,\ldots\ . \tag{3.2}$$

In other words, for any $1\leq i\leq n$, $z_m^{(i)}$ is the point in $\mathbb{C}^n$ whose i-th component is $z_m^i\varepsilon\mathbb{C}$, $m=1,2,\ldots$, and the rest of its $n-1$ components coincide with those of $z\varepsilon\mathbb{C}^n$. For any other point $\zeta=(\zeta^1,\ldots,\zeta^n)\varepsilon\mathbb{C}^n$, $z-\zeta$ will always stand for

$$z-\zeta = \prod_{i=1}^{n}(z^i-\zeta^i)$$

and <u>not</u> for $(z^1-\zeta^1,\ldots,z^n-\zeta^n)$. It follows that

$$\frac{1}{z-\zeta} - \frac{1}{z-\tau} = \sum_{j=1}^{n}\{\prod_{i=1}^{n}(z^i-\zeta^i)\cdot\prod_{i=1}^{n}(z^i-\tau^i)\}^{-1}(\zeta^j-\tau^j)$$

where $\tau=(\tau^1,\ldots,\tau^n)$ is another point in $\mathbb{C}^n$. In particular, using Cauchy-Schwarz inequality,

$$\left\|\frac{1}{z-\zeta} - \frac{1}{z-\tau}\right\| \leq B_n(z:\zeta,\tau)\,\|\zeta-\tau\| \tag{3.3}$$

where

$$B_n(z:\zeta,\tau) = \left\{ \sum_{j=1}^{n} \left| \prod_{i=1}^{n}(z^i-\zeta^i)\cdot\prod_{i=1}^{j}(z^i-\tau^i)\right|^{-2}\right\}^{1/2} \tag{3.4}$$

is a non-negative function which is symmetric in ζ and τ.

Theorem 1. *Let* $k(\cdot,\cdot)$ *be a sesqui-holomorphic scalar-valued kernel on* $D\times D$, D_o *is a dense subset of* D *and* $T_o(\cdot)\,\varepsilon(D_o;\mathcal{B}(U:W))$. *Assume that the operator-valued kernel*

$$K_o(z,\zeta) = k(z,\zeta)I_U - T_o(\zeta)^* T_o(z) \quad ; \quad z,\zeta\varepsilon D_o \tag{3.5}$$

belonging to $(D_o\times D_o;\mathcal{B}(U:U))$ *is (weakly) almost positive-definite of order* 3, *i.e.* $K_o\varepsilon A\mathcal{P}_3(D_o:U)$. *Then there exists a unique strongly continuous* $T(\cdot)\,\varepsilon(D;\mathcal{B}(U:W))$ *such that* $T(z)=T_o(z)$ *for all* $z\varepsilon D_o$. *Moreover* $T(\cdot)\,\varepsilon H(D;\mathcal{B}(U:W))$. *In particular, the operator-valued kernel* $K(z,\zeta)=k(z,\zeta)I_U-T_o(\zeta)^*T(z)$; $z,\zeta\varepsilon D$, *belongs to* $H(D\times\bar{D};\mathcal{B}(U:U))$ *and is of class* $A\mathcal{P}_3(D:U)$, *and it extends* $K_o(\cdot,\cdot)$ *to* $D\times D$.

Proof. We first prove the uniqueness assertion. Assume that there are two such strongly continuous extensions $T_1(\cdot)$ and $T_2(\cdot)$. Let z be any point in D and let $\{z_m\}$ be a sequence of points in D_o converging to z. Then

$$\|(T_o(z_m) - T_j(z))u\|_W \xrightarrow[m\to\infty]{} 0 \quad (j=1,2)$$

for every $u\varepsilon U$. In particular,

$$\|(T_1(z)-T_2(z))u\|_W \leq \|(T_1(z)-T_o(z_m))u\|_W + \|(T_2(z)-T_o(z_m)u\|_W \xrightarrow[m\to\infty]{} 0,$$

showing that $T_1(z)=T_2(z)$. As for the existence of a strongly continuous extension $T(\cdot)$, it is sufficient to show that $T_o(\cdot)$ is strongly uniformly continuous on the intersection of D_o with the closure $\hat{P}_o$ of any polydisk P_o such that $\hat{P}_o\subset D_o$. Let P_o be such a polydisk and consider another polydisk P, $\hat{P}_o\subset P\subset D$ with $\hat{P}_o\cap\partial_o P=\emptyset$, where $\partial_o P$ is the distinguished boundary of P. Now, as $k(\cdot,\cdot)$ is sesqui-holomorphic on $D\times D$, $k\varepsilon H(D\times\bar{D})$. In particular, by Cauchy's integral formula and using multinomial notation

(3.6) $$k(\omega,\tau) = \frac{1}{(2\pi)^{2n}} \int_{\partial_o P} \int_{\partial_o P} \frac{1}{(z-\omega)(\bar{\zeta}-\bar{\tau})} k(z,\zeta)\,dz\,d\bar{\zeta}$$

for every $\omega,\tau\varepsilon P$. Here we have also used the multinomial abbreviation of $dz=dz^1\ldots dz^n$ for $z=(z^1,\ldots,z^n)\varepsilon\mathbb{C}^n$. The fact that the kernel $K_o(\cdot,\cdot)$ of (3.5) is in $AP_3(D_o:U)$ implies that

(3.7) $$\sum_{i,j=1}^{3} k(z_i,z_j)\alpha_i\overline{\alpha_j} \geq \left|\left|\sum_{i=1}^{3}\alpha_i T_o(z_i)u\right|\right|_W^3$$

for any unit vector u of U, and for all $z_1,z_2,z_3\varepsilon D_o$ and for all $\alpha_1,\alpha_2,\alpha_3\varepsilon\mathbb{C}$ with $\alpha_1+\alpha_2+\alpha_3=0$. Substituting (3.6) in (3.7), we obtain

(3.8) $$K\left[\sum_{m=1}^{3}\frac{\alpha_m}{z-z_m}\right] \geq \left|\left|\sum_{m=1}^{3}\alpha_m T_o(z_m)u\right|\right|_W^3$$

for any $z_1,z_2,z_3\varepsilon\hat{P}_o\bigcap D_o$. Here $u,\alpha_1,\alpha_2,\alpha_3$ are as before, and for $f\varepsilon L_\infty(\partial_o P)$, $K[f]$ is the quadratic form

$$K[f] = \frac{1}{(2\pi)^{2n}} \int_{\partial_o P} \int_{\partial_o P} f(z)\overline{f(\zeta)}k(z,\zeta)\,dz\,d\bar{\zeta}.$$

Therefore

$$|K[f]| \leq \frac{1}{(2\pi)^{2n}} [L(\partial_o P)]^2 ||k||_{L_\infty(\partial_o P\times\partial_o P)} ||f||^2_{L_\infty(\partial_o P)}$$

where $L(\partial_o P)$ is the "length" of the distinguished boundary $\partial_o P$, namely the product of the lengths of the n one-dimensional components of $\partial_o P$. The last inequality is equivalent to

$$|K[f]| \leq C_n^2 ||f||^2_{L_\infty(\partial_o P)}$$

where $C_n=C_n(k,\partial_o P)$ is a positive constant denending only on $k(\cdot,\cdot)$ and $\partial_o P$. It follows from (3.8) that

$$\left|\left|\sum_{m=1}^{3}\alpha_m T_o(z_m)u\right|\right|_W^2 \leq K\left[\sum_{m=1}^{3}\frac{\alpha_m}{z-z_m}\right] = \left|K\left[\sum_{m=1}^{3}\frac{\alpha_m}{z-z_m}\right]\right| \leq C_n^2\left|\left|\sum_{m=1}^{3}\frac{\alpha_m}{z-z_m}\right|\right|^2_{L_\infty(\partial_o P)}$$

and thus, for any unit vector $u\varepsilon U$,

(3.9) $$\left\| \sum_{m=1}^{3} \alpha_m T_o(z_m) u \right\|_W \leq C_n \max_{z\in\partial_o P} \left| \sum_{m=1}^{3} \frac{\alpha_m}{z-z_m} \right|$$

for all $z_1,z_2,z_3 \in \hat{P}_o \bigcap D_o$ and all $\alpha_1,\alpha_2,\alpha_3 \in \mathbb{C}$ with $\alpha_1+\alpha_2+\alpha_3=0$. In (3.9) we choose $\alpha_1=1$, $\alpha_2=-1$, $\alpha_3=0$, $z_1=\zeta$ and $z_2=\tau$, where $\zeta,\tau \in \hat{P}_o \bigcap D_o$. Then, in the notation of (3.1) and (3.3)-(3.4),

$$\|(T_o(\zeta)-T_o(\tau)u\|_W \leq C_n \max_{z\in\partial_o P} \left|\frac{1}{z-\zeta} - \frac{1}{z-\tau}\right|$$

$$\leq C_n B_n(\partial_o P:\zeta,\tau)\|\zeta-\tau\|,$$

where

$$B_n(\partial_o P:\zeta,\tau) \equiv \max_{z\in\partial_o P} B_n(z:\zeta,\tau).$$

Since $\hat{P}_o$ and $\partial_o P$ are disjoint compact sets, the function $B_n(\partial_o P:\zeta,\tau)$ is uniformly bounded in $\zeta,\tau\in\hat{P}_o$. Therefore, $\|(T_o(\zeta)-T_o(\tau))u\|_W$ tends to zero uniformly as $\zeta\to\tau$ in $\hat{P}_o \bigcap D_o$, for any $u\in U$. It follows that $T_o(\cdot)$ is strongly uniformly continuous on $\hat{P}_o \bigcap D_o$ and hence $T_o(\cdot)$ has a strongly uniformly continuous extension $T(\cdot)$ to all D with $T(\cdot)\varepsilon(D;\mathcal{B}(U:W))$. We now show that $T(\cdot)\in H(D;\mathcal{B}(U:W))$. Let P_o and P be the previously considered polydisks. Since the strongly continuous extension $T(\cdot)$ is obtained from $T_o(\cdot)$ by a pointwise limiting process, inequality (3.9) is also valid for $T(\cdot)$ on all of $\hat{P}_o$, i.e. for any unit vector $u\in U$,

(3.10) $$\left\| \sum_{m=1}^{3} \alpha_m T(z_m) u \right\|_W \leq C_n \max_{z\in\partial_o P} \left| \sum_{m=1}^{3} \frac{\alpha_m}{z-z_m} \right|$$

for all $z_1,z_2,z_3\in\hat{P}_o$ and all $\alpha_1,\alpha_2,\alpha_3\in\mathbb{C}$ with $\alpha_1+\alpha_2+\alpha_3=0$. Let $\zeta=(\zeta^1,\ldots,\zeta^n)$ be a fixed point in P_o, and fix an index j, $1\leq j\leq n$, and a unit vector $u\in U$. Consider, following the notation of (3.2), the points $\zeta_m^{(j)}=(\zeta^1,\ldots,\zeta_m^j,\ldots,\zeta^n)$, $m=1,2,\ldots$, which are in P_o. We must show that

$$(W) - \lim_{\zeta_1^j\to\zeta^j} \frac{(T(\zeta_1^{(j)}) - T(\zeta))u}{\zeta_1^j - \zeta^j}$$

exists in W for any unit vector $u\in U$. For this purpose, it is sufficient to show that the above difference-quotients form a Cauchy-sequence in W, that is

$$||(\frac{T(\zeta_1^{(j)}) - T(\zeta)}{\zeta_1^j - \zeta^j} - \frac{T(\zeta_2^{(j)}) - T(\zeta)}{\zeta_2^j - \zeta^j})u||_W \to 0$$

uniformly as $\zeta_1^j \to \zeta^j$ and $\zeta_2^j \to \zeta^j$. In order to prove this, we consider (3.10) with the choices of

$$z_1 = \zeta \ , \ z_2 = \zeta_1^{(j)} \ , \ z_3 = \zeta_2^{(j)}$$

and

$$\alpha_1 = \frac{1}{\zeta_2^j - \zeta^j} - \frac{1}{\zeta_1^j - \zeta^j} \ , \ \alpha_2 = \frac{1}{\zeta_1^j - \zeta^j} \ , \ \alpha_3 = -\frac{1}{\zeta_2^j - \zeta^j} .$$

This gives, using multinomial notation,

$$||(\frac{T(\zeta_1^{(j)}) - T(\zeta)}{\zeta_1^j - \zeta^j} - \frac{T(\zeta_2^{(j)}) - T(\zeta)}{\zeta_2^j - \zeta^j})u||_W$$

$$\leq C_n \max_{z \varepsilon \partial_o P} |\frac{1}{z-\zeta} (\frac{1}{z^j - \zeta_1^j} - \frac{1}{z^j - \zeta_2^j})|$$

$$= C_n |\zeta_1^j - \zeta_2^j| \max_{z \varepsilon \partial_o P} |(z-\zeta)(z^j - \zeta_1^j)(z^j - \zeta_2^j)|^{-1} .$$

This upper bound tends to zero as $\zeta_1^j \to \zeta^j$ and $\zeta_2^j \to \zeta^j$ and thus $T(\cdot)$ is holomorphic in the j-component variable, $1 \leq j \leq n$, at any point ζ of the polydisk P_o. Since this is true for any $j=1,\ldots,n$ and P_o is an arbitrary polydisk inside D, we deduce that $T(\cdot) \varepsilon H(D;\mathcal{B}(U:W))$. The last statement of the theorem concerning the extension $K(\cdot,\cdot)$ of $K_o(\cdot,\cdot)$ is deduced at once from the inequality (3.10). This concludes the proof of the theorem.

4. Other Generalizations

The following theorem constitutes the most general form of the contractive version of the problems considered here.

Theorem 2. *Let* $k(\cdot,\cdot)$ *be a sesqui-holomorphic scalar-valued kernel on* $D\times D$ *which is also of class* $P^+(D)$. *Let* D_o *be a dense subset of* D *and let* $T_o(\cdot)\varepsilon(D_o;\mathcal{B}(U:W))$. *Assume that the operator-valued kernel*

$$L_o(z,\zeta) = k(z,\zeta)[I_U - T_o(\zeta)^* T_o(\zeta)] \quad ; \quad z,\zeta\varepsilon D_o, \tag{4.1}$$

belonging to $(D_o\times D_o;\mathcal{B}(U:U))$ *is (weakly) almost positive-definite of order* 3, *i.e.* $L_o\varepsilon AP_3(D_o:U)$. *Then there exists a unique* $T(\cdot)\varepsilon H(D;\mathcal{B}(U:W))$ *extending* $T_o(\cdot)$ *to* D. *In particular, the operator-valued kernel* $L(z,\zeta)=k(z,\zeta)[I_U-T(\zeta)^*T(z)]$; $z,\zeta\varepsilon D$, *belongs to* $H(D\times\bar{D};\mathcal{B}(U:U))$ *and is of class* $AP_3(D:U)$, *and it extends* $L_o(\cdot,\cdot)$ *to* $D\times D$. *Moreover, if also* $L_o\varepsilon P_1(D_o:U)$ *then* $T(\cdot)\varepsilon H(D;\mathcal{C}(U:W))$ *and, in particular,* $L\varepsilon P_1(D:U)$.

Proof. By assumption, $k(\cdot,\cdot)$ admits the bilinear expansion (2.1) in terms of any orthonormal basis $\{\psi_m\}_{m=1}^{\infty}$ of $H_k(D)\subset H(D)$. Moreover, since $k\varepsilon P^+(D)$, it is also assumed that $k(z,z)>0$ for every $z\varepsilon D$. This is completely equivalent to the requirement that for any $z\varepsilon D$, there exists an integer $m\geq 1$ so that $\psi_m(z)\neq 0$. Next, the requirement that the kernel $L_o(\cdot,\cdot)$ in (4.1) belongs to class $AP_3(D_o:U)$ means that for any unit vector $u\varepsilon U$,

$$\sum_{i,j=1}^{3} k(z_i,z_j)\alpha_i\bar{\alpha}_j \geq \sum_{i,j=1}^{3} k(z_i,z_j)(T_o(z_i)u,\ T_o(z_j)u)_W\alpha_i\overline{\alpha_j} \tag{4.2}$$

for all $z_1,z_2,z_3\varepsilon D_o$ and for all $\alpha_1,\alpha_2,\alpha_3\varepsilon\mathbb{C}$ with $\alpha_1+\alpha_2+\alpha_3=0$. Using the expansion (2.1), the right hand side of (4.2) admits the expression

$$\sum_{m=1}^{\infty} \Big|\Big| \sum_{i=1}^{3} \alpha_i\psi_m(z_i)T_o(z_i)u\Big|\Big|_W^2.$$

Consequently, for every integer $m\geq 1$.

$$\sum_{i,j=1}^{3} k(z_i,z_j)\alpha_i\overline{\alpha_j} \geq \Big|\Big| \sum_{i=1}^{3} \alpha_i T_o^{(m)}(z_i)u\Big|\Big|_W^2$$

where $T_o^{(m)}(\cdot)\varepsilon(D_o;\mathcal{B}(U:W))$ is defined by

$$T_o^{(m)}(z) = \psi_m(z)T_o(z) \quad , \quad z\varepsilon D_o. \tag{4.3}$$

This shows that the operator-valued kernel

$$K_o^{(m)}(z,\zeta) \equiv k(z,\zeta)I_U - T_o^{(m)}(\zeta)^* T_o^{(m)}(z) \quad ; \quad z,\zeta\varepsilon D_o,$$

is of class $AP_3(D_o:U)$ for every integer $m\geq 1$. According to Theorem 1, there exists a unique $T^{(m)}(\cdot)\varepsilon H(D;\mathcal{B}(U:W))$ extending $T_o^{(m)}(\cdot)$ to D. In view of (4.3) and since $\psi_m \varepsilon H(D)$, $T^{(m)}(\cdot)$ is of the form

$$T^{(m)}(z) = \psi_m(z)T(z) \quad , \quad z\varepsilon D,$$

and thus $\psi_m(\cdot)(T(\cdot)u,w)_W \varepsilon H(D)$ for any $(u,w)\varepsilon U\times W$ and all integers $m\geq 1$. This implies that $(T(\cdot)u,w)_W$ is meromorphic in D. We claim that this function is in fact holomorphic in D. Indeed, if this function has a pole at some $\zeta\varepsilon D$ then $\psi_m(\zeta)=0$ for all integers $m\geq 1$; a contradiction to our assumption. It follows that $T(\cdot)\varepsilon H(D;\mathcal{B}(U:W))$ and it is the unique holomorphic extension of $T_o(\cdot)$ to D. If also $L_o \varepsilon P_1(D_o:U)$, then in (4.2) we may choose $\alpha_1=1$ and $\alpha_2=\alpha_3=0$, i.e.

$$k(z_1,z_1) \geq k(z_1,z_1)||T_o(z_1)U||_W^2$$

for every $z_1\varepsilon D_o$ and every unit vector $u\varepsilon U$. Since $k(z_1,z_1)>0$ we also conclude that $||T_o(z_1)u||_W \leq 1$. Now, let u be any unit vector in U, z any point in D and $\{z_m\}$ any sequence of points in D_o, converging to z. Then

$$||T(z)u||_W = \lim_{n\to\infty}||T_o(z_n)u||_W \leq 1$$

which means that $T(\cdot)$ is actually in $H(D;\mathcal{C}(U:W))$. The theorem now follows and the proof is complete.

Theorem 3. Let $k(\cdot,\cdot)$ be a sesqui-holomorphic scalar-valued kernel on $D\times D$ which is also of class $P^+(D)$. Let D_o be a dense subset of D and let $S_o(\cdot)\varepsilon(D_o;\Omega(U:U))$, where Ω is a hyperbolic simply connected domain in $\mathbb{C}$. Assume that, for any unit vector $u\varepsilon U$, the scalar-valued kernel

$$(4.4) \qquad M_o^{(u)}(z,\zeta) = k(z,\zeta)[k_\Omega((S_o(z)u,u)_U, (S_o(\zeta)u,u)_U)]^{-1} \quad ; \quad z,\zeta\varepsilon D_o,$$

where $k_\Omega(\cdot,\cdot)$ is the Szegö kernel of Ω, is in $P_3(D_o)$. Then there exists a unique $S(\cdot)\varepsilon H(D;\Omega(U:U))$, extending $S_o(\cdot)$ to D.

Proof. Fix a unit vector $u \in U$ and define

$$s_o^{(u)}(z) \equiv (S_o(z)u,u)_U \quad , \quad z \in D_o. \tag{4.5}$$

It follows, since $S_o(\cdot) \in (D_o;\Omega(U:U))$, that $s_o^{(u)}(z) \in \Omega$ for every $z \in D_o$ and that (4.4) can be written as

$$M_o^{(u)}(z,\zeta) = k(z,\zeta)[k_\Omega(s_o^{(u)}(z), s_o^{(u)}(\zeta))]^{-1} \quad ; \quad z,\zeta \in D_o. \tag{4.6}$$

Let $\phi:\Omega \to \Delta$ be a Riemann mapping of Ω onto the unit disk Δ and define $t_o^{(u)} = \phi o s_o^{(u)}$. Therefore, $t_o^{(u)}(z) \in \Delta$ for every $z \in D_o$. Consider the kernel

$$L_o^{(u)}(z,\zeta) \equiv k(z,\zeta)[1-t_o^{(u)}(\zeta)\, t_o^{(u)}(z)] \quad ; \quad z,\zeta \in D_o.$$

In view of (2.2) and (4.6), this kernel admits the form

$$L_o^{(u)}(z,\zeta) = \{\phi'(s_o^{(u)}(z))\overline{\phi'(s_o^{(u)}(\zeta))}\}^{1/2} M_o^{(u)}(z,\zeta) \quad ; \quad z,\zeta \in D_o.$$

It follows, since $M_o^{(u)}(\cdot,\cdot)$ is of class $P_3(D_o)$ by assumption, that $L_o^{(u)}(\cdot,\cdot)$ is also of class $P_3(D_o)$. Therefore, using the scalar version of Theorem 2, we conclude the unique existence of an holomorphic function $t^{(u)} \in H(D)$ which extends $t_o^{(u)}$ to D, and furthermore $|t^{(u)}(z)| \leq 1$ for every $z \in D$. Since $t^{(u)}(z) = t_o^{(u)}(z) \in \Delta$ for every $z \in D_o$, the maximum modulus principle shows that, in fact, $t^{(u)}(z) \in \Delta$ for all $z \in D$. We now define $s^{(u)} = \phi^{-1} o t^{(u)}$. It follows easily that $s^{(u)} \in H(D)$, $s^{(u)}(z) = s_o^{(u)}(z)$ for every $z \in D_o$, and that $s^{(u)}(z) \in \Omega$ for all $z \in D$. Now, let z be a point of D and let $\{z_m\}$ be a sequence of points of D, converging to z. Since, using (4.5),

$$s^{(u)}(z) = \lim_{m\to\infty} s_o^{(u)}(z_m) = \lim_{m\to\infty} (S_o(z_m)u,u)_U$$

and since u is an arbitrary unit vector in U, we conclude that $\{(S_o(z_m)u,u)_U\}$ is a Cauchy-sequence for any vector $u \in U$. In particular,

$$\{(S_o(z_m)(u+\lambda w), u+\lambda w)_U\}$$

is a Cauchy-sequence for any $\lambda \in \mathbb{C}$ and any $u,w \in U$. Choosing $\lambda=1$ and $\lambda=i$, successively, we conclude that

$$\{(S_o(z_m)w,\ u)_U\}$$

is a Cauchy-sequence for any $u,w\varepsilon U$. This implies, since every Hilbert space is weakly complete, that $S_o(z_m)$, being an element of $\Omega(U:U)$ $\mathcal{B}(U:U)$, converges weakly in U to an operator $S(z)\varepsilon\mathcal{B}(U:U)$, $z\varepsilon D$. Thus $S(\cdot)\varepsilon(D;\mathcal{B}(U:U))$ and clearly $S(\zeta)=S_o(\zeta)$ for any $\zeta\varepsilon D_o$. Moreover, since for any unit vector $u\varepsilon U$, the function

$$s^{(u)}(z) = \lim_{m\to\infty} (S_o(z_m)u,u)_U = (S(z)u,u)_U$$

is holomorphic in $z\varepsilon D$, we conclude that $S(\cdot)\varepsilon H(D;\mathcal{B}(U:U))$. Finally, since also $s^{(u)}(z)\varepsilon\Omega$, $S(\cdot)$ is, in fact, in $H(D;\Omega(U:U))$ and the proof is complete.

The last theorem can be extended somewhat by letting Ω to be any hyperbolic domain in $\mathbb{C}$ (or a Riemann surface). In this case the Szegö-kernel $k_\Omega(\cdot,\cdot)$ in (2.2) should be modified by using any holomorphic cover map $\pi:\Delta\to\Omega$ instead of a Riemann mapping $\phi:\Omega\to\Delta$. We shall not pursue this point here. Instead, however, we shall strengthen the accretive version of Theorem 3 by establishing the following result:

Theorem 4. Let $k(\cdot,\cdot)$ *be a sesqui-holomorphic scalar-valued kernel on* $D\times D$ *which is also of class* $P^+(D)$. *Let* D_o *be a dense subset of* D *and let* $S_o(\cdot)\varepsilon(D_o;\mathcal{B}(U:U))$. *Assume that the operator-valued kernel*

$$M_o(z,\zeta) = k(z,\zeta)[S_o(z)+S_o(\zeta)^*] \quad ; \quad z,\zeta\varepsilon D_o,$$

belonging to $(D_o\times D_o;\mathcal{B}(U:U))$ *is positive-definite of order* 3, *i.e.* $M_o\varepsilon P_3(D_o:U)$. *Then there exists a unique holomorphic and accretive operator* $S(\cdot)\varepsilon H(D;A(U:U))$ *extending* $S_o(\cdot)$ *to* D.

Proof. Fix an arbitrary unit vector $u\varepsilon U$ and define, as in (4.5),

$$s_o^{(u)}(z) \equiv (S_o(z)u,u)_U \quad , \quad z\varepsilon D_o.$$

Consider the scalar-valued kernel

$$M_o^{(u)}(z,\zeta) = (M_o(z,\zeta)u,u)_U \quad ; \quad z,\zeta\varepsilon D_o.$$

It follows that

$$(4.7)\qquad M_o^{(u)}(z,\zeta) = k(z,\zeta)[s_o^{(u)}(z)+\overline{s_o^{(u)}(\zeta)}] \quad ; \quad z,\zeta\varepsilon D_o.$$

By assumption, this kernel belongs to $P_3(D_o)$ which means that

$$\sum_{i,j=1}^{3} k(z_i,z_j)[s_o^{(u)}(z_i)+\overline{s_o^{(u)}(z_j)}]\alpha_i\overline{\alpha_j} \geq 0$$

for every $z_1,z_2,z_3\varepsilon D_o$ and all $\alpha_1,\alpha_2,\alpha_3\varepsilon\mathbb{C}$. In this inequality we choose $\alpha_1=1$, $\alpha_2=\alpha_3=0$ and $z_1=z\varepsilon D_o$. This gives

$$k(z,z)[s_o^{(u)}(z)+\overline{s_o^{(u)}(z)}] \geq 0.$$

and since $k(z,z)>0$ we deduce that $Re\{s_o^{(u)}(z)\}\geq 0$ or that $s_o^{(u)}(z)\varepsilon\hat{R}$ for every $z\varepsilon D_o$. Since also $S_o(\cdot)\varepsilon(D_o;\mathcal{B}(U:U))$ we further deduce that $s_o^{(u)}(z)$ is finite for all $z\varepsilon D_o$. Now, the Möbius mapping

$$\phi(\omega) = \frac{\omega-1}{\omega+1}$$

is a Riemann mapping of the right half-plane $\mathcal{R}$ onto the unit disk Δ. Therefore, in view of (2.2), the Szegö kernel of $\mathcal{R}$ is

$$k_{\mathcal{R}}(\omega,\tau) = (\omega+\bar{\tau})^{-1}.$$

This permits writing $M_o^{(u)}(\cdot,\cdot)$ of (4.7) as

$$M_o^{(u)}(z,\zeta) = k(z,\zeta)[k_{\mathcal{R}}(s_o^{(u)}(z),\ s_o^{(u)}(\zeta))]^{-1} \quad ; \quad z,\zeta\varepsilon D_o.$$

The proof of the theorem follows now from Theorem 3 with $\mathcal{R}$ in place of Ω.

References

[1] ARONSZAJN, N.: Theorey of reproducing kernels, Tans. Amer. Math. Soc. 68 (1950), 337-404.

[2] BURBEA, J.: Pick's theorem with operator-valued holomorphic functions, Kōdal Math. J. 4 (1981), 495-507.

[3] BURBEA, J.: Operator-valued Pick's conditions and holomorphicity, Pacific J. Math. 98 (1982), 295-311.

[4] DONOGHUE, W.F.: Monotone Matrix Functions and Analytic Continuation, Springer-Verlag, New York, 1974.

[5] FITZGERALD, C.H., and HORN, R.A.: On quadratic and bilinear forms in function theory, Proc. London Math. Soc. (3) 44 (1982), 554-576.

[6] HINDMARSH, A.C.: Pick's conditions and analyticity, Pacific J. Math. 27 (1968), 527-531.

Department of Mathematics
University of Pittsburgh
Pittsburgh, Pennsylvania 15260,U.S.A.

ABOUT THE EQUALITY BETWEEN THE p-MODULE AND THE p-CAPACITY IN R^n

Petru Caraman (Iaşi)

C o n t e n t s

Introduction

In this paper, we establish that

$$M_p\Gamma(E_o,E_1,D) = \text{cap}_p(E_o,E_1,D), \tag{1}$$

where D is an open set in the Euclidean n-space R^n, $E_o,E_1 \subset R^n$ are two sets such that the distance $d(E_o,E_1) > 0$ and $E_o \cap \partial D$ or $E_1 \cap \partial D$ (∂D = the boundary of D) satisfies some additional conditions, $\Gamma(E_o,E_1,D)$ is the family of arcs joining E_o and E_1 in D, $M_p\Gamma$ is the p-module of an arc family Γ, and $\text{cap}_p(E_o,E_1,D)$ is the p-capacity of E_o,E_1 relative to D. In order to do this, we use a very recent result established by us in [9] about the completion of the class of admissible functions involved in the definition of M_p, when they are supposed to be bounded in R^n, continuous in D and 0 in the complement CD of D. In some of the cases considered in this paper, (1) still holds if E_o,E_1 and D are supposed to be contained in the one-point compactification $\overline{R^n}$ of R^n. In the particular case $n = 2$, we obtain

$$M_pZ = \text{cap}_pZ \quad (p > 1), \tag{2}$$

where $M_p Z = M_p \Gamma(B_o, B_1, Z)$ is the p-module of a topological cylinder Z with respect to the euclidean metric. This result represents a generalization of the case of (2) for topological cylinders with respect to the relative metric (obtained by us [9] in R^n). We show that (2) still holds in R^n if the corresponding topological cylinder with respect to the euclidean metric is quasiconformally equivalent to the unit cylinder. Finally, we make some historical remarks pointing out certain errors. As an application, we show that a certain exceptional set E^o of the unit sphere (corresponding to a quasiconformal mapping of the unit ball B) is of conformal capacity zero.

1. p-module and p-capacity in the case of open sets $D \subset R^n$

Now, let us precise the concepts contained in this paper.

Let χ be a family of continua γ and $F(\chi)$ the class of admissible functions ρ characterized by the following conditions: $\rho \geq 0$ in R^n is Borel measurable and $\int_\gamma \rho dH^1 \geq 1$, $\forall \gamma \in \Gamma$ ($\forall$ means "for every"), where H^1 is the Hausdorff linear measure. Then the p-module of χ is given as

$$M_p \chi = \inf_{\rho \in F(\chi)} \int \rho^p dm,$$

where dm is the volume element with respect to the Lebesgue n-dimensional measure and the integration is taken over the whole space R^n. If $F(\chi) = 0$, then $M_p \chi = \infty$, while if $\chi = \emptyset$, then $M_p \chi = 0$. In the particular case $p = n$, we write $M_n = M$ and call it the module.

Next, let us remind several definitions of the p-capacity of two closed sets $C_o, C_1 \subset \bar{D}$ and then, let us give a generalization of these definitions.

The p-capacity of two closed sets $C_o, C_1 \subset \bar{D}$ relative to a domain D is defined as

$$\mathrm{cap}_p(C_o, C_1, D) = \inf_u \int_{D-(C_o \cup C_1)} |\nabla u|^p dm,$$

where $\nabla u = (\partial u/\partial x^1, \ldots, \partial u/\partial x^n)$ is the gradient of u and the infimum is taken over all u, which are continuous in $D \cup C_o \cup C_1$, locally Lipschitzian in $D - (C_o \cup C_1)$ and assume the boundary values 0 on C_o and 1 on C_1. If D is contained in a fixed ball, then $\mathrm{cap} = \mathrm{cap}_n$ is called conformal capacity.

A function $u : D \to R$ is said to be ACL (_absolutely continuous on lines_) in D if $\forall\, I = \{x;\ \alpha^i < x^i < \beta^i \ (i = 1,\dots,n)\}$, $I \subset\subset D$ (i.e. $\bar{I} \subset D$), u is AC (absolutely continuous) on a.e. (almost every) line segment parallel to the coordinate axes, which means that if $I_i = \{x;\ x \in \bar{I},\ x^i = \alpha^i\}$ is a face of I and E is the set of the points $\xi \in I_i$ such that f is not AC on the segment $J_\xi = \{x;\ x = \xi + \lambda e_i,\ 0 < \lambda < \beta^i - \alpha^i\}$, then $m_{n-1}E = 0$, where m_{n-1} is the $(n-1)$-dimensional Lebesgue measure.

Now, we obtain two other definitions of $\operatorname{cap}_p(C_0, C_1, D)$ if we change the condition on u of being in $D - (C_0 \cup C_1)$ locally Lipschitzian by being ACL and of class C^1 (continuously differentiable), respectively. The equivalence of the last two definitions is established by us in [7] (Lemma 10). Since the condition that u is locally Lipschitzian in $D - (C_0 \cup C_1)$ is stronger than u is ACL and weaker than $u \in C^1$, it follows that also the first definition of $\operatorname{cap}_p(C_0, C_1, D)$ is equivalent to the other two.

Now, let us remind

PROPOSITION 1. _If_ $C_0, C_1 \subset D$ _are two disjoint closed sets and_ u _is admisible for_ $\operatorname{cap}_p(C_0, C_1, D)$, _then_ $\nabla u(x) = 0$ _a.e. in_ $C_0 \cup C_1$ _so that_

$$\int_{D-(C_0 \cup C_1)} |\nabla u|^p dm = \int_D |\nabla u|^p dm$$

and then

$$\operatorname{cap}_p(C_0, C_1, D) = \inf_u \int_{D-(C_0 \cup C_1)} |\nabla u|^p dm = \inf_u \int_D |\nabla u|^p dm$$

(see Proposition 2 of our paper [7], the proof is similar as in Lemma 3 of Gehring's paper [11]).

Arguing as in [11], we obtain

LEMMA 1. _In the hypotheses of the preceding proposition,_ u _is ACL in_ D.

From the preceding proposition and lemma, we deduce the following

COROLLARY. _We obtain for_ $\operatorname{cap}_p(C_0, C_1, D)$ $(C_0, C_1 \subset \bar{D}$ _closed disjoint sets) an equivalent definition by_

$$\operatorname{cap}_p(C_0, C_1, D) = \inf_u \int_D |\nabla u|^p dm, \tag{3}$$

where the functions u _admissible for_ $\operatorname{cap}_p(C_0, C_1, D)$ _are supposed to_ be ACL in D, not only in $D-(C_o \cup C_1)$.

PROPOSITION 2. If D is bounded, $C_o, C_1 \subset \bar{D}$ are two disjoint closed sets, then

$$\mathrm{cap}_p(C_o, C_1, D) = \inf_u \int_{D-(C_o \cup C_1)} |\nabla u|^p dm,$$

where the infimum is taken over all $u \in C^1$ in $D-(C_o \cup C_1)$ and with the boundary values 0 on C_o and 1 on C_1 (our paper [7], Lemma 10).

Following the general line of the argument of the preceding proposition, we have

LEMMA 2. If $C_o, C_1 \subset \bar{D}$ are two disjoint closed sets, then (3) holds, where the infimum is taken over all $u \in C^1$, $0 \leq u(x) \leq 1$ in D and with boundary values 0 on C_o and 1 on C_1.

If we denote the new infimum by $\mathrm{cap}_p^*(C_o, C_1, D)$ then, evidently, on account of the preceding corollary,

$$\mathrm{cap}_p(C_o, C_1, D) \leq \mathrm{cap}_p^*(C_o, C_1, D).$$

In order to obtain the opposite inequality, it is enough to prove that

$$\mathrm{cap}_p^*(C_o, C_1, D) \leq \int_D |\nabla u|^p dm \tag{4}$$

$\forall$ u admissible for $\mathrm{cap}_p(C_o, C_1, D)$. Given such a u, we may assume that $|\nabla u| \in L^p(D)$ (i.e. $\int_D |\nabla u|^p dm < \infty$), for otherwise there is nothing to prove. Next, fix $0 < a < 1/2$ and let

$$v(x) = \begin{cases} 0 & \text{if} \quad u(x) < a, \\ \dfrac{u(x)-a}{1-2a} & \text{if} \quad a \leq u(x) \leq 1-a, \\ 1 & \text{if} \quad u(x) > 1-a. \end{cases} \tag{5}$$

The set $E_a = \{x;\ a \leq u(x) \leq 1-a\}$ is a relatively closed subset of D, lying at a distance b>0 from $C_o \cup C_1$. Let $\varepsilon < \min[1, b, d(C_o, C_1)]$ and extend v to be 0 on a b-neighbourhood $C_o(b)$ of C_o and 1 on a b-neighbourhood $C_1(b)$ of C_1. Next, set $\delta(x) \equiv 1$ if $\partial D - (C_o \cup C_1) = \emptyset$ (where ∂D means the boundary of D) and $\delta(x) = \min\{1, d[x, \partial D - (C_o \cup C_1)]\}$ otherwise. By means of δ, let us define $y = x + \xi\delta(x)$, $\xi \in B(\varepsilon)$, where $B(\varepsilon)$ is a ball of radius ε centred at O. For a fixed ξ, y maps E_a

into $D-(C_o \cup C_1)$. The function

$$w(x,\varepsilon) = \frac{1}{\omega_n \varepsilon^n} \int_{B(\varepsilon)} v[x+\xi\delta(x)]dm,$$

where ω_n is the volume of the unit ball, is clearly continuous in $C_o \cup C_1$, taking boundary values 0 on C_o and 1 on C_1. To see that w is continuous in $D-(C_o \cup C_1)$ too, let $x, x' \in D-(C_o \cup C_1)$; then, since $0 \leq v(x) \leq 1$ in $D-[C_o(\varepsilon) \cup C_1(\varepsilon)]$ and, arguing as in the preceding proposition, it follows that

$$|w(x',\varepsilon) - w(x,\varepsilon)|$$

$$\leq \left|\frac{1}{\omega_n[\varepsilon\delta(x')]^n} - \frac{1}{\omega_n[\varepsilon\delta(x)]^n}\right| \int_{B[x'-x,\varepsilon\delta(x')]\cap B[\varepsilon\delta(x)]} v(x+y)\,dm$$

$$+ \frac{1}{\omega_n[\varepsilon\delta(x')]^n} \int_{B[x'-x,\varepsilon\delta(x')]-B[\varepsilon\delta(x)]} v(x+y)\,dm$$

$$+ \frac{1}{\omega_n[\varepsilon\delta(x)]^n} \int_{B[\varepsilon\delta(x)]-B[x'-x,\varepsilon\delta(x')]} v(x+y)\,dm$$

$$\leq \omega_n \left|\frac{1}{\omega_n[\varepsilon\delta(x')]^n} - \frac{1}{\omega_n[\varepsilon\delta(x)]^n}\right|$$

$$+ \frac{1}{\omega_n[\varepsilon\delta(x')]} \int_{B[x'-x,\varepsilon\delta(x')]-B[\varepsilon\delta(x)]} dm$$

$$+ \frac{1}{\omega_n[\varepsilon\delta(x)]^n} \int_{B[\varepsilon\delta(x)]-B[x'-x,\varepsilon\delta(x')]} dm,$$

which, on account of Radon-Nikodym's theorem, becomes arbitrarily small for $|x-x'|$ small enough.

From above, v is bounded, continuous and ACL in $D \cup C_o(\varepsilon) \cup C_1(\varepsilon)$, so that ∇v exists a.e. in $D \cup C_o(\varepsilon) \cup C_1(\varepsilon)$. Now, let us extend $\partial v/\partial y^i$ in $D \cup C_o(\varepsilon) \cup C_1(\varepsilon)$ by

$$\frac{\partial v(y)}{\partial y^i} = \begin{cases} \dfrac{\partial v(y)}{\partial y^i} & \text{if } \dfrac{\partial v(y)}{\partial y^i} \text{ exists,} \\ 0 & \text{otherwise.} \end{cases}$$

And now, $|\nabla u| \in L^p(D)$ implies, by (5), that $|\nabla v| \in L^p(D)$. Then, for every compact set $F \subset D$, by Hölder's inequality, we have

$$\int_F |\nabla v|\,dm \leq \left(\int_F |\nabla v|^p dm\right)^{1/p} (mF)^{1-1/p},$$

i.e. $|\nabla v|$ and then ∇v too (see S. Saks [21], Theorem 12, p.66) and a fortiori $|\partial v/\partial x^i|$ and $\partial v/\partial x^i$ are integrable over every compact set $F \subset D$.

Finally, arguing as in the preceding proposition, we conclude that

$$|\nabla w| \leq \frac{1+\varepsilon}{\omega_n \varepsilon^n} \int_{B(\varepsilon)} |\nabla v|\,dm.$$

Hence, applying Minkowski's inequality,

$$\begin{aligned}[\int_D |\nabla w(x,\varepsilon)|^p dm(x)]^{1/p} &= \frac{1+\varepsilon}{\omega_n \varepsilon^n} \left(\int_D \{\int_{B(\varepsilon)} |\nabla v[x+\xi\delta(x)]\,dm(\xi)\}^p dm(x)\right)^{1/p} \\ &\leq \frac{(1+\varepsilon)}{\omega_n \varepsilon^n} \int_{B(\varepsilon)} \{\int_D |\nabla v[x+\xi\delta(x)]|^p dm(x)\}^{1/p} dm(\xi),\end{aligned} \tag{6}$$

where we write $dm(x)$ and $dm(\xi)$ instead of dm in order to point out the variable of integration. But, it is easy to see that δ is Lipschitzian with Lipschitz constant 1. Hence, for instance,

$$\left|\frac{\partial v[x^1+\xi^1\delta(x),\ldots,x^n+\xi^n\delta(x)]}{\partial x^1}\right| = \left|\frac{\partial v(y^1,\ldots,y^n)}{\partial y^1}\right| \cdot \left|\frac{\partial [x^1+\xi^1\delta(x)]}{\partial x^1}\right|$$

$$= \left|\frac{\partial v(y)}{\partial y^1}\right| \cdot \left|1+\xi^1 \frac{\partial \delta(x)}{\partial x^1}\right| \leq \left|\frac{\partial v(y)}{\partial y^1}\right| (1+|\xi^1|) \leq \left|\frac{\partial v(y)}{\partial y^1}\right| (1+\varepsilon) < \frac{\left|\frac{\partial v(y)}{\partial y^1}\right|}{1-\varepsilon}.$$

Hence

$$|\nabla v[x+\xi\delta(x)]| < \frac{|\nabla v(x)|}{1-\varepsilon},$$

so that, $\forall \xi \in B(\varepsilon)$,

$$\int_D |\nabla v[x+\xi\delta(x)]|^p dm < \frac{1}{(1-\varepsilon)^p} \int_D |\nabla v(x)|^p dm \leq \frac{1}{[(1-\varepsilon)(1-2a)]^p} \int_D |\nabla u|^p dm,$$

which, by (6) and the fact that $w(x,\varepsilon)$ is admissible for $\mathrm{cap}_p^*(C_o, C_1, D)$, yields

$$\mathrm{cap}_p^*(C_o, C_1, D) \leq \int_D |\nabla w(x,\varepsilon)|^p dm <$$

$$\frac{1}{[(1-\varepsilon)^2(1-2a)\omega_n\varepsilon^n]^p}\{\int_{B(\varepsilon)}[\int_D|\nabla u(x)|^p dm(x)]^{1/p}dm(\xi)\}^p$$

$$= \frac{1}{[(1-\varepsilon)^2(1-2a)]^p}\int_D|\nabla u|^p dm,$$

whence

$$\operatorname{cap}_p^*(C_o,C_1,D) < \frac{1}{[(1-\varepsilon)^2(1-2a)]^p}\int_D|\nabla u|^p dm$$

and, letting $\varepsilon \longrightarrow 0$ and $a \rightarrow 0$, we obtain (4), as desired.

From this lemma and the preceding corollary, we deduce

COROLLARY 1. <u>The three definitions of the p-capacity $\operatorname{cap}_p(C_o,C_1,D)$, where $C_o,C_1 \subset \bar{D}$ are 2 disjoint closed sets and the admissible functions u involved in the definitions are supposed to be in D: ACL, of class C^1 and locally Lipschitzian, respectively, are equivalent and also equivalent to the corresponding definitions, where the properties of being ACL, of class C^1 or locally Lipschitzian are supposed to hold</u> only in $D-(C_o \cup C_1)$.

This corollary suggests the following generalization of the preceding definitions:

The p-capacity of two sets E_o,E_1 relative to a domain D, where $E_o,E_1 \subset \bar{D}$ and $d(E_o,E_1)>0$, is defined by

$$(7) \qquad \operatorname{cap}_p(E_o,E_1,D) = \inf_u \int_D |\nabla u|^p dm,$$

where the infimum is taken over all u, $0 \leq u(x) \leq 1$, which are continuous in $D \cup E_o \cup E_1$, locally Lipschitzian in D and assume the boundary values 0 on E_o and 1 on E_1.

We obtain the other two generalizations of $\operatorname{cap}_p(E_o,E_1,D)$ if, instead of being locally Lipschitzian in D, functions u are ACL or of class C^1 there respectively.

<u>Remarks</u>. 1. Clearly, in the particular case in which E_o,E_1 are closed, the preceding three new definitions come to the corresponding previous ones.

2. We had to suppose, in the last three definitions that u is ACL, locally Lipschitzian or of class C^1 in D and not in $D-(E_o \cup E_1)$ since the properties of being ACL, locally Lipschitzian or of class C^1 are meaningless in $D-(E_o \cup E_1)$ if this set is not

open. Of course, it is possible to try to extend the corresponding concepts for more general sets, but we prefered this way.

3. We had to introduce the condition $d(E_o, E_1) > 0$ since u is supposed to be continuous in $D \cup E_o \cup E_1$ and to have boundary values 0 on E_o and 1 on E_1 and then, if $\bar{E}_o \cap \bar{E}_1 \cap D \neq \emptyset$, $\bar{E}_o \cap E_1 \neq \emptyset$ or $E_o \cap \bar{E}_1 \neq \emptyset$, at the points of such a set, u has to be at the same time equal to 0 and to 1.

Arguing as in the preceding lemma, we obtain

COROLLARY 2. *If $E_o, E_1 \subset \bar{D}$ with $d(E_o, E_1) > 0$, then for $\mathrm{cap}_p(E_o, E_1, D)$ in (7), we have the same value, nc matter if the admissible functions u involved in the definition are supposed to be ACL, locally Lipschitzian or of class C^1 in D.*

From each of the above definitions for the p-capacity, we obtain the corresponding definition for the conformal capacity if we take $p = n$ and suppose that $\bar{D}$ is contained in a fixed ball.

Another generalization may be obtained if we get rid of the condition $E_o, E_1 \subset \bar{D}$. In the particular cases $E_1 \cap \bar{D} = \emptyset$ or $E_o \cap \bar{D} = \emptyset$, we assume, obviously that $\mathrm{cap}_p(E_o, E_1, D) = 0$ because, in the first case, the function u defined by $u|_{D \cup E_o} = 0$ and $u|_{E_1} = 1$ is admissible, while in the other case, the function u such that $u|_{D \cup E_1} = 1$ and $u|_{E_o} = 0$ id adnissible too and, in the both cases, $|\nabla u|\,|_D = 0$.

Finally let us mention a generalization by supposing that D is only open. In this case, we precise that, if there are components D_o with $\bar{D}_o \cap E_1 = \emptyset$, it is enough to consider only admissible functions u such that the restriction $u|_{\bar{D}_o} = 0$ and if there are components D_1 such that $\bar{D}_1 \cap E_o = \emptyset$, then it is enough to consider only admissible functions such that $u|_{\bar{D}_1} = 1$. In these two cases, $\nabla u|_{\bar{D}_o} = \nabla u|_{\bar{D}_1} = 0$, so that, if we eliminate from the open set D all the components of these two kinds, the value of $\mathrm{cap}_p(E_o, E_1, D)$ remains unchanged.

LEMMA 3. *If an open set D is a union of domains of the form $D = (\bigcup_k D_k) \cup (\bigcup_m D_m^o) \cup (\bigcup_q D_q^1)$, where D_m^o and D_q^1 are of the type D_o, D_1 introduced above, while $\bar{D}_k \cap E_o, \bar{D}_k \cap E_1 \neq \emptyset$ $(k = 1, 2, \ldots)$, then*

$$\mathrm{cap}_p(E_o, E_1, D) = \sum_k \mathrm{cap}_p(E_o, E_1, D_k).$$

Remark. From this Lemma 3 it follows that it does not matter if two different components of D have common boundary points and if some of these common boundary points belong to E_o or to E_1.

An important role in the generalization of Ziemer's relation (1) is played by

PROPOSITION 3. *If D is a domain and $\Gamma = \Gamma(E_o, E_1, D)$ is the family of arcs joining two disjoint sets $E_o, E_1 \subset \partial D$, then*

$$(8) \qquad M_p\Gamma = M_p^{oD}\Gamma = \inf_{\rho\in F^{oD}(\Gamma)} \int \rho^p dm,$$

where $F^{oD}(\Gamma)$ is the class of admissible functions $\rho \in F(\Gamma)$ bounded in R^n, continuous in D and 0 in CD (see our note [9]).

Arguing as in the preceding proposition, we have also

LEMMA 4. *Under the hypotheses of the preceding proposition, where D is only and open set, (8) still holds.*

LEMMA 5. *Let D be an open set, $\Gamma = \Gamma(E_o, E_1, D)$, $\Gamma' = \Gamma(\bar{E}_o, \bar{E}_1, D)$, and $\{\rho_m\}$ a sequence of functions $\rho_m \in F(\Gamma)$ such that*

$$(9) \qquad M_p\Gamma = \lim_{m\to\infty} \int \rho_m^p dm.$$

Let further

$$\tilde{\Gamma}_m = \{\gamma\in\Gamma;\ \int_\gamma \rho_m ds = \infty\}, \quad \Gamma_o = \bigcup_{m=1}^{\infty} \tilde{\Gamma}_m, \quad \Gamma'' = \Gamma - \Gamma_o,$$

$$\Gamma_1 = \{\gamma' \in \Gamma';\ \gamma' \subset \gamma'' \in \Gamma''\},$$

and

$$\Delta = D - (\bar{E}_o \cup \bar{E}_1) \neq \emptyset, \quad M_p^*\Gamma = \inf_{\rho\in F^*(\Gamma)} \int \rho^p dm,$$

where $F^*(\Gamma)$ is the class of admissible functions $\rho \in F(\Gamma)$ satisfying the additional condition $\int_\gamma \rho ds < \infty$ $\forall\ \gamma \in \Gamma$ rectifiable. Then

$$M_p\Gamma = M_p\Gamma'' = M_p^*\Gamma'' = M_p^*\Gamma_1 = M_p^{o\Delta}\Gamma_1.$$

Without loss of generality, we may suppose that all the arcs of Γ are rectifiable. Next, by Theorem 1 and 2 in Fuglede's paper [10],

$$M_p\Gamma_o \le \sum_{m=1}^{\infty} M_p\tilde{\Gamma}_m = 0 \quad \text{and} \quad M_p\Gamma'' \le M_p\Gamma \le M_p\Gamma'' + M_p\Gamma_o = M_p\Gamma'', \quad \text{whence}$$

$$M_p\Gamma = M_p\Gamma''.$$

Clearly, $M_p\Gamma'' \le M_p^*\Gamma''$, but $\rho_m \in F^*(\Gamma'')$ so that, taking into account (9),

$$M_p^*\Gamma'' \le \lim_{m\to\infty} \int \rho_m^p dm = M_p\Gamma = M_p\Gamma''.$$

Hence $M_p\Gamma'' = M_p^*\Gamma''$.
Since $\Gamma'' \subset \Gamma_1$, we have

(10) $\quad M_p^*\Gamma'' \le M_p^*\Gamma_1$.

In order to prove the opposite inequality, let $\rho \in F(\Gamma'')$. Then, $\forall\gamma_1 \in \Gamma_1$, there is an arc $\gamma'' \in \Gamma''$ such that $\gamma_1 \subset \gamma''$. Hence

$$1 \le \int_{\gamma''} \rho ds = \int_{\gamma_1} \rho ds + \int_{\gamma_2} \rho ds, \quad \text{where} \quad \gamma_2 = \gamma'' - \gamma_1 \quad \text{and} \quad \int_{\gamma_2} \rho ds \le \int_{\gamma''} \rho ds < \infty.$$

Since γ'' is rectifiable, we have $\gamma'' : (a,b) \to D, \gamma_1 : (c,d) \to D$ and $\gamma_2 : (a,c) \cup (c,d) \to D$, where $a < c < d < b$. Then, by Radon-Nikodym's theorem,

$$1 \le \lim_{\substack{a\to c\\ b\to d}} \int_{\gamma''} \rho ds = \int_{\gamma_1} \rho ds + \lim_{\substack{a\to c\\ b\to d}} \int_{\gamma_2} \rho ds = \int_{\gamma_1} \rho ds.$$

Hence $\rho \in F(\Gamma_1)$ and then $F^*(\Gamma'') \subset F^*(\Gamma_1)$. Consequently, $M_p^*\Gamma_1 \le M_p^*\Gamma''$, which, together with (10), yields $M_p^*\Gamma'' = M_p^*\Gamma_1$, as desired.

Finally, since $M_p^*\Gamma_1 \le M_p^{o\Delta}\Gamma_1$, γ is rectifiable, and $\rho \in F^{o\Delta}(\Gamma_1)$ (and then bounded in R^n), we have $\int_\gamma \rho ds < \infty$, so it remains to prove the opposite inequality. In order to do this, it is enough to show that $M_p^{o\Delta}\Gamma_1 \le \int \rho^p dm$ for an arbitrary $\rho \in F^*(\Gamma_1)$. Let

$$\tilde{\rho}(x) = \begin{cases} \dfrac{1+\varepsilon}{\omega_n} \displaystyle\int_B \rho[x + \varepsilon r(x) y] dm(y) & \text{if} \quad x \in \Delta, \\ 0 & \text{if} \quad x \in C\Delta, \end{cases}$$

where $\varepsilon > 0$ and $r(x) = d(x, \partial D \cup E_o \cup E_1)$. Arguing as in Theorem 4.2 of Hesse's Ph.D. thesis [15], we establish that $\tilde{\rho}$ is continuous in Δ

and, as in the case of the preceding proposition, we show that $\tilde{\rho}$ is bounded in R^n. Thus we have only to prove that $\tilde{\rho} \in F^*(\Gamma_1)$, or at least that $\int_{\gamma_1} \tilde{\rho} ds \geq 1$, where $\gamma_1 \in \Gamma_1$. Indeed, arguing as in the case of Hesse's Theorem 4.2, we obtain

$$\int_{\gamma_1} \tilde{\rho} ds = \frac{1+\varepsilon}{\omega_n} \int_{\gamma_1} \int_B [x+\varepsilon r(x) y] dm(y) ds(x)$$
$$= \frac{1+\varepsilon}{\omega_n} \int_B \int_{\gamma_1} \rho[x+\varepsilon r(x) y] ds(x) dm(y) = \frac{1+\varepsilon}{\omega_n} \int_B \int_{\gamma_1} \rho[z(x,y)] ds(x) dm(y)$$
$$= \frac{1+\varepsilon}{\omega_n} \int_B \int_{\gamma_1'} \rho(z) \frac{ds(x)}{ds(z)} ds(z) dm(y),$$

where $z(x,y) = x + \varepsilon r(x) y$ is a homeomorphism of Δ onto itself, $\gamma_1' = z(\gamma_1)$, and $ds(z)/ds(x) \leq 1 + \varepsilon$. Hence

$$\int_{\gamma_1} \tilde{\rho} ds = \frac{1+\varepsilon}{\omega_n} \int_B \int_{\gamma_1'} \rho(z) \frac{ds(x)}{ds(z)} ds(z) dm(y)$$
$$\geq \frac{1}{\omega_n} \int_B \int_{\gamma_1'} \rho(z) ds(z) dm(y) \geq \frac{1}{\omega_n} \int_B dm(y) = 1,$$

since $\int_{\gamma_1'} \rho(z) ds(z) \geq 1$ if $\gamma_1' \in \Gamma_1$ and $\int_{\gamma_1'} \rho(z) ds(z) = \infty > 1$ if $\gamma_1' \in \Gamma - \Gamma_1 \subset \Gamma_0$.

PROPOSITION 4. <u>If χ is the set of all continua in R^n that intersect two disjoint closed sets C_0 and C_1, where C_0 is assumed to contain the complement of a ball then</u> $M_p \chi \leq \text{cap}_p(C_0, C_1, R^n)$ (Ziemer [24], Lemma 3.1).

Now, we recall that the families Γ_m $(m = 1,2,\ldots)$ are said to be <u>separate</u> if there exist disjoint Borel sets E_m $(m = 1,2,\ldots)$ such that $\gamma \in \Gamma_m$ implies $H^1(\gamma - E_m) = 0$.

THEOREM 1. <u>Suppose that D is open, E_0, E_1 are two sets such that $d(E_0, E_1) > 0$ and for each component D_k of D with $\bar{D}_k \cap E_i \neq \emptyset$ $(i = 0,1)$, for $i = 0$ or $i = 1$, $\forall \xi \in \partial D_k \cap E_i$ we have</u>

$$\liminf_{\substack{x \to \xi \\ x \in D_k}} \inf_{\gamma} H^1[\gamma(E_i, x)] = 0, \tag{11}$$

<u>where the infimum is taken over all $\gamma = \gamma(E_i, x)$ joining x to E_i in D_k. Then (1) holds.</u>

Using the notation of Lemma 3, $D = (\bigcup_k D_k) \cup (\bigcup_m D_m^0) \cup (\bigcup_q D_q^1)$ and, clearly, the arc families $\Gamma_k = \Gamma(E_0, E_1, D_k)$, $\Gamma_m^0 = \Gamma(E_0, E_1, D_m^0)$,

$\Gamma_q^1 = \Gamma(E_o, E_1, D_q^1)$ (k,m,q = 1,2,...) are separate so that, on account of Lemma 2.1(c) of Väisälä's paper [22], we have

$$M_p\Gamma = \sum_k M_p\Gamma_k + \sum_m M_p\Gamma_m^o + \sum_q M_p\Gamma_q^1,$$

where

$$\Gamma = (\bigcup_k \Gamma_k) \cup (\bigcup_m \Gamma_m^o) \cup (\bigcup_q \Gamma_q^1) \quad \text{and} \quad \Gamma_m^o = \Gamma_q^1 = \emptyset \quad \text{yields} \quad M_p\Gamma_m^o = M_p\Gamma_q^1 = 0$$

$$(m,q = 1,2,...).$$

Then, taking into account Lemma 3, it follows that we may suppose, without loss of generality, that D itself is a domain, since otherwise, we can establish (11) for each component of D, separately.

The inequality

$$(12) \qquad M_p\Gamma \leq \text{cap}_p(E_o, E_1, D)$$

can be proved by the same argument as that used by Ziemer [24] for the preceding proposition since the additional condition "C_o, C_1 closed and C_o containing the complement of a ball" is not involved in the proof, while the use of Γ instead of χ rather simplifies things.

Next, in order to prove also the opposite inequality, it is sufficient to establish that

$$\text{cap}_p(E_o, E_1, D) \leq \int \rho^p dm,$$

where, on account of the preceding lemma and Lemma 2.3 of Ziemer [25], ρ may be supposed to belong to $F^{o\Delta}(\Gamma_1)$, with $\Delta = D - (\bar{E}_o \cup \bar{E}_1)$ and $\Gamma_1 = \{\gamma' \in T(\bar{E}_o, \bar{E}_1, D); \gamma' \subset \gamma \in \Gamma\} \supset \Gamma$ and then, $M_p\Gamma = M_p^{o\Delta}\Gamma_1 = M_p^{o\Delta}\Gamma$.

Now, assume that (11) holds for E_o and $\forall x \in \bar{D}$. Let

$$u(x) = \inf_\gamma \int_{\gamma(E_o,x)} \rho \, dH^1,$$

where the infimum is taken over all $\gamma = \gamma(E_o, x)$ joining x to E_o in $D \cup E_o \cup E_1$. Extend u to be 0 and 1 on the components of E_o and E_1, respectively, which are disjoint of $\bar{D}$. Then, clearly, $u_{|E_1} \geq 1$. Next, since $\rho \in F^{o\Delta}(\Gamma_1)$, it is easy to see that u has a boundary value 0 on $E_o \cap \bar{D}$. Indeed, if $\xi \in E_o \cap \bar{D}$ and $\sup_{x \in R^n} \rho(x) = M < \infty$, then by (11)

$$\lim_{\substack{x\to\xi\\x\in D}} u(x) = \liminf_{\substack{x\to\xi\\x\in D}} \gamma \int_{\gamma(E_o,x)} \rho dH^1 \leq \lim_{\substack{x\to\xi\\x\in D}} M \inf_{\gamma} \int_{\gamma(E_o,x)} dH^1$$

$$= M \liminf_{\substack{x\to\xi\\x\in D}} {}_{\gamma} H^1[\gamma(E_o,x)] = 0$$

and taking $u(x) = 0$ also for $x \in E_o - \bar{D}$, if follows that $u_{|E_o} = 0$.

It is easy to see that $u(x) < \infty$ in D. Indeed, from $u(x) \to 0$ as $x \to \xi$ and $x \in D$, it follows that, for $x_o \in D$ sufficiently close to ξ, $u(x_o) < \infty$. Hence, there exists an open arc $\gamma_o \subset D$, joining x_o to E_o, such that

$$\int_{\gamma_o} \rho dH^1 < \infty.$$

If $\tilde{x}$ is an arbitrary point of D, then there is a rectifiable arc $\gamma_1 = \gamma(x_o,\tilde{x}) \subset D$. Indeed, $\bar{\gamma}_1 = \gamma_1 \cup \{x_o,\tilde{x}\} \subset D$ is compact and then $d(\bar{\gamma}_1,\partial D) = r_o > 0$. Next, let us consider a covering of $\bar{\gamma}_1$ by balls $B(x,r)$ with $x \in \bar{\gamma}_1$, $r < r_o$. Since $\bar{\gamma}_1$ is compact, we can extract a finite covering $\{B(x_k,r_k)\}$ of $\bar{\gamma}_1$ by balls with $x_k \in \bar{\gamma}_1$, $r_k < r_o$ $(k = 1,\ldots,q)$. Clearly, these q balls can be ordered along $\bar{\gamma}_1$ form x_o towards $\tilde{x}$ and the intersection Δ_k of two successive balls be an open non-empty domain. Assuming that the indices of x_k,r_k are ordered accordingly and that, in each $\Delta_k = B(x_k,r_k) \cap B(x_{k+1},r_{k+1})$, we choose a point y_k, the polygonal line λ corresponding to the points $x_o,y_1,\ldots,y_{q-1},\tilde{x}$, is contained in D and $H^1(\lambda) < 2(q+1)r_o < \infty$. Next, let $\gamma \subset \gamma_o \cup \lambda$ be an arc joining E_o to $\tilde{x}_1$. Then, since $\rho(x) < M$, we obtain that

$$u(\tilde{x}) \leq \int_{\gamma} \rho dH^1 \leq \int_{\gamma_o} \rho dH^1 + \int_{\lambda} \rho dH^1 \leq \int_{\gamma_o} \rho dH^1 + 2(q+1) r_o M < \infty,$$

as desired.

Let us show that u is locally Lipschitzian in D. Indeed, if $x_o \in D$, $x_1,x_2 \in B(x_o,r) \subset D$, $\gamma_1 \in \Gamma(E_o,x_1,D \cup E_o)$ and λ is the line segment joining x_1 to x_2, then

$$u(x_2) \leq \int_{\gamma_1} \rho dH^1 + \int_{\lambda} \rho dH^1,$$

and, taking the infimum over all the arcs joining x_1 and E_o in $D \cup E_o$, we obtain

$$u(x_2) \leq \inf_{\gamma_1} \int_{\gamma_1} \rho dH^1 + \int_{\lambda} \rho dH^1 \leq u(x_1) + M \int_{\lambda} dH^1 = u(x_1) + M|x_1 - x_2|;$$

but then, $|u(x_1) - u(x_2)| \leq M|x_1 - x_2|$, i.e. u is locally Lipschitzian in D, as desired.

Next, let us prove that

$$(13) \qquad |\nabla u(x)| \leq \rho(x)$$

a.e. in Δ. Firstly, we observe that u is differentiable a.e. in D and, at a point of differentiability,

$$(14) \qquad |\nabla u(x)| = \sup_s \left|\frac{\partial u(x)}{\partial s}\right|,$$

where

$$\frac{\partial u(x)}{\partial s} = \lim_{t \to 0} \frac{u(x+te_s) - u(x)}{t}$$

is the directional derivative of u. Clearly,

$$u(x + te_s) \leq \int_{\gamma} \rho dH^1 + \int_{\lambda_t} \rho dH^1,$$

where $\gamma = \gamma(E_0, x)$, while λ_t is the line segment joining x and $x + te_s$. This inequality yields

$$u(x + te_s) \leq \inf_{\gamma} \int_{\gamma(E_0,x)} \rho dH^1 + \int_{\lambda_t} \rho dH^1 = u(x) + \int_{\lambda_t} \rho dH^1.$$

Hence, in the case $u(x) \leq u(x + te_s)$, we have

$$\left|\frac{u(x+te_s) - u(x)}{t}\right| = \frac{u(x+te_s) - u(x)}{t} \leq \frac{1}{t} \int_{\lambda_t} \rho dH^1.$$

Yet, arguing as above, we also get

$$u(x) \leq \inf_{\gamma} \int_{\gamma(E_0, x+te_s)} \rho dH^1 + \int_{\lambda_t} \rho dH^1 = u(x + te_s) + \int_{\lambda_t} \rho dH^1,$$

so that, for $u(x + te_s) \leq u(x)$, we deduce

$$\left|\frac{u(x+te_s) - u(x)}{t}\right| = \frac{u(x) - u(x+te_s)}{t} \leq \frac{1}{t} \int_{\lambda_t} \rho dH^1,$$

and then, in both cases,

$$\left|\frac{u(x+te_s)-u(x)}{t}\right| \leq \frac{1}{t}\int_{\lambda_t}\rho dH^1,$$

whence

$$\left|\frac{\partial u(x)}{\partial s}\right| = \lim_{t\to 0}\left|\frac{u(x+te_s)-u(x)}{t}\right| \leq \lim_{t\to 0}\frac{1}{t}\int_{\lambda_t}\rho dH^1 = \rho(x)$$

a.e. in Δ, since ρ is supposed to be continuous in Δ and the points of continuity are Lebesgue points implying the last part of the preceding relation, which, by (14), yields (13), as desired.

Now, we remark that $h = \inf_{E_1} u(x) \geq 1$. Let us consider the truncation

$$u^*(x) = \begin{cases} h & \text{for } u(x) > h, \\ u(x) & \text{for } 0 \leq u(x) \leq h, \end{cases}$$

and observe that u^*/h is admissible for $\mathrm{cap}_p(E_0, E_1, D)$. Indeed, $0 \leq u^*(x)/h \leq 1$, $u^*/h_{|E_0} = 0$ since $u_{|E_0} = 0$ and $u^*/h_{|E_1} = 1$ because, by the definition of h, $u(x) \geq h$ $\forall x \in E_1$ and u^* is the truncation of u at the level h. Yet, since u is locally Lipschitzian in D with Lipschitz constant M (as it was proved above), it is easy to see that u^*/h is also locally Lipschitzian in D with Lipschitz constant M/h. Indeed, if $x, y \in D_h = \{x \in D;\ u(x) < h\}$, then, clearly,

$$\left|\frac{u^*(x)}{h} - \frac{u^*(y)}{h}\right| = \frac{1}{h}|u(x) - u(y)| \leq \frac{M}{h}|x-y|;$$

if $x, y \in D \cap CD_h$, then $u^*(x) = u^*(y) = h$, so that

$$\left|\frac{u^*(x)}{h} - \frac{u^*(y)}{h}\right| = 0 \leq \frac{M}{h}|x-y|;$$

and if $x \in D_h$, $y \in D \cap CD_h$, then

$$\left|\frac{u^*(x)}{h} - \frac{u^*(y)}{h}\right| = \left|\frac{u(x)}{h} - 1\right| \leq \left|\frac{u(x)}{h} - \frac{u(y)}{h}\right| \leq \frac{M}{h}|x-y|.$$

Now, let us also verify that

$$(15) \qquad |\nabla u^*(x)| = \sup_s\left|\frac{\partial u^*(x)}{\partial s}\right| \leq \rho(x)$$

at any point of differentiability of u^*, i.e. a.e. in D. Indeed, suppose that x is such a point differentiability. If $x \in D_h$, then, evidently,

$$\frac{\partial u^*(x)}{\partial s} = \lim_{t\to 0} \frac{u^*(x+te_s)-u^*(x)}{t} = \lim_{t\to 0} \frac{u(x+te_s)-u(x)}{t} = \frac{\partial u(x)}{\partial s} \quad \forall s.$$

Hence

$$|\nabla u^*(x)| = |\nabla u(x)|.$$

If $x \in C\bar{D}_h \cap D$ for t sufficiently small, $u^*(x+te_s) = u^*(x) = h$, yielding $\left|\frac{\partial u^*(x)}{\partial s}\right| = 0$ $\forall s$ and then, $|\nabla u^*(x)| = 0$. Now, if $x \in \partial D_h \cap D$, then

$$\left|\frac{\partial u^*(x)}{\partial s}\right| = \lim_{t\to 0} \left|\frac{u^*(x+te_s)-u^*(x)}{t}\right| = \lim_{t\to 0} \left|\frac{u^*(x+te_s)-u(x)}{t}\right|$$
$$\leq \lim_{t\to 0} \left|\frac{u(x+te_s)-u(x)}{t}\right| \leq \left|\frac{\partial u(x)}{\partial s}\right|.$$

Hence, $|\nabla u^*(x)| \leq |\nabla u(x)|$ a.e. in D and, on account of (13)

(16) $\quad |\nabla u^*(x)| \leq \rho(x)$

a.e. in Δ.

Next, since $\rho \in F^{o\Delta}(\Gamma_1)$, it follows that $\rho_{|C\Delta} = 0$ and then, a fortiori, $\rho_{|\bar{E}_o \cup \bar{E}_1} = \rho_{|(\bar{E}_o \cup \bar{E}_1)\cap D} = 0$. In order to show that $|\nabla u^*(x)| \leq \rho(x)$ a.e. in D, it remains to prove that $|\nabla u^*(x)| = 0$ a.e. in $(\bar{E}_o \cup \bar{E}_1) \cap D$. Yet, u^* is continuous in $D \cup E_o \cup E_1$, and then, in particular, in $D \cap \bar{E}_o$ and in $D \cap \bar{E}_1$, so that $u^*_{|E_o} = 0$ and $u^*_{|E_1} = 1$ imply $u^*_{|\bar{E}_o \cap D} = 0$ and $u^*_{|\bar{E}_1 \cap D} = 1$, respectively. And now, since u^* is differentiable a.e. in D, ∇u^* exists a.e. in D and then a.e. in the sets $\bar{E}_o \cap D$ and $\bar{E}_1 \cap D$ which are measurable, so that almost all their points are of linear density in the direction of the coordinate axes (see Saks [21], p.298), implying that $|\nabla u^*(x)| = 0$ a.e. in $\bar{E}_o \cap D$ and $\bar{E}_1 \cap D$. Hence $|\nabla u^*(x)| = 0 = \rho(x)$ a.e. in $(\bar{E}_o \cup \bar{E}_1) \cap D$ which, together with (16), allows us to conclude that (15) holds.

We recall that $x_o \in R^n$ is said to be a _point of linear density in the direction of the coordinate axes_ if

$$\lim_{r\to 0} \frac{m_1\{[B(x_o,r)\cap E];\ x_o^1,\dots,x_o^{i-1},x_o^{i+1},\dots,x_o^n\}}{2r} = 1 \quad (i=1,..,n),$$

where $(E;x_o^1,\dots,x_o^{i-1},x_o^{i+1},\dots,x_o^n)$ denotes the intersection of E with the axis X_i and m_1 is the linear Lebesgue measure.

Finally, since u^*/h is admissible for $\operatorname{cap}_p(E_o,E_1,D)$ with $h\geq 1$ and taking into account (15), we obtain

$$\operatorname{cap}_p(E_o,E_1,D) \leq \frac{1}{h^p}\int_D |\nabla u^*|^p dm \leq \frac{1}{h^p}\int \rho^p dm \leq \int \rho^p dm.$$

Since ρ is an arbitrary function of $F^{o\Delta}(\Gamma_1)$, taking the infimum over all such ρ, we deduce that

$$(17)\qquad \operatorname{cap}_p(E_o,E_1,D) \leq M_p^{o\Delta}\Gamma_1 = M_p\Gamma,$$

which, together with (12), yields (1) if (11) holds for E_o.

When E_1 satisfies the condition (11), we repeat the above argument for E_1 instead of E_o and finally consider the function $v=1-u^*/h$, which is admissible for $\operatorname{cap}_p(E_o,E_1,D)$.

R e m a r k. We shall provide an example to show that it is possible to have a bounded simply connected domain D with two disjoint compact sets $E_o,E_1\subset\partial D$ such that all the points of $E_o\cup E_1$ are accessible by rectifiable arcs, but they do not verify the condition (11). Indeed, let $Q\subset R^2$ (Fig.1) be a square with the side $l=2$, E_o and E_1 be two closed segment of length 1, parallel to a side of Q with the endpoints (a,a') and (b,b'), respectively, and $\{\alpha_k\}$, $\{\beta_k\}$ be two sequence of segments parallel to E_o, with the endpoints a_k,b_k converging to a and b, respectively. Finally, let $D=Q-[E_o\cup E_1\cup(\bigcup_k\alpha_k)\cup(\bigcup_k\beta_k)]$. Then, all the points of $E_o\cup E_1$ are accessible from D by rectifiable arcs, but for each of them, except for a' and b', (11) does not hold. Indeed, if $\{x_k\}$ is, for instance, a sequence of points at a distance $1/2$ from the side of Q containing a_k and lying in D between α_{k-1} and α_k, then $x_k \to \xi_o\in E_o$, but

Fig.1

$$\varliminf_{k\to\infty}\ \inf_\gamma H^1[\gamma(E_o,x_k)]\geq 1,$$

while if a sequence $y_k\in D$ converges to the same ξ_o from E_o, then,

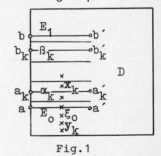

clearly,

$$\liminf_{k\to\infty} \inf_{\gamma} H^1[\gamma(E_o,y_k)] = 0.$$

Now, let D be a domain. For any two points $x,y \in D$, we define the <u>relative distance</u> $d_D(x,y)$ as the greatest lower bound of the lengths of all polygonal lines joining x to y in D. It is clear that $d_D(x,y)$ is a metric and that $d_D(x,y) \geq |x-y|$, with equality iff x,y lie in some convex subset of D. If $x \in D$ and $\xi \in \partial D$, we define $d_D(x,\xi)$ as the infimum of $\lim_{m\to\infty} d_D(x,x_m)$ on all sequences $\{x_m\}$ tending to ξ, with $x_m \in D$ $(m=1,2,\ldots)$.

<u>COROLLARY 1</u>. <u>If</u> D <u>is open</u>, $d(E_o,E_1)>0$ <u>and</u> $\forall D_k$ <u>with</u> $\bar{D}_k \cap E_i \neq \emptyset$ $(i=0,1)$ <u>for</u> $i=0$ <u>or</u> $i=1$, $\forall \xi \in \partial D \cap E_i$, $\lim d_{D_k}(\xi,x)=0$ (<u>i.e.</u> <u>the relative distance is continuous with respect to</u> D <u>on the corresponding set</u> $\partial D_k \cap E_i$), <u>then</u> (1) <u>holds</u>.

From the inequality (17) and Fuglede's [10] Theorem 1 quoted above, we deduce

<u>COROLLARY 2</u>. <u>If</u> χ <u>is the set of all continua in</u> R^n <u>that intersect the sets</u> E_o,E_1, <u>where</u> $d(E_o,E_1)>0$, <u>then</u>

$$M_p\chi = \mathrm{cap}_p(E_o,E_1,R^n).$$

<u>COROLLARY 3</u>. <u>Under the hypotheses of the proceding theorem</u>,

$$M\Gamma = \mathrm{cap}(E_o,E_1,D). \tag{18}$$

Now, in order to establish (1) under more general hypotheses, let us prove some properties of the p-capacity.

<u>LEMMA 6</u>. <u>The</u> p-<u>capacity</u> $\mathrm{cap}_p(E_o,E_1,D)$, <u>where</u> $D \subset R^n$ <u>is a domain and</u> $E_o=\emptyset$ <u>or</u> $E_1=\emptyset$ <u>or</u> $E_o, E_1 \subset R^n$ <u>are such that</u> $d(E_o,E_1)>0$, <u>satisfies the following conditions</u>:

(i) $\mathrm{cap}_p(\emptyset,E_1,D) = 0$.

(ii) $E_o \subset E_o' \Rightarrow$ ($\Rightarrow$ <u>means "implies"</u>) $\mathrm{cap}_p(E_o,E_1,D) \leq \mathrm{cap}_p(E_o',E_1,D)$.

(iii) $E_o \subset \bigcup_{k=1}^{m} E_o^k \Rightarrow \mathrm{cap}_p(E_o,E_1,D) \leq \sum_{k=1}^{m} \mathrm{cap}_p(E_o^k,E_1,D)$.

(i') $\mathrm{cap}_p(E_o,\emptyset,D) = 0$.

(ii') $E_1 \subset E_1' \Rightarrow \mathrm{cap}_p(E_o,E_1,D) \leq \mathrm{cap}_p(E_o,E_1',D)$.

(iii') $E_1 \subset \bigcup_{k=1}^{m} E_1^k \Rightarrow \mathrm{cap}_p(E_o,E_1,D) \leq \sum_{k=1}^{m} \mathrm{cap}_p(E_o,E_1^k,D)$.

The condition (i) is trivial since $u=1$ is an admissible function. It is easy to see that (ii) also holds since, if $E_o \subset E'_o$ and $\mathcal{U},\mathcal{U}'$ are the two corresponding classes of admissible functions, then $\mathcal{U}' \subset \mathcal{U}$. Hence

$$\mathrm{cap}_p(E_o,E_1,D) = \inf_{U} \int_D |\nabla u|^p dm \le \inf_{U'} \int_D |\nabla u|^p dm = \mathrm{cap}_p(E'_o,E_1,D).$$

In order to establish (iii), let $\mathcal{U}_k$ be the class of admissible functions for $\mathrm{cap}_p(E_o^k,E_1,D)$ $(k=1,\dots,q)$ and let $u(x) = \min[u_1(x),\dots,u_q(x)]$, where $u_k \in \mathcal{U}_k$ $(k=1,\dots,q)$. Clearly, $0 \le u(x) \le 1$, $u_{|E_o} = 0$ and $u_{|E_1} = 1$.

Now, let us prove that u is also locally Lipschitzian in D. Indeed, given a point $x \in D$, let V_k be a neighbourhood of x where u_k is Lipschitzian and let us show that u is Lipschitzian in any neighbourhood $W \subset \bigcap_{k=1}^{m} V_k$. To this purpose, let $y \in W$. Since u_k is Lipschitzian (let us precise: with Lipschitz constant M_k) in W, then it follows, in particular, that $|u_k(x) - u_k(y)| < M_k|x-y|$ $(k=1,\dots,q)$. Hence, if for instance $u(x) \le u(y)$, then, since, by definition, there is an integer $k \in [1,q]$ such that $u(x) = u_k(x)$, it follows that

$$|u(x) - u(y)| = |u_k(x) - u(y)| \le |u_k(x) - u_k(y)| \le M_k|x-y| \le M|x-y|,$$

where $M = \max(M_1,\dots,M_q)$. Therefore, there exists a neighbourhood W_x of x, $\forall x \in D$, where the preceding inequality holds, i.e. u is locally Lipschitzian in D, so we can conclude that $u \in \mathcal{U}$, which means that u is admisible for $\mathrm{cap}_p(E_o,E_1,D)$.

Next, let us show that

(19) $$|\nabla u(x)|^p \le \sum_{k=1}^{q} |\nabla u_k(x)|.$$

Let us consider a unit vector e_s of direction s and suppose first that

(20) $$u(x) \le u(x + |\Delta x| e_s).$$

Then, if $u_k(x) = \min_{1 \le i \le q} u_i(x) = u(x)$, we have

$$[\frac{|u(x)-u(x+|\Delta x|e_s)|}{|\Delta x|}]^p = [\frac{|u_k(x)-u(x+|\Delta x|e_s)|}{|\Delta x|}]^p$$

$$\leq [\frac{|u_k(x)-u_k(x+|\Delta x|e_s)|}{|\Delta x|}]^p \leq \sum_{k=1}^{q} [\frac{|u_k(u)-u_k(x+|\Delta x|e_s)|}{|\Delta x|}]^p .$$

Yet, since u and u_k are Lipschitzian with Lipschitz constant M in an n-dimensional neighbourhood W_x of x, then they are Lipschitzian also in a linear neighbourhood of x contained in the axis X_s passing through x and having the direction s, so that u and u_k $(k = 1,\dots,q)$ (considered as functions of a real variable) have a directional derivative $\partial u/\partial s$ and $\partial u_k/\partial s$ a.e. in $W_x \cap X_s$. Assume that the point considered above is such a point. Then, letting $|\Delta x| \to 0$ in the preceding inequality, we obtain

$$|\frac{\partial u(x)}{\partial s}|^p = \lim_{|\Delta x|\to 0} [\frac{|u(x)-u(x+|\Delta x|e_s)|}{|\Delta x|}]^p$$

$$\leq \overline{\lim_{|\Delta x|\to 0}} [\sum_{k=1}^{q} \frac{|u_k(x)-u_k(x+|\Delta x|e_s)|}{|\Delta x|}]^p$$

$$\leq \sum_{k=1}^{q} \overline{\lim_{|\Delta x|\to 0}} \frac{|u_k(x)-u_k(x+|\Delta x|e_s)|}{|\Delta x|}^p = \sum_{k=1}^{q} |\frac{\partial u_k(x)}{\partial s}|^p .$$

Next, since u and u_k $(k = 1,\dots,q)$ are Lipschitzian in W_x, then they are differentiable a.e. in W_x. Let us suppose that x is such a point at which u and u_k $(k = 1,\dots,q)$ are differentiable, i.e. at which the relation (14) holds. Yet then, the preceding inequality yields

$$|\nabla u(x)|^p = \sup_s |\frac{\partial u(x)}{\partial s}|^p \leq \sup_s \sum_{k=1}^{q} |\frac{\partial u_k(x)}{\partial s}|^p \leq \sum_{k=1}^{q} \sup_s |\frac{\partial u_k(x)}{\partial s}|^p$$

$$= \sum_{k=1}^{q} |\nabla u_k(x)|^p .$$

Thus, we have established (19) under the hypothesis (20).

Now, assume that the opposite inequality holds, i.e. that $u(x) > u(x+|\Delta x|e_s)$. Then, if $u_k(x+|\Delta x|e_s) = \min_{1\leq i\leq q} u_i(x+|\Delta x|e_s) = u(x+|\Delta x|e_s)$, we have

$$[\frac{|u(x)-u(x+|\Delta x|e_s)}{|\Delta x|}]^p = [\frac{|u(x)-u_k(x+|\Delta x|e_s)|}{|\Delta x|}]^p$$

$$\leq [\frac{|u_k(x)-u_k(x+|\Delta x|e_s)}{|\Delta x|}]^p \leq \sum_{k=1}^{q} [\frac{|u_k(x)-u_k(x+|\Delta x|e_s)|}{|\Delta x|}]^p$$

and, as above, in the hypothesis (20), we obtain (19) in this case as well.

Finally, since $u \in \mathcal{U}$, we deduce by (19) that

$$\operatorname{cap}_p(E_o,E_1,D) \leq \int_D |\nabla u|^p dm \leq \sum_{k=1}^{q} \int_D |\nabla u_k|^p dm.$$

Hence, since each u_k is an arbitrary function of $\mathcal{U}_k$,

$$\operatorname{cap}_p(E_o,E_1,D) \leq \sum_{k=1}^{q} \operatorname{cap}_p(E_o^k,E_1,D),$$

as desired.

The same argument still holds for (iii'), but with $u(x) = \max[u_1(x),\ldots,u_q(x)]$.

(i') is trivial since $u = 0$ is an admissible function; arguing as for (ii), we also establish (ii').

R e m a r k. In order to be able to obtain the subadditivity of the p-capacity in the cases (iii) and (iii') (i.e. the corresponding inequality with $q = \infty$), we have, to suppose for instance, that all u_k $(k = 1,2,\ldots)$ are locally Lipschitzian with a fixed Lipschitz constant $K < \infty$, or at least that the set of all these constants is bounded (which is equivalent).

Let $q(x,y) = |x-y|/(1+|x^2|)^{1/2}(1+|y|^2)^{1/2}$ be the spherical distance between x and y. If E_1,E_2 are two sets, then $q(E_1/E_2) = \inf\{q(x,y) : x \in E_1, y \in E_2\}$.

PROPOSITION 5. For each $p > 0$, p-almost every bounded curve is rectifiable (Väisäla [23], Theorem 2.3).

This proposition means that if Γ_o is the family of all bounded curves, which are not rectifiable, then $M_p\Gamma_o = 0$.

THEOREM 2. Suppose that D is open, E_o,E_1 are such that $q(E_o,E_1) > 0$ (hence one of the sets E_o,E_1 is bounded) and for each component D_k of D with $\bar{D}_k \cap E_i \neq \emptyset$ $(i = 0,1)$, $\forall \xi \in \partial D_k \cap E_i$ (where E_i is the bounded set) either (11) is satisfied or ξ is not accessible from D_k by rectifiable arcs. Then (1) holds.

Suppose that E_o is bounded. As we have stated in the proof of the preceding theorem, we may assume, without loss of generality, that D is a domain and $E_o \cap \bar{D}$, $E_1 \cap \bar{D} \neq \emptyset$. Next, let us write $D \cap E_o = E' \cup E''$, where E' is the set of the points of $E_o \cap \partial D$ inaccessible from D by rectifiable arcs. Since E_o - and a fortiori E' - is supposed to be bounded, then $E'(r)$ is also bounded, where $E(r)$ is assumed to be the open set of points (of R^n), which lie within a distance r from E. Then, for $r < d(E_o, E_1)$, $\Gamma_1 = \Gamma[E', E'(r), D \cap E'(r)] < \Gamma(E', E_1, D)$ - i.e. Γ_1 is minorised by $\Gamma(E', E_1, D)$ (see Fuglede [10]) - where, evidently, $D \cap E'(r) \subset E'(r)$ is bounded. But then, Theorem 1 of Fuglede's paper [10], combined with the preceding proposition yields

$$M_p\Gamma(E', E_1, D) \leq M_p\Gamma[E', E'(r), D \quad E'(r)] = 0$$

and

$$M_p\Gamma(E'', E_1, D) \leq M_p\Gamma(E_o, E_1, D) \leq M_p\Gamma(E', E_1, D) + M_p\Gamma(E'', E_1, D)$$

$$= M_p\Gamma(E'', E_1, D).$$

Hence

$$M_p\Gamma(E_o, E_1, D) = M_p\Gamma(E'', E_1, D). \tag{21}$$

Next, let us show that

$$\operatorname{cap}_p(E', E_1, D) = 0. \tag{22}$$

Firstly, let us denote $E(r_1, r_2) = \{x; r_1 < d(E,x) < r_2\}$ and $E(r_1, \infty) = \{x; d(E,x) > r_1\}$. Again, on account of the preceding proposition, we have

$$M_p\Gamma[E', E'(r_1, r_2), D \cap E'(r_2)] = 0. \tag{23}$$

Yet, all the points $\xi \in \partial[D \cap E'(r_2)] \cap E'(r_1, r_2)$ verify a condition of the form (11), i.e.

$$\lim_{\substack{x \to \xi \\ x \in D \cap E'(r_2)}} \inf_{\gamma} H^1\{\gamma[E'(r_1, r_2), x]\} = 0,$$

since, for a given $\xi \in \partial[D \cap E'(r_2)] \cap E'(r_1, r_2) \subset E'(r_1, r_2)$, every $x \in D \cap E'(r_2)$ sufficiently close to ξ belongs to $E'(r_1, r_2)$ (which is an open set) so that such an x may be joined to $E'(r_1, r_2)$ by an arc of length zero. Thus, the hypotheses of the preceding theorem, are satisfied what, by (23), yields

$$\operatorname{cap}_p[E', E'(r_1, r_2), D \cap E'(r_2)] = M_p\Gamma[E', E'(r_1, r_2), D \cap E'(r_2)] = 0.$$

Hence

$$(24)\qquad \operatorname{cap}_p[E',E'(r_1,\infty),D] = \inf_u \int_D |\nabla u|^p dm = \inf_u \int_{D\cap E'(r_2)} |\nabla u|^p dm$$

$$= \operatorname{cap}_p[E',E'(r_1,r_2),D\cap E'(r_2)] = 0.$$

If we denote by $\mathcal{U}(E_0,E_1,D)$ the class of admissible functions for $\operatorname{cap}_p(E_0,E_1,D)$, then for $r_1 < d(E_0,E_1)$ we have $\mathcal{U}[E',E'(r_1,\infty),D] \subset \mathcal{U}(E',E_1,D)$, since $E_1 \subset E'(r_1,\infty)$. Hence, taking into account the condition (ii′) of the preceding lemma and (24), we obtain

$$\operatorname{cap}_p(E',E_1,D) \leq \operatorname{cap}_p[E',E'(r_1,\infty),D] = 0,$$

and this implies (22).

Now, by the preceding lemma, arguing as above for the p-module, we deduce

$$\operatorname{cap}_p(E'',E_1,D) \leq \operatorname{cap}_p(E_0,E_1,D) \leq \operatorname{cap}_p(E',E_1,D) + \operatorname{cap}_p(E'',E_1,D)$$

$$= \operatorname{cap}_p(E'',E_1,D),$$

so that

$$(25)\qquad \operatorname{cap}_p(E_0,E_1,D) = \operatorname{cap}_p(E'',E_1,D).$$

Now, Theorem 1 assures us that

$$\operatorname{cap}_p(E'',E_1,D) = M_p\Gamma(E'',E_1,D),$$

which, together with (25) and (21), yields

$$\operatorname{cap}_p(E_0,E_1,D) = \operatorname{cap}_p(E'',E_1,D) = M_p\Gamma(E'',E_1,D) = M_p\Gamma(E_0,E_1,D),$$

as desired.

<u>COROLLARY 1</u>. <u>If</u> D <u>is open</u> $E_0 \cap E_1 = \emptyset$, $q(D\cap E_0, D\cap E_1) > 0$ <u>and for</u> i <u>corresponding to the bounded set</u> $\partial D \cap E_i$, <u>the conditions, of the preceding theorem are satisfied, then</u> (1) <u>holds</u>.

<u>COROLLARY 2</u>. <u>If</u> D <u>is open</u>, $\partial D \cap E_0$ <u>is bounded</u>, $E_0 \cap E_1 = \emptyset$, $d(E_0 \cap \bar{D}, E_1 \cap \bar{D}) > 0$ <u>and</u> $E_0 \cap \partial D$ <u>verifies the conditions of the preceding theorem, then</u> (1) <u>holds</u>.

About the Equality between the p-Module and the p-Capacity in R^n

In order to obtain (1) under some other hypotheses, let us recall first some concepts and preliminary results.

A domain $D \subset R^n$ is said to be m-connected at a boundary point $\xi \in \partial D$ if m is the least integer, for which there is an arbitrary small neighbourhood U_ξ of ξ such that $U_\xi \cap D$ consists of m components.

We say that a domain D is m-smooth at a boundary pint ξ if

1° D is m-connected at ξ;

2° there exists $\lambda_o > 0$ and a neighbourhood U_ξ with the property that $U_\xi \cap D$ consists of m components $\Delta_1, \dots, \Delta_m$, and if V_ξ is any neighbourhood of ξ contained in U_ξ, there is a neighbourhood $V'_\xi \subset V_\xi$ of ξ so that $M\Gamma(E_1, E_2, V_\xi \cap \Delta_i) \geq \lambda_o$, whenever E_1, E_2 are disjoint connected sets in Δ_i $(i = 1, \dots, m)$, which meet both V_ξ and V'_ξ.

If D is m-smooth at each point of a set $E \subset \partial D$, it is said to be m-smooth on E.

R e m a r k. The above definition of the m-smoothness is a modified version of J. Hesses's definition, in which D is not supposed to be m-connected. Yet then, the m-smoothness is no more a characteristic of the boundary point ξ, but depends on the neighbourhood U_ξ. Thus, for instance, in Fig.2 ξ is 1-smooth according to our definition and 1-, 2-, 3- and 4-smooth at the same time according to Hesse's definition, depending on the fact that the selected neighbourhood U_ξ of his definition is U_1, U_2, U_3 or U_4.

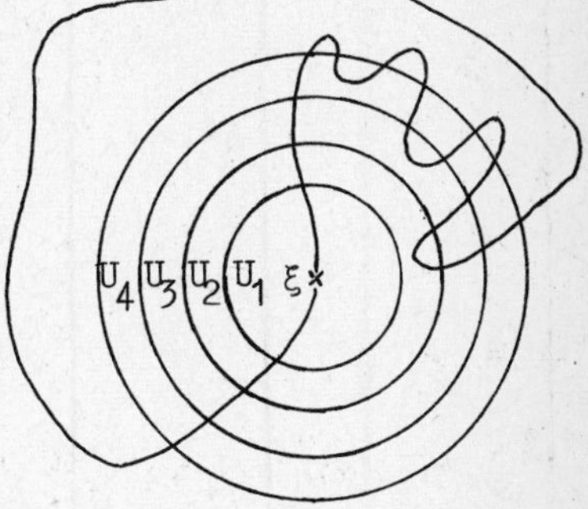

Fig.2

Let $E_o, E_1 \subset \bar{D}$ be two sets with $d(E_o, E_1) > 0$. For $r \in (0,1)$, let $E_i(r) = \{x;\ d(x, E_i) \leq r\}$ $(i = 0, 1)$. If $\rho \in F[\Gamma(E_o, E_1, D)]$, let $L(\rho, r) = \inf_\gamma \int_\gamma \rho\, dH^1$, where the infimum is taken over all locally rectifable $\gamma \in \Gamma[E_o(r), E_1(r), D]$. It is easy to verify that $0 \leq r_1 \leq r_2 \leq 1$ implies $L(\rho, r_2) \leq L(\rho, r_1)$. Then, let us define (following Hesse) $L(\rho) = \lim_{r \to 0} L(\rho, r)$.

PROPOSITION 6. *Let $D \subset \overline{R^n}$ be a domain and $\rho \in F[\Gamma(E_o, E_1, D)]$, where E_o, E_1 are two disjoint compact nonempty sets. Then, $L(\rho) \geq 1$ iff $\forall \varepsilon > 0$ there exists a $\delta = \delta(\varepsilon) \in (0,1)$ such that $\frac{\rho}{1-\varepsilon} \in F\{\Gamma[E_o(r), E_1(r), D]\}$ for all $0 < r \leq \delta$* (Hesse [15], Theorem 4.16).

PROPOSITION 7. *Let* D *be a domain in* $\overline{R^n}$, E *and* F *compact disjoint non-empty sets in* $\bar{D}$. *Suppose that at each point of* $(E \cup F) \cap \partial D$, D *is m-smooth for some* m. *Let* $\Gamma = \Gamma(E,F,D)$ *and* $\mathfrak{a}_o = \{\rho + \varepsilon\theta;\ \rho \in F(\Gamma) \cap L^n(R^n)$ *and* $\varepsilon \in (0,1)\}$, *where*

$$\theta(x) = \begin{cases} 1/|x|\log|x| & \text{if } |x| \geq e, \\ 1/e & \text{if } |x| \leq e. \end{cases}$$

Then, $L(\rho) \geq 1 \quad \forall\rho \in \mathfrak{a}_o$ (Hesse [15], Theorem 4.27).

R e m a r k. In our opinion, the proof contains some inaccuracies. For instance, the author asserts that "since for all $i = 1,2,\ldots$, $\int_{\gamma_i} \rho ds < 2$ and since $\rho \geq \varepsilon\theta$ for some $\varepsilon \in (0,1)$, it follows that all the curves γ_i lie in some fixed closed euclidean ball and they are all rectifiable." However, the preceding two inequalities can give

$$\frac{H^1(\gamma_i)\varepsilon}{|x_i|\log|x_i|} \leq \theta(x_i)\varepsilon\int_{\gamma_i} ds \leq \int_{\gamma_i}\theta\varepsilon ds \leq \int_{\gamma_i} \rho ds < 2,$$

where $|x_i| = \sup\{|x| : x \in \gamma_i\}$. Yet, since $\gamma_i \subset \bar{D} \subset \overline{R^n}$, where γ_i are supposed to be locally rectifiable, then it is possible to have $x_i = \infty$ for some i or, at least, $|x_i| \to \infty$ as $i \to \infty$, contradicting the boundedness of the family of curves and even their rectifiability. Then, as a consequence of this mistake, Hesse does not consider the case $F = \{\infty\}$, which is compatible with the hypotheses of the theorem, because his erroneous conclusion that all γ_i lie in a closed ball implies that such a case cannot happen. We establish the preceding proposition in R^n changing slightly the hypotheses and use some parts of the proof of the preceding proposition, but, for the sake of completeness - since (as far as we know) the mentined proof has not been published and it is to be found only in J. Hesse's Ph.D.Thesis (1972) - we shall give a complete proof.

Let us denote first $L_1(\rho,r) = \inf_{\gamma}\int_{\gamma}\rho dH^1$, where the infimum is taken over all $\gamma \in \Gamma[E_o(r),E_1,D]$ and $L_1(\rho) = \lim_{r\to 0} L_1(\rho,r)$.

Arguing as in Proposition 6, we have

LEMMA 7. *If* $D \subset \overline{R^n}$ *is a domain and* $\rho \geq 0$ *is Borel measurable, then* $L_1(\rho) \geq 1$ *iff* $\forall\varepsilon \in (0,1)$ *there exists a* $\delta(\varepsilon) \in (0,1)$ *such that* $\frac{\rho}{1-\varepsilon} \in F\{\Gamma[E_o(r),E_1,D]$ *for all* $0 < r \leq \delta(\varepsilon)$, *where* $E_o(r) = \{x \in R^n;\ d(x,E_o) < r\}$.

Proposition 8. Let $A=\{x\in R^n;\ r_1<|x|<r_2\}$ *and* $E_1,E_2\subset A$ *be two disjoint sets such that each sphere* $S(r)$ $(r_1<r<r_2)$ *contains at least one point of each* E_i $(i=1,2)$. *Then*

$$M\Gamma[E_1,E_2,A-(E_1\cup E_2)]\geq\frac{2^n}{A_o}\log\frac{r_2}{r_1},$$

where A_o *is a constant depending only on the dimension* n (Väisälä [23], Theorem 3.9 or our monograph [5], Proposition 11, chapt.5, part I).

LEMMA 8. *Let* D *be a domain,* $E_o,E_1\subset\bar{D}\subset R^n$ *with* $d(E_o,E_1)>0,E_o$ *compact and suppose that* D *is* m-*smooth on* $E_o\cap\partial D$ *for some* m. *Then* $L_1(\rho)\geq 1$ $\forall\rho\in\mathcal{O}\!\iota'_o=\{\rho\in F(\Gamma)\cap L^n(R^n);\ \rho(x)\geq\alpha_F>0\ \ \forall F\subset R^n$ *compact*$\}$.

Assume first that $E_o=\{\xi\}$. Then, suppose, (in order to prove it is false), that $L_1(\rho)<1$. Fix any sequence $\{\eta_i\}$, $\eta_i\in(0,1)$ $(i=1,2,\ldots)$ so that

$$\sum_{i=1}^{\infty}\eta_i<\infty.$$

Since $\rho\in L^n$, the absolute continuity of the integral allows us to choose a strictly decreasing sequence $r_i\in(C,1)$ such that $\lim_{i\to\infty}r_i=0$,

$$(26)\qquad\int_{B(\xi,r_i)}\rho^n dm<\eta_i^n\lambda_o\qquad(i=1,2,\ldots).$$

We may also assume that $E_o(r_i)\cap E_1=\emptyset$. We notice that $E_o(r_i)=B(\xi,r_i)$ $(i=1,2,\ldots)$. Choose a sequence of arcs $\gamma_i\in\Gamma[E_o(r_i),E_1,D]$ so that

$$\int_{\gamma_i}\rho dH^1<L_1(\rho,r_i)+\eta_i\qquad(i=1,2,\ldots).$$

Since, $\forall i\in N$, we have

$$\int_{\bar{\gamma}_i}\rho dH^1=\int_{\gamma_i}\rho dH^1<L_1(\rho,r_1)+\eta_i\leq L_1(\rho)+\eta_i<1+1=2$$

and since ρ is supposed to be bounded away from zero on compact sets and the closure $\bar{\gamma}_i$ of γ_i (obtained by adding to γ_i its endpoints) (see our monograph [5], p. 51) is compact, we get

$$0<a_{\bar{\gamma}_i}H^1(\bar{\gamma}_i)\leq\int_{\bar{\gamma}_i}\rho dH^1<2.$$

Hence

$$H^1(\bar{\gamma}_i) < 2/a_{\bar{\gamma}_i} < \infty,$$

i.e. γ_i is rectifiable. Since one of its endpoints lies in $B(\xi,r_i)$, we can decompose γ_i into $\gamma_i = \alpha_i \cdot \alpha_i' \cdot \chi_i$, where $\alpha_i \in \Gamma[B(\xi,r_i), S(\xi,r_{i-1}), B(\xi,r_{i-1})]$, $\alpha_i' \in \Gamma[S(\xi,r_{i-1}), S(\xi,r_{i-2}), B(\xi,r_{i-2})]$, $\chi_i \in \Gamma[S(\xi,r_{i-2}),E_1,D]$, If ξ_i is the endpoint of γ_i contained in $E_o(r_i) = B(\xi,r_i)$, then, clearly, $\xi_i \to \xi$ as $i \to \infty$. Assume $\xi \in \partial D$. Since D is supposed to be m-smooth at ξ for some m, let U_ξ be the neighbourhood of ξ, involved in the definition of the smoothness. Then, it follows that

$$U_\xi \cap D = \bigcup_{k=1}^{m} \Delta_k.$$

We may assume (choosing, if necessary, a subsequence) that all ξ_i are contained in one of the Δ_k, which we denote by Δ. The m-smoothness of D at ξ implies the existence of a constant $\lambda_o > 0$ such that

$$\text{(27)} \qquad M\Gamma[\overline{\alpha_i'}, \overline{\alpha_{i-1}}, B(\xi,r_{i-2})] \geq \lambda_o \qquad (i = 1,2,\dots).$$

If $\xi \in D$, then the preceding inequality still holds on account of Proposition 8 (we have to choose, if necessary, a subsequence of $\{\gamma_i\}$). Next, since (26) yields

$$\text{(28)} \qquad \int_{B(\xi,r_i-2)} \left(\frac{\rho}{\eta_{i-2}}\right)^n dm < \lambda_o,$$

it follows form (27) that there is a rectifiable arc β_i in $B(\xi,r_{i-2})$ connecting $\overline{\alpha_i'}$ and $\overline{\alpha_{i-1}}$, such that

$$\text{(29)} \qquad \int_{\beta_i} \rho dH^1 < \eta_{i-2},$$

because otherwise, $\rho/\eta_{i-2} \in F\{\Gamma[\overline{\alpha_i'}, \overline{\alpha_{i-1}}, B(\xi,r_{i-2})]\}$ and (28) would contradict (27). The arc $\alpha_i \cdot \alpha_i'$ contains a subarc α_i'' joining an endpoint of β_{i+1} to an endpoint of β_i $(i = 3,4,\dots)$. Yet then

$$\int_{\alpha_i''} \rho dH^1 + L_1(\rho,r_{i-2}) \leq \int_{\alpha_i \cdot \alpha_i'} \rho dH^1 + \int_{\chi_i} \rho dH^1 = \int_{\gamma_i} \rho dH^1 < L_1(\rho,r_i) + \eta_i.$$

Hence

$$(30) \qquad \int_{\alpha_i''} \rho\, dH^1 \le L_1(\rho, r_i) - L_1(\rho, r_{i-2}) + \eta_i \qquad (i = 3, 4, \dots).$$

Define a locally rectifiable arc

$$\tilde{\gamma}_k = \dots (\alpha_{i+1}'' \cdot \beta_{i+1}) \cdot (\alpha_i'' \cdot \beta_i) \dots (\alpha_k'' \cdot \beta_k) \cdot \tau_{k-1},$$

where τ_{k-1} is a subarc of γ_{k-1} joining an endpoint of β_k to E_1. We get $\tilde{\gamma}_k \in \Gamma(E_0, E_1, D)$ and

$$1 \le \int_{\tilde{\gamma}_k} \rho\, dH^1 \le \sum_{i=k}^{\infty} \int_{\alpha_i''} \rho\, dH^1 + \sum_{i=k}^{\infty} \int_{\beta_i} \rho\, dH^1 + \int_{\tau_{k-1}} \rho\, dH^1 \qquad (k = 3, 4, \dots),$$

where

$$\int_{\tau_{k-1}} \rho\, dH^1 \le \int_{\gamma_{k-1}} \rho\, dH^1 \le L_1(\rho, r_{k-1}) + \eta_{k-1} \qquad (k = 3, 4, \dots).$$

Hence, taking into account (29) and (30), we get

$$1 \le \int_{\tilde{\gamma}_k} \rho\, dH^1 \le \sum_{i=k}^{\infty} [L_1(\rho, r_i) - L_1(\rho, r_{i-2})] + \sum_{i=k}^{\infty} \eta_i + \sum_{i=k}^{\infty} \eta_{i-2} +$$
$$+ L_1(\rho, r_{k-1}) + \eta_{k-1},$$

whence

$$1 \le \int_{\tilde{\gamma}_k} \rho\, dH^1 \le L_1(\rho) + [L_1(\rho) - L_1(\rho, r_{k-2})] + 2 \sum_{i=k-2}^{\infty} \eta_i \qquad (k = 3, 4, \dots).$$

For a large k, the last part of the preceding inequality is less than 1, which is absurd.

Now, let us consider the general case of E_0 compact in R^n (and then bounded). Let $\rho \in \mathfrak{A}_0'$ and suppose again (to prove it is false) that $L_1(\rho) < 1$. By Lemma 7, there exists an $\varepsilon \in (0, 1)$, a strictly decreasing sequence $\{r_i\}$, $r_i \in (0, 1)$ $(i = 1, 2, \dots)$ with $r_i \to 0$ as $i \to \infty$, and a sequence of locally rectifiable arcs $\gamma_i \in \Gamma[E_0(r_i), E_1, D]$ such that

$$\int_{\gamma_i} \rho\, dH < 1 - \varepsilon \qquad (i = 1, 2, \dots).$$

Hence, since ρ is bounded away from zero on compact sets, we deduce (arguing as above) that all γ_i are rectifiable. Next, since the endpoints x_i and y_i of γ_i lie in $E_0(r_i)$ and E_1, respectively, and E_0, E_1 are compact, it follows, by considering a subsequence, $x_i \to \xi_0 \in E_0$, $y_i \to \xi_1 \in E_1$ as $i \to \infty$ and, by the same argument as in the first part of the proof, we obtain again a contradiction, as desired.

PROPOSITION 9. Let Γ be any curve family in $\overline{R^n}$, let $p \in (1,\infty)$, and assume $M_p\Gamma < \infty$. Let $\mathfrak{A}_o'' = \{\rho \in F(\Gamma);\ \rho$ be bounded away from zero on compact sets and $\rho \in L^p\}$. Then $\mathfrak{A}_o''$ is a complete family for $M_p\Gamma$ (Hesse [15], Lemma 4.40).

LEMMA 9. Let $D \subset R^n$ be a domain, $\Gamma = \Gamma(E_0, E_1, D)$, where $d(E_0, E_1) > 0$, $M_p\Gamma < \infty$, $p \in (1,\infty)$ and let $\mathfrak{A}_p^\Delta$ be a family of admissible functions $\rho \in F(\Gamma) \cap L^p$, bounded in R^n, continuous in $\Delta = D - (\bar{E}_0 \cup \bar{E}_1)$, bounded away from zero on compact sets $F \subset \bar{\Delta}$ and 0 in $C\bar{\Delta}$. Then $\mathfrak{A}_p^\Delta$ is a complete family for $M_p\Gamma$.

Indeed, if $\rho \in F(\Gamma)$ is supposed to be only bounded in R^n, continuous in Δ and 0 in $C\Delta$, then, by Lemma 5 and Ziemer's Lemma 2.3 of [25], the corresponding subfamily of $F(\Gamma)$ is complete (i.e. yields the same value for $M_p\Gamma$). Next, since $M_p\Gamma < \infty$, suppose $\rho \in L^p$ and, arguing as in the preceding proposition, $\forall \varepsilon > 0$, let us consider $\tilde{\rho} = \rho + \varepsilon\psi$, where $\psi(x) = 1/[1 + |x|^{(n+1)/p}]$. Clearly, $\tilde{\rho}$ is bounded away from zero on compact sets in R^n and then a fortiori in $\bar{\Delta}$. By Minkowski's inequality

$$(\int \tilde{\rho}^p dm)^{\frac{1}{p}} = [\int (\rho+\varepsilon\psi)^p dm]^{\frac{1}{p}} \leq (\int \rho^p dm)^{\frac{1}{p}} + \varepsilon(\int \psi^p dm)^{\frac{1}{p}}$$

$$= (\int \rho^p dm)^{\frac{1}{p}} + \varepsilon(n\omega_n)^{\frac{1}{p}}\{\int_0^\infty [r^{n-1}/(1+r^{\frac{n+1}{p}})^p]dr\}^{\frac{1}{p}}$$

$$= (\int \rho^p dm)^{\frac{1}{p}} + \varepsilon(n\omega_n)^{\frac{1}{p}}\{(\int_0^1 + \int_1^\infty)[r^{n-1}/(1+r^{\frac{n+1}{p}})^p]dr\}^{\frac{1}{p}}$$

$$\leq (\int \rho^p dm)^{\frac{1}{p}} + \varepsilon(n\omega_n)^{\frac{1}{p}}(\int_0^1 dr + \int_1^\infty \frac{r^{n-1}dr}{r^{n+1}})^{\frac{1}{p}} = (\int \rho^p dm)^{\frac{1}{p}} + \varepsilon(n\omega_n)^{\frac{1}{p}} 2^{\frac{1}{p}} < \infty,$$

where $n\omega_n$ is the area of the unit sphere in R^n. Hence $\tilde{\rho} \in L^p$, and then $\tilde{\rho} \in \mathfrak{A}_o''$. If $M = \inf_{\mathfrak{A}_o''} \int \tilde{\rho}^p dm$, then

$$M^{\frac{1}{p}} \leq (\int \tilde{\rho}^p dm)^{\frac{1}{p}} = [\int (\rho+\varepsilon\psi)^p dm]^{\frac{1}{p}} \leq (\int \rho^p dm)^{\frac{1}{p}} + \varepsilon(\int \psi^p dm)^{\frac{1}{p}},$$

where $\psi \in L^p$. Letting $\varepsilon \to 0$, we obtain $M \leq \int \rho^p dm$. Since ρ is an arbitrary admissible function for $M_p^{o\Delta}\Gamma$, then taking the infimum over all such admissible ρ, we conclude, taking also into account our Lemma 5 and Ziemer's Lemma 2.3, that $M \leq M_p^{o\Delta}\Gamma = M_p\Gamma$. Since, evidently, $M \geqq M_p\Gamma$, then $M = M_p\Gamma$, as desired.

Finally, let us consider the function

$$\rho_0(x) = \begin{cases} \tilde{\rho}(x) & \text{for } x \in \bar{\Delta}, \\ 0 & \text{otherwise.} \end{cases}$$

Clearly,

$$\int_\gamma \rho_0(x)\, dH^1 = \int_\gamma \tilde{\rho}(x)\, dH^1 \geq 1,$$

since each open arc $\gamma \in \Gamma(E_0, E_1, \Delta)$ is contained in $\bar{\Delta}$.

LEMMA 10. *If a domain* D *is* m-*smooth at a point* $\xi \in \partial D$, *then for* $p \in [n, \infty)$ *there exists* $\lambda_p > 0$ *and a neighbourhood* U_ξ *with the property that* $U_\xi \cap D$ *consists of* m *components* $\Delta_1, \dots, \Delta_m$ *and if* V_ξ *is any neighbourhood of* ξ *contained in* U_ξ, *there is a neighbourhood* $V_\xi' \subset V_\xi$ *such that* $M_p\Gamma(E_1, E_2, V_\xi \cap \Delta_i) \geq \lambda_p$, *where* E_1, E_2 *are two disjoint connected sets in* Δ_i *which meet both* ∂V_ξ *and* $\partial V_\xi'$.

Let U_ξ be the neighbourhood involved in the definition of the m-smoothness of D at ξ and suppose that $V_\xi \subset U_\xi$. If $M_p\Gamma(E_1, E_2, V_\xi \cap \Delta_i) = \infty$, then, a fortiori, $M_p\Gamma(E_1, E_2, V_\xi \cap \Delta_i) \geq \lambda_p$ for any $\lambda_p > 0$. If $M_p\Gamma(E_1, E_2, V_\xi \cap \Delta_i) < \infty$, then, arguing as in the preceding lemma, ρ may be assumed to be bounded away for zero on compact sets of $\bar{U}_\xi$, and 0 on $C\bar{U}_\xi$. For such a ρ, since $\bar{U}_\xi$ is compact, we have

$$\begin{aligned} (31) \quad \int_{V_\xi \cap \Delta_i} \rho^p dm &= \int_{E_i'} \rho^p dm + \int_{E_i''} \rho^p dm \geqq \alpha^p_{\bar{U}_\xi} \int_{E_i'} dm + \int_{E_i''} \rho^n dm \\ &\geqq \alpha^p_{\bar{U}_\xi} \int_{E_i'} \rho^n dm + \int_{E_i''} \rho^n dm \geqq \alpha^p_{\bar{U}_\xi} \int_{V_\xi \cap \Delta_i} \rho^n dm \geqq \alpha^p_{\bar{U}_\xi} M\Gamma(E_1, E_2, V_\xi \cap \Delta_i) \geqq \alpha^p_{\bar{U}_\xi} \lambda_o, \end{aligned}$$

where

$$E_i' = \{x \in V_\xi \cap \Delta_i;\ \rho(x) \leq 1\}, \quad E_i'' = \{x \in V_\xi \cap \Delta_i;\ \rho(x) > 1\} \text{ and } \alpha^p_{\bar{U}_\xi} \leq 1.$$

Taking the infimum in (31) over all admissible ρ, we obtain

$$M_p\Gamma(E_1,E_2,V_\xi\cap\Delta_i) \ge \alpha^p_{U_\xi}\lambda_o,$$

so that we may denote $\lambda_p = \alpha^p_{U_\xi}\lambda_o$.

Arguing as in Lemma 8, on account of the preceding lemma, we deduce

LEMMA 11. *If the hypotheses of Lemma 8, are satisfied, then* $L_1(\rho) \ge 1$ $\forall\rho \in \mathfrak{a}_p = \{\rho \in F(\Gamma)\cap L^p;$ ρ *is bounded away from zero on compact sets in* $R^n\}$, *where* $p \in [n,\infty)$.

LEMMA 12. *If the hypotheses of Lemma 8, are satisfied and if* $p \in (1,\infty)$ *and* E_o *is compact, then* $L_1(\rho) \ge 1$ $\forall\rho \in \mathfrak{a}^\Delta_p$.

We have only to show that each $\rho \in \mathfrak{a}^\Delta_p$ belongs to $L^n(R^n)$. Indeed, let $E' = \{x \in \bar{\Delta};\ \rho(x) \le 1\}$, $E'' = \{x \in \bar{\Delta};\ \rho(x) > 1\}$. Then, if $p \in (1,n)$, from the condition $\rho \in L^p$ it follows that

$$\infty > \int \rho^p dm = \int_{\bar{\Delta}} \rho^p dm = \int_{E'} \rho^p dm + \int_{E''} \rho^p dm \ge mE''.$$

Hence

$$\int \rho^n dm = \int_{\bar{\Delta}} \rho^n dm = \int_{E'} \rho^n dm + \int_{E''} \rho^n dm \le \int_{E'} \rho^p dm + M^n mE'' < \infty,$$

where $M = \sup_{x\in R^n} \rho(x)$.

Finally, if $p \ge n$, the conclusion of the lemma follows on account of the preceding lemma.

PROPOSITION 10. *Suppose that* D *is an open set;* E_o, E_1 *are disjoint bounded continua in* D, *and* $\Gamma = \Gamma(E_o,E_1,D)$, $\Gamma_r = \Gamma[E_o(r), E_1(r),D]$. *Then* $M = \lim_{r\to 0} M\Gamma_r$ (Gehring and Väisälä [13], Lemma 3.4).

Arguing as in the preceding proposition and taking into account the preceding lemma combined with Lemma 7, we obtain

LEMMA 13. *Suppose that* E_1 *is a set;* E_o *is compact with* $d(E_o,E_1) > 0$, D *is a domain* m*-smooth on* $E_o \cap \partial D$ *for some* m, $\Gamma = \Gamma(E_o,E_1,D)$, *and* $\Gamma'_r = \Gamma[E_o(r),E_1,D]$. *Then*

$$M_p\Gamma = \lim_{r\to 0} M_p\Gamma'_r. \tag{32}$$

PROPOSITION 11. *Let* $E_1 \supset E_2 \supset \ldots$ *and* $F_1 \supset F_2 \supset \ldots$ *be disjoint sequences of nonempty compact sets in the closure of a domain* $D \subset \overline{R^n}$. *Let* $E = \bigcap_{m=1}^\infty E_m$, $F = \bigcap_{m=1}^\infty F_m$. *Then*

(33) $\lim_{m\to\infty} \operatorname{cap}_p(E_m, F_m, D) = \operatorname{cap}_p(E,F,D)$.

(For the proof, see Hesse [16], Theorem 3.3.)

Arguing as in the preceding propositon, we obtain

LEMMA 14. *Let* $E_1 \supset E_2 \supset \dots$ *and* $F_1 \supset F_2 \supset \dots$ *be two sequences of sets in the closure of a domain* $D \subset R^n$ *such that* $E = \bigcap_{k=1}^{\infty} E_k \neq \emptyset$, $F = \bigcap_{k=1}^{\infty} F_k \neq \emptyset$ *and* $d(E_1, F_1) > 0$. *Then* (33) *holds.*

THEOREM 3. *Suppose that* D *is open, and* E_0, E_1 *are two sets such that* $d(E_0, E_1) > 0$. *Let each component* D_k *of* D *with* $\bar{D}_k \cap E_i \neq \emptyset$ $(i = 0, 1)$ *for* $i = 0$ *or* $i = 1$ *be* m-*smooth on the set* $\partial D_k \cap E_i$, *where* $\bar{D}_k \cap E_i$ *is compact. Then* (1) *holds.*

As we have observed in the proof of the preceding theorem, we may suppose - without loss of generality - that D is a domain m-smooth on $\partial D \cap E_0$ for some m. Next, let us denote $C_0 = E_0 \cap \bar{D}$ and consider the sequence $C_0(r_1) \supset C_0(r_2) \supset \dots$, where $\lim_{k\to\infty} r_k = 0$, $\bigcap_{k=1}^{\infty} C_0(r_k) = C_0$ and $d[(E_1, C_0(r_1)] > 0$. Clearly, $\forall \xi \in C_0(r_k) \cap \bar{D}$,

$$\liminf_{x\to\xi}\ \inf_{\gamma} H^1\{\gamma[C_0(r_k), x]\} = 0 \qquad (k = 1, 2, \dots),$$

where the infimum is taken over all the arcs $\gamma[C_0(r_k), x]$ joining $C_0(r_k)$ to x in D, since if $\xi \in C_0(r_k) \cap \bar{D}$, then any $x \in D$ sufficiently close to ξ belongs to $C_0(r_k)$ so it can be joined to $C_0(r_k)$ by an arc of lenght zero, yielding

$$\inf_{\gamma} H^1\{\gamma[C_0(r_k), x]\} = 0 \qquad (k = 1, 2, \dots).$$

Then, the hypotheses of Theorem 1 are satisfied, and we conclude that

$$M_p\Gamma[C_0(r_k), E_1, D] = \operatorname{cap}_p[C_0(r_k), E_1, D] \qquad (k = 1, 2, \dots).$$

Hence, by the preceding two lemmas,

$$M_p\Gamma(E_0 \cap \bar{D}, E_1, D) = \lim_{k\to\infty} M_p\Gamma[C_0(r_k), E_1, D]$$

$$= \lim_{k\to\infty} \operatorname{cap}_p[C_0(r_k), E_1] = \operatorname{cap}_p(E_0 \cap \bar{D}, E_1, D).$$

Finaly, extending all the admissible functions for $\operatorname{cap}_p(E_0 \cap \bar{D}, E_1, D)$

to be 0 on $E_o - \bar{D}$, we conclude that

$$M_p\Gamma(E_o,E_1,D) = M_p\Gamma(E_o \cap \bar{D},E_1,D)$$

$$= \mathrm{cap}_p(E_o \cap \bar{D},E_1,D) = \mathrm{cap}_p(E_o,E_1,D),$$

as desired.

LEMMA 15. *If* D *is a domain and* E_i $(i=0,1)$ *are two sets such that for at least one of them (say for* E_o) $E_o \cap \bar{D} = E' \cup E''$, *where*,

$$\liminf_{\substack{x\to\xi \\ x\in D}} \inf_{\gamma} H^1[\gamma(E',x)] = 0 \qquad \forall \xi \in E',$$

if D *is* m-*smooth on* $E'' \cap \partial D$ *for some* m *and* $E'' \cap \partial D$ *is compact, then*

(34) $\quad M_p\Gamma(E_o,E_1,D) = \lim_{k\to\infty} M_p\Gamma[D,E' \cup E''(r_k),E_1]$, *where* $\lim_{k\to\infty} r_k = 0$.

Indeed, on account of Lemma 8, if $L_1'(\rho,r_k) = \inf_{\gamma} \int_{\gamma} \rho dH^1$, where the infimum is taken over all $\gamma \in \Gamma[E''(r_k),E_1]$, then

$$L_1'(\rho) = \lim_{k\to\infty} L_1'(\rho,r_k) \geq 1.$$

Clearly, if $L_1(\rho,r_k) = \inf_{\gamma} \int_{\gamma} \rho dH^1$, where the infimum is taken over all

$$\gamma \in \Gamma[D,E' \cup E''(r_k),E_1] = \Gamma[D,E''(r_k),E_1] \cup \Gamma(D,E',E_1),$$

then

$$L_1(\rho) = \lim_{k\to\infty} L_1(\rho,r_k) \geq 1,$$

since all the additional arcs $\gamma \in \Gamma(D,E',E_1)$, appearing in the infimum in $L_1(\rho,r_k)$, satisfy the condition $\int_{\gamma} \rho dH^1 \geq 1$, so that

$$\lim_{k\to\infty} L_1'(\rho,r_k) \geq 1 \quad \text{implies} \quad \lim_{k\to\infty} L_1(\rho,r_k) \geq 1$$

and, arguing as in Lemma 13, we obtain (34), as desired.

THEOREM 4. *Suppose that* D *is open,* $d(E_o,E_1) > 0$, $E_o \cap \bar{D}$ *is bounded and for every component* D_k *of* D *with* $\bar{D}_k \cap E_i \neq \emptyset$ $(i = 0,1)$ *the set* $\partial D_k \cap E_o$ *can be expressed as* $\partial D_k \cap E_o = E' \cup E'' \cup E'''$, *where* E', E'' *are as in the preceding lemma (with* $D = D_k$*), and all the points of* E''' *are not accessible from* D_k *by rectifiable arcs. Then* (1) *holds.*

This theorem is a consequence of the preceding lemma and Corollary 2 of Theorem 2.

Now, in order to consider another case, let us recall (following Näkki [18]) the concept of quasiconformal m-collardeness.

A domain D is *quasiconformally* m-*collared at a boundary point* $\xi \in \partial D$, if there exists a neighbourhood U_ξ such that $U_\xi \cap D$ consists of m components $\Delta_1,\dots,\Delta_m$, and for each of them there is a quasiconformal mapping $g_i : \Delta_i \to B_+$ with

$$\lim_{x \to \xi} g_i(x) = 0, \quad \lim_{y \to 0} g_i^{-1}(y) = \xi,$$

and where $B_+ \doteq \{x \in R^n;\ |x| < 1,\ x^n > 0\}$. D is said to be *quasiconformally* m-*collared* on $E \subset \partial D$ if it is quasiconformally m-collared at each point of E. In particular, a domain quasiconformally 1-collared is simply called *quasiconformally collared*.

PROPOSITION 12. *If a domain* D *is quasiconformally* m-*collared at* $\xi \in \partial D$, *then it is* m-*smooth at* ξ (according to Hesse's definition - Hesse [15], Theorem 4.23).

PROPOSITION 13. *For a given domain* D *and a boundary point* $\xi \in \partial D$, *the following statements are equivalent*:

(i) D *is quasiconformally* m-*collared at* ξ.

(ii) *There is a neighbourhood* U_ξ *such that* $U_\xi \cap D$ *consists of* m *components and each of them is quasiconformally collared at* ξ.

(iii) *There exist arbitrarily small neighbourhood* U_ξ *such that* $U_\xi \cap D$ *consist of* m *components and each* U_ξ *is quasiconformally collared at* ξ (Näkki [18], Theorem 1.12).

COROLLARY. *If* D *is* m-*collared at* $\xi \in \partial D$, *then* D *is* m-*smooth at* ξ (according to the above definition).

THEOREM 5. *Suppose that* D *is open,* E_o, E_1 *are two sets such that* $d(E_o,E_1) > 0$, *and each component* D_k *of* D *with* $\bar{D}_k \cap E_i \neq \emptyset$ $(i = 0,1)$ *is* m-*collared on* $\partial D_k \cap E_o$, *which is supposed to be compact. Then* (1) *holds.*

This is a direct consequence of Theorem 3 and of the preceding corollary.

2. p-module and p-capacity in the case of open sets $D \subset \overline{R^n}$

Now, in order to generalize some of the above results to $\overline{R^n}$, we define the p-capacity in this case.

The p-capacity of two sets $E_o, E_1 \subset \overline{R^n}$ with $q(E_o, E_1) > 0$ relative to a domain $D \subset \overline{R^n}$ is given by the formula

$$\operatorname{cap}_p(E_o, E_1, D) = \inf_u \int_{D-\{\infty\}} |\nabla u|^p dm,$$

where the infimum is taken over all u which are continuous in $D \cup E_o \cup E_1$, locally Lipschitzian in $D - \{\infty\}$ and $u|_{E_o} = 0$, $u|_{E_1} = 1$.

Arguing as for the p-capacity in R^n, we have

LEMMA 6'. *If the hypotheses of Lemma 6 and satisfied, then the p-capacity in* $\overline{R^n}$ *satisfies the conditions* (i) - (iii) *and* (i') - (iii').

Next, we define the p-module of an arc family Γ of $\overline{R^n}$ by the formula

$$M_p \Gamma = \inf_{\rho \in F(\Gamma)} \int_{R^n} \rho^p dm,$$

where $F(\Gamma)$ is the class of Borel measurable functions $\rho \geq 0$ such that $\int_{\gamma-\{\infty\}} \rho^p dH^1 \geq 1 \quad \forall \gamma \in \Gamma$.

An open arc $\gamma \subset \overline{R^n}$ is called *extended locally rectifiable* if γ is locally rectifiable or $\gamma - \{\infty\}$ is the union of two locally rectifiable arcs.

R e m a r k. Hesse [15] defined the extended locally rectifiable curves as the open curves $\alpha : (a,b) \to \overline{R^n}$ with the property that there exists a finite set $S \subset (a,b)$ such that the restriction of α to any closed subinterval in $(a,b) - [S \cup \alpha^{-1}(\infty)]$ is rectifiable. Evidently, this definition is more general that the preceding one, which has the advantege of being closer to the concept of locally rectifiable arcs. In this paper, the expression "extended locally rectifiable" will be used only in the sense of our definition.

LEMMA 16. *If* Γ_o *is the family of the arcs* $\gamma \subset \overline{R^n}$, *which are not extended locally rectifiable, then* $M_p \Gamma_o = 0 \quad (1 < p < \infty)$.

By the definition of extended locally rectifiable arcs, it follows that each $\gamma_o \in \Gamma_o$ contains at least a closed subarc $\gamma_1 \subset \gamma_o - \{\infty\}$, which is not rectifiable. Next, if Γ_1 is the family of all such arcs γ_1, then Proposition 5 yields $M_p \Gamma_1 = 0$. Yet, Γ_1 is minorized by Γ_o, so that (Fuglede [10], Theorem 1, p.176) $M_p \Gamma_o \leq M_p \Gamma_1 = 0$, as desired.

R e m a r k. The definition of the p-module $\hat{M}_p\Gamma$, given by Hesse [15], in which the condition $\int_\gamma \rho\, dH^1 \geq 1$ is supposed to hold only for extended locally rectifiable arcs of Γ, is equivalent, if the finite set $S = \emptyset$, to our definition of $M_p\Gamma$, by the preceding lemma.

PROPOSITION 14. *If* $D \subset \overline{R^n}$ *is a domain,* $E, F \subset \bar{D}$ *are compact disjoint nonempty sets and* $p \in (1,\infty)$, *then* $\hat{M}_p\Gamma(E,E,D) \leq \mathrm{cap}_p(E,F,D)$ (for the proof, see Hesse [15], Theorem 5.7, p.82; the proof is similar to that given by him in [16], Lemma 5.2.)

COROLLARY. *If* $D \subset \overline{R^n}$ *is open,* $q(E_0,E_1) > 0$, $\Gamma = \Gamma(E_0,E_1,D)$ *and* $p \in (1,\infty)$, *then* (12) *still holds.*

Next, by the preceding lemma we can conclude that Lemma 5 holds in $\overline{R^n}$ as well.

On the basis of the above considerations, we obtain

THEOREM 1'. *Suppose that* $D \subset \overline{R^n}$ *is open,* $E_0, E_1 \subset \overline{R^n}$ *are such that* $q(E_0,E_1) > 0$, E_0 *is bounded and* $\forall D_k$ *of* D *with* $\bar{D}_k \cap E_i \neq \emptyset$ $(i = 0,1)$, $\forall \xi \in \partial D_k \cap E_0$, *the relation* (11) *is verified with* $i = 0$. *Then* (1) *holds.*

COROLLARY 1. *If* χ *is the set of all continua in* $\overline{R^n}$ *that meet* E_0, E_1, *where* $q(E_0,E_1) > 0$, *then* $M_p\chi = \mathrm{cap}_p(E_0,E_1,\overline{R^n})$.

COROLLARY 2. *If the hypotheses of the preceding theorem are satisfied, then* (18) *holds.*

Now, arguing exactly as in Theorem 2, we get

THEOREM 2'. *Suppose that* $D \subset \overline{R^n}$ *is a domain,* $q(E_0,E_1) > 0$, E_0 *is bounded and* $E_0 \cap \partial D = E' \cup E''$, *where* $\forall \xi \in E'$, (11) *is verified, while every point of* E'' *is inaccessible from* D *by rectifiable arcs. Then* (1) *holds.*

3. p-module and p-capacity of topological cylinders

Now, let us recall three definitions of a topological cylinder: two of them with respect to the euclidean metric and the third with respect to the relative metric.

I. A triple (B_0,B_1,Z), in which $Z \subset R^n$ is a domain and $B_0, B_1 \subset \partial Z$, is called *a topological cylinder with respect to the euclidean metric* if there exists a homeomorphism $\varphi : \bar{Z}_0 \rightleftarrows \bar{Z}$ such that $\varphi(B_k^0) = B_k$ $(k = 0,1)$, where $Z_0 = \{x; (x^1)^2 + \ldots + (x^{n-1})^2 < 1;\ 0 < x^n < 1\}$ is the unit cylinder and B_k^0 $(k = 0,1)$ its bases. B_0, B_1 are the *bases* of the topological cylinder.

II. A triple (B_0, B_1, Z) is said to be a <u>topological cylinder</u> <u>with respect to the euclidean metric</u> if there exists a homeomorphism $\psi : Z_o \cup B_o^o \cup B_1^o \rightleftarrows Z \cup B_0 \cup B_1$ such that $\psi(B_k^o) = B_k$ $(k = 0, 1)$.

III. A triple (B_0, B_1, Z) is said to be a <u>topological cylinder</u> <u>with respect to the relative metric</u> if there exists a bijection $\psi : Z_o \cup B_o^o \cup B_1^o \rightleftarrows Z \cup B_0 \cup B_1$ with the following property: given $\varepsilon > 0$ and a point $x_0 \in Z_o \cup B_o^o \cup B_1^o$, there is $\delta = \delta(\varepsilon, x_0) > 0$ such that $x \in Z_0$ with $|x - x_0| < \delta$ imply $d_Z[\psi(x_0), \psi(x)] < \varepsilon$.

<u>Remarks</u>. 1. Clearly, the bijection ψ of the preceding definition is a homeomorphism (with respect to the euclidean metric) of $Z_o \cup B_o^o \cup B_1^o$ onto $Z \cup B_0 \cup B_1$. Hence a topological cylinder with respect to the relative metric is also a topological cylinder with respect to the euclidean metric according to Definition II, but not generally, according to Definition I.

2. All the points of the bases B_0, B_1 of a topological cylinder with respect to the relative metric are accessible from Z by rectifiable arcs.

The p-module M_pZ of a topological cylinder (B_0, B_1, Z) according to Definition I, II, III is given by $M_pZ = M_p\Gamma_Z$, where $\Gamma_Z = \Gamma(B_0, B_1, Z)$.

In our note [9], we have established

<u>PROPOSITION 15</u>. <u>If</u> $Z = (B_0, B_1, Z)$ <u>is a topological cylinder with respect to the relative metric, then</u> (2) <u>holds</u>.

In the proof of this proposition, we have established that $\forall \xi \in B_0$, $u(x) \to 0$ as $x \to \xi$ and $x \in Z$. It seems that it would be worth-while to prove this assertion in detail. It is enough to show that $u(x_k) \to 0$ as $k \to \infty$ for any sequence $\{x_k\}$ with $x_k \in Z$ and $x_k \to \xi \in B_0$ as $k \to \infty$. Indeed, let $y_k = \psi^{-1}(x_k)$ $(k = 1, 2, \ldots)$ and $\eta = \psi^{-1}(\xi) \in B_0^o$, where ψ is the homeomorphism involved in the definition of Z. Since $\psi : Z_o \cup B_o^o \cup B_1^o \rightleftarrows Z \cup B_0 \cup B_1$ is a homeomorphism, it follows from $x_k \to \xi$, that $y_k \to \eta$. Hence $d(y_k, \eta) \to 0$ as $k \to \infty$. Since ψ is continuous with respect to the relative metric in $Z_o \cup B_o^o \cup B_1^o$ then, in particular, it follows that at $\eta \in B_0^o$, we have $d_Z(x_k, \xi) \to 0$ as $k \to \infty$, i.e. in other words, $\liminf_{k\to\infty} \inf_\gamma H^1[\gamma(x_k, \xi)] = \lim_{k\to\infty} d_Z(x_k, \xi) = 0$. Hence

$$\liminf_{k\to\infty}\ \inf_{\gamma} H^1[\gamma(x_k, B_0)] \le \liminf_{k\to\infty}\ \inf_{\gamma} H^1[\gamma(x_k, \xi)] = 0,$$

and since this relation holds for any sequence $\{x_k\}$ with $x_k \in Z$ and

$x_k \to \xi$, we conclude that the condition (11) is satisfied for any $\xi \in B_o$. This implies (1) in our case, so we arrive at (2).

Now, we establish that, in the particular case $n=2$, the relation (2) is also true for a topological cylinder with respect to the euclidean metric, i.e. for topological quadrilaterals.

THEOREM 6. *If $Z \subset R^2$ is a topological quadrilateral with respect to the euclidean metric (according to Definition I), then (2) holds.*

Let us show first that $\forall \varepsilon > 0$ there exists an $r > 0$ such that, $\forall \rho \in F^{oZ}(\Gamma)$, we have $\frac{\rho}{1-\varepsilon} \in F^{oZ}(\Gamma'_r)$, where $\Gamma = \Gamma_Z$ and $\Gamma'_r = \Gamma[B_o(r), B_1, Z]$. Let $\dot{B}_o$ denote the basis B_o without its endpoints; let $\xi_o \in \dot{B}_o$ and $\gamma_o \in \Gamma$ be an arc joining ξ_o to B_1. $\partial Z - \dot{B}_o$ is a closed set. Let $r < d(\xi_o, \partial Z - B_o)$. The circumference $C(\xi_o, r)$ is disjoint with $\partial Z - B_o$ and meets $\dot{B}_o$ in at least two points. Hence there is a circular subarc γ_r of $C(\xi_o, r)$ of length $l < \pi r$ joining the first point x_o of the intersection $\gamma_o \cap C(\xi_o, r)$ (taken along γ_o from B_1 toward B_o) to B_o in Z. Since $\rho \leq M$ in R^2, it follows that

$$\int_{\gamma_r} \rho ds \leq M\pi r < \varepsilon \quad \text{for} \quad r < \frac{\varepsilon}{M\pi}.$$

Therefore, if γ'_r is the subarc of γ_o joining B_1 and x_o, then $\gamma''_r = \gamma_r \cup \gamma'_r$ belongs to Γ so that

$$1 \leq \int_{\gamma''_r} \rho dH^1 \leq \int_{\gamma_r} \rho ds + \int_{\gamma'_r} \rho dH^1 < \int_{\gamma'_r} \rho dH^1 + \varepsilon.$$

Hence

$$(35) \qquad \int_{\gamma'_r} \rho dH^1 > 1 - \varepsilon.$$

Since any arc joining B_1 to $C(\xi_o, r)$ can be considered as a subarc of an arc joining B_1 to ξ_o, and belonging to Γ'_r, then if we consider a closed subarc $\tilde{B}_o \subset \dot{B}_o$, we can take $r_o = \min[d(\tilde{B}_o, \partial Z - B_o), \frac{\varepsilon}{\pi M}, d(B_o, B_1)]$ and $\frac{\rho}{1-\varepsilon}$ is admissible for all the arcs of Γ'_{r_o} joining $\tilde{B}_o(r_o)$ to B_1.

Now, let us consider a point $\xi \in B_o - \tilde{B}_o$ and the circumference $C(\xi, r_o)$. If $r_o \leq d(\xi, \partial Z - B_o)$ with $C(\xi, r_o) \cap (\partial Z - B_o) = \emptyset$, then, arguing as above we conclude that any arc joining B_1 to $C(\xi, r_o)$ satisfies (35) with $r = r_o$. Finally, if $d(\xi, \partial Z - B_o) < r_o < d(B_o, B_1)$, then one of the subarcs of $C(\xi, r_o)$ contained in Z joins $\partial Z - (B_o \cup B_1)$ to B_o.

This subarc is contained in the boundary of a simply connected subdomain of Z whose boundary also contains ξ. Yet then, all the arcs γ_0 joining B_1 to ξ cross this arc. Arguing as above we obtain (35) in this case as well, and that, $\frac{\rho}{1-\varepsilon} \in F^{oZ}(\Gamma'_{r_0})$. Consequently, arguing as in Proposition 10, we get (32).

Next, since $\Gamma[B_0(r), B_1, Z]$ satisfies the relation (11), it follows, on account of Theorem 1, that

$$M_p\Gamma[B_0(r), B_1, Z] = \mathrm{cap}_p[B_0(r), B_1, Z], \quad \forall r < r_0.$$

In particular, the preceding relation is verified for a sequence $\{r_k\}$ with $r_k \to 0$ as $k \to \infty$. Yet then, from (32) and Lemma 14, we deduce that

$$M_p\Gamma(B_0, B_1, Z) = \lim_{k\to\infty} M_p\Gamma[B_0(r_k), B_1, Z]$$

$$= \lim_{k\to\infty} \mathrm{cap}_p[B_0(r_k), B_1, Z] = \mathrm{cap}_p(B_0, B_1, Z),$$

as desired.

As a consequence of Theorem 4, we have

THEOREM 7. *If $Z \subset R^n$ is a topological cylinder with respect to the euclidean metric (Definition I or II) and B_0 or B_1 can be expressed as the union $E' \cup E'' \cup E'''$ (with the same meaning as in Theorem 4), then (2) holds.*

Now, in order to obtain another case of a topological cylinder verifying the condition (2), let us recall

PROPOSITION 16. *Let D be a domain, which is locally connected on ∂D and quasiconformally equivalent to a ball. Then D is a Jordan domain, which is quasiconformally collared on ∂D (Näkki [18], Theorem 4.5).*

THEOREM 8. *If (E_0, E_1, Z) is a topological cylinder with respect to the euclidean metric according to Definition I and the homeomorphism φ involved in its definition is quasiconformal in Z_0, then (2) holds.*

Let us show first that D is locally connected on ∂D. Indeed, if $\xi \in \partial D$, since φ is a homeomorphism, then, given $\varepsilon > 0$, there is a connected neighbourhood U_0 of $\xi_0 = \varphi^{-1}(\xi)$ such that $\varphi(U_0 \cap \overline{Z_0}) \subset B(\xi, \varepsilon)$ is connected as well. Next, on account of the preceding proposition, Z is quasiconformally collared on ∂D and, in particular, on the basis E_0 of Z, so that we have checked the hypotheses of Theorem 5, and this enables us to conclude that (2) holds.

4. Historical remarks

For the first time, Hersch [14] established the relation (1) for the harmonic capacity of a ring in R^3 and the relation (2) for the harmonic capacity of a domain $D \subset R^3$ homeomorhic to a ball, with two distinguished continua $B_0, B_1 \subset \partial D$. He defined the harmonic capacity by the relation

$$\operatorname{cap}_p(B_0, B_1, D) = \inf_u \int_\sigma \frac{\partial u}{\partial \nu} d\sigma,$$

where $\sigma \subset D$ was a surface separating B_0 from B_1 and $\partial u/\partial \nu$ was the normal derivative of the admissible function u. Fuglede [10] established something similar to (1) in the particular case $p = 2$ for the harmonic capacity $\operatorname{cap}_2(K, \infty, D)$, where Γ was the family of the arcs joining the point at infinity of R^n to the compact set K within the unbounded domain D. Gehring [12] obtained (1) for the conformal capacity of a ring (according to Loewner's definition, by means of Dirichlet integral used also in this paper). Bagby [2] showed that $\operatorname{cap}(C_0, C_1, \overline{R^n}) = M\Gamma(C_0, C_1, \overline{R^n})$, where C_0, C_1 were disjoint compact sets. Ziemer indicated in [24] how to verify (1) in the case $p = n$ if $D \subset R^n$ was a bounded domain and $C_0, C_1 \subset D$ were disjoint closed sets. He asserted that this result was also valid for $C_0, C_1 \subset \overline{D}$ if certain conditions were imposed on the tangential behaviour of $\partial D \cap (C_0 \cup C_1)$. In [25], he established the equality $\operatorname{cap}_p(C_0, C_1, R^n) = M_p\chi(C_0, C_1, R^n)$, where C_0 contained the complement of a ball and χ was the family of continua joining C_0 to C_1. In [26], he showed the equality between the p-module $(1 \leq p < n)$ of the family of all continua that join a Suslin set E to the point at infinity and the p-capacity of E, where the infimum involved in the definition was taken over all u being ACL in $R^n - E$, $u_{|E} = 1$ and with compact support. In the case $p \geq n$, the support of each u was required to lie in some fixed ball containing E and the corresponding family of continua was supposed to join E to the complement of that ball.

Krivov [17] (Theorem 3) tried to prove the equality

$$M\Gamma(E_0, E_1, D) = \operatorname{cap}(E_0, E_1, D)$$

in the case of a simply connected domain $D \subset R^n$ with $E_0, E_1 \subset \partial D$ being continua, $\partial D - (E_0 \cup E_1)$ connected and $\rho \geq 0$ defined in D, satisfying the following conditions:

1°. ρ is bounded on each compact subset $F \subset D$,

2°. $\lim\limits_{x\to\xi} \rho(x) < \infty \quad \forall \xi \in \partial D - (E_o \cup E_1)$,

3°. $\int\limits_\gamma \rho dH^1 < \infty \quad \forall \gamma \subset D$ for γ rectifiable,

4°. $\int\limits_\sigma \rho dH^{n-1} < \infty \quad \forall \sigma \subset D$.

In the last inequality, G was supposed to be measurable with respect to the (n-1)-dimensional Hausdorff measure H^{n-1},

$$M\Gamma = \inf_\rho \frac{V(\rho)}{h(\rho)^n}, \quad V(\rho) = \int_D \rho^n dm, \quad h(\rho) = \inf_{\gamma(E_o,E_1)} \int_\gamma \rho dH^1$$

and the infimum involved in the definition of the conformal capacity was supposed to be taken over all $u \in C^1$, $u_{|E_o} = 0$, $u_{|E_1} = 1$. Moreover, he tacitly supposed that

$$u_o = \frac{u}{h(\rho)}, \quad \text{where} \quad u(x) = \inf_\gamma \int_{\gamma(E_o,E_1)} \rho dH^1 \quad \text{(the potential of } \rho),$$

is admissible for the corresponding conformal capacity, i.e $u_o \in C^1$, $u_{o|E_o} = 0$, $u_{o|E_1} = 1$. Yet, he only proved that u satisfied a Lipschitz condition on any compact subset $F \subset D$. Since the Lipschitz constant depends on the compact F, it follows that, in general, u is locally Lipschitzian in D (it satisfies a Lipschitz condition in D if, for instance, D is convex). However u_o, and then also u, have to be of the class C^1 (according to his definition of admissible functions for the conformal capacity) and the property of u to be locally Lipschitzian is not enough to assure that $u \in C^1$.

In this paper we have established Lemma 2 and Corollary 1 which assure the infimum in the definition of the p-capacity to be the same if we assume that u is locally Lipshitzian or $u \in C^1$. Thus, the above inaccuracy can be easily corrected. Yet, there is still something else: he tacitly supposed that the potential u of ρ satisfied the condition $u(P) \to 0$ as $P \to P_o \in E_o$ with $P \in D$, which, in general, is not true: it follows from the counter-example corresponding, in the case $n \geq 3$, to that given in Fig.1 (where D and ρ satisfy all his hypotheses mentioned above). Thus, the correctness of his Theorem 3 is still an open problem. Yet then, also his assertion "It is easy to show also that $\sigma(\alpha) = \{x;\ u(x) = \alpha\}$ for any $0 < \alpha < h(\rho)$ separates E_o form E_1 and then that

$$\inf_{\sigma\in\Sigma} \int_\sigma \rho^{n-1} d\sigma \leq \int_{\sigma(\alpha)} \rho^{n-1} d\sigma,$$

where Σ is the family of Hausdorff measurable surfaces $\sigma \subset D$ which spearate E_o from E_1" is false. Similarly the corresctness of the Krivov's lemma and his Theorems 1,2 (together with corollaries), and 4 still remain an open problem because the proofs are based on the fact that the potential u of ρ has the boundary value 0 on E_o.

Syčev's [22] proof for (1), where $p=n=3$, D is a ring and $E_o, E_1 \subset \partial D$ are two simply connected sets of the two boundary components of ∂D, respectively, is also incorrect: he claims that if $\rho \in F(\Gamma)$ then "according to Gehring [12], the function

$$u(x) = \min(1, \inf_\beta \int_\beta \rho ds),$$

where β is an arbitrary curve joining x and E_o, is admissible for $\text{cap}(E_o, E_1, D)$". However Syčev did not notice that his hypotheses were different. Indeed, Gehring [12] considered a more particular case in which E_o, E_1 were not only subsets of the boundary components of the ring D, but also coincided with them. Hence, any rectifiable curve β joining E_o to E_1 (no matter if $\beta \subset D$ or not) satisfies the condition $\int_\beta \rho ds \geq 1$ $\forall \rho \in F(\Gamma)$ since every such β contains a subcurve $\beta_o \subset D$ joining its boundary components. However, it is easy to see that this is not any more true, in general, if E_o or E_1 are the only subsets of the corresponding boundary component. Then we are not sure any longer that $u|_{E_1} = 1$. Nevertheless, this inaccuracy can be easily corrected if we suppose, in addition, that the curves involved in Syčev's definition of u have to be contained in D. In this new case $u|_{E_1} = 1$, however, a new inconvenience arises. Namely u is, in general, no longer continuous on E_o (as for instance if we consider a bidimensional ring obtained from the simply connected doamin D given in Fig.1, by taking off a square $Q_o \subset D$, and E_o being as in Fig.1, while $E_1 = \partial Q_o$).

A mistake of the same kind has been made also by Reimann [20] (for more detailed comments, see our papers [7, 9]). However, in a latter of March 23[d] 1971 he asserted that (1) can be established, in the particular case where $p=n$ and $D=(E_o, E_1, Z)$ is a topological cylinder with respect to the euclidean metric (according to definition I). He stated that arguing as Gehring [12] for rings, it is possible to show what follows:

<u>Given</u> $a>0$, <u>there is a</u> $b>0$ <u>such that if</u> P_o, P_1 <u>are the end-points of a rectifiable curve</u> $\beta \subset Z$ <u>such that</u> $P_o \in E_o$, $d(P_1, E_1) < k \leq n$

<u>and</u> $d(x,\partial Z - E_1) > 2k$ <u>for</u> <u>any</u> $k \leq b$, <u>then</u> $\int_\beta \rho ds \geq 1 - a$ $\forall \rho \in F[\Gamma(E_0, E_1, Z)]$.

This preliminary result allows him to obtain the functions

$$\rho_1(x) = \begin{cases} 0 & \text{if } d(x,E_1) < k \leq b \text{ and } d(x,\partial Z - E_1) > 2k, \\ \dfrac{\rho(x)}{1-a} & \text{elsewhere,} \end{cases}$$

$$\rho_2(x) = \begin{cases} 0 & \text{if } d(x,E_0) < k \leq b \text{ and } d(x,\partial Z - E_0) > 2k, \\ \dfrac{\rho_1(x)}{1-a} & \text{elsewhere,} \end{cases}$$

where $\rho_1, \rho_2 \in F[\Gamma(E_0, E_1, Z)]$. By means of ρ_2, he constructs

$$g(x,\varepsilon) = \frac{1}{mB(\varepsilon)} \int_{B(\varepsilon)} \rho_2[y_t(x)]\, dm(t),$$

where $y_t(x) = x + t\delta(x)$, $\delta(x) = \min[1, d(x,\partial Z)]$. Next, given $x_0 \in E_0$, there exist $r > 0$ and $M < \infty$ such that $v(x,\varepsilon) = g(x,\varepsilon)(1 + \frac{1}{2}\varepsilon) < M$ $\forall x \in Z$ with $|x - x_0| < r$. Next Reimann claims that for

$$u(x,\varepsilon) = \min[1, \inf_\beta \int_\beta v(x,\varepsilon) ds(y)], \quad \text{we have} \quad \lim_{x \to x_0} u(x,\varepsilon) = 0.$$

We observe that, according to the lemma of our note [9], g and v can be supposed to be bounded even in the whole R^n, but this does not seem to us to be enough to assure that

$$\lim_{x \to \xi} u(x) = 0 \quad \text{if} \quad \xi \in E_0 \cap \overline{\partial Z - E_0}.$$

If $x \to \xi$ along a rectifiable arc $\gamma \subset Z$ with an endpoint at ξ, then, clearly, $u(x,\varepsilon) \to 0$ since

$$u(x,\varepsilon) \leq \int_{\gamma(x,\varepsilon)} \rho dH^1 \leq M \int_{\gamma(x,\varepsilon)} dH^1 = MH^1[\gamma(x,\xi)],$$

which tends to 0 as $x \to \xi$ along γ. Moreover, $\lim u(x,\varepsilon) \to 0$ as $x \to \xi$, $x \in Z$ for $\xi \in E_0 \cap \overline{\partial Z - E_0}$ in the case $n = 2$, as it was established above (Theorem 5) or if Z is a topological cylinder with respect to the relative metric (Proposition 15). Yet, in the case of a topological cylinder $Z \subset R^n$ $(n > 2)$ with respect to the euclidean metric according to definition I, it seems that the continuity of u in $E_o \cap \overline{\partial Z - E_o}$ is still an open problem, while in the case corresponding to definition II, it may not be true, as it is possible to see considering the counterexample provided by Fig.1.

Later on, Aseev [1] asserted that (1) still holds for $p=n$ in the more general case of a domain $D \subset R^n$ with $E_o, E_1 \subset \partial D$ being two arbitrary disjoint sets, "as it may be proved by an argument analogous to that given by Syčev [22]". First of all, we shall show that the definitions of the conformal cappacity and the p-capacity become meaningless in such a general case. This is evident in the case of the definitions given in this paper, where the admissible functions u are supposed to be continuous on $D \cup E_o \cup E_1$, since if, for instance, $D \subset R^2$ is a disk and $E_o \subset \partial D$ is a closed arc, while $E_1 = \partial D \cap CE_o$, then, at the endpoints of E_o, u has to be simultaneosly 0 and 1, which is absurd. In fact, Aseev does not give in [1] a definition of the conformal capacity, but since he mentions Syčev's argument of [22], it is natural to suppose that he tacitly accepted Syčev's corresponding definition, where u is continuous in D and has the boundary values 0 and 1 on E_o and E_1, repectively. Yet, considering the above counterexample even in the case of Syčev's definition, we deduce that $E_o, E_1 \subset \partial D$ cannot be arbitrary. Indeed, let ξ_o be an endpoint of E_o. Since $u(x) \to 0$ as $x \to \xi_o$, then for $0 < \varepsilon < \frac{1}{2}$, there exists a neighbourhood U_{ξ_o} such that $|u(x)| = |u(\xi_o) - u(x)| < \varepsilon \ \forall x \in D \cap U_{\xi_o}$. Yet, since $\overline{D \cap U_{\xi_o}} \cap E_1 \neq \emptyset$, let $\xi_1 \in \overline{D \cap U_{\xi_o}} \cap E_1$ and $\{x_k\}$ be a sequence, where $x_k \in D \cap U_{\xi_o}$ $(k=1,2,\ldots)$, and $x_k \to \xi_1$ as $k \to \infty$. This implies

$$\lim_{k\to\infty} u(x_k) \leq \varepsilon < \frac{1}{2} \quad \text{and,} \quad \lim_{k\to\infty} u(x_k) = \lim_{x\to\xi_1} u(x) = 1,$$

where the latter equality follows from the fact that, according to Syčev's definition, u has boundary value 1 on E_1. Yet, even in a more particular case where E_o, E_1 are two disjoint closed sets of the boundary of an arbitrary domain, Aseev's assertion remains still an open problem since Syčev's argument on which it is based is incorrect.

In our paper [7] we have established (2) for topological cylinders with respect to the relative metric if $\rho \in F(\Gamma)$ satisfies the additional condition $\int_\gamma \rho \, dH^1 < \infty \ \forall \gamma \in \Gamma_Z$ rectifiable, but, according to Proposition 3 of this paper (established in [9]), it follows that the value of $M_p Z$ is not influenced by this condition.

And now, let us mention the extension of the equality between the p-module and the p-capacity in $\overline{R^n}$ considered by Hesse [16] (Theorem 5.5) in the case where $E_o, E_1 \subset D$ are compact, disjoint, non-empty sets. In his Ph.D. thesis [15], he proved also that $M_p\Gamma(E_o, E_1, \overline{R^n}) = \text{cap}_p(E_o, E_1, \overline{R^n})$, where $E_o, E_1 \subset \overline{R^n}$ are again supposed to be dosjoint, compact and non-empty. He asserted also that (18) holds if $D \subset \overline{R^n}$ and $E_o, E_1 \subset \bar{D}$ are disjoint, compact, non-empty sets and D is m-smooth on $(E_o \cup E_1) \cap \partial D$ for some m. However, this result is based on his Theorem 4.27 (quoted in this paper as Proposition 7), which is incorrect (see our comment on Proposition 7).

Finally, let us observe that we have tried to establish (18) in [8] in the particular case in which E_o is a closed set containing the complement of a closed ball and $D \subset R^n$, but the corresponding proof contains a mistake. Indeed, given $\rho \in F(\Gamma)$, we considered its truncation

$$\rho_i^k(x) = \begin{cases} k & \text{if } \rho(x) \geq k, \\ \rho(x) & \text{if } i^{-1} < \rho(x) < k, \\ i^{-1} & \text{if } \rho(x) \leq i^{-1}, \end{cases}$$

and the corresponding potential

$$u_i^k(x) = \inf_\beta \int_\beta \rho_i^k dH^1,$$

where β was an arc joining x to E_o in R^n. It is easy to see that, in this case, $u^k_{i|E_o} = 0$, and the difficulty is to show that

$$\underline{\lim}_{k\to\infty} m_i^k \geq 1, \quad \text{where} \quad m_i^k = \inf_{E_1} u_i^k(x).$$

In order to prove it is false, we have assumed that

$$\underline{\lim}_{k\to\infty} m_i^k < 1, \quad \text{or even} \quad \lim_{k\to\infty} m_i^k < 1$$

(appealling to a subsequence, if necessary). Hence, $\forall m_i^k$ with k sufficiently large, there exists an $x_k \in E_1$ such that $u_i^k(x_k) < 1$, implying the existence of an arc γ_k such that

$$u_i^k(x_k) = \int_{\gamma_k} \rho_i^k dH^1$$

(see Ziemer [25], p.47). Next, it is easy to see that it is possible to choose the sequence $\{x_k\}$ in such a way that

$$\lim_{k\to\infty} u_i^k(x_k) < 1, \quad \text{whence} \quad \lim_{k\to\infty} \int_{\gamma_k} \rho_i^k dH^1 < 1.$$

Yet, since

$$1 > \int_{\gamma_k} \rho_i^k dH^1 \geq \int_{\gamma_k} \rho_i^q dH^1 \quad \text{for} \quad k \geq q,$$

it follows that

$$\lim_{k\to\infty} \int_{\gamma_k} \rho_i^k dH^1 \geq \lim_{k\to\infty} \int_{\gamma_k} \rho_i^q dH^1 \qquad (q = 1, 2, \ldots).$$

Hence

$$1 > \lim_{k\to\infty} \int_{\gamma_k} \rho_i^k dH^1 \geq \lim_{q\to\infty} \lim_{k\to\infty} \int_{\gamma_k} \rho_i^q dH^1.$$

On the other hand,

$$\lim_{q\to\infty} \int_{\gamma_k} \rho_i^q dH^1 = \int_{\gamma_k} \rho_i dH^1 \geq \int_{\gamma_k} \rho dH^1 \geq 1 \qquad (k = 1, 2, \ldots),$$

whence

$$\lim_{k\to\infty} \lim_{q\to\infty} \int_{\gamma_k} \rho^q dH^1 \geq 1.$$

Now, in order to show that the hypothesis $\lim_{q\to\infty} m_i^q < 1$ is false, it remains to show that

$$\lim_{q\to\infty} \lim_{k\to\infty} \int_{\gamma_k} \rho_i^q dH^1 = \lim_{k\to\infty} \lim_{q\to\infty} \int_{\gamma_k} \rho_i^q dH^1,$$

which, on account of the above considerations, is equivalent to

$$\lim_{k\to\infty} \sup_q \int_{\gamma_k} \rho_i^q dH^1 = \sup_q \lim_{k\to\infty} \int_{\gamma_k} \rho_i^q dH^1. \tag{36}$$

We have tried to establish this relation by using the following minimax theorem (cf. Barbu and Precupanu [3], Ch. 2, Theorem 3.4, p.141):

If $\{f_k\}$ *is a non-increasing sequence of real-valued upper semi-continuous functions on a compact set* A, *then*

(37) $$\lim_{k\to\infty} \max_{A} f_k(x) = \max_{A} \lim_{k\to\infty} f_k(x).$$

But first, we have had to make some changes in order that the hypotheses of the preceding minimax theorem be satisfied. Thus, we have denoted $x_q = 1/q$ $(q = 1,2,\ldots)$, $A = \{0\} \cup \{x_1, x_2, \ldots\}$, and

$$f_k(x) = \begin{cases} \int_{\gamma_k} \rho_i^q dH^1 & \text{if} \quad x = x_q, \\ \int_{\gamma_k} \rho_i dH^1 & \text{if} \quad x = 0 \end{cases} \qquad (k = 1,2,\ldots).$$

If, for an infinite number of indices q, the sequences $\{f_k(x_q)\}$ are non-decreasing, then it is easy to prove directly that (36) holds. If, for an infinite number of indices q, $\{f_k(x_q)\}$ are non-increasing, then, on account of the preceding minimax theorem, (37) holds indeed. However, we have deduced in [8] that this relation implies (36), which is false. Indeed, it is easy to see that

$$\sup_q \lim_{k\to\infty} \int_{\gamma_k} \rho_i^q dH^1 \leq \lim_{k\to\infty} \sup_q \int_{\gamma_k} \rho_i^q dH^1.$$

Next, since $\forall k \in N$, the numerical sequences $\{f_k(x_q)\}$ are non-decreasing, then we have

$$f_k(0) = \max_A f_k(x) = \sup_q f_k(x_q) = \sup_q \int_{\gamma_k} \rho_i^q dH^1.$$

Hence

$$\lim_{k\to\infty} \sup_q \int_{\gamma_k} \rho_i^q dH^1 = \lim_{k\to\infty} f_k(0) = \lim_{k\to\infty} \max_A f_k(x).$$

On the other hand, the numerical sequences $\{\lim_{k\to\infty} f_k(x_q)\}$ are non-decreasing as well, so that

$$\lim_{k\to\infty} f_k(x_1) \leq \lim_{k\to\infty} f_k(x_2) \leq \ldots \leq \lim_{k\to\infty} f_k(0) = \lim_{k\to\infty} \max_A f_k(x),$$

but we are not allowed to conclude that

$$\lim_{q\to\infty} \lim_{k\to\infty} f_k(x_q) = \sup_q \lim_{k\to\infty} f_k(x_q) = \lim_{k\to\infty} f_k(0) = \lim_{k\to\infty} \max_A f_k(x).$$

It is easy to construct counterexamples showing that, in general, the implication is wrong. For instance, if

$$f_k(x_q) = 1 \text{ if } k \leq q, \quad f_k(x_q) = 0 \text{ for } k > q,$$

then $\forall q \in N$,

$$\lim_{k\to\infty} f_k(x_q) = 0, \quad \text{whence} \quad \sup_q \lim_{k\to\infty} f_k(x_q) = 0.$$

On the other hand, $\forall k \in N$,

$$\sup_q f_k(x_q) = 1, \quad \text{whence} \quad \lim_{k\to\infty} \sup_q f_k(x_q) = 1.$$

We also remark that in our paper [4], we have used Reimann's [20] assertion that (2) holds for a topological cylinder with respect to the euclidean metric according to definition I. It is true that, by Theorem 6, (2) holds, in this case only for $n = 2$, while for $n > 2$ its validity remains an open problem. However, in our paper [4], we have considered a definition of a topological cylinder Z with respect to the euclidean metric slightly different from I, i.e. the bases B_i $(i = 0,1)$ of Z are the images of the open bases $\dot{B}_i^o = \{x;\ (x^1)^2 + \ldots + (x^{n-1})^2 < 1,\ x^n = i\}$ $(i = 0,1)$ of the unit cylinder by the homeomorphism φ (involved in the definition). Yet, in such a case, (2) can be proved by the same argument as in the first part of the proof of Theorem 6. In this way, all our results of [4] are correct.

5. Evaluation of an exceptional set for quasiconformal mappings

In this section, we are going to give an application (mentioned in the introduction) of Theorem 1 to the theory of quasiconformal mappings. We need however some preliminaries.

A homeomorphism $f : D = D^*$ is said to be K-quasiconformal $(1 \leq K < \infty)$ if

$$(38) \qquad (1/K)M\Gamma \leq M\Gamma^* \leq KM\Gamma$$

$\forall \Gamma$ of D, where $\Gamma^* = f(\Gamma)$. If the constant K has not to be specified, f is said to be quasiconformal.

Let $f : B = D^*$ be a quasiconformal mapping and E^o the exceptional set of points of S such that the image $f(\gamma)$ of any endcut γ of B from an arbitrary point $\xi \in E^o$ is unrectifiable (we recall that an endcut γ of B from $\xi \in S$ is an open arc $\gamma \subset B$ with an endpoint at ξ and the other one at a point of B).

LEMMA 17. *Let* $R_+^n = \{x; x_n > 0\}$, $E_0 \subset R_+^n$, $E_1 \subset \partial R_+^n = \{x; x^n = 0\}$, $d(E_0, E_1) > 0$, $\Gamma = \Gamma(E_0, E_1, R_+^n)$, *and* $\tilde{\Gamma} = \Gamma(E_0, E_1, \overline{R_+^n})$. *Then*

(39) $M_p \tilde{\Gamma} = M_p \Gamma$.

Clearly, $\Gamma \subset \tilde{\Gamma}$ implies

(40) $M_p \Gamma \leq M_p \tilde{\Gamma}$,

so that it remains to prove only that

(41) $M_p \tilde{\Gamma} \leq M_p \Gamma$.

Since the case $M_p \Gamma = \infty$ is trivial, we may suppose $M_p \Gamma < \infty$. Let $\rho \in F^{o\Delta}(\Gamma)$, where $\Delta = R_+^n - (\bar{E}_0 \cup \bar{E}_1)$ and let $E_i(t) = \{x \in \overline{R_+^n}; d(x, E_i) \leq t\}$ $(i = 0, 1)$ ($\overline{R_+^n}$ being the closure of R_+^n in $\overline{R^n}$), γ_t an arc joining $E_0(t)$ to $E_1(t)$ in Δ, λ_i a segment of length t joining E_i to γ_t $(i = 0, 1)$, and $\gamma = \gamma_t \cup \lambda_0 \cup \lambda_1$. Given $\varepsilon > 0$, let $t < \varepsilon/2M$, where $M = \sup_{R^n} \rho(x)$, so that

$$1 \leq \int_\gamma \rho dH^1 \leq \int_{\gamma_t} \rho dH^1 + \int_{\lambda_0} \rho dH^1 + \int_{\lambda_1} \rho dH^1 \leq \int_{\gamma_t} \rho dH^1 + 2Mt < \int_{\gamma_t} \rho dH^1 + \varepsilon.$$

Hence $\int_{\gamma_t} \rho dH^1 > 1 - \varepsilon$ and then $\int_{\gamma_t} \frac{\rho}{1-\varepsilon} dH^1 > 1$. Clearly,

$$\rho_1(x) = \frac{\rho(x + te_n)}{1-\varepsilon} \quad \text{belongs to} \quad F(\tilde{\Gamma})$$

since $\forall \tilde{\gamma} \in \tilde{\Gamma}$ we have

$$\int_{\tilde{\gamma}} \rho_1 dH^1 = \int_{\tilde{\gamma}} \frac{\rho(x + te_n)}{1-\varepsilon} dH^1(x) = \int_{\gamma_t} \frac{\rho}{1-\varepsilon} dH^1 > 1.$$

Consequently,

$$M_p \tilde{\Gamma} \leq \int \rho_1^p dm = \frac{1}{(1-\varepsilon)^p} \int \rho^p dm,$$

and, since ρ is an arbitrary function of $F^{o\Delta}(\Gamma)$ and $M_p^{o\Delta} \Gamma = M_p \Gamma$, by taking the infimum over all these ρ we arrive at $M_p \tilde{\Gamma} \leq M_p \Gamma/(1-\varepsilon)$. Hence, letting $\varepsilon \to 0$ we obtain (41) which, together with (40), yields (39), as desired.

LEMMA 18. *If* $E \subset S$, $E \neq S$, $\Gamma = \Gamma[E, B(r), B]$, *and* $\tilde{\Gamma} = \Gamma[E, B(r), \bar{B}]$, *then* $M\tilde{\Gamma} = M\Gamma$.

Let $x_o \in S-E$, φ be the inversion with respect to a sphere centred at x_o, $C_o = \varphi[B(r)]$, $C_1 = \varphi(E)$, $\Gamma' = \varphi(\Gamma)$, and $\tilde{\Gamma}' = \varphi(\tilde{\Gamma})$. Then $\varphi(S) = \Pi$ is a plane, $C_1 \subset \Pi$ and C_o is contained in one of the half spaces of R^n determined by Π, so we are able to apply the preceding lemma. Taking into account the conformal invariance of the module, we deduce that $M\Gamma = M\Gamma' = M\tilde{\Gamma}' = M\tilde{\Gamma}$, as desired.

PROPOSITION 17. Let $\tilde{\Gamma}$ be the family of arcs $\tilde{\gamma} \subset \bar{B}$ having an endpoint in a set $E \subset S$ and let Γ'' be the family of arcs $\gamma'' \subset R^n$ which meet E. Then $M\Gamma'' = 2M\tilde{\Gamma}$ (Zorič [27]).

LEMMA 19. If Γ_o is the family of arcs $\gamma \subset R^n$ with an endpoint belonging to the exceptional set $E^o \subset S$ (defined above), then $M\Gamma_o = 0$.

Clearly, $\Gamma_o \subset \Gamma''$, so, by the preceding proposition and lemma (with the corresponding notation), we obtain

(42) $$M\Gamma_o \leq M\Gamma'' = 2M\tilde{\Gamma} = 2M\Gamma.$$

Yet, the points of $f(E^o)$ are inaccessible from $f(B)$ by rectifiable arcs, where f is a quasiconformal mapping, so if $\Gamma_1 = f(\Gamma)$, then, by (38), we get $M\Gamma \leq KM\Gamma_1 = 0$. Hence, by (42), we conclude that Γ_o is exceptional, as desired.

Finally we are going to prove that the conformal capacity of E^o is zero:

THEOREM 9. $\operatorname{cap} E^o = 0$.

Clearly,

(43) $$\operatorname{cap}[CE^o(r), E^o, R^n] \geq \operatorname{cap}[CB(R), E^o, R^n] \equiv \operatorname{cap} E^o,$$

where $B(R)$ is a fixed ball sufficiently large containing $E^o(r)$. This follows from the fact that the class of functions admissible for $\operatorname{cap}[CE^o(r), E^o, R^n]$ is contained in that of $\operatorname{cap} E^o$. Next, let $\Gamma_r = \Gamma[E^o, CE^o(r), R^n]$ and let Γ_o be the arc family of the preceding lemma. Then, evidently, $\Gamma_r \subset \Gamma_o$ and that lemma implies $M\Gamma_r \leq M\Gamma_o = 0$ $\forall r > 0$. Hence, by (43) and Corollary 2 to Theorem 1, we obtain

$$\operatorname{cap} E^o \leq \operatorname{cap}[CE^o(r), E^o, R^n] = M\Gamma_r \leq M\Gamma_o = 0,$$

as desired.

References

[1] АСЕЕВ, В.В.: Об одном свойстве модуля, Докл. Акад. Наук СССР 200 (1971), 513-514.

[2] BAGBY, T.: Ph.D. thesis, Harvard Univ., Cambridge, Massachusets 1966.

[3] BARBU, V. and T.PRECUPANU: Convexity and optimization in Banach spaces, Edit. Acad. Bucereşti România and Sijthoff & Noordhoff, International Publishers 1978, 316 pp.

[4] CARAMAN, P.: Quasiconformality and extremal length, Proc. 1st Romanian-Finnish Seminar on Teichmüller spaces and quasiconformal mappings, Braşow 1969, Edit. Acad. Bucureşti România 1971, pp. 111-145.

[5] —— : n-dimensional quasiconformal mappings, Edit. Acad. Bucureşti România and Abacus Press, Tunbridge Wells, Kent, England 1974, 553 pp.

[6] —— : Quasiconformality and boundary correspondence, Rev. Anal. Numer. Théorie Approximation 5(1976), 117-126 pp.

[7] —— : p-capacity and p-modulus, Symposia Math. 18(1976), 455-484 pp.

[8] —— : Estimate of an exceptional set for quasiconformal mappings in R^n, Komplexe Analysis und ihre Anwendung auf partielle Differentialgleichungen, Teil 3, Martin Luther Univ. Halle-Wittenberg Wissenschaftliche Beiträge 1980/41 (M18), Halle (Saale) 1980, pp. 210-221.

[9] —— : Le p-module et la p-capacité du cylindre, C. R. Acad. Sci. Paris 290(1980), 171-219.

[10] FUGLEDE, B.: Extremal length and functional completion, Acta Math. 9(1957), 171-219.

[11] GEHRING, F.: Rings and quasiconformal mappings in space, Trans. Amer. Math. Soc. 103(1962), 353-393.

[12] —— : Extremal length definition for conformal capacity in space, Michigan Math. J. 9(1962), 137-150.

[13] —— and J.VÄISÄLÄ: The coefficient of quasiconformality of domains in space, Acta Math. 114(1965), 1-70.

[14] HERSCH, J.: Longueurs extremales dans l'espace, résistence électrique et capacité, C. R. Acad.Sci. Paris 238(1954), 1693-1641.

[15] HESSE, J.: Modulus and capacity, Ph. D. thesis, Univ. of Michigan, Ann Arbor, Michigan 1972, 117 pp.

[16] —— : A p-extremal length and a p-capacity equality, Ark. Mat. 13(1975), 131-144.

[17] КРИВОВ, В.В.: Некоторые свойства модулей в пространстве, Докл. Акад. Наук СССР 154 (1964), 510-513.

[18] NÄKKI, R.: Boundary behavior of quasiconformal mappings in n-space, Ann. Acad. Sci. Fenn. Ser. A I Math. 484(1970), 50 pp.

[19] RADO, T. and P.V.REICHELDERFER: Continuous transformations in analysis, Springer, Berlin-Heidelberg-New York 1955, 442 pp.

[20] REIMANN, H.M.: Über harmonische Kapazität und quasikonforme Abbildungen im Raum, Comment, Math. Helv. 44(1969), 284-304.

[21] SAKS, S.: Theory of the integral, Second revised edition with two Notes by Prof.Stefan Banach. Hafner Publishing Company, New York 1955, 347 pp.

[22] СЫЧЁВ, А.В.: О некоторых свойствах модулей, Сибирский Мат. Ж. 6 (1965), 1108-1119.

[23] VÄISÄLÄ, J.: On quasiconformal mappings in space, Ann. Acad. Sci. Fenn. Ser. A I Math. 298 (1961), 36 pp.

[24] ZIEMER, W.: Extremal length and conformal capacity, Trans. Amer. Math. Soc. 126 (1967), 460-473.

[25] ——— : Extremal length and p-capacity, Michigan Math. J. 16 (1969), 43-51.

[26] ——— : Extremal length as a capacity, Michigan Math. J. 17 (1970), 117-123.

[27] ЗОРИЧ, В.А.: Об угловых граничных значениях квазиконформных отображений шара, Докл. Акад. Наук СССР 177 (1967), 771-773.

Institute of Mathematics
University "Al.I.Cuza"
Iaşi, România

AN ESTIMATE OF THE COMPLEX MONGE-AMPÈRE OPERATOR

Urban Cegrell (Uppsala)*)

C o n t e n t s

1. Introduction

In this note we prove an estimate for the complex Monge-Ampere operator. This estimate implies the Chern-Levine-Nirenberg estimate [5] and generalizes the estimates obtained for $\mathbb{C}^2$ in Cegrell [3,4].

2. Notation

Let Ω be an open subset of $\mathbb{C}^n$. Then $PSH(\Omega)$ denotes the plurisubharmonic functions on Ω and δ-$PSH(\Omega)$ is the Fréchet space $PSH(\Omega) - PSH(\Omega)$ (cf. Cegrell [3]). If $\varphi_\nu \in C^2(\Omega)$, $1 \leq \nu \leq n$ we define the n-linear operator T by

$$T(\varphi_1, \ldots, \varphi_n) = i\partial\bar{\partial}\varphi_1 \wedge \ldots \wedge i\partial\bar{\partial}\varphi_n$$

and if $\chi \in C_0(\Omega)$ we have by Stokes' theorem (cf. Bedford and Taylor [2]):

$$(1) \qquad \int_\Omega \chi T(\varphi_1, \ldots, \varphi_n)dw = \int_\Omega \varphi_p T(\varphi_1, \ldots, \chi, \ldots, \varphi_n)dw$$

If furthermore $\varphi_\nu \in PSH(\Omega)$, $1 \leq \nu \leq n$ then $T(\varphi_1, \ldots, \varphi_n) \geq 0$.

An example by Shiffman which can be found in Siu [6] proves that T

---

*) Partially supported by the Swedish Natural Science Research Council, Contract no. 4145-101.

cannot be extended to a positive measure-valued n-linear operator on all n-tuples of plurisubharmonic functions. However, in [1] Bedford gave the extension of T to all bounded and plurisubharmonic functions. The following proposition will give an asymmetric extension of T.

PROPOSITION. *Assume that* $\varphi_\nu^j \in PSH \cap C^2(\Omega)$, $1 \le \nu \le n$, $j \in \mathbf{N}$ *and that* $\varphi_1^j \longrightarrow \varphi \in PSH(\Omega)$, $j \to +\infty$ *in the sense of distributions*, $\varphi_\nu^j \longrightarrow \varphi_\nu \in PSH \cap C(\Omega)$, $j \longrightarrow +\infty$, $2 \le \nu \le n$, *uniformly on compact subsets of* Ω. *Then*

$$\{T(\varphi_1^j, \dots, \varphi_n^j)\}_{j=1}^{\infty}$$

is a weakly convergent sequence of positive measures. The limit is independent of the particular choices of the sequences.

Proof. The proof is by induction. If the sequences φ_ν^j, $2 \le \nu \le n$, $j \in \mathbf{N}$, are independent of j then the proposition is clearly true.

So let us assume that the proposition is true if φ_ν^j, $m \le \nu \le n$, does not depend on j. Assume that φ_ν^j, φ_ν^{1j} does not depend on j for $m+1 \le \nu \le n$ where $\lim_{j\to+\infty} \varphi_\nu^j = \lim \varphi_\nu^{1j}$, $1 \le \nu \le n$. If $\psi \in C_0^\infty(\Omega)$ we have, by (1),

$$\int \psi T(\varphi_1^{1j}, \dots \varphi_m^{1j}, \varphi_{m+1}, \dots, \varphi_n) - \int \psi T(\varphi_1^j, \dots, \varphi_m^j, \varphi_{m+1}, \dots, \varphi_n)$$

$$= \int \varphi_m^{1j} T(\varphi_1^{1j}, \dots, \psi, \varphi_{m+1}, \dots) - \int \varphi_m^j T(\varphi_1^j, \dots, \psi, \varphi_{m+1}, \dots) =$$

$$= \int \varphi_m [T(\varphi_1^{1j}, \dots, \psi, \varphi_{m+1}, \dots) - T(\varphi_1^j, \dots, \psi, \varphi_{m+1}, \dots)] +$$

$$+ \int (\varphi_m^{1j} - \varphi_m) T(\varphi_1^{1j}, \dots, \psi, \varphi_{m+1}, \dots) + \int (\varphi_m - \varphi_m^j) T(\varphi_1^j, \dots, \psi, \dots).$$

Now, every function in $C_0^\infty(\Omega)$ is in δ-$PSH \cap C(\Omega)$ so by assumption the first integral converges to zero, $j \longrightarrow \infty$, and the measures converge weakly. By the Banach-Steinhaus theorem they have uniformly bounded mass and since $\varphi_m^{1j} - \varphi_m$ and $\varphi_m^j - \varphi_m$ converge uniformly to zero, $j \longrightarrow \infty$, we see that all the integrals converge to zero, $j \longrightarrow \infty$, and the proposition is proved.

3. The estimate

THEOREM. *To every compact subset* K *of* Ω *there is a constant* C *and a compact subset* L *of* Ω *so that*

$$(2) \qquad \int_K T(\varphi_1, \ldots, \varphi_n) \leq C\left(\int_L |\varphi_1| dw\right) \prod_{\nu=2}^{n} \sup_{\xi \in L} |\varphi_\nu(\xi)|,$$

$$\forall \varphi_1 \in PSH(\Omega), \ \varphi_\nu \in PSH \cap C(\Omega), \ 2 \leq \nu \leq n.$$

Proof. By the proposition we can extend T to $PSH(\Omega) \overset{n}{\underset{\nu=2}{\times}} PSH \cap C(\Omega)$ and so by n-linearity to the Frechet space $\delta - PSH(\Omega) \overset{n}{\underset{\nu=2}{\times}} \delta - PSH \cap C(\Omega)$. We thus have

$$\delta - PSH(\Omega) \overset{n}{\underset{\nu=2}{\times}} \delta - PSH \cap C(\Omega) \xrightarrow{T} \delta - B(\Omega)$$

where $B(\Omega)$ are the positive Borel measures on Ω. We note that $T(PSH(\Omega) \overset{n}{\underset{\nu=2}{\times}} PSH \cap C(\Omega))$ $B(\Omega)$ so by Cegrell [3, Theorem 2:3,1] T is separately continuous and therefore continuous.

For plurisubharmonic functions the seminorms on $\delta - PSH$ and $\delta - PSH \cap C$ are of the form $\int_L |f| dw$ and $\sup_{\xi \in L} |\varphi(\xi)|$ which proves the theorem.

References

[1] BEDFORD, E.:Envelopes of continuous plurisubharmonic functions, Math. Ann. 251 (1980), 175-183.

[2] —— and B.A. TAYLOR,: The Dirichlet problem for a complex Monge-Ampere equation., Invent. Math. 37 (1976), 1-44.

[3] CEGRELL, U.:Delta-plurisubharmonic functions, Math. Scand. 43 (1978), 343-352.

[4] ——., Capacities and extremal plurisubharmonic functions on subsets of C^n, Ark. Mat. 18 (1980), 199-206.

[5] CHERN, S.S., H.I. LEVINE and L. NIRENBERG; Intrinsic norms on a complex manifold, Global analysis, Papers in honor of K.Kodaira, ed. by D.C.Spencer and S.Iynaga, Univ. of Tokyo Press and Princeton Univ. Press, Tokyo 1969, pp. 119-439.

[6] SIU, Y.-T.: Extension of meromorphic maps, Ann. of Math. 102 (1975), 421-462.

Department of Mathematics
Uppsala University
Thunbergsvgägen 3
S-752 38 Uppsala, Sweden

ON THE PARAMETRIC AND ALGEBRAIC MULTIPLICITIES OF AN ISOLATED ZERO OF A HOLOMORPHIC MAPPING

Jacek Chądzyński, Tadeusz Krasiński
and Wojciech Kryszewski (Łódź)

Contents

Summary

In this paper there has been given a constructive proof of the equivalence of the parametric and algebraic multiplicities of an isolated zero of a holomorphic mapping.

Introduction

An important part in algebraic geometry, algebraic topology, complex analysis and the theory of singularities is played by the notion of multiplicity. Several definitions of multiplicity are well-known: algebraic, covering, gradient and parametric ones. This last definition predominates over the others in that, according to it, the multiplicity can be calculated, in general, in a simple manner . For instance, the calculation of the parametric multiplicity of the zero at the point $(0,0)$ of the mapping $H(x,y) = (x^3-y^5 , x^4-y^7)$ presents no serious difficulties (see sec. 5) , while that of the other multiplicities is not simple in this case.

Consequently, there arises a question whether these definitions are equivalent in the case of an isolated zero of a holomorphic mapping. This question was answered in the affirmative by Orlik (see [8]), as regards the first three definitions.

In this paper we examine holomorphic mappings in $\mathbb{C}^2$. We show the equivalence of the definitions of the algebraic and parametric multiplicities. Thus, all the definitions are equivalent in this case, and the last one gives a simple criterion for calculating multiplicities. Besides, the parametric multiplicity is very useful when one investigates the order of decrease of a mapping in a neighbourhood of an isolated zero (see sec. 5).

1. Notation

In this paper $\mathbb{C}$, $\mathbb{C}^m$ will stand for the field of complex numbers and the m-dimensional complex vector space. If $z = (x,y) \in \mathbb{C}^2$, then $|z| = \max(|x|, |y|)$.

By $\mathcal{O}_w^m$ we shall denote the ring of germs of holomorphic functions at a point $w \in \mathbb{C}^m$. If w is the origin, then, instead of $\mathcal{O}_o^m$, we shall write $\mathcal{O}^m$, and if $m = 1$ - simply $\mathcal{O}$. If f is a holomorphic function at the point w, we shall denote by $\hat{f}$ the germ from $\mathcal{O}_w^m$ generated by the function f. If f is of a more complicated form, say $f = g \circ h$, then, instead of $\widehat{(g \circ h)}$, we shall write $(g \circ h)\hat{}$. Further definitions and notations concerning germs of holomorphic functions will be taken after Hervé (see [5]).

If $\mathcal{P}$ is a commutative ring and A - its arbitrary subset, then by (A) we shall denote the smallest ideal in $\mathcal{P}$ containing the set A. When $A = \{\alpha_1, \ldots, \alpha_s\}$, we simply write $(\alpha_1, \ldots, \alpha_s)$. If $\alpha \in \mathcal{P}$ and $\mathfrak{a}$ is an ideal of this ring, then by $[\alpha]$ we denote the element of the residue-class ring $\mathcal{P} / \mathfrak{a}$, generated by α.

Let h be a holomorphic function at the point $0 \in \mathbb{C}$, not vanishing identically. Then, in some neighbourhood of $0 \in \mathbb{C}$, we have

$$h(t) = \sum_{k=\rho}^{\infty} c_k t^k, \quad c_\rho \neq 0 .$$

The number ρ is called the order of h at the point 0 and denoted by $O(h)$. In addition, we assume that $O(h) = \infty$ when h vanishes identically.

2. The parametric multiplicity

Let f be a holomorphic function in a neighbourhood Ω of the point $z_0 \in \mathbb{C}^2$, $f(z_0) = 0$ and $\Gamma = \{z \in \Omega: f(z) = 0\}$. If there exist a neighbourhood $U \subset \Omega$ of the point z_0, a neighbourhood V of the point $0 \in \mathbb{C}$ and a holomorphic mapping $\Psi: V \to U$ such that $\Psi(0) = z_0$ and Ψ maps V bijectively onto $\Gamma \cap U$, then the triple (V,Ψ,U) is called a parametrization of the set of zeros of the function f in a neighbourhood of the point z_0. Let (V_1,Ψ_1,U_1) and (V_2,Ψ_2,U_2) be two such parametrization. We say that these parametrizations are equivalent when there exist a neighbourhood $U \subset U_1 \cap U_2$ of the point z_0 and a conformal mapping α of the set $V_1' = \Psi_1^{-1}(U)$ onto the set $V_2' = \Psi_2^{-1}(U)$ such that $\alpha(0) = 0$ and $\Psi_1|_{V_1'} = \Psi_2|_{V_2'} \circ \alpha$.

In the sequel, we shall deal with the question of the existence of parametrizations and with their equivalence. However, we shall first give a definition convenient for further considerations. If f is a holomorphic function in a neighbourhood of the point $0 \in \mathbb{C}^2$, $f(0) = 0$ and the function $x \mapsto f(x,0)$ does not vanish identically, then f is called normed in the direction of x.

Let f be a holomorphic function in a neighbourhood of the point $0 \in \mathbb{C}^2$, normed in the direction of x, and let $x = 0$ be the n - fold zero of the equation $f(x,0) = 0$. Then it follows from the Weierstrass preparation theorem that, in some neighbourhood Ω of the point $0 \in \mathbb{C}^2$, there exists a distinguished pseudopolynomial of the form

$$(1) \qquad P(x,y) = x^n + a_1(y)x^{n-1} + \ldots + a_n(y)$$

whose set of zeros is identical with that of the function f. Denote this set by Γ. Let us additionally assume that $\hat{f}$ is an irreducible germ in $\mathcal{O}^2$. Then, applying Rouche's theorem and the theory of analytic continuations (see [9], pp. 275 and 262, cf. [5], p. 72), we easily show that there exist: a bicylinder $B = \{z \in \mathbb{C}^2: |x| < \rho, |y| < \eta\} \subset \Omega$ and a function φ holomorphic in the disc $K = \{t \in \mathbb{C} : |t| < \eta^{1/n}\}$, such that (K,Φ,B), where $\Phi(t) = (\varphi(t),t^n)$, is a parametrization of Γ in a neighbourhood of the point $0 \in \mathbb{C}^2$. Moreover, for any $\varepsilon > 0$, there exists a bicylinder $B' = \{z \in \mathbb{C}^2: |x| < \rho' < \varepsilon, |y| < \eta' < \varepsilon\} \subset \Omega$ such that (K',Φ',B'), where $K' = \{t \in \mathbb{C}: |t| < \eta'^{1/n}\}$ and

$\Phi' = \Phi_{|K'}$, is a parametrization of Γ in a neighbourhood of the point $0 \in \mathbb{C}^2$. The above parametrizations will be called _canonical parametrizations of the set of zeros of the function_ f _in a neighbourhood of the point_ $0 \in \mathbb{C}^2$. Summing up, we obtain

PROPOSITION 1. _If_ f _is a holomorphic function in a neighbourhood of the point_ $0 \in \mathbb{C}^2$ _, normed in the direction of_ x _, and_ $\hat{f}$ _is the irreducible germ in_ $\mathcal{O}^2$ _, then, in an arbitrarily small neighbourhood of the point_ 0 _, there exists a canonical parametrization of the set of zeros of the function_ f .

Under the above assumptions, we also show easily that any parametrization (V,Ψ,U) of the set Γ is equivalent to a canonical parametrization (K,Φ,B) . Consequently, we have

PROPOSITION 2. _All parametrizations of the set of zeros of the function_ f _are equivalent._

Let now f be an arbitrary holomorphic function in a neighbourhood of the point $z_o \in \mathbb{C}^2$, not vanishing identically there, and let $\hat{f}$ be the irreducible and non-invertible germ in $\mathcal{O}^2_{z_o}$. By means of a non-singular linear change of variables l , one can reduce the function f to a function $\tilde{f} = f \circ l$ normed in the direction of x . Of course, $\tilde{f}$ generates the irreducible germ in $\mathcal{O}^2$, too. By proposition 1, there exists a canonical parametrization (K,Φ,B) of the set of zeros of $\tilde{f}$ in a neighbourhood of the point 0 . Then the triple $(K,l^{-1} \circ \Phi,l^{-1}(B))$ is a parametrization of the set of zeros of f in a neighbourhood of z_o . From proposition 2 we get that all parametrizations of the set of zeros of f in a neighbourhood of z_o are equivalent. Hence we obtain

PROPOSITION 3. _If_ f _is a holomorphic function in a neighbourhood of the point_ $z_o \in \mathbb{C}^2$ _, not vanishing identically, and_ $\hat{f}$ _is the irreducible and non-invertible germ in_ $\mathcal{O}^2_{z_o}$ _, then there exist parametrizations of the set of zeros of_ f _in a neighbourhood of the point_ z_o _and they are equivalent._

Let f and g be holomorphic functions at the point $z_o \in \mathbb{C}^2$, having there a common isolated zero. Assume additionally that $\hat{f}$ is the irreducible germ in $\mathcal{O}^2_{z_o}$ and (V,Ψ,U) is a parametrization of zeros of the function f in a neighbourhood of z_o . Let f' be another holomorphic function in a neighbourhood of z_o , such that $(f')^{\wedge} = \hat{f}$, and let (V',Ψ',U') be a parametrization of zeros of f' in a neighbourhood of the point z_o. Since f and f' are identical in some neighbourhood of the point z_o, from proposition 3 it follows that there exists a holomorphic function α in some neighbourhood of $0 \in \mathbb{C}$, such that $\alpha(0) = 0$, $\alpha'(0) \neq 0$ and $\Psi' = \Psi \circ \alpha$. Hence $0(g \circ \Psi) = 0(g \circ \Psi')$. The number $0(g \circ \Psi)$ is called the <u>order of the function</u> g <u>at the irreducible germ</u> $\hat{f}$ and denoted by $0_{\hat{f}}(g)$.

Directly from the definition we have

$$(2) \qquad 0_{\hat{f}}(g_1 \cdot g_2) = 0_{\hat{f}}(g_1) + 0_{\hat{f}}(g_2) .$$

It is also easily seen that if l is a non-singular linear change of variables in $\mathbb{C}^2$, then it induces an isomorphism $\mathcal{O}^2_{z_o} \to \mathcal{O}^2_{l(z_o)}$, and $(f \circ l^{-1})^{\wedge}$ is the irreducible germ in $\mathcal{O}^2_{l(z_o)}$. It is easy to check that

$$(3) \qquad 0_{\hat{f}}(g) = 0_{(f \circ l^{-1})^{\wedge}}(g \circ l^{-1}) .$$

Let now f and g be holomorphic functions in a neighbourhood of the point $z_o \in \mathbb{C}^2$, having a common isolated zero at this point. Let us factorize $\hat{f}$ into irreducible and non-invertible elements in $\mathcal{O}^2_{z_o}$. Let $\hat{f} = \hat{f}_1 \dots \hat{f}_r$. From the theorem on the uniqueness of factorization in $\mathcal{O}^2_{z_o}$ and the remark preceding the definition of the order of a function at an irreducible germ it follows that the number

$$\sum_{i=1}^{r} 0_{\hat{f}_i}(g)$$

depends on the germ $\hat{f}$ and the function g only. The above number is called the <u>parametric multiplicity of the common isolated zero</u> z_o <u>of</u> f <u>and</u> g and denoted by $\mu_p(f,g;z_o)$. If $H = (f,g)$, then $\mu_p(f,g;z_o)$ is also called the <u>parametric multiplicity of the isolated zero</u> z_o <u>of the mapping</u> H.

The above definition is non-symmetric with respect to f and g. It can be shown directly (cf. [11], pp. 109-110) that $\mu_p(f,g;z_o) = \mu_p(g,f;z_o)$; however, we shall not make use of this fact. It will follow from the symmetry of the algebraic multiplicity and from the equivalence of both definitions.

From the definition of the parametric multiplicity as well as from (2) we obtain

PROPOSITION 4. $\mu_p(f_1 \cdot f_2\ , g; z_o) = \mu_p(f_1, g; z_o) + \mu_p(f_2, g; z_o)$ and $\mu_p(f, g_1 \cdot g_2; z_o) = \mu_p(f, g_1; z_o) + \mu_p(f, g_2; z_o)$.

Let l be a non-singular linear transformation of $\mathbb{C}^2$ onto $\mathbb{C}^2$. Then the mapping $\nu : \mathcal{O}^2_{z_o} \to \mathcal{O}^2_{l^{-1}(z_o)}$ defined by the formula $\nu(\hat{h}) = (h \circ l)$ is an isomorphism of the above rings. Thus, it assign irreducible germs to irreducible germs. From (3) we get

PROPOSITION 5. $\mu_p(f, g; z_o) = \mu_p(f \circ l\ , g \circ l\ ; l^{-1}(z_o))$.

3. The algebraic multiplicity

Let $H = (f,g)$ be, as before, a holomorphic mapping in a neighbourhood Ω of the point z_o , having an isolated zero at this point. Consider the residue-class ring $\mathcal{O}^2_{z_o} / (\hat{f}, \hat{g})$. From the analytical version of Hilbert's Nullstellensatz (see [5], p. 78) it follows that this ring is a finite-dimensional vector space over $\mathbb{C}$.

The number $\mu_a(f, g; z_o) = \dim_{\mathbb{C}} \mathcal{O}^2_{z_o} / (\hat{f}, \hat{g})$ is called the algebraic multiplicity of the isolated zero z_o of the mapping H .

Directly from the definition we have

PROPOSITION 6. $\mu_a(f, g; z_o) = \mu_a(g, f; z_o)$.

It is not hard, either, to show (cf. [10], pp. 185-186)

PROPOSITION 7. $\mu_a(f_1 \cdot f_2, g; z_o) = \mu_a(f_1, g; z_o) + \mu_a(f_2, g; z_o)$.

Let l be a non-singular linear transformation of $\mathbb{C}^2$ onto $\mathbb{C}^2$ and ν - the isomorphism defined in the preceding section, before proposition 5. From this isomorphism follows

PROPOSITION 8. $\mu_a(f, g; z_o)) = \mu_a(f \circ l\ , g \circ l\ ; l^{-1}(z_o))$.

4. The equivalence of the parametric and algebraic multiplicities

We shall first prove

LEMMA 1. *Let $P \subset \mathcal{O}$ be a subring having the property that there exists a non-negative integer N such that, for each germ $\hat{\lambda} \in \mathcal{O}$ with $O(\lambda) > N$, we have $\hat{\lambda} \in P$. If $\hat{\mu} \in P$ and*

$0(\mu) = \rho < \infty$, then $\dim_{\mathbb{C}} P/(\hat{\mu}) = \rho$.

P r o o f. First, note that if $\hat{\lambda} \in O$ and $0(\lambda) > N + \rho$, then $\hat{\lambda} \in (\hat{\mu})$. Indeed, the function $\lambda' = \lambda/\mu$ is holomorphic in some vicinity of the point 0 and has there a removable singularity. Moreover, $0(\lambda') > N$. Consequently, by assumption, $\hat{\lambda}' \in P$, and since $\hat{\lambda} = \hat{\lambda}' \cdot \hat{\mu}$, therefore $\hat{\lambda} \in (\hat{\mu})$.

For any $i = 1,\dots,\rho$, let us now define the set

$$\Lambda_i = \{0(\lambda) : \hat{\lambda} \in P,\ 0(\lambda) \equiv i \,(\mathrm{mod}\ \rho)\} .$$

From the assumption it follows that every such set contains the minimal element. Thus, for every i , there exists an element $\hat{\lambda}_i \in P$ such that $0(\lambda_i) = \min \Lambda_i$. Let $l_i = 0(\lambda_i)$ and $\Lambda = \{l_1 ,\dots, l_\rho\}$

We shall now show that the elements $[\hat{\lambda}_1] ,\dots, [\hat{\lambda}_\rho] \in P/(\hat{\mu})$ are linearly independent over $\mathbb{C}$. Let σ be a permutation of the set of numbers $\{1 ,\dots, \rho\}$, such that $l_{\sigma(1)} < \dots < l_{\sigma(\rho)}$.

Let $\alpha_1 ,\dots, \alpha_\rho$ be any complex numbers such that

$$\sum_{i=1}^{\rho} \alpha_i [\hat{\lambda}_{\sigma(i)}] = 0 ,$$

i.e. there exists a germ $\hat{\lambda} \in P$ such that

$$\sum_{i=1}^{\rho} \alpha_i \hat{\lambda}_{\sigma(i)} = \hat{\lambda}\hat{\mu} .$$

Consequently, in some neighbourhood of the point $0 \in \mathbb{C}$, we have

$$\sum_{i=1}^{\rho} \alpha_i \lambda_{\sigma(i)}(t) = \lambda(t)\mu(t) .$$

Suppose that not all α_i are equal to zero. Let i_0 be a smallest number such that $\alpha_{i_0} \neq 0$. Then

$$l_{\sigma(i_0)} = 0\Big(\sum_{i=1}^{\rho} \alpha_i \lambda_{\sigma(i)}\Big) = 0(\lambda\cdot\mu) = 0(\lambda) + 0(\mu) = 0(\lambda) + \rho .$$

Hence $0(\lambda) \equiv l_{\sigma(i_0)} \equiv \sigma(i_o) \,(\mathrm{mod}\ \rho)$ and $0(\lambda) < l_{\sigma(i_0)}$, which contradicts the definition of $l_{\sigma(i_0)}$.

Now, we shall show that the elements $[\hat{\lambda}_1] ,\dots, [\hat{\lambda}_\rho]$ form a basis in $P/(\hat{\mu})$. Let $[\hat{\lambda}]$ be any element of $P/(\hat{\mu})$. Then it

nical parametrization of the set of zeros of f in a neighbourhood of the point $0 \in \mathbb{C}^2$. Let us now take any function h holomorphic in some in some neighbourhood Ω' of the point $0 \in \mathbb{C}^2$. From proposition 1 it follows that there exists a parametrization (K',Φ',B') where $\Phi' = \Phi_{|K'}$, such that $B' \subset \Omega'$. Consequently, for every such function h, we shall find a disc K' with centre at the point $0 \in \mathbb{C}$, such that $h \circ \Phi$ is well defined in K', and further understand the superposition $h \circ \Phi$ in this sense.

Define now the set $\mathcal{Q} \subset \mathcal{O}$ in the following way:

$$(4) \qquad \mathcal{Q} = \{\hat{\lambda} \in \mathcal{O} : \lambda = h \circ \Phi,\ \hat{h} \in \mathcal{O}^2\}.$$

It is easily seen that $\mathcal{Q}$ is a subring in $\mathcal{O}$. We shall show

PROPOSITION 9. $\mathcal{O}^2 / (\hat{f}) \cong \mathcal{Q}$.

Proof. Let us define the homomorphism $\nu: \mathcal{O}^2/(\hat{f}) \to \mathcal{Q}$ by the formula $\nu([\hat{h}]) = (h \circ \Phi)^{\wedge}$. Since (K,Φ,B) is a parametrization of the set of zeros of f in a neighbourhood of the origin, the mapping ν is well defined. This mapping is, of course, a surjection. It suffices to prove that it is an injection. Let $\nu([\hat{h}_1]) = \nu([\hat{h}_2])$. Hence $(h_1 \circ \Phi)^{\wedge} = (h_2 \circ \Phi)^{\wedge}$, that is, in some neighbourhood V of the origin in $\mathbb{C}$, $h_1 \circ \Phi = h_2 \circ \Phi$. Let (K',Φ',B') be a parametrization such that h_1 and h_2 are defined in B' and $K' \subset V$. From the injectivity of Φ' we have $h_1 = h_2$ on $\Gamma \cap B'$. From the analytical version of Hilbert's Nullstellensatz (see [4], p. 78) we get that $\hat{h}_1 - \hat{h}_2 \in (\hat{f})$, which gives $[\hat{h}_1] = [\hat{h}_2]$.

Remark 1. It follows from the above that if $(h \circ \Phi)^{\wedge} = 0$ in $\mathcal{O}$, then $\hat{h} \in (\hat{f}) \subset \mathcal{O}^2$

Let us now assume that g is an arbitrary holomorphic function in a neighbourhood Ω of the point $0 \in \mathbb{C}^2$, having at this point an isolated zero in common with the function f. Put $\mu = g \circ \Phi$.

PROPOSITION 10. $\mathcal{O}^2 / (\hat{f},\hat{g}) \cong \mathcal{Q}/(\hat{\mu})$.

Proof. It is easy to notice (see [1], p. 19) that

$$\mathcal{O}^2/(\hat{f},\hat{g}) \cong (\mathcal{O}^2/(\hat{f}))/([\hat{f}],[\hat{g}]) \cong (\mathcal{O}^2/(\hat{f}))/([\hat{g}])$$

follows from the definition of Λ that there exist numbers $l_{i_1} \in \Lambda$ and a non-negative integer n_1, such that $O(\lambda) = l_{i_1} + n_1 \cdot \rho$. Thus, $O(\lambda) = O(\lambda_{i_1} \mu^{n_1})$ and, in consequence, there exists a number $\beta_1 \in \mathbb{C}$ such that $O(\lambda) < O(\lambda - \beta_1 \lambda_{i_1} \mu^{n_1})$. Applying an analogous reasoning to $\lambda - \beta_1 \lambda_{i_1} \mu^{n_1}$, we find an element $\beta_2 \lambda_{i_2} \mu^{n_2}$ such that $O(\lambda - \beta_1 \lambda_{i_1} \mu^{n_1}) < O(\lambda - \sum_{k=1}^{2} \beta_k \lambda_{i_k} \mu^{n_k})$. By simple induction we find a sequence $\{\beta_k \lambda_{i_k} \mu^{n_k}\} \subset P$ such that the sequence of numbers $\{O(\lambda - \sum_{k=1}^{p} \beta_k \lambda_{i_k} \mu^{n_k})\}$ is increasing. So, there exists a number m such that $O(\lambda - \sum_{l=1}^{m} \beta_l \lambda_{i_l} \mu^{n_l}) > N + \rho$. In virtue of the remark made at the beginning of the proof, $(\lambda - \sum_{l=1}^{m} \beta_l \lambda_{i_l} \mu^{n_l})^{\wedge} \in (\hat{\mu})$. Hence $\hat{\lambda} - \sum_{l=1}^{m} \beta_l \hat{\lambda}_{i_l} \hat{\mu}^{n_l} \in (\hat{\mu})$. Let us decompose this last sum into the sum of two addends $\sum' + \sum''$, the first one consisting of those terms for which $n_l = 0$, and the second - of those for which $n_l > 0$. The first addend can be written in the form

$$\sum{}' \beta_l \hat{\lambda}_{i_l} = \sum_{i=1}^{\rho} \gamma_i \hat{\lambda}_i .$$

Since the second addend belongs to $(\hat{\mu})$, therefore

$$\hat{\lambda} - \sum_{i=1}^{\rho} \gamma_i \hat{\lambda}_i \in (\hat{\mu}) .$$

Hence

$$[\hat{\lambda}] - \sum_{i=1}^{\rho} \gamma_i [\hat{\lambda}_i] = 0 ,$$

which was to be shown.

This concludes the proof of the lemma.

Let f still satisfy the assumptions of proposition 1, and let Γ be the set of zeros of f and (K,Φ,B) - a fixed cano-

because $[\hat{f}] = 0$. From proposition 9 we have $\mathcal{O}^2 / (\hat{f}) \cong \mathcal{Q}$, with that the isomorphism ν , defined in the proof of proposition 9, assigns $\hat{\mu}$ to $[\hat{g}]$. So, $(\mathcal{O}^2 / (\hat{f})) / ([\hat{g}]) \cong \mathcal{Q} / (\hat{\mu})$. This ends the proof.

Let f still satisfy the assumptions of proposition 1, and let P be the distinguished pseudopolynomial of form (1), corresponding to f in Ω . Let now h be an arbitrary holomorphic function in some neighbourhood Ω' of the point $0 \in \mathbb{C}^2$. Then, from the Späth-Cartan division theorem (see [5], p. 13) it follows that there exist: a bicylinder Δ with centre at the origin , $\Delta \subset \Omega \cap \Omega'$, a holomorphic function q and a pseudopolynomial H of degree less than n , defined in Δ , such that

$$(5) \qquad h = P \cdot q + H \quad \text{in} \quad \Delta$$

and decomposition (5) is unique. Moreover, from remark 1 and from (5) it follows that if $(H \circ \Phi)\hat{} = 0$ in $\mathcal{O}$, then $\hat{H} = 0$ in $\mathcal{O}^2$ because $(P \circ \Phi)\hat{} = 0$. It also follows from (5) that the function h in (4) can be replaced by the pseudopolynomial H . Summing up, we obtain

PROPOSITION 11. $\mathcal{Q} = \{\hat{\lambda} \in \mathcal{O} : \lambda = H \circ \Phi$, H -a pseudopolynomial of degree $< n\}$. Besides, $\hat{H} = 0$ in $\mathcal{O}^2$ if and only if $(H \circ \Phi)\hat{} = 0$ in $\mathcal{O}$.

We shall now prove

LEMMA 2. There exists a non-negative integer N such that, for each $\hat{\lambda} \in \mathcal{O}$ with $O(\lambda) > N$, we have $\hat{\lambda} \in \mathcal{Q}$.

P r o o f. Let (K,Φ,B) , where $\Phi(t) = (\varphi(t),t^n)$, $B = \{z \in \mathbb{C}^2 : |x| < \rho , |y| < \eta\}$, $K = \{t \in \mathbb{C}: |t| < \eta^{1/n}\}$, be a canonical parametrization of the set of zeros of f in a neighbourhood of the origin. It is easy to notice that, for any $i = 1,\dots,n$, the function φ^{n-i} can be represented in the form

$$(6) \qquad [\varphi(t)]^{n-i} = t^{n-1} A_{1i}(t^n) + \dots + A_{ni}(t^n)$$

where A_{ji} are holomorphic functions in $\Delta = \{y: |y| < \eta\}$.

We shall first show that $W(y) = \det [A_{ij}(y)]$ does not vanish identically in Δ . Suppose the contrary. Then the system of equations

$$(7) \qquad \hat{A}_{j1}x_1 + \dots + \hat{A}_{jn}x_n = 0 \ , \quad j = 1,\dots,n \ ,$$

would possess a non-zero solution $(\hat{c}_1', \dots, \hat{c}_n')$ where $\hat{c}_i'$ belong to the quotient field of $\mathcal{O}$. Hence it easily follows that there would exist a non-zero solution $(\hat{c}_1, \dots, \hat{c}_n)$, $\hat{c}_i \in \mathcal{O}$, of system (7). Consequently, there would exist a neighbourhood $\Delta' = \{y: |y| < \eta' < \eta\}$ and functions $c_1, \dots, c_n$ holomorphic in Δ', not all vanishing identically, such that

$$A_{j1}(y)c_1(y) + \dots + A_{jn}(y)c_n(y) = 0 \quad \text{for} \quad y \in \Delta' .$$

Let us now define in $\mathbb{C} \times \Delta'$ a pseudopolynomial H of degree less than n, not vanishing identically, by the formula

$$H(x,y) = c_1(y)x^{n-1} + \dots + c_n(y) .$$

From (6) we have in $K' = \{t: |t| < \eta'^{1/n}\}$

$$H \circ \Phi(t) = H(\varphi(t), t^n) = \sum_{i=1}^{n} c_i(t^n)\left[\varphi(t)\right]^{n-i} =$$

$$= \sum_{i=1}^{n} c_i(t^n) \cdot \sum_{j=1}^{n} A_{ji}(t^n)t^{j-1} = \sum_{j=1}^{n} t^{j-1} \sum_{i=1}^{n} A_{ji}(t^n)c_i(t^n) \equiv 0,$$

which contradicts proposition 11.

Let $\mathcal{O}(W) = s$. We put $N = n(s+1)$. Let $\hat{\lambda} \in \mathcal{O}$ and $\mathcal{O}(\lambda) > N$. Then, for t from some neighbourhood of $0 \in \mathbb{C}$, we have

$$\lambda(t) = t^{n-1}\lambda_1(t^n) + \dots + \lambda_n(t^n) .$$

Hence

$$n(s+1) < \mathcal{O}(t^{n-j}\lambda_j(t^n)) = (n-j) + n\mathcal{O}(\lambda_j) < n + n\mathcal{O}(\lambda_j)$$

and, in consequence, $s < \mathcal{O}(\lambda_j)$. Consider the system of equations

$$(8) \qquad \hat{A}_{j1}x_1 + \dots + \hat{A}_{jn}x_n = \hat{\lambda}_j \ , \quad j = 1,\dots,n .$$

Since $\hat{W} \neq 0$ in $\mathcal{O}$, system (8) possesses exactly one solution $(\hat{c}_1, \dots, \hat{c}_n)$ in the quotient field of the ring $\mathcal{O}$, of the form $(\hat{W}_1/\hat{W}, \dots, \hat{W}_n/\hat{W})$ where $\hat{W}_i$ is the determinant of the matrix formed by replacing the i-th column of $[\hat{A}_{ji}]_{1 \le j,\, i \le n}$ by the column $(\hat{\lambda}_j)_{1 \le j \le n}$. Since $0(\lambda_j) > s$, therefore $0(W_i) > s$ and $(W_i/W)^{\wedge} \in \mathcal{O}$ for $i = 1,\dots,n$. So, $\hat{c}_i \in \mathcal{O}$ and, similarly as before, in some neighbourhood $\Delta' = \{y: |y| < \eta' < \eta\}$,

$$A_{j1}(y)c_1(y) + \dots + A_{jn}(y)c_n(y) = \lambda_j(y) \quad \text{for} \quad y \in \Delta',$$

where $c_1, \dots, c_n$ are holomorphic functions in Δ'. In $\mathbb{C} \times \Delta'$ we define a pseudopolynomial H by the formula

$$H(x,y) = c_1(y)x^{n-1} + \dots + c_n(y) .$$

We easily verify that $H \circ \Phi(t) = \lambda(t)$ for $|t| < \eta'^{1/n}$. Consequently, $\hat{\lambda} \in \mathcal{Q}$

This completes the proof of the lemma.

Now, we shall prove the fundamental theorem of this paper.

<u>THEOREM 1</u>. <u>If</u> $H = (f,g)$ <u>is a holomorphic mapping in a neighbourhood of the point</u> $z_o \in \mathbb{C}^2$, <u>having an isolated zero at this point, then</u>

$$\mu_p(f,g;z_o) = \mu_a(f,g;z_o) .$$

<u>P r o o f</u>. Since a non-singular linear change of variables does not alter the parametric and algebraic multiplicities (propositions 5 and 8), we may assume, without loss of generality, that $z_o = 0$ and f is normed in the direction of x. Let $\hat{f} = \hat{f}_1 \cdot \dots \cdot \hat{f}_r$ be a factorization of the germ $\hat{f}$ into irreducible and non-invertible elements in $\mathcal{O}^2$, and let $f = f_1 \cdot \dots \cdot f_r$. Then f_i are normed in the direction of x and, in virtue od proposition 4 and 7,

$$\mu_p(f,g;z_o) = \sum_{i=1}^{r} \mu_p(f_i,g;z_o); \quad \mu_a(f,g;z_o) = \sum_{i=1}^{r} \mu_a(f_i,g;z_o) .$$

Consequently, it is enough to prove the equality $\mu_p(f,g;z_o) = \mu_a(f,g;z_o)$ when f is a function normed in the direction of x, $\hat{f}$ is an irreducible germ in $\mathcal{O}^2$ and (K,Φ,B) - a canonical parametrization of the set of zeros of f in a neighbourhood of the point $0 \in \mathbb{C}^2$. In this case, from proposition 10 we have

$$\mu_a(f,g;z_o) = \dim_{\mathbb{C}} \mathcal{O}^2 / (\hat{f},\hat{g}) = \dim_{\mathbb{C}} \mathcal{Q} / (\hat{\mu}) .$$

From lemma 2 it follows that the ring $\mathcal{Q}$ satisfies the assumptions of lemma 1, and so, $\dim_{\mathbb{C}} \mathcal{Q} / (\hat{\mu}) = \rho$ where $\rho = 0(g \circ \Phi)$. Hence $\mu_a(f,g;z_o) = 0(g \circ \Phi)$. On the other hand, $0(g \circ \Phi)$ is equal to the order of g at the irreducible germ $\hat{f}$, that is, to $\mu_p(f,g;z_o)$.

This ends the proof.

5. Concluding remarks

(i) From theorem 1 and proposition 6 it follows at once that $\mu_p(f,g;z_o) = \mu_p(g,f,z_o)$ (cf. sec. 2) .

(ii) Let us now calculate the parametric multiplicity of the isolated zero $z_o = (0,0)$ of the mapping $H = (f,g)$ where $f(x,y) = x^3 - y^5$, $g(x,y) = x^4 - y^7$. It is easily noticed that $(\mathbb{C},\Phi,\mathbb{C}^2)$, where $\Phi(t) = (t^5,t^3)$, is a parametrization of the set of zeros of the function f in a neighbourhood of the point z_o . Thus $\mu_p(f,g;z_o) = 0(g \circ \Phi) = 0(t^{20} - t^{21}) = 20$. From the theorem just proved it follows that the algebraic multiplicity, i.e. $\dim_{\mathbb{C}} \mathcal{O}^2 / (\hat{f},\hat{g})$, is also equal to 20 in this case.

(iii) The equivalence of definitions of the algebraic and parametric multiplicities of an isolated zero of a polynomial mapping can also be obtained from the axiomatic definition of multiplicity (see [4]).

(iv) Let $H = (f,g)$ be a holomorphic mapping in a neighbourhood of the point z_o , and let z_o be an isolated zero of this mapping. Then, from the separation theorem (see [6], [7] and cf. [2]) it follows that there exist positive numbers D and r and a positive integer α , such that

(9) $$|H(z)| \geq D|z - z_o|^{\alpha} \quad \text{for} \quad |z - z_o| < r .$$

The smallest positive integer satisfying (9) will be called the <u>order of the isolated zero</u> z_o <u>of the mapping</u> H and denoted by $\nu(f,g;z_o) = \nu$.

One can prove (see [3])

THEOREM 2. *If z_0 is an isolated zero of a holomorphic mapping* $H = (f,g)$, μ_p - *the parametric multiplicity of this zero, and* f,g *have, respectively, at the point* z_0 *zeros of orders* m *and* n, *then*

(10) $\nu \leq \mu_p - mn + \max(m,n)$,

with that there exist mappings for which equality holds in (10).

From theorem 1 it follows that, in the above theorem, μ_p can be replaced by any other multiplicity mentioned in the introduction. This fact inspired us to write the present paper.

References

[1] ATIYAH, M.F. and I.G. MACDONALD: Introduction to Commutative Algebra, Addison-Wesley Publ. Co., 1969.

[2] CHARZYŃSKI, Z.: Sur certaines évaluations générales d'une fonction analytique de deux variables dans le voisinage d'un zéro isolé, Colloq. Math. 26 (1962), No. 8.

[3] CHĄDZYŃSKI, J.: On the order of an isolated zero of a holomorphic map (to appear).

[4] FULTON, W.: Algebraic Curves, Benjamin Publ. Co., 1969.

[5] HERVÉ, M.: Several Complex Variables. Local Theory, Oxford Univ. Press, 1963.

[6] HÖRMANDER, L.: Differentiability properties of solutions of differential equations, Ark. Math., 3 (1956).

[7] ŁOJASIEWICZ, S.: Sur le problème de la division, Studia Math., 18 (1959).

[8] ORLIK, P.: The multiplicity of a holomorphic map at an isolated critical point, Proc. of Nordic Summer School/NAVF, Symp. of Math., Oslo 1976.

[9] SAKS, S. and A. ZYGMUND: Analytic Functions, P.W.N., 1965.

[10] SHAFAREVICH, J.R.: Basic Algebraic Geometry, Springer-Verlag, 1977.

[11] WALKER, R.J.: Algebraic Curves, Springer-Verlag, 1978.

Institute of Mathematics
University of Łódź
Banacha 22, PL-90-258 Łódź, Poland
(J.Chądzyński, T. Krasiński and W. Kryszewski)

PROPRIÉTÉS LOCALES DES FONCTIONS PRESQUE-HOLOMORPHES

Stancho Dimiev (Sofia)

Table des matières   page

Introduction

C'est un exposé de quelques resultats recents [1,2,5] pour les fonctions presque holomorphes correspondantes aux structures presque--complexes analytiques*. On verra que les propriétés locales des fonctions presque-holomorphes correspondantes peuvent être étudier d'une manière plus detaillé. Le but est de characteriser les algèbres des germes $\mathcal{AH}_a(2n,J)$ des fonctions presque-holomorphes correspondantes à la structure presque-complexe analytique J dans le point $a \in M$, (M,J) étant une variété presque-complexe de dimension réelle égale à 2n. A l'aide d'une notion naturelle (appelee systeme de Lie) on reusit de separer un cas particulier important dans lequel on a localement

$$\mathcal{AH}_a(2n, J) \cong \mathcal{O}_q, \quad 0 \leq q \leq n,$$

$\mathcal{O}_q$ étant l'algebre des germes des fonctions de q variables complexes dans l'origine de $\mathbb{C}^q$, $\mathcal{O}_0 = \mathbb{C}$, ($\mathcal{AH}$ = almost-holomorphic). Il est à remarquer qu'en général aux différents points $a \in M$ corespondent differents nombres entiers q. Plus précisément, le nombre q est le même sur certains sous-ensembles ouverts de M et aussi sur certaines

* i.e. analytiques réelles

sous-variétés analytiques de M.Cela montre qu'une variété presque-complexe (M,J), admettant des fonctions presque-holomorphes non-triviales, peut être considerer comme une variété analytique complexe avec des singularités, i.e. comme un espace analytique complexe.

1. Rappels des fonctions presque-holomorphes

On rappelera d'abord quelques propriétés élementaires de l'algèbre $\mathcal{O}_U$ de fonctions de plusieurs variables complexes $f(z_1,\ldots,z_n) = f(x_1,\ldots,x_n,x_{n+1},\ldots,x_{2n})$, $z_k = x_k + ix_{n+k}$, $k=1,\ldots,n$, $(z_1,\ldots,z_n) \in U$, U étant un ouvert de $\mathbb{C}^n$. La fonction f est holomorphe sur U, $f \in \mathcal{O}_U$, si et seulement si

$$(1.1) \quad (S^* - iE_{2n})df = 0,$$

où E_{2n} est la $(2n \times 2n)$-matrice unité et S^* est la matrice transposé de la matrice suivante S

$$S = \begin{bmatrix} 0 & -E_n \\ E_n & 0 \end{bmatrix}.$$

Il est a remarquer encore qu'on a $S^2 = (S^*)^2 = -E_{2n}$.

L'équation (1.1) est équivalent au système surdéterminé

$$(1.2) \quad Z_p(f) = 0, \quad p = 1,\ldots,n,$$

où $Z_p = \partial/\partial x_p + i\partial/\partial x_{n+p}$ est le champ vectoriel indiqué, $Z_p : \mathcal{E}_U^\infty \otimes \mathbb{C} \to \mathcal{E}_U^\infty \otimes \mathbb{C}$. Donc l'algèbre des fonctions holomorphes sur U, $\mathcal{O}_U$, coicide avec l'intersection des annulateurs des champs vectoriels Z_p. Les crochets de Lie $[Z_p, Z_q]$, $p,q = 1,\ldots,2n$, sont tous égals à zéro.

Soit maintenant $J = \| J_q^p(x) \|$, $x = (x_1,\ldots,x_{2n}) \in \mathbb{R}^{2n}$, une $(2n \times 2n)$--matrice de coefficients $J_q^p(x)$, qui sont des fonctions indéfiniment differentiables sur l'ouvert U de $\mathbb{R}^{2n}$. On suppose qu'on a $J^2 = -E_{2n}$. Considérons l'équation suivant

(1.3) $(J^* - iE_{2n})df = 0, \quad f \in \mathcal{E}_U^\infty \otimes \mathbb{C},$

ou bien le système surdéterminé a coefficients non-constants indéfiniment differentiables suivant

(1.4) $J_p^1 \partial f/\partial x_1 + \ldots + J_p^{2n} \partial f/\partial x_{2n} - i\, \partial f/\partial x_p = 0, \quad p = 1, \ldots, 2n.$

Toute solution de l'équation (1.3) ou bien du système (1.4) est dite fonction presque-holomorphe par rapport à la structure J ou plus brievement: fonction J-presque-holomorphe. Les propriétés fondamentales des fonctions presque-holomorphes sont semblables à celles des fonctions holomorphes de plusieurs variables: le principe du maximum est vrai pour J générales et aussi le principe d'unité de la prolongation pour J à coéfficients analytiques réelles, (i.e. $J_q^p(x)$ sont développables localement en series convergentes des variables réelles $x_1, \ldots, x_{2n}$). Dans le dernier cas on dit que la structure J est une structure presque-complexe analytique. Pour une telle structure J les solutions de (1.3) ou bien de (1.4) sont aussi des fonctions analytiques réelles à valeurs complexes, i.e. développables en séries convergentes des variables réelles $x_1, \ldots, x_{2n}$, mais à coefficients complexes [3.4].

Les fonctions presque-holomorphes apparaissent naturellement sur les variétés presque-complexes. Pour la première fois elles ont été considerées par Ehresmann et Hermann, voir [6]. Les structures presque--complexes analytiques sont étudier d'abord dans [2] et puis dans [1,5].

2. Système de Lie attaché a l'équation fondamental des fonctions presque-holomorphes

Notons $\mathcal{A}_{\mathbb{R}}(U)$ l'anneau des fonctions analytiques réelles definies sur l'ouvert U de $\mathbb{R}^{2n}$. Le champ des quotients de l'anneau $\mathcal{A}_{\mathbb{R}}(U)$ est note $\mathcal{M}_{\mathbb{R}}(U)$ (ses élements sont les fonctions meromorphes réelles sur U). Dans la suite on dira que $\mathcal{A}_{\mathbb{R}}(U)$ est l'aneau de coefficients et respectivement $\mathcal{M}_{\mathbb{R}}(U)$ est le champ de coefficients. On considerera encore l'anneau des coefficients complexifié $\mathcal{A}_{\mathbb{R}}(U) \otimes \mathbb{C}$ et son champ des quotients qui n'est que $\mathcal{M}_{\mathbb{R}}(U) \otimes \mathbb{C}$.

Soit $\mathfrak{X}$ l'ensemble de tous les champs vectoriels sur U

$$p_1(x)\partial/\partial x_1 + \ldots + p_{2n}(x)\partial/\partial x_{2n},\ x \in U,$$

à coéfficients dans le champ des coéfficients complexifié, i.e.

$$p_k = p_k(x) \in \mathcal{M}_{\mathbb{R}}(U) \otimes \mathbb{C},\ k = 1,\ldots,2n.$$

Par definition, les champs vectoriels

$$(2.1)\qquad Z_p = J_p^1 \partial/\partial x_1 + \ldots + J_p^{2n}\partial/\partial x_{2n} - i\partial/\partial x_p$$

appartienent à $\mathfrak{X}$. On a aussi $[Z_p, Z_q] \in \mathfrak{X}$, mais maintenant les crochets ne s'annulent pas en general (voir §1).

L'ensemble $\mathfrak{X}$ est un espace vectoriel sur $\mathbb{C}$, mais ce n'est pas une algebre de Lie sur le champ des coéfficients complexfié $\mathcal{M}_{\mathbb{R}}(U) \otimes \mathbb{C}$. Cependant les propriétés suivantés sont remplies, si $X, Y, Z \in \mathfrak{X}$ et $p,q \in \mathcal{M}_{\mathbb{R}}(U) \otimes \mathbb{C}$, on a

1. $[X,Y] = -[Y,X]$,
2. L'identité d'Jacobi,
3. $[X+Y, Z] = [X,Z] + [Y,Z]$, $[\lambda X, Y] = \lambda[X,Y], \lambda \in \mathbb{C}$,
4. $[pX, qY] = pq[X,Y] + pX(q)Y - qY(p)X$.

Si $Q \subset \mathfrak{X}$, posons $[Q,Q] = \{[X,Y] : X,Y \in Q\}$, i.e. $[Q,Q]$ est lensemble des touts les crochets de Lie possibles d'elements de Q.

Soit H_1 l'ensemble des champs vectoriels Z_p, $p = 1,\ldots,2n$, mentionés ci-dessus (2.1). On considerera encore les ensembles $H_2 = [H_1, H_1]$, $H_3 = [H_1, H_2]$, $H_4 = [H_1, H_3] \cup [H_2, H_2]$ etc.

$$H_j = \bigcup_{q=1}^{[J/2]+1} [H_q, H_{j-q}] .$$

On dira que la suite d'ensembles suivants

$$H_1, H_2, \ldots, H_j, \ldots$$

est le système de Lie correspondant a l'équation (1.3).

Soit H l'enveloppe vectoriel sur le champ $\mathcal{M}_{\mathbb{R}}(U) \otimes \mathbb{C}$ de la réunion $\bigcup_{j \geq 1} H_j$, i.e.

$$H = \mathrm{Lin}_{\mathcal{M}_{\mathbb{R}}(U)\otimes\mathbb{C}}(\bigcup_{j\geq 1} H_j).$$

Il est clair qu'on a $H \subset \mathcal{X}$. La dimension de H comme espace vectoriel sur $\mathcal{M}_{\mathbb{R}}(U)\otimes\mathbb{C}$ est notée par **k**. Evidemment,on a $k \leq 2n$.

Dans le cas classique des fonctions holomorphes (le cas où $J = S$) on a $H_2 = \{0\}$ (voir § 1). Plus généralement le lemme suivant est vrai.

LEMME. Pour que le structure presque-complexe analytique soit integrable il est necessaire et suffisant qu'on ait $H_2 \subset \mathrm{Lin}_{\mathcal{M}_{\mathbb{R}}(U)\otimes\mathbb{C}}(H_1)$, c'est-à-dire tout crochet $[Z_p, Z_q]$ s'exprime linéairement par les champs Z_p à coefficients de $\mathcal{M}_{\mathbb{R}}(U)\otimes\mathbb{C}$.

Démonstration. On à

$$[(J - iE)X,\ (J - iE)Y] = -[(E + iJ)X,\ (E + iJ)Y],$$

X et Y étant des champs vectoriels réelles. On rappelera [7] que la structure presque-complexe J est integrable, si et seulement si les crochets de Lie de tous les champs vectoriels de type (0,1) sont aussi des champs vectoriels de type (0,1). Comme pour tous champs vectoriels réels X et Y les champs vectoriels $(E + iJ)X$ et $(E + iJ)Y$ sont de type (0,1), on peut affirmer que J est integrable si et seulement si

$$(2.2)\quad [(J - iE)X,\ (J - iE)Y] = (J - iE)Z.$$

En prenant les champs vectoriels réels correspondants au df, on voit que (2.2) signifie que $H_2 \subset \mathrm{Lin}_{\mathcal{M}_{\mathbb{R}}(U)\otimes\mathbb{C}}(H_1)$.

Une autre forme de la condition d'integrabilité indiquée est la suivante: pour touts champs vectoriels réels X et Y on a

$$(2.3)\quad (J + iE)[(J - iE)X,\ (J - iE)Y] = 0.$$

En effet, on a evidemment $(J + iE)(J - iE)Z = 0$. Il est à remarquer qu'on a

$$(J + iE)[(J - iE)X,\ (J - iE)Y] = \tfrac{1}{2}(J + iE)N(X,Y),$$

où $N(X,Y)$ est le tenseur de Nijenhuis pour la couple (X,Y).

On developpera encore la version locale de la notion du système de Lie attaché à (1.3). En effet, soit a un point de U, $a \in U$. L'ensemble $H_j(a)$ est constitué par les éléments de l'ensemble H_j dont les coefficients sont remplacés par ses valeurs au point a. La suite

$$H_1(a),\ H_2(a),\ \ldots\ ,H_j(a),\ \ldots$$

est dite système de Lie attaché a l'equation (1.3) au point a de U. C'est un système d'ensembles des champs vectoriels à coefficients constants.

La définition de $H(a)$ est completement analogue:

$$H(a) = \mathrm{Lin}_{\mathbb{C}}\Big(\bigcup_{j\geq 1} H_j(a)\Big).$$

Evidemment, $H(a)$ est un espace $\mathbb{C}$-vectoriel. Sa dimension est noté $k(a)$, $\dim_{\mathbb{C}} H(a) = k(a)$. On a pour tout $a \in U$

$$k(a) \leq k_U,\ k_U = k.$$

Dans le paragraphe suivant on verra que pour tout $a \in U$ on a $n \leq k(a)$. Ainsi: $n \leq k(a) \leq k_U \leq 2n$.

3. Une forme localement equivalent du système fondamental

On reprend le systeme (1.4) ou bien l'equation (1.3). Pour la matrice $J^*(0)$ on peut supposer qu'elle coincide avec S^*. En effet, il existe une matrice non-dégénerée à coefficients constants T telle que $T^{-1}J^*(0)T = S^*$. Alors après le changement des variables $y = Tx$, $x = (x_1,\ldots,x_{2n})$, $y = (y_1,\ldots,y_{2n})$, on obtient le cas desiré.

Par la suite le mot "localement" signifie "dans un voisinage de l'origine".

THÉORÈME 1. *Le système* (1.4) *ou l'équation* (1.3) *sont équivalents localement au système de* n *équations suivants*:

(3.1) $(E_nP + iQ)df = 0,$

où P *et* Q *sont des* $(n \times n)$-*matrices de la même classe que* J.

D é m o n s t r a t i o n. Posons

$$(3.2)\quad J^*(x) = \begin{bmatrix} A(x) & B(x)+E_n \\ C(x)-E_n & D(x) \end{bmatrix},$$

où $A(x)$, $B(x)$, $C(x)$, et $D(x)$ sont des $(n\times n)$-matrices à coefficients des fonctions indéfiniment differentiables ou analytiques de $x = =(x_1,\dots,x_{2n})$. En vue de $J^*(0) = S^*$, on a

$$A(Q) = B(Q) = C(Q) = D(Q) = 0.$$

Comme on a $(J^*(x))^2 = -E_{2n}$ pour tout $x \in U$, il suit de (3.2) que les quatres égalités ci-desous sont remplies

$$(3.3)\quad \begin{aligned} & A^2 + (B+E_n)(C-E_n) = -E_n, \\ & A(B+E_n) + (B+E_n)D = 0, \\ & (C-E_n)A + D(C-E_n) = 0, \\ & D^2 + (C-E_n)(B+E_n) = -E_n, \end{aligned}$$

où $A = A(x)$, $B = B(x)$, $C = C(x)$ et $D = D(x)$.

Les deux dernieres égalités de (3.3) impliquent localement

$$(3.4)\quad \begin{aligned} & B + E_n = -(C-E_n)^{-1}(D^2+E_n), \\ & A = -(C-E_n)^{-1}D(C-E_n), \end{aligned}$$

A l'aide de (3.3) et (3.4) on voit qu'on a

$$(A-iE_n)(C-E_n)^{-1}(C-E_nD-iE_n) = (A-iE_nB+E_n),$$

Ainsi (1.3) est equivalent localement au système suivant:

$$(C-E_nD - iE_n)df = 0,$$

et après la multiplication par $(C-E)^{-1}$, le système (1.3) est équivalent localement au

$$(E_n(C-E_n)^{-1}(D-iE_n))df = 0.$$

Il ne nous reste plus qu'a poser

$$P = (C-E_n)^{-1}D \quad \text{et} \quad Q = -(C-E_n)^{-1}.$$

R e m a r q u e s. Ayant en vue que les coefficients de la matrice inverse $(C-E_n)^{-1}$ sont localement de la même classe (indéfiniment differentiables ou analytiques réelles) que ceux de la matrice C, on conclut que les coéfficients des matrices P et Q sont de la même classe que A, B, C et D.

Le matrice Q est inversible. Ayant P et Q on peut obtenir C et D directement par des formules

$$C = E - Q^{-1}, \quad D = -Q^{-1}P.$$

Pour A et B on a

$$A = PQ^{-1} \quad \text{et} \quad B = PQ^{-1}P + Q - E_n.$$

COROLLAIRE. *Le système correspondant à l'équation* (3.1) *est le suivant:*

$$(3.5) \quad \partial f/\partial x_p + e_1^p \partial f/\partial x_{n+1} + \ldots + e_n^p \partial f/\partial x_{2n} = 0, \quad p = 1, \ldots, n,$$

où e_q^p *sont les coefficients de la matrice* $P + iQ$.

Il est clair que les champs vectoriels

$$(3.6) \quad \partial/\partial x_p + e_1^p \partial/\partial x_{n+1} + \ldots + e_n^p \partial/\partial x_{2n}, \quad p = 1, \ldots, n,$$

sont $\mathbb{C}$-linéarement independants, donc $n \leq k(a)$ pour tout $a \in U$.

4. Changements des variables

On précisera ici coment employer le théorème de Cauchy-Kovalevskaya sur le système (3.5) pour réaliser certains changements de variables.

Considérons l'équation

(4.1) $\partial f/\partial x_1 + e_1^1 \partial f/\partial x_{n+1} + \ldots + e_n^1 \partial f/\partial x_{2n} = 0.$

D'après le théorème de Cauchy-Kovalevskaya l'équation (4.1) a des solutions dans l'anneau $\mathcal{R}_{\mathbb{R}}(U) \otimes \mathbb{C}$. Soient

(4.2) $f_p = -e_p^1(0)x_1 + x_{n+p} + \xi_p(x_1,\ldots,x_{2n}), \quad p = 1,\ldots,n,$

n solutions differentes de (4.1) avec $\omega(\xi_p) \geq 2$, $\omega(\xi_p)$ étant l'ordre de la série f_p.

Considerons le changement des variables suivant

$$\begin{aligned} y_1 &= x_1, \\ &\ldots\ldots \\ y_n &= x_n, \\ y_{n+1} &= f_1(x_1,\ldots,x_{2n}), \\ &\ldots\ldots\ldots\ldots\ldots\ldots\ldots \\ y_{2n} &= f_n(x_1,\ldots,x_{2n}), \end{aligned}$$

On remarquera que les premières variables nouvelles $y_1,\ldots,y_n$ sont réelles, tandis que les variables $y_{n+1},\ldots,y_{2n}$ sont complexes. En vue de choix des parties linéaires dans (4.2), on voit que le déterminant de la matrice de Jacobi de la transformation (4.3) est égal à 1.

Le champ vectoriel

$$L_1 = \partial/\partial x_1 + e_1^1 \partial/\partial x_{n+1} + \ldots + e_n^1 \partial/\partial x_{2n}$$

en coordonnées nouvelles $(y_1,\ldots,y_{2n})$ est le suivant:

(4.4) $L_1(y_1)\partial/\partial y_1 + \ldots + L_1(y_n)\partial/\partial y_n + L_1(y_{n+1})\partial/\partial y_{n+1} + \ldots$

$$+ L_1(y_{2n})\partial/\partial y_{2n}.$$

En vue du choix de $y_{n+1},\dots,y_{2n}$ et des égalités évidentes $L_1(y_1)=1$, $L_1(y_2)=\dots=L_1(y_n)=0$, le champ vectoriel L_1 est transformé en un champ de la forme suivante: $L_1=\partial/\partial y_1$.

Les autres champs vectoriels

$$L_q=\partial/\partial x_q+e_1^q\partial/\partial x_{q+1}+\dots+e_n^q\partial/\partial x_{2n},\quad q=2,\dots,n,$$

sont transformés par la même formule (4.4). On a

$$L_q=\partial/\partial y_q+L_q(y_{n+1})\partial/\partial y_{n+1}+\dots+L_q(y_{2n})\partial/\partial y_{2n},$$

$$L_q(y_{n+1})(0,\dots,0)=e_1^q(0,\dots,0),$$

$$\dots\dots\dots\dots\dots\dots$$

$$L_q(y_{2n})(0,\dots,0)=e_n^q(0,\dots,0).$$

Pour L_2 on peut supposer que $L_2(y_{n+1})(0,\dots,0)=e_1^2(0,\dots,0)\neq 0$ etc.

Maintenant on prend l'équation

$$L_2(f(y))=\partial f/\partial y_2+L_2(y_{n+1})\partial f/\partial y_{n+1}+\dots+L_2(y_{2n})\partial f/\partial y_{2n}=0, \tag{4.5}$$

dont les coefficients sont considérés comme fonctions de $y_1,y_2,\dots,y_{2n}$, avec $y_1=0$.

Avec cet équation on peut répéter la même procedure de changement des variables

$$z_2=y_2,$$

$$\dots\dots$$

$$z_n=y_n$$

$$z_{n+1}=\varphi_1(y_2,\dots,y_{2n}),$$

$$\dots\dots\dots\dots$$

$$z_{2n}=\varphi_n(y_2,\dots,y_{2n}),$$

en ajoutant encore $z_1=y_1$. Les fonctions $\varphi_1,\dots,\varphi_n$ sont choisies comme étant des solutions fonctionelement indépendantes de l'équation (4.5).

Dans les coordinnées $(z_1, z_2, \dots, z_{2n})$ on a

$$L_1 = \partial/\partial z_1$$

$$L_2 = \partial/\partial z_2,$$

$$L_3 = \partial/\partial z_3 + L_3(z_{n+1})\partial/\partial z_{n+1} + \dots + L_3(z_{2n})\partial/\partial z_{2n} \quad \text{etc.}$$

5. Système characteristique local attaché à l'équation fondamental

On peut prendre une base pour l'espace H, contenant un système maximal de champs vectoriels linéairement indépendants de H_1. Comme on a vu dans §3, pour la forme localement équivalente(3.1) on peut prendre les n champs vectoriels correspondants (c'est dans H_1). On peut donner un algorithme pour construire une telle base de H, mais on ne s'occupera pas ici de cette question.

Soit donné une base de H:

$$\begin{aligned} &\alpha_1^1 \partial/\partial x_1 + \dots + \alpha_{2n}^1 \partial/\partial x_{2n}, \\ (5.1) \qquad &\dots\dots\dots\dots\dots\dots\dots \\ &\alpha_1^k \partial/\partial x_1 + \dots + \alpha_{2n}^k \partial/\partial x_{2n}. \end{aligned}$$

On peut supposer que les coefficients α_j^i appartient à l'anneau complexifié des coefficients $\mathcal{A}_{\mathbb{R}}(U) \otimes \mathbb{C}$. En effet, il suffit de multiplier convenablement tout les champs de (5.1) (tuer les dénominateurs des α_j^i).

Si a est un point de U, soit $r = k(a)$ (voir §2) et supposons qu'on peut resoudre les premières r équations correspondantes à (5.1) par rapport à $\partial/\partial x_1, \dots, \partial/\partial x_r$. Ainsi on obtien localement (autour du point a) la forme suivant pour la base (5.1):

$$\partial/\partial x_1 + \alpha_{r+1}^1 \partial/\partial x_{r+1} + \dots + \alpha_{2n}^1 \partial/\partial x_{2n},$$

$$\partial/\partial x_2 + \alpha_{r+1}^2 \partial/\partial x_{r+1} + \dots + \alpha_{2n}^2 \partial/\partial x_{2n},$$

$$\dots\dots\dots\dots\dots\dots\dots\dots\dots\dots$$

$$\partial/\partial x_r + \alpha_{r+1}^r \partial/\partial x_{r+1} + \dots + \alpha_{2n}^r \partial/\partial x_{2n},$$

$$\alpha^{r+1}_{r+1}\partial/\partial x_{r+1} + \ldots + \alpha^{r+1}_{2n}\partial/\partial x_{2n},$$

$$(5.2) \qquad \ldots\ldots\ldots\ldots\ldots\ldots\ldots\ldots$$

$$\alpha^{k}_{r+1}\partial/\partial x_{r+1} + \ldots + \alpha^{k}_{2n}\partial/\partial x_{2n}.$$

On peut supposer que β^i_j sont de l'anneau de coefficients complexifié. A l'aide de changements successifs de variables du type decrit dans §4 on obtient une autre forme de la base (5.2).

THÉORÈME 2. *La base de l'espace* H *peut être présentée localement sous la forme canonique suivante:*

$$\partial/\partial z_1,$$

$$\ldots\ldots$$

$$\partial/\partial z_r,$$

$$\gamma^{r+1}_{r+1}\partial/\partial z_{r+1} + \ldots + \gamma^{r+1}_{r+1}\partial/\partial z_{k+1} + \ldots + \gamma^{2n}_{2n}\partial/\partial z_{2n},$$

$$(5.3) \qquad \ldots\ldots\ldots\ldots\ldots\ldots\ldots\ldots\ldots\ldots\ldots\ldots$$

$$\gamma^{k}_{k}\partial/\partial z_{k} + \ldots + \gamma^{k}_{2n}\partial/\partial z_{2n},$$

où $\gamma^i_j(a) = 0$ et $\gamma^i_j = \gamma^i_j(z_{r+1},\ldots,z_{2n})$.

On obtient la forme méntionné ei-dessus par une sorte de "nettoyage" de coefficients, mais comme les derniers $2n-r$ éléments de (5.3) ne sont pas importants pour les corollaires presentés ici on n'éxposera pas la demonstration en détails.

Il est à remarquer encore que dans (5.3) les variables $z_1,\ldots,z_r$ sont "separés" des variables $z_{r+1},\ldots,z_{2n}$ dans le sens que $z_1,\ldots,z_r$ n'entrent pas dans les $2n-r$ dernières équations de (5.3) et $z_{r+1}, \ldots, z_{2n}$ n'entrent pas dans les premières r équations de (5.3).

Deux cas particuliers: 1.) $k(a) = k_U = 2n$. Dans ce cas on a pour (5.3) $\partial f/\partial z_1 = 0,\ldots,\partial f/\partial z_{2n} = 0$, qui signifie que toute solution f est constante. On a $\mathcal{AH}_a(2n,J) = \mathbb{C}$.

2.) $k(a) = k_U = n$. Dans ce cas d'après le lemme du §2 on a que J est integrable, et f est holomorphe sur une carte locale convenable. On a $\mathcal{AH}_a(2n,J) = \mathcal{O}_n$.

Le cas important est le cas où ils existent de points a de U pour lesquels on a $k(a) < k_U$ (inégalité stricte). Dans ce cas on introduit la notion du point singulier.

D é f i n i t i o n. Tout élément de l'ensemble

$$\Gamma = \{a \in U : k(a) < k_U\}$$

est dit point singulier de la structure presque-complexe analytique J. L'ensemble Γ est dit l'ensemble des points singuliers de la structure J.

COROLLAIRE. Si $a \in U \setminus \Gamma$, pour toute solution locale on a

$$f = f(z_{k+1}, \ldots, z_{2n}).$$

Donc f est une fonction holomorphe des $2n - k$ variables complexes $z_{k+1}, \ldots, z_{2n}$, Le nombre de ces variables ne depasse pas $n (2n-k \leq n$ parce qu'on a $n \leq k)$.

On voit que l'ensemble des points reguliers (i.e. nonsinguliers) est ouvert. D'autre part comme tout ensemble

$$\{a \in U : k(a) = \text{const}\}$$

est un sous-ensemble analytique de U, on conclut que localement Γ est une reunion de sous-ensembles analytiques de U.

6. Exemples et autres remarques.

On peut donner une classification complete pour les petites dimensions 4 et 6. Voici une breve description. Soit U un ouvert de $\mathbb{R}^4$, alors on a $2 \leq k(a) \leq k_U \leq 4$. A l'exception de deux cas extremes $k(a) = 4$ et $k = 2$, on considerera encore les cas suivants:

1.) $k(a) = k_U = 3$ et 2.) $k(a) = 2$ et $k_U = 3$.

Dans le premier cas la forme canonique locale (5.3) est la suivante: $\partial f/\partial z_1 = \partial f/\partial z_2 = \partial f/\partial z_3 = 0$. Alors on a $f = f(z_4)$ et $\mathcal{AH}_a(2n,J) = \mathcal{O}_1$.

Dans le deuxieme cas la forme canonique locale est la suivante:

$$\partial f/\partial z_1 = \partial f/\partial z_2 = 0,$$

$$c(z_3,z_4)\partial f/\partial z_3 + d(z_3,z_4)\partial f/\partial z_4 = 0,$$

où les séries $c(z_3,z_4)$ et $d(z_3,z_4)$ sont sans termes constants. Il y a deux possiblités: ou il n'existe pas de solution analytique, ou il y a une telle solution φ et toute autre solution est de la forme $f = f(\varphi)$.

Soit U un ouvert de $\mathbb{R}^6$. A l'exception de deux cas extremes on a les suivants. (Ici on a $3 \leq k(a) \leq k_U \leq 6$)

$k_U = 5$, $k(a) = 5$. On a $f = f(z_6)$ et $\mathcal{AH}_a(6,J) = Q_1$.

$k_U = 5$, $k(a) = 4$. La forme cononique locale est

$$\partial f/\partial z_1 = \partial f/\partial z_2 = \partial f/\partial z_3 = 0,$$

$$a\partial f/\partial z_4 + b\partial/\partial z_4 = 0,$$

$$c\partial f/\partial z_5 + d\partial/\partial z_6 = 0,$$

où les séries $a = a(z_4,z_5,z_6)$, b,b et d sont sans des termes constants. Comme dans le deuxième cas mentionné ci-dessus ($U \subset \mathbb{R}^4$) on a où bien qu'il n'existe pas de solutions analytiques, ou bien que $f = f(\psi)$, ψ étant une solution analytique.

$k_U = 4$, $k(a) = 4$. On a $f = f(z_5,z_6)$, i.e. $\mathcal{AH}_a(6,J) = Q_2$.

$k_U = 4$, $k(a) = 3$. La forme cannonique locale est

$$\partial f/\partial z_1 = \partial f/\partial z_2 = \partial f/\partial z_3 = 0$$

$$a\partial f/\partial z_4 + b\partial f/\partial z_5 + c\partial f/\partial z_6 = 0,$$

où les séries $a = a(z_1,z_2,z_3)$, b,c sont sans des termes constants. Ici il y a trois cas différents: il n'existe pas de solution analytique, touts les deus solutions analytiques sont fonctionellement dépendantes et il y a deux solutions analytiques qui sont fonctionellement indépendantes (voir [1]).

On remarquera encore que dans [2] on a demontré que pour certaines structures presque-complexes analytiques J sur $\mathbb{R}^4$ les fonctions presque-holomorphes correspondantes sont de la forme suivante $f(x) = \tilde{f}(\xi+i\eta)$, avec f holomorphe, $x\in\mathbb{R}^4$, où $\xi = a_1x_1 + a_2x_2$, $\eta = b_1x_1 + b_2x_2$, les constantes réelles a_1, a_2, b_1, b_2 étant uniquement determinée par J. Dans ce théorème on suppose qu'il existe des fonctions J-presque-holomorphes linéaires, i.e. J adment des fonctions presque-holomorphes linéaires.

L.Apostolova a proposé une courte démonstration de ce théorème en remplaçant le supposition mentionné ci-dessus par exigence d'avoir k fonctions J-presque-holomorphes qui forment un système maximal avec la propriété que leurs differentielles sont $\mathbb{C}$-linéairement indépendants ($k < n$, comme J est non-integrable).

En effet, soient $\varphi_1,\dots,\varphi_k$ les fonctions en question. Posons $\varphi_q = \xi_q + i\eta_q$, $q = 1,\dots,k$. On fait le changement des coordonnées suivant:

$$y_1 = \xi_1,\ y_2 = \eta_2,\dots,\ y_{2k-1} = \xi_k,\ y_{2k} = \eta_k,$$

$$y_{2k+1} = x_{2k+1},\dots,y_{2n} = x_{2n}.$$

Le fait que $d\varphi_1,\dots,d\varphi_k$ sont $\mathbb{C}$-linéairement indépendants implique que c'est un changement de coordennées regulièr.

Pour df dans les coordonnées nouvelles on a

$$df = \partial f/\partial y_1 d\xi_1 + \partial f/\partial y_2 d\eta_1 + \dots + \partial f/\partial y_{2k-1} d\xi_k + \partial f/\partial y_{2k} d\eta_k$$

$$+ \partial f/\partial y_{2k+1} dx_{2k+1} + \dots + \partial f/\partial y_{2n} dx_{2n}.$$

D'autre part, on a

$$df = \lambda_1 d\varphi_1 + \lambda_2 d\varphi_2 + \dots + \lambda_k d\varphi_k.$$

Le comparaison de deux expressions pour df montre que

$$\partial f/\partial y_2 - i\partial f/\partial y_1 = 0, \ldots, \partial f/\partial y_{2k} - i\partial f/\partial y_{2k-1} = 0,$$

$$\partial f/\partial y_{2k+1} = \ldots = \partial f/\partial y_{2n} = 0.$$

Cela signifie que $f = f(y_1, \ldots, y_{2k})$ et après la complexification des variables réelles $y_1, \ldots, y_{2k}$ on voit que f est une fonction holomorphe de k variables complexes.

Il est à remarquer que la démonstration exposée n'exige pas que la structure presque-complexe J soit analytique.

References

[1] APOSTOLOVA, S. et S.DIMIEV: Structures presque-quaternioniques (à paraitre).

[2] ARNAUDOVA, E., S.DIMIEV et T.VITANOV: Some almost complex structures and their almost holomorphic functions, Serdika, 7 (1981), 234-242.

[3] DIMIEV, S. et O.MUSHKAROV: Fonctions presque-pluriharmoniques, C.R. Acad. Bulg. Sci. 33 , 1 (1980).

[4] ——, Fonctions presque-holomorphes, Banach Center Publications (à paraitre).

[5] —— et N.MILEV: Structures presque-complexes analytiques (à paraitre).

[6] HERMANN, R.: Complex homogeneous almost complex spaces of positive characteristic, Trans. Amer. Math. Soc. 83 (1956).

[7] KOBAYASHI, S. and K.NOMIZU:Foundations of Differential Geometry II, Interscience Publishers, New-York-London 1963.

Institute of Mathematics of the
Bulgarian Academy of Sciences
BG-1090 Sofia, P.O.Box 373, Bulgaria

ON HOLOMORPHIC CHAINS WITH GIVEN BOUNDARY IN $\mathbb{P}^n(\mathbb{C})$

Pierre Dolbeault (Paris)

Contents page

Résumé. Etant donné une sous-variété M analytique réelle, orientée, maximalement complexe, de dimension $(2p-1)$ de l'espace projectif complexe $X=\mathbb{P}^n(\mathbb{C})$, on donné une construction d'une p-chaîne holomorphe T de $X\smallsetminus M$ admettant une extension simple à X de bord le courant d'intégration sur M moyennant une restriction sur M. La démonstration repose sur la construction, dans un hyperplan réel de $\mathbb{C}^n$, d'une chaîne unique, maximalement complexe, à support relativement compact, de bord un cycle donné, analytique réel à support compact, de dimension $(2p-2)$, de dimension holomorphe $(p-2)$ et sur l'extension de ce résultat à une donnée non compacte dans un cas particulier.

Un résultat partiel a été annonce dans "Symposia Mathematica", Ist. Naz. Alta Mat. 24 (1981), 205-213; référence principale: R. Harvey, Proc. Symp. in Pure Math. 30 (1977), 309-382, A.M.S.

Summary. Consider a maximal complex oriented $(2p-1)$-dimensional real-analytic subvariety M of the complex projective space $X=\mathbb{P}^n(\mathbb{C})$. We construct a holomorphic p-chain T in $X\smallsetminus M$ admitting a simple extension to X whose boundary is the current of integration over M, in terms of a restriction to M. The proof is based upon the construction, on a real hyperplane in $\mathbb{C}^n$, of a unique maximal complex chain, with a relatively compact support and a given (2p - 2)-dimensional real analytic boundary of compact support, whose holomorphic dimension is (p - 2). The proof explores also an extension of this result to a given non-compact support in a particular case.

A partial result has been announced in "Symposia Mathematica", Ist. Naz. Alta Mat. 24 (1981), 205-213; the principal reference being: R. Harvey, Proc. Symp. in Pure Math. 30 (1977), 309-382, A.M.S.

INTRODUCTION

Let Z be a complex analytic manifold, a holomorphic chain in Z is a current $T = \sum_j n_j [W_j]$ where W_j is an irreducible, closed, complex analytic subvariety of pure complex dimension p, where $[W_j]$ is the integration current on W_j and $n_j \in \mathbb{Z}$, the family $(W_j)_{j \in \mathbb{N}}$ being locally finite.

In a complex manifold X, let M be a cycle of dimension (2p-1) defining an integration current [M] ; in $Z = X \setminus M$ we look for an holomorphic p- chain T having, as an integration current, a simple extension to X such that dT = [M].

Example : Let Ω be a bounded open set of $\mathbb{C}^n$ with smooth connected boundary $b\Omega$; let f be a CR function on $b\Omega$, then there exists a continuous function F on $\bar{\Omega}$, holomorphic on Ω, such that $F|b\Omega = f$ (theorem of Hartogs-Bochner) ; then, in $\mathbb{C}^{n+1}$, the graph of F is an analytic set in the complement of gr f , having gr f as boundary.

1. DEFINITIONS AND PRELIMINARIES

1.1. A closed set U of X is called a C^k subvariety ($k \in \mathbb{N} \cup \{\infty\} \cup \{\omega\}$), of dimension q, with negligible singularities if there exists a closed set τ of U of Hausdorff q-dimensional measure zero ($\mathcal{H}_q(\tau) = 0$) such that $U \setminus \tau$ is a closed C^k submanifold of dimension q of $X \setminus \tau$ with locally finite q-dimensional volume ; τ is said the singular set of U. We say that $U \setminus \tau$ is of holomorphic dimension r in $x \in U \setminus \tau$ if the tangent space $T_x(U)$ contains a maximal complex vector subspace of complex dimension r ; we set $dh_x(U \setminus \tau) = r$. We call holomorphic dimension $dh(U)$ of U the integer $\inf_{x \in U \setminus \tau} dh_x(U \setminus \tau)$.

Let $V = \sum_j n_j U_g$ be a linear combination of such subvarieties U_j of dimension q, of singular sets τ_j with $n_j \in Z$, we call $|V| = \bigcup_j U_j$ the support of V, then if $|V|$ is of locally finite q-dimensional volume, V is called a q-chain. For every j, the integration current on $U_j \setminus \tau_j$ is defined and denoted by $[U_j]$. We also call q-chain the current $\sum_j n_j [U_j]$ and, usually, we omit [ ]. Then, if $|V|$ is compact, $|V|$ is of finite volume.

We say that V is closed or is a cycle if $dV = 0$. We set $dh(V) = \inf_j dh(U_j)$. In particular, a q-chain V is said to be maximally complex if dhV is the highest integer $\leqslant \frac{1}{2}$ dim V.

A necessary condition to solve the problem of the Introduction is : M is maximally complex ; it is sufficient if M is compact, in $\mathbb{C}^n$ for $p > 2$ and in $\mathbb{P}^n \setminus \mathbb{P}^{n-q}$ for $p > q + 1$ (Harvey-Lawson) [4], [5], [6]. For $p = 1$ in $\mathbb{C}^n$ and $p = q$ in $\mathbb{P}^n \setminus \mathbb{P}^{n-q}$, M has to satisfy the moment condition.

1.2. In what follows all chains are assumed to be C^∞. Let E be a real hyperplane of $\mathbb{C}^n$, $p \in \mathbb{N}$, $p \geqslant 2$, N a cycle of dimension $(2p-2)$, with $dhN = p-2$, with compact support in E.* Then, it is proved that there exists a $(2p-1)$ - maximally complex cycle M in $\mathbb{C}^n \setminus |N|$ having a simple extension also denoted by M, with compact support in E, such that $dM = N$ (Besnault-Dolbeault for $p = n-1$ [2], [3] ; Benlarabi for any $p \geqslant 2$ [1]).

1.3. A 1-cycle γ with compact support in a complex analytic manifold satisfies the moment condition if, for every holomorphic 1-form α, we have $\int_\gamma \alpha = 0$.

*N is supposed to be transversal, almost everywhere, to maximal complex affine subspaces of E.

Let F be a vector subspace of the $\mathbb{R}$-vector space $\mathbb{C}^m = \mathbb{R}^{2m}$; π the orthogonal projection parallel to F. Let M be a chain of $\mathbb{C}^m$ · when the slicing of M relatively to π is defined at 0, we call intersection of M and F, the current $I(M,F) = \langle M, \pi, 0 \rangle$; [recall : when M is the integration current defined by an oriented submanifold M, when $\pi|M$ is of maximal rank, then $\langle M, \pi, 0\rangle$ is the intersection $M \cap F$ with convenient orientation].

Let E be a real hyperplane of $\mathbb{C}^n$; a 0-cycle ν of $\mathbb{C}^n$ with support in E is said special if there exists an integer $q \geq 1$ and a maximally complex cycle M of $\mathbb{C}^n \times \mathbb{C}^q$ of dimension $2q+1$ such that $\nu = I(M, E \times \{0\})$.

Let Y be a real submanifold of $\mathbb{C}^n$, a 1-cycle γ of $\mathbb{C}^n \setminus Y$, with support in a real hyperplane E of $\mathbb{C}^n$ satisfies the moment condition if there exist an integer $q \geq 1$, a holomorphic $(q+1)$-chain T of $(\mathbb{C}^n \setminus Y)\times \mathbb{C}^q$ in a neighbourhood of $E \times \mathbb{C}^q$ having a simple extension also denoted by T to $\mathbb{C}^n \times \mathbb{C}^q$ in a neighbourhood of $E \times \mathbb{C}^q$ such that $\gamma = I(T, E \times \{0\})$. [This definition is compatible with the previous one].

Then, given a special 0-cycle ν with compact support in E, there exists a unique 1-cycle γ of finite lenght, with support in $E \setminus |\nu|$ satisfying the moment condition and having a simple extension γ such that $d\gamma = \nu$ (Benlarabi [1]).

2. SOLUTION IN $\mathbb{P}^n(\mathbb{C})$; the problem has been set up by J. King in 1978 [7].

For simplicity, we consider the case $p = n-1 \geq 2$. We shall consider a condition (B) on M which will be explicited in 2.4.

2.1. Theorem. In $\mathbb{P}^n(\mathbb{C})$, let M be a maximally complex cycle, of dimension $(2n-3)$ satisfying the condition (B), then there exists a holomorphic $(n-1)$-chain T in $\mathbb{P}^n(\mathbb{C})\setminus |M|$ having a simple extension to $\mathbb{P}^n(\mathbb{C})$, still denoted T, such that $dT = M$.

Using a convenient subdivision of M, we shall come back to the problem in $\mathbb{C}^n$.

2.2. Covering $\mathbb{P}^n(\mathbb{C})$ by $(n+1)$ affine blocks. In $\mathbb{P}^n(\mathbb{C})$, consider homogeneous coordinates $\zeta_0\ \zeta_1, \ldots, \zeta_n$; the affine spaces $A_k = \{x \in \mathbb{P}^n(C) ;\ \zeta_k(x) \neq 0\}$; $\{k = 0, 1, \ldots, n\}$ are the elements of a covering of $\mathbb{P}^n$ by open sets isomorphic to $\mathbb{C}^n$.

Let $(\xi_1, \ldots, \xi_{2n})$ be real coordinates in $\mathbb{C}^n \simeq \mathbb{R}^{2n}$. An open block in $\mathbb{C}^n$ is deduced by dilatation and translation from the set

$$\{\xi = (\xi_1, \ldots, \xi_{2n}) \in \mathbb{C}^n\ ,\ |\xi_j| < 1,\ j = 1, \ldots 2n\}$$

Then, there exists an open covering $(D_k)_{k=0,\ldots,n}$ of $\mathbb{P}^n(\mathbb{C})$ by open blocks D_k of A_k ; $k = 0, 1, \ldots, n$. The property remains true after small perturbations of the affine spaces A_k and small linear changes of the D_k.

2.3. Let A be one of the affine spaces of the previous covering of $\mathbb{P}^n(\mathbb{C})$. Let H be an oriented real hyperplane of A ; H is determined by a point $a \in A$ and a point b of the grassmannian $G_{\mathbb{R}}(2n-1, 2n)$. We call small perturbation of such an H, a variation of the points a and b in arbitrarily small open sets of A and of $G_{\mathbb{R}}(2n-1, 2n)$ respectively.

Lemma 1. Let H_o be a given real hyperplane of A, then there exists a small perturbation of H_o into a hyperplane H such that $N = I(M,H)$ has the following properties : N is a cycle C^ω (with non compact support in general), with $\dim N = 2n-4$; $\mathrm{dh}\ N = n-3$.

This results from the definition of $I(M,H)$ and a result of Shiffman on the Hausdorff measure of the projection of a set [8].

Lemma 2. Let (A_o, H_o) be given, there exist small perturbations (A,M) of (A_o, H_o)

such that the following properties are true : there exists a sequence of real hyperplanes $(H_1^{p_1})$ of H going to infinity when $|p_1| \to \infty$ where induced structure is complex and such that : $I(N,H_1^{p_1})$ is a C^ω cycle, of dimension $(2n-5)$ and holomorphic dimension $(n-3)$. More generally, there exists a sequence $(H_k^{p_1 \cdots p_k})$ of hyperplanes of $H_{k-1}^{p_1 \cdots p_{k-1}}$ such that :

Cycles	M	$N = N_1$	N_2		N_{2n-4}	N_{2n-3}
dim	$2n-3$	$2n-4$	$2n-5$		1	0
dh	$n-2$	$n-3$	$n-3$		0	0
Sub-vector space containing cycle	A	H	$H_1^{p_1}$		$H_{2n-5}^{p_1 \cdots p_{2n-5}}$	$H_{2n-4}^{p_1 \cdots p_{2n-4}}$
dim	$2n$	$2n-1$	$2(n-1)$		4	3
dh	n	$n-1$	$n-1$		2	1

where $N_2 = \bigcup_{p_1 \in \mathbb{Z}} I(N,H_1^{p_1}),\ldots,N_{2n-3} = \bigcup_{(p_1 \cdots p_{2n-4}) \in \mathbb{Z}^{2n-4}} N_{2n-3}^{p_1 \cdots p_{2n-4}}$

$N_{2n-3}^{p_1 \cdots p_{2n-4}} = I(N,H_{2n-4}^{p_1 \cdots p_{2n-4}})$: the support of this last cycle having a finite number of points.

Then, we shall say that the affine subspaces intersect M generically.

Denote by H_k the vector space which is the direction of $H_2^{p_1 \cdots p_2}$. Consider :

a) complex linearly independent generators of H_{2n-4} and of the complex spaces $H_{2n-5}, H_{2n-7}, \ldots, H_1$;

b) a basis of the $\mathbb{R}$-vector space H given by the real and imaginary parts of the former vectors and a real vector of H linearly independent of them ;

c) all the real hyperplanes of H generated by the different combinations of the vectors of the basis and sequences of affine hyperplanes of H having the directions of the vector hyperplanes ; finally, choose the affine subspaces of lemma 2 so that they are intersections of some of these hyperplanes. After a small perturbation, we may assume that any intersection of these hyperplanes intersect M generically.

2.4. Condition (B)

Let K be an affine subspace of (real) dimension 4 from the family described in 2.3. and let $\tilde{K}$ be the strip of K between two consecutive affine subspaces of the family, of dimension 3 contained in K. We assume that (B) the 1-chain $\tilde{K} \cap M$(*) can be complemented by well defined 1-chains in the boundary of $\tilde{K}$ into a compact 1-cycle satisfying the moment condition. Moreover, we assume that M is not contained in a complex projective variety.

2.5. Proposition

Let M be a maximally complex cycle (C^{ω}) of dimension (2n-3) of $\mathbb{P}^n(\mathbb{C})$ satisfying the condition (B) ; let A' be an affine subspace of $\mathbb{P}^n(\mathbb{C})$, H' a real hyperplane of A', then for a small perturbation (A,H) of (A',H') the following properties are true :

a) $N = I(M,H)$ is a (non compact) cycle of A, C^{ω} with $\dim N = 2n-4$; $dh\, N = n-3$;

b) there exists a unique maximally complex C^{ω} cycle M', with $\dim M' = 2n-3$, in $A \setminus |N|$ having a simple extension to A still denoted by M', with (non compact) support in H, such that $dM' = -N$.

(*) In the following, for simplicity, we use $\cap$ for the intersection of chains.

Proof : a) This is Lemma 1 ;

b) (α) we shall subdivide the cycle N of H into a locally finite sum of cycles $({}_jN)$; $j \in \mathbb{N}$, with compact supports where $\dim {}_jN = (2n-4)$ and $dh\, {}_jN = n-3$. From 1.2., for every j, there exists a maximally complex chain ${}_jM'$ of dimension (2n-3), with support in H such that $d_jM' = -{}_jN$.

Then $M' = \sum_j {}_jM'$ is a maximally complex (2n-3)-chain of H, because, the ${}_jM'$ beeing relatively compact in $H \setminus |N|$, the sum is locally finite ; moreover $dM' = d(\sum_j {}_jM') = \sum_j d_jM' = -\sum_j N_j = -N$.

(β) Existence of the ${}_jN$: The cycle

$$N^{p_1 \cdots p_{2n-4}}_{2n-3} = N_{2n-4} \cap H^{p_1 \cdots p_{2n-4}}_{2n-4} = M \cap H^{p_1 \cdots p_{2n-4}}_{2n-4}$$

is of dimension 0, finite from Lemma 2, and special ; then (1.3), there exists a unique 1-chain $\gamma^{p_1 \cdots p_{2n-4}}_{2n-4}$ whose boundary is $N^{p_1 \cdots p_{2n-4}}_{2n-3}$ satisfying the moment condition and whose support is in $H^{p_1 \cdots p_{2n-4}}_{2n-4}$. Let $\gamma^{p_1 \cdots p_{2n-4}^+}_{2n-4}$ be the intersection of N (i.e. of M) with the strip of $H^{p_1 \cdots p_{2n-5}}_{2n-5}$ between $H^{p_1 \cdots p_{2n-4}}_{2n-4}$ and $H^{p_1 \cdots (p_{2n-4}+1)}_{2n-4}$; it is a chain whose support is in the support of N_{2n-4} and whose boundary is $N^{p_1 \cdots (p_{2n-4}+1)}_{2n-3} - N^{p_1 \cdots p_{2n-4}}_{2n-3}$; then the chain $\delta = \delta^{p_1 \cdots p_{2n-4}^+}_{2n-4} = \gamma^{p_1 \cdots p_{2n-4}^+}_{2n-4} + \gamma^{p_1 \cdots p_{2n-4}}_{2n-4} - \gamma^{p_1 \cdots (p_{2n-4}+1)}_{2n-4}$ is a compact 1-cycle which, from condition (B), satisfies the moment condition.

From Harvey-Lawson theorem in $\mathbb{C}^2$ (1.1) there exists a 1-holomorphic chain $\gamma^{p_1 \cdots p_{2n-4}}_{2n-5}$ in $H^{p_1 \cdots p_{2n-5}}_{2n-5}$ with relatively compact support in the complement of $|\delta|$, whose boundary is δ.

We consider the strip of $H^{p_1\cdots p_{2n-6}}_{2n-6}$ between $H^{p_1\cdots p_{2n-5}}_{2n-5}$ and $H^{p_1\cdots (p_{2n-5}+1)}_{2n-5}$. From 2.4. c), the affine subspaces $H^{p_1\cdots p_{2n-5}\,p_{2n-4}}_{2n-4}$ and $H^{p_1\cdots (p_{2n-5}+1)p_{2n-4}}_{2n-4}$ lie in an affine space $'H^{p_1\cdots p_{2n-4}}_{2n-4}$ of dimension 4 and holomorphic dimension 1 contained in $H^{p_{2n}\cdots p_{2n-6}}_{2n-6}$ which intersects M along $'\gamma^{p_1\cdots p_{2n-4}}_{2n-4}$.

From condition (B) the 1-cycle $'\gamma^{p_1\cdots p_{2n-4}}_{2n-4} - \gamma^{p_1\cdots p_{2n-5}}_{2n-4} + \gamma^{p_1\cdots (p_{2n-5}+1)p_{2n-4}}_{2n-4}$ satisfies the moment condition ; so in $H^{p_1\cdots p_{2n-7}}_{2n-7} (\simeq \mathbb{C}^3)$ there exists a compact holomorphic 1-chain whose boundary is this 1-cycle ; the orthogonal projection of it being a maximally complex 2-chain of $H^{p_1\cdots p_{2n-6}}_{2n-6}$ with same boundary ; $M \cap H^{p_1\cdots p_{2n-6}}_{2n-6}$ has holomorphic dimension 0.

Finally we get a uniquely determined compact 2-cycle $\delta^{p_1\cdots p_{2n-5}p_{2n-4}^{+}}_{2n-5}$ with holomorphic dimension 0 in the affine space $H^{p_1\cdots p_{2n-6}}_{2n-6}$ of dimension 5 and holomorphic dimension 2. $\delta^{p_1\cdots p_{2n-4}^{+}}_{2n-5}$ is the boundary of maximally complex 3-chain in $H^{p_1\cdots p_{2n-6}}_{2n-6}$. Using alternatively Harvey-Lawson theorem in $\mathbb{C}^n$(1.1) and results of 1.2 and 1.3 we construct a sequence $({}_jN)$ of uniquely determined (2n-4)-cycles of holomorphic dimension (n-3) and a sequence of uniquely determined maximally complex (2n-3)-chains ${}_jM'$ when the family of hyperplanes of H is chosen.

(γ) Unicity of $M' = \sum_j {}_jM'$.

Let $\mathcal{H}_1$ and $\mathcal{H}_2$ be two families of hyperplanes of H difining the chains M_1' and M_2'. One considers the sections of M_1' by the hyperplanes of $\mathcal{H}_2$; this defines a solution for the family $\mathcal{H}_2$; because of the unicity when the family $\mathcal{H}_2$ is chosen, one has $M_1' = M_2'$.

2.6. Proof of Theorem 2.1.

Let $(D_j)_{j=0,1,\dots,n}$ be a covering of $\mathbb{P}^n(\mathbb{C})$ by affine blocks (2.2.). Consider $D_o \subset A_o$:
Let F_o^q be the oriented faces of D_o, then the boundary of D_o, $bD_o = \sum_q F_o^q$.
Let H_o^q be the real hyperplane of A_o which contains the support of F_o^q and let $N_o^q = M \cap H_o^q = I(M,H_o^q)$, then $d\, N_o^q = 0$.

From Proposition 2.5., after an eventual small perturbation, there exists a unique maximally complex chain $M_q'^o$ in A_o, with support contained in H_o^q, and boundary $-N_o^q$. Let M_o be the restriction of the current M to D_o ; it is a maximally complex (2n-3)-chain of A_o ; let N_o be the (clearly defined) intersection of M with $b\,D_o$, then $N_o = bM_o$.

2.7. Lemma : In the notations of 2.6, there exists a unique maximally complex chain M_o' such that $M_o'' = M_o + M_o'$ is a maximally complex (2n-3)-cycle whose support is contained in $\bar{D}_o$.

Proof : $M_o' = \sum_q (M_q'^o \cap F_o^q)$ is a maximally complex chain whose support is contained in the support of bD_o. If M_q' meets an edge G_o common to F_o^q and another face $F_o^{q'}$, then by the unicity of $M_q'^o$ in H_o^q, one has $M_q'^o \cap G_o = M_{q'}'^o \cap G_o$, hence $dM_o' = -N_o$. Let $M_o'' = M_o + M_o'$, then $dM_o'' = 0$; moreover, the supports of M_o and M_o' are contained in $\bar{D}_o$.

2.8. End of proof of Theorem 2.1.

M_o'' (Lemma 2.7) is a maximally complex, compact (2n-3)-cycle with support in $\bar{D}_o$, hence with compact support in A_o.

The support of the chain M_o' is contained in $bD_o \subset D_1 \cup \dots \cup D_n$ and $\mu_o = M - M_o'' = M - M_o - M_o'$ is a cycle with support in $D_1 \cup \dots \cup D_n$. By induction on k, we construct a maximally complex compact (2n-3)-cycle M''_k with support in $\bar{D}_k$, hence in A_k, such that $M = \sum_{j=0}^{n} M''_j$.

From the theorem of Harvey-Lawson (1.1) in A_k, there exists a holomorphic (n-1)-chain T_k in $A_k \setminus |M''_k|$ whose simple extension, still denoted by T_k, satisfies $dT_k = M''_k$.

Moreover T_k and T_l $(k \neq l)$ are disjoint or have as common boundary a C^ω chain which is, up to a negligible singular set, a C^ω submanifold. From a lemma of Harvey-Lawson ([5], Lemma 9.3), T_k is an extension of T_l through the common boundary, then $T = \sum_{k=0}^{n} T_k$ is a holomorphic chain of $\mathbb{P}^n(\mathbb{C}) \setminus |M|$ whose boundary is M.

3. REMARKS

3.1. The following example of Harvey-Lawson [6] shows that the hypothesis of maximal complexity for M is not sufficient : let M be a real hypersurface of a p-dimensional complex submanifold X of $\mathbb{P}^n(\mathbb{C})$ such that M is not homologous to zero in X ; M, as a hypersurface of X, is maximally complex ; if there exists a holomorphic chain T in $\mathbb{P}^n$ such that $dT = M$, then from a theorem of Bishop, it is known that $|T|$ is contained in X which contradicts the fact that M is not homologous to zero in X.

3.2. In the proof of the Theorem 2.1., T_k has support in A_k, and boundary in $\bar{D}_k$, then it is possible that $|T_k|$ meets M again outside $\bar{D}_k$; hence $|T|$ may intersect M outside its boundary.

3.3. In the same proof, the chain T is completely determined by the given construction, but this one depends on the choice of the blocks D_k ; let T' be another holomorphic chain such that $dT' = M$, then $T - T'$ is a holomorphic (n-1)-cycle, i.e. an algebric variety of $\mathbb{P}^n(\mathbb{C})$.

3.4. In case T in $\mathbb{P}^n$ comes from solution in $\mathbb{P}^n - \mathbb{P}^{n-q}$ [6], comparison between the hypothesis in [6] and in this paper should be done.

Université de Paris VI
Mathématiques
4, Place Jussieu
F-75 230 Paris, France

REFERENCES

[1] A. BENLARABI : Thèse du 3ème cycle, Université de Paris VI, 1982.

[2] J. BESNAULT and P. DOLBEAULT : Sur les bords d'ensembles analytiques complexes dans $\mathbb{P}^n(\mathbb{C})$, Symposia Mathematica, vol. 24 (1981), 205-213.

[3] J. BESNAULT et P. DOLBEAULT : Sur les chaînes maximalement complexes de bord donné, to be published.

[4] R. HARVEY : Holomorphic chains and their boundaries, Proc. of Symposia in Pure Math., 30, Part 1, A.M.S. (1977), 309-382.

[5] R. HARVEY and B. LAWSON, On boundaries of complex analytic varieties, I, Ann. of Math., 102 (1975), 233-290.

[6] R. HARVEY and B. LAWSON : On boundaries of complex analytic varieties, II, Ann. of Math., 106 (1977), 213-238.

[7] J. KING : Open problems in Geometric Function Theory, Proceeding of the fifth international symposium, division of Math., the Taniguchi foundation (1978), p.4.

[8] B. SHIFFMANN : On the removal of singularities of analytic sets, Michigan Jl. of Math. 15 (1968), 111-120.

THE JENKINS' TYPE INEQUALITY FOR BAZILEVIČ FUNCTIONS

Wiesława Drozda (Olsztyn),
Anna Szynal and Jan Szynal (Lublin)

1. Let S denote the class of holomorphic and univalent functions f in the unit disk $K = \{z : |z| < 1\}$ which have the form

(1) $$f(z) = z + a_2 z^2 + \dots, \quad z \in K.$$

By $B(\alpha,\beta;g,h_a)$ we denote the class of Bazilevič functions. Namely, we say that a holomorphic function f of the form (1) is in $B(\alpha,\beta;g,h_a)$ if it has the representation

(2) $$f(z) = \left\{ \frac{\alpha + i\beta}{1 + ia} \int_0^z g^{\alpha}(s) h_a(s)\, s^{i\beta - 1}\, ds \right\}^{1/(\alpha + i\beta)}, \quad z \in K.$$

The parameters α and β satisfy the conditions: $\alpha > 0$, $\beta \in (-\infty, +\infty)$, whereas g is a starlike and univalent function in K of the form (1) and $h_a = h_a(z) = 1 + ai + 2h_1 z + 2h_2 z^2 + \dots$ (a is real) is holomorphic in K and satisfies the condition $\operatorname{Re} h_a(z) > 0$, $z \in K$.

Bazilevič [1] (see also Pommerenke [6]) showed that the functions f given by (2) are univalent in K.

From (2) one can get the differential equation for $f \in B(\alpha,\beta;g,h_a)$. After some calculations we obtain from (2):

(3) $$1 + \frac{zf''(z)}{f'(z)} + (\alpha - 1 + i\beta)\frac{zf'(z)}{f(z)} = i\beta + \frac{zh_a'(z)}{h_a(z)} + \alpha\frac{zg'(z)}{g(z)}.$$

The class $B(\alpha,\beta;g,h_a)$ with particular choices of α, β, g and h_a yields several classes of univalent functions.

We quote below some of them:

(a) $B(1,0;g,h_a) \equiv L$ - the class of close-to-convex functions,

(b) $B(1,\mathrm{tg}\,\gamma;g,1) \equiv S^c_\gamma$ $(|\gamma|<\frac{1}{2}\pi)$ - the class of spiral-convex functions [10],

(c) $S^c_0 \equiv S^c$ - the class of convex functions,

(d) $\lim_{\alpha\to+\infty} B(\alpha, \alpha\,\mathrm{tg}\,\gamma; g,1) \equiv S^*_\gamma$ $(|\gamma|<\frac{1}{2}\pi)$- the class of spiral-like functions,

(e) $S^*_0 \equiv S^*$ - the class of starlike functions,

(f) $B(\frac{1}{\alpha},0;g,1) \equiv M_\alpha$ - the class of α-convex functions in the sense of Mocanu.

The coefficient problem for the class S has been attacked for many years and the results concerning that problem are well known.

However, as for the class of Bazilevič functions, very little is known about coefficients and even for initial coefficients there are only partial results [2] and [4].

In the estimates of some functionals concerning a_2 and a_3 for the class S the important role plays so-called Jenkins' inequality.

This inequality has the form

$$\text{(4)} \qquad \mathrm{Re}\{(a_3 - a_2^2) - \lambda a_2\} \le 1 + \tfrac{3}{8}\lambda^2 - \tfrac{1}{4}\lambda^2 \log \frac{|\lambda|}{4},$$

where $\lambda \in [-4,4]$ and f is an arbitrary function in S. (For a short proof of (4) we refer to [3] or [8]).

The extension of (4) for complex λ and bounded univalent functions is much more complicated but, as it was pointed out in [9], leads also to the determination of the coefficient region (a_2,a_3) within the considered classes.

In this paper we are going to prove the Jenkins' type inequality for the class $B(\alpha,\beta;g,h_a)$ and show some applications of it.

The obtained results will have quite complicated forms, however in some special cases it will be possible to find concise and clear estimates.

In what follows we denote by $B^r(\alpha,\beta;g,h_a)$ the subclass of Bazilevič functions consisting of functions with real coefficients. In the same way we indexed all the other classes which appear in the paper and which consist of functions with real coefficients .

2. In order to get the Jenkins' type inequality for the class

$B(\alpha,\beta;g,h_a)$ we should form the appropriate expressions.

Let P denote the class of holomorphic functions p in K which have the form

$$(5)\qquad p(z) = 1 + 2p_1 z + 2p_2 z^2 + \dots, \qquad z \in K,$$

and satisfy the conditions $\operatorname{Re} p(z) > 0$, $z \in K$.

If h_a is a function appearing in (2), then the function

$$(6)\qquad h(z) = h_a(z) - ai = 1 + 2h_1 z + 2h_2 z^2 + \dots \in P.$$

From the definition of the class S^* it also follows that for the function g in (2) we have

$$(7)\qquad \frac{zg'(z)}{g(z)} = 1 + 2q_1 z + 2q_2 z^2 + \dots = q(z) \in P.$$

Let denote

$$(8)\qquad \sin\delta/2 = \frac{-a}{(1+a^2)^{1/2}}, \quad \cos\delta/2 = \frac{1}{(1+a^2)^{1/2}}, \quad \operatorname{tg}\delta/2 = t,$$

$$\eta = 1 + e^{i\delta}, \quad \delta \in (-\pi, \pi).$$

By the comparison of the coefficients in (3) and taking into account (6), (7) and (8) we get the formulae

$$(9)\qquad \begin{cases} \varkappa a_2 = \eta h_1 + 2\alpha q_1, \\ 2(\varkappa+1)a_3 - (\varkappa+2)a_2^2 = 2\eta h_2 - \eta^2 h_1^2 + 2\alpha q_2, \\ \varkappa = (\alpha+1) + i\beta . \end{cases}$$

Let $\lambda = \lambda_1 + i\lambda_2$ be an arbitrary fixed complex number. We form the following functional of the Jenkins' type within the class $B(\alpha,\beta;g,h_a)$:

$$(10)\qquad I(f) = \operatorname{Re}\{[2(\varkappa+1)a_3 - (\varkappa+2)a_2^2] - \lambda\varkappa a_2\}$$

$$= \operatorname{Re}\{\eta[2h_2 - \eta h_1^2 - \lambda h_1]\} + 2\alpha \operatorname{Re}(q_2 - \lambda q_1)$$

$= I_1(h) + 2\alpha I_2(q)$.

From (10) we can see that in order to find $\max I(f)$ over the class $B(\alpha,\beta;g,h_a)$ we should find $\max I_1(h)$ and $\max I_2(q)$, where h and q have the form (6) and (7), respectively, and run independently over the class P.

3. In this section we prove some lemmas which will lead to the main result.

LEMMA 1. *Let $p \in P$ and has the form (5). Then the following sharp inequalities hold:*

(11) $$|p_1| \le 1, \qquad |p_2 - p_1^2| \le 1 - |p_1|^2.$$

The inequalities (11) are well known as the first two Carathéodory inequalities.

LEMMA 2. *Let $q \in P$ and has the form (7). For any complex number $\lambda = \lambda_1 + i\lambda_2$ the following sharp inequalities hold:*

(12) $$\mathrm{Re}(q_2 - \lambda_1 q_1) \le 1 + |\lambda_1| \qquad \text{if} \quad \lambda_2 = 0,$$

(13) $$\mathrm{Re}(q_2 - i\lambda_2 q_1) \le \begin{cases} 1 + \frac{1}{8}\lambda_2^2 & \text{if} \quad |\lambda_2| \le 4 \quad \text{and} \quad \lambda_1 = 0, \\ |\lambda_2| - 1 & \text{if} \quad |\lambda_2| \ge 4 \quad \text{and} \quad \lambda_1 = 0, \end{cases}$$

(14) $$\mathrm{Re}(q_2 - q_1) \le 1 - \tfrac{1}{2}\eta^{-1}\lambda_1^2 + \tfrac{1}{2}(1-\eta)(2-\eta)^{-2}\lambda_2^2 \quad \text{in other cases},$$

where $\eta < 0$ is the root of the equation

(15) $$\eta^{-2}\lambda_1^2 + (2-\eta)^{-2}\lambda_2^2 = 4.$$

The extremal functions have the form:

$$q(z) = \frac{1+z}{1-z}\ (\lambda_1 > 0), \quad q(z) = \frac{1-z}{1+z}\ (\lambda_1 < 0) \qquad \text{in (12);}$$

$$q(z) = \frac{1+iz}{1-iz}\ (\lambda_2 > 0), \quad q(z) = \frac{1-iz}{1+iz}\ (\lambda_2 < 0) \qquad \text{in (13) if } |\lambda_2| \ge 4;$$

$$q(z) = \mu \frac{1+ze^{-i\varphi}}{1-ze^{-i\varphi}} + (1-\mu)\frac{1-ze^{i\varphi}}{1+ze^{i\varphi}}, \quad 0 \le \mu \le 1, \quad \varphi = \arcsin\frac{\lambda_2}{4}$$

in (13) if $|\lambda_2| \le 4$;

$$q(z) = \frac{1+ze^{-i\varphi_0}}{1-ze^{-i\varphi_0}}, \quad \cos\varphi_0 = \frac{\lambda_1}{2\eta}, \qquad \sin\varphi_0 = \frac{\lambda 2}{2(2-\eta)} \qquad \text{in (14).}$$

P r o o f. Let $q_1 = x + iy$. Then, according to (11) we have

$$x^2 + y^2 \le 1 \quad \text{and} \quad \mathrm{Re}\, q_2 \le 1 - |q_1|^2 + \mathrm{Re}\, q_1^2 = 1 - 2y^2.$$

Now we have

$$\text{(16)} \qquad \mathrm{Re}(q_2 - \lambda q_1) \le 1 - 2y^2 - \lambda_1 x + \lambda_2 y.$$

We can see that in order to find $\max\limits_{q \in P} \mathrm{Re}\,(q_2 - \lambda q_1)$ we should find $\sup u(x,y) = \max\,(1 - 2y^2 - \lambda_1 x + \lambda_2 y)$ in the disk $x^2 + y^2 < 1$.

We consider the following cases:

1° $\lambda_1 = 0$ and $\lambda_2 = 0$. Then $u(x,y) = 1 - 2y^2$. We have $\max u(x,y) = 1$ which is attained for $y = 0$ and arbitrary $x \in [-1,1]$.

2° $\lambda_1 = 0$ and $\lambda_2 \ne 0$. Then $u(x,y) = 1 - 2y^2 + \lambda_2 y$. We have

$$\max u(x,y) = \begin{cases} 1 + \frac{1}{8}\lambda_2^2 & \text{if} \quad |\lambda_2| \le 4, \\ |\lambda_2| - 1 & \text{if} \quad |\lambda_2| \ge 4. \end{cases}$$

The maximum is attained for $y = \pm 1$ and arbitrary $x \in [-1,1]$ if $|\lambda_2| \ge 4$ as well as for $y = \lambda_2/4$ and arbitrary $x \in [-1,1]$ if $|\lambda_2| \le 4$.

3° $\lambda_1 \ne 0$ and $\lambda_2 = 0$. Then $u(x,y) = 1 - 2y^2 + \lambda_1 x$. We have $\max u(x,y) = 1 + |\lambda_1|$ which is attained for $y = 0$ and $x = \pm 1$.

4° $\lambda_1 \ne 0$ and $\lambda_2 \ne 0$. Then $\max u(x,y)$ in the disk $x^2 + y^2 \le 1$ is attained when $x^2 + y^2 = 1$. We introduce the function

$$\text{(17)} \qquad \hat{u}(x,y;\eta) = 1 - 2y^2 - \lambda_1 x + \lambda_2 y + \eta(x^2 + y^2 - 1),$$

which gives

$$(\partial/\partial x)\hat{u} = -\lambda_1 + 2\eta x = 0,$$

$$(\partial/\partial y)\hat{u} = -4y + \lambda_2 + 2\eta y = 0.$$

The solution of the above system is given by the formulae: $\hat{x} = \frac{1}{2}\eta^{-1}\lambda_1$, $\hat{y} = \frac{1}{2}(2-\eta)^{-1}\lambda_2$, so according to the condition $x^2 + y^2 = 1$ the number η is the root of the equation

$$\frac{1}{4}\eta^{-2}\lambda_1^2 + \frac{1}{4}(2-\eta)^{-2}\lambda_2^2 = 1.$$

If we consider the function: $F(\eta) = \frac{1}{4}\eta^{-2}\lambda_1^2 + \frac{1}{4}(2-\eta)^{-2}\lambda_2^2 - 1$, then we find that $F'(\eta) > 0$ for $\eta < 0$ and $F'(\eta) < 0$ for $\eta > 2$. Moreover $F(\eta) = 0$ if $\eta = \eta_1 < 0$, and $\eta = \eta_2 > 2$, and has no zeros for $0 < \eta < 2$.

We conclude that $\max \hat{u}(x,y;\eta) = \hat{u}(\hat{x},\hat{y};\eta_1)$. Thus all the estimates (12), (13) and (14) have been proved. The statements about extremal functions follow from the proof.

Before we formulate Lemma 3 we introduce some notation. We have: $\eta = 1 + e^{i\delta}$, $t = \mathrm{tg}\,\delta/2$, $\delta \in (-\pi, \pi)$, $\tau = t/(1+t^2)^{\frac{1}{2}} \in (-1, 1)$, $\lambda = \lambda_1 + i\lambda_2$, $\lambda_1 = 2^{-\frac{1}{2}}(\lambda' - \lambda'')$, $\lambda_2 = 2^{-\frac{1}{2}}(\lambda' + \lambda'')$. Furthermore,

$$(18)\qquad \bar{E}_\tau = \{(\lambda', \lambda''): (1+\tau)^{-2}\lambda'^2 + (1-\tau)^{-2}\lambda''^2 \le 16\};$$

$\theta = \theta(\tau, \lambda', \lambda'') > |\tau|$ is the root of the equation

$$(19)\qquad (\theta+\tau)^{-2}\lambda'^2 + (\theta-\tau)^{-2}\lambda''^2 = 16;$$

$$(20)\qquad \Phi_1(\tau, \lambda', \lambda'') = 4(1-\tau^2)^{\frac{1}{2}} + 4^{-1}(1-\tau^2)^{-\frac{1}{2}}\{(1-\tau)\lambda'^2 + (1+\tau)\lambda''^2\},$$

$$(21)\qquad \Phi_2(\tau, \lambda', \lambda'') = 4(1-\tau^2)^{\frac{1}{2}} + \frac{1}{4}(1-\tau^2)^{\frac{1}{2}}\left\{\frac{2\theta+\tau-1}{(\theta+\tau)^2}\lambda'^2 + \frac{2\theta-\tau-1}{(\theta-\tau)^2}\lambda''^2\right\},$$

$$(22_1)\qquad x_\theta = \frac{1}{4}\cdot 2^{-\frac{1}{2}}\left\{\frac{-[(1-\tau^2)^{\frac{1}{2}} - \tau]\lambda'}{\theta+\tau} + \frac{[(1-\tau^2)^{\frac{1}{2}} + \tau]\lambda''}{\theta-\tau}\right\},$$

$$(22_2)\qquad y_\theta = \frac{1}{4}\cdot 2^{-\frac{1}{2}}\left\{\frac{[(1-\tau^2)^{\frac{1}{2}}+\tau]\lambda'}{\theta+\tau} + \frac{[(1-\tau^2)^{\frac{1}{2}}-\tau]\lambda''}{\theta-\tau}\right\}.$$

LEMMA 3. *Let* $h\in P$ *and have the form* (6). *For any complex number* $\lambda = \lambda_1 + i\lambda_2$ *under the notation* (18) - (21) *the following sharp estimate holds:*

$$(23)\qquad I_1(h) \le \begin{cases} \Phi_1(\tau,\lambda',\lambda'') & \text{if } (\lambda',\lambda'')\in \bar{E}_\tau, \\ \Phi_2(\tau,\lambda',\lambda'') & \text{if } (\lambda',\lambda'')\notin E_\tau. \end{cases}$$

The extremal functions have the form:

$$h(z) = \mu(1+ze^{-i\varphi_0})/(1-ze^{-i\varphi_0}) + (1-\mu)(1-ze^{i\varphi_0})/(1+ze^{i\varphi_0}),$$

$$0\le\mu\le 1,\quad \varphi_0 = \operatorname{arc\,tg}(x_0/y_0),$$

where x_0, y_0 *are given by* (22) *if* $(\lambda',\lambda'')\in\bar{E}_\tau$; $h(z) = (1+ze^{-i\varphi_\theta})/(1-ze^{-i\varphi_\theta})$, $\varphi_\theta = \operatorname{arc\,tg}(y_\theta/x_\theta)$, *where* x_θ, y_θ *are given by* (22) *if* $(\lambda',\lambda'')\notin E_\tau$.

Proof. Let $h\in P$ and $h_1 = x+iy$, $h_2 = u+iv$, $\eta = \eta_1 + i\eta_2 = 1+e^{i\delta}$. Then $\operatorname{Re}[\eta h_2] = \eta_1 u - \eta_2 v$. According to Lemma 1 we have

$$x^2+y^2\le 1 \quad\text{and}\quad [u-(x^2-y^2)]^2 + [v-2xy]^2 \le (1-x^2-y^2)^2.$$

It is not difficult to check that the quantity $\eta_1 u - \eta_2 v$ attains its maximal value when $u = u_0 = (x^2-y^2) + (1-x^2-y^2)\cos\delta/2$, $v = v_0 = 2xy - (1-x^2-y^2)\sin\delta/2$. Calculating this maximal value we find that

$$(24)\qquad \max_{h\in P} 2\operatorname{Re}[\eta h_2] = 4(1+t^2)^{-\frac{1}{2}} - 4(1+t^2)^{-1}\{[(1+t^2)^{\frac{1}{2}}-1]x^2 + 2txy + [(1+t^2)^{\frac{1}{2}}+1]y^2\},$$

where $t = \operatorname{tg}\delta/2$.

By simple calculations we find also that

$$\operatorname{Re}[\eta h_1]^2 = 4(1+t^2)^{-2}\{(1-t^2)(x^2-y^2) - 4txy\},$$

$$\mathrm{Re}\,[\eta\lambda q_1] = 2^{\frac{1}{2}}(1+t^2)^{-1}\{[(1-t)\lambda' - (1+t)\lambda'']x - [(1+t)\lambda' + (1-t)\lambda'']y\}.$$

Now we have

$$I_1(h) = \mathrm{Re}\{\eta(2h_2 - \eta h_1^2 - \lambda h_1)\} \le 4\{(1+t^2)^{-\frac{1}{2}} - (1+t^2)^{-2}G(x,y)\},$$

where

$$(25)\qquad G(x,y) = [(1+t^2)^{\frac{3}{2}} - 2t^2]x^2 - 2t(1-t^2)xy + [(1+t^2)^{\frac{3}{2}} + 2t^2]y^2 + 2^{-\frac{3}{2}}(1+t^2)[(1-t)\lambda' - (1+t)\lambda'']x - 2^{-\frac{3}{2}}(1+t^2)[(1+t)\lambda' + (1-t)\lambda'']y.$$

It is clear that in order to find $\max_{h\in P} I_1(h)$ we should determine $\min G(x,y)$ in the disk $x^2+y^2 \le 1$ for the fixed $t \in (-\infty, +\infty)$.

The system of the equations $\partial G/\partial x = 0$ and $\partial G/\partial y = 0$ has a unique solution (x_0, y_0) given by the formula (22). At this point, as it is not very difficult to check, the function $G(x,y)$ has minimum unless it satisfies the relation $x_0^2 + y_0^2 \le 1$.

The last condition: $x_0^2 + y_0^2 \le 1$ is equivalent to the fact that the point $(\lambda', \lambda'') \in \bar{E}_\tau$. In this case, if we calculate the value of $G(x_0, y_0)$, we obtain for the functional $I_1(h)$ the estimate by $\Phi_1(\tau, \lambda', \lambda'')$ given in (20).

If the point (x_0, y_0) is such that $x_0^2 + y_0^2 \ge 1$, then it is clear that the function $G(x,y)$ attains its minimal value on the circle $x^2 + y^2 = 1$.

In this case we consider the function

$$(26)\qquad \hat{G}(x,y;\theta) = G(x,y) + \hat{\theta}(x^2+y^2-1), \qquad \hat{\theta} = (1+t^2)^{\frac{3}{2}}(\theta - 1).$$

The tedious but straightforward considerations lead to the conclusion that the function $G(x,y)$ attains its minimum at the point (x_θ, y_θ) given by the formulae (22). The corresponding maximal value for $I_1(h)$ is equal to $\Phi_2(\tau, \lambda', \lambda'')$. The forms of the extremal functions follow from the proof.

COROLLARY 1. *If* $h \in P$ *and* $\eta = 2$ *i.e.* $\delta = 0$ *then for any com-*

plex λ the following sharp inequality holds:

$$(27)\qquad I_1(h) \le \begin{cases} 4 + \frac{1}{4}|\lambda|^2 & \text{if } |\lambda| \le 4, \\ 2|\lambda| & \text{if } |\lambda| \ge 4. \end{cases}$$

COROLLARY 2. *If* $h \in P$ *and* λ *is real or purely imaginary then the following sharp inequality holds*:

$$(28)\qquad I_1(h) \le \begin{cases} 4(1-\tau^2)^{\frac{1}{2}} + \frac{1}{4}|\lambda|^2/(1-\tau^2)^{\frac{1}{2}} & \text{if } |\lambda| \le 4(1-\tau^2)/(1+\tau^2)^{\frac{1}{2}}, \\ \frac{1}{2}|\lambda|^2\theta^3(1-\tau^2)^{\frac{1}{2}}/(\theta^2-\tau^2)^2 & \text{if } |\lambda| \ge 4(1-\tau^2)/(1+\tau^2)^{\frac{1}{2}}, \end{cases}$$

where $\theta = \frac{1}{4}\{16\tau^2 + \frac{1}{2}|\lambda|[(|\lambda|^2 + 128\tau^2)^{\frac{1}{2}} + |\lambda|]\}^{\frac{1}{2}}$.

In particular we have the sharp inequality

$$(29)\qquad I_1(h) \le 4 + \frac{1}{4}|\lambda|^2 \quad \text{if } |\lambda| \le 4.$$

4. So far we have found $\max_{h\in P} I_1(h)$, $\max_{q\in P} I_2(q)$. Now we can state the following

THEOREM 1. *If* $f \in B(\alpha,\beta;g,h_a)$ *then for any complex* λ *the following sharp inequality holds*:

$$(30)\qquad I(f) = \operatorname{Re}\{[2(\varkappa+1)a_3 - (\varkappa+2)a_2^2] - \lambda\varkappa a_2\}$$

$$\le \max_{h\in P} I_1(h) + 2\alpha \max_{q\in P} I_2(q),$$

where $\max_{h\in P} I_1(h)$ *and* $\max_{q\in P} I_2(q)$ *are given by* (23) *and* (12) - (14), *respectively*.

COROLLARY 3. *If* $f \in B(\alpha,\beta;g,h_a)$ *then for any real* λ *the following sharp inequalities hold*:

$$(31)\qquad I(f) \le \begin{cases} \frac{1}{4}|\lambda|^2 + 2\alpha|\lambda| + 2\alpha + 4 & \text{if } |\lambda| \le 4, \\ 2(1+\alpha)|\lambda| + 2\alpha & \text{if } |\lambda| \ge 4. \end{cases}$$

COROLLARY 4. *If* $f \in B(\alpha,\beta;g,h_a)$ *then for any purely imaginary*

λ the following sharp inequality holds:

$$(32)\qquad I(f) \le \begin{cases} \frac{1}{4}(1+\alpha)|\lambda|^2 + 2\alpha + 4 & \text{if } |\lambda| \le 4 \\ 2(1+\alpha)|\lambda| - 2\alpha & \text{if } |\lambda| \ge 4. \end{cases}$$

5. At the end we show one more application of Lemmas 2 and 3.

Let the parameters γ and δ satisfy the conditions: $\gamma \ge 0$, $\delta \in (-\pi/2, \pi/2)$.

The family of functions f holomorphic in K which have the form (1) and satisfy the condition

$$(33)\qquad |\arg\{e^{i\delta} f'(z)/g'(z)\}| \le \gamma\pi/2, \qquad z \in K,$$

for some univalent and convex function g, we call (after Pommerenke [7] or Goodman [5]) the close-to-convex functions of the order $\gamma (L_\gamma)$.

It is known that the class L_γ consists of univalent functions if $\gamma \in [0,1]$ and is a linearly invariant family in the sense of Pommerenke for all $\gamma \ge 0$. For the subclass $\hat{L}^r_\gamma \subset L_\gamma$ consisting of functions with real coefficients and $\delta = 0$ in (33), we have

THEOREM 2. *If $f \in \hat{L}^r_\gamma$, $\gamma > 0$, then the set $V = \{(a_2, a_3) : f \in \hat{L}^r_\gamma\}$ is given by the inequality*

$$(34)\qquad \frac{2}{3}\left(\frac{\gamma+2}{\gamma+1}\right)a_2^2 - \frac{2\gamma+1}{3} \le a_3 \le \begin{cases} \frac{1}{3}(2a_2^2 + 2\gamma + 1) & \text{if } a_2 \in [-1, 1], \\ \frac{1}{3}[2\frac{\gamma-1}{\gamma} a_2^2 + \frac{4}{\gamma} a_2 + 2(\gamma - \frac{1}{\gamma}) + 1] & \text{if } a_2 \in [1, \gamma+1], \\ \frac{1}{3}[2\frac{\gamma-1}{\gamma} a_2^2 - \frac{4}{\gamma} a_2 + 2(\gamma - \frac{1}{\gamma}) + 1 & \text{if } a_2 \in [-(\gamma+1), -1], \end{cases}$$

The inequalities (34) describe precisely the region of variability (a_2, a_3) for $f \in \hat{L}^r_\gamma$.

Proof. Let $f \in \hat{L}^r_\gamma$ and λ be an arbitrary real number. Using the condition (33) and the definition of convex function, after some calculations we obtain the equality

(35) $$(6a_3 - 4a_2^2) - 2\lambda a_2 = \gamma \cdot 2(2h_2 - 2h_1^2 - \lambda h_1) + (q_2 - \lambda q_1),$$

where $h, q \in P^r$.

It is clear that

(36) $$\max_{f \in \hat{L}_\gamma^r} \{(6a_3 - 4a_2^2) - 2\lambda a_2\} \le \gamma \max_{h \in P^r} I_1(h) + \max_{q \in P^r} I_2(q),$$
$$\min_{f \in \hat{L}_\gamma^r} \{(6a_3 - 4a_2^2) - 2\lambda a_2\} \ge \gamma \min_{h \in P^r} I_1(h) + \min_{q \in P^r} I_2(q).$$

From (12) and (27) we have:

(37) $$\max_{q \in P^r} I_2(q) = 1 + |\lambda|,$$
$$\max_{h \in P^r} I_1(q) = \begin{cases} 4 + \frac{1}{4}|\lambda|^2 & \text{if } |\lambda| \le 4, \\ 2|\lambda| & \text{if } |\lambda| \ge 4. \end{cases}$$

Using Lemma 1 in similar way as in Lemma 3 we can prove that

(38) $$\min_{q \in P^r} I_2(q) = \begin{cases} -1 - \frac{\lambda^2}{8} & \text{if } |\lambda| \le 4, \\ 1 - |\lambda| & \text{if } |\lambda| \ge 4; \end{cases}$$
$$\min_{h \in P^r} I_1(h) = \begin{cases} -4 - \frac{\lambda^2}{4} & \text{if } |\lambda| \le 4, \\ -2|\lambda| & \text{if } |\lambda| \ge 4. \end{cases}$$

All the relations (37)-(38) are sharp. From the relations (37)-(38) we obtain (34).

Remark. Lemmas 2 and 3 can be also used for estimating other functionals, e.g. $\mathrm{Re}(a_3 - \mu a_2^2)$ and $|a_3| - |a_2|$.

References

[1] BAZILEVIČ, I.E.: On a case of integrability in quadratures of the Loewner-Kufarev equation, Mat. Sb. 37 (1955), 471-476.

[2] CAMPBELL, D.M. and K. PEARCE: Generalized Bazilevič functions, Rocky Mounatain Journ. of Math. 9, 2 (1979), 197-226.

[3] DeTEMPLE, D.W.: On coefficient inequalities for bounded univalent functions, Ann. Acad. Sci. Fenn. Ser. A I, No. 469 (1970).

[4] EENIGENBURG, P.J. and E.M. SILVIA: A coefficient inequality for Bazilevič functions, Ann. Univ. Mariae Curie-Skłodowska Sectio A 27 (1973), 5-12.

[5] GOODMAN, A.W.: On close-to-convex functions of higher order, Ann. Univ. Sci. Budapest. Eötvös Sect. Math. 15 (1972), 17-30.

[6] POMMERENKE, Ch.: Über die Subordination analytischer Funktionen, J. Reine Angew. Math. 218 (1965), 159-173.

[7] ——: On close-to-convex analytic functions, Trans. Amer. Math. Soc. 114 (1961), 176-186.

[8] SZYNAL, A. and J. SZYNAL: The extension of Jenkins inequality, Ann. Univ. Mariae Curie-Skłodowska Sectio A, to appear.

[9] TAMMI, O.: On optimized inequalities in connection with coefficient bodies of bounded univalent functions, Ann. Acad. Sci. Fenn. Ser. A I. Math. 4 (1978/1979), 45-52.

[10] YOSHIKAWA, H.: On a subclass of spiral-like functions, Mem. Fac. Sci. Kyushu Univ. Ser. A 25 (1971), 271-279.

Institute of Mathematics
Maria Curie-Skłodowska University
Pl. M. Curie-Skłodowskiej 1
PL-20-031 Lublin, Poland

DIVISION OF CAUCHY-RIEMANN FUNCTIONS ON HYPERSURFACES

Roman Dwilewicz (Warszawa)

Contents | page

Summary

In this paper we discuss the problem of division of two CR functions defined on a smooth hypersurface having the same zero sets. We prove that if the CR differentials of the functions f,g are not singular on the hypersurface, then the quotients f/g, g/f can be extended to smooth CR functions on the whole hypersurface.

Introduction

In this paper we discuss the problem of division of two functions defined on a manifold and having the same zero sets. In the case where the manifold has a differential structure, a real analytic structure or a complex analytic structure the problems become obvious and have already been solved in the classical analysis. We intend to extend those results to the case of the so-called Cauchy-Riemann structures, CR structures for short. We begin by formulating the problem for the differential structures.

Let U be an open subset of $\mathbb{R}^n$ and let $f,g : U \longrightarrow \mathbb{R}$ be smooth functions whose differentials df, dg are not singular and let

$$\{p \in U ; f(p) = 0\} = \{p \in U ; g(p) = 0\}.$$

It is well-known that in this situation the quotients f/g, g/f extend to smooth functions on U. The same is true for both real and complex analytic structures.

The division problem for functions is very important and has numerous applications. For example the notion of the derivative of a function is actually based on the division of functions. Also the explicit definition of the normal bundle to a submanifold makes use of the division problem.

The CR structures are very natural and often used in complex analysis. In this paper we present the results concerning the division of CR functions on a special kind of CR manifolds, namely on hypersurfaces. Some results about division of CR functions on an arbitrary CR manifold and with weakened assumptions on the differentials of functions will be published elsewhere.

The main theorem of this paper reads

Theorem 0.1. *Let* M *denotes a smoothly embedded hypersurface of a complex manifold* X, $\dim_{\mathbb{C}} X = n \geq 2$, $\dim_{\mathbb{R}} M = 2n-1$. *Let* f,g *be* CR *functions on* M *such that*

$$\{p \in M ; f(p) = 0\} \supset \{p \in M ; g(p) = 0\}$$

and the CR *differential of* g *is not singular at each point of* M ($\partial_{CR} g(p) \neq 0$, $p \in M$; *see* §1c). *Then the quotient* f/g *can be extended to a smooth* CR *function on* M.

It can easily be observed that the theorem is not true for $n = 1$ since the requirement that $\partial_{CR} g(p) \neq 0$ is then void.

The contents of the paper is the following. After introducing in Section 1 some notions and definitions we reduce the main theorem to a special case (Section 2). In Section 3 we prove the reduced version by using the vanishing of some partial derivatives of CR functions. The proof of this fact (Section 4) is the essential step in the proof of the main theorem. It uses the theorem of Hans Lewy (see [4,Th.2.6.13],[6]) on the extension of CR functions and the theorem on local parametrization of Levi flat CR manifolds in the form given by Wells [10,Th.2.2] based on Sommer paper [9] or in the form given by Rea [8,Prop.4.3] based on the "complex Frobenius theorem" of Newlander - Nirenberg [7]. In Section 5 we generalize the main theorem to systems of CR functions and obtain a version which is suitable for the explicit definition of the normal bundles to CR submanifolds (Section 6).

§1. Some notion and definitions

In the sequel we need the notion of CR structures only in the case of differential submanifolds of complex manifolds and therefore our definitions are reduced to this special situation. Concerning the general case see for example papers [1,§1],[11,§4,§5].

All differential manifolds, appearing in this paper,are smooth and by differential submanifolds we mean smoothly embedded submanifolds.

a) Let X be a complex manifold of complex dimension n. Let $\mathbb{C}T(X)$ be the complexified tangent bundle to X. The complex structure on X gives a natural decomposition of $\mathbb{C}T(X)$ into the holomorphic tangent bundle and the antiholomorphic tangent bundle

$$\mathbb{C}T(X) = H(X) \oplus \overline{H(X)}.$$

Given local holomorphic coordinates at the point p, say $z_1,\dots,z_n$, the fibres $H_p(X)$, $\overline{H_p(X)}$ assume the form

$$\overline{H_p(X)} = \{a_1\frac{\partial}{\partial \bar z_1} + \dots + a_n\frac{\partial}{\partial \bar z_n} \in \mathbb{C}T_p(X)\ ;\ a_1,\dots,a_n \in \mathbb{C}\},$$

$$H_p(X) = \{b_1\frac{\partial}{\partial z_1} + \dots + b_n\frac{\partial}{\partial z_n} \in \mathbb{C}T_p(X)\ ;\ b_1,\dots,b_n \in \mathbb{C}\} .$$

Let M be a C^∞ differential submanifold of X, $\dim_{\mathbb{R}} M = m$, $1 \le m \le 2n$. There is a natural inclusion $\mathbb{C}T(M) \subset \mathbb{C}T(X)$. For each point $p \in M$ we define $H_p(M)$ as follows

$$(1.1) \qquad H_p(M) = \mathbb{C}T_p(M) \cap H_p(X).$$

It is not difficult to compute that

(1.2) $\max(0, m-n) \leq \dim_{\mathbb{C}} H_p(M) \leq m/2$.

If $\dim_{\mathbb{C}} H_p(M)$ is constant on M, then we put

$$H(M) = \bigcup_{p \in M} H_p(M).$$

Simple computations show that $H(M)$ is a smooth subbundle of the bundle $\mathbb{C}T(X)|_M$. We then say that $(M, H(M))$ is a CR submanifold of the complex manifold X. In what follows we often denote by M a CR submanifold if it does not lead to a confusion. Notice that in particular the complex manifold X is a CR submanifold.

Denote by $\dim_{CR} M$ the common complex dimension of the fibres $H_p(M)$. Using the inequalities (1.2) we obtain

(1.3) $\max(0, m-n) \leq \dim_{CR} M \leq m/2$.

If $\dim_{CR} M$ is minimal, i.e. $\dim_{CR} M = \max(0, m-n)$, then M is called a generic CR submanifold of the complex manifold X.

b) Assume that M is a real hypersurface of a complex manifold X. It is easy to see that for each point $p \in M$, $\dim_{\mathbb{C}} H_p(M) = n-1$. So, M is a natural Cauchy-Riemann submanifold of X. Suppose that the hypersurface M is given in coordinates $(z_1, \ldots, z_n)$ on a neighbourhood $U \subset X$ by the equation $\rho = 0$, where ρ is a real C^∞ function such that grad $\rho \neq 0$ at each point of U. Then

$$H_p(M) = \{ \sum_{\alpha=1}^{n} \xi_\alpha \frac{\partial}{\partial z_\alpha} \in H_p(X) ; \sum_{\alpha=1}^{n} \xi_\alpha \frac{\partial \rho(p)}{\partial z_\alpha} = 0 \},$$

$$\overline{H_p(M)} = \{ \sum_{\alpha=1}^{n} \eta_\alpha \frac{\partial}{\partial \bar{z}_\alpha} \in \overline{H_p(X)} ; \sum_{\alpha=1}^{n} \eta_\alpha \frac{\partial \rho(p)}{\partial \bar{z}_\alpha} = 0 \}.$$

We introduce the Levi form only for hypersurfaces, because only this case is needed later. The definition for an arbitrary CR manifold can be found in [10, Def.2.1] or [11, pp.427-428]. The main properties of the Levi form are invariant under biholomorphic transformations, therefore it is enough to define the form for hypersurfaces in a coordinate neighbourhood U in X. Assume that M is given, as above, by the function ρ. The Levi form is the usual quadratic form L_ρ obtained from the complex Hessian of ρ acting on the complex tangent space

$$L_\rho(p)(\xi,\bar{\eta}) = \sum_{\alpha,\beta=1}^{n} \frac{\partial^2\rho(p)}{\partial z_\alpha \partial \bar{z}_\beta} \xi_\alpha \bar{\eta}_\beta,$$

where

$$\xi = \sum_{\alpha=1}^{n} \xi_\alpha \frac{\partial}{\partial z_\alpha} \in H_p(M), \quad \eta = \sum_{\alpha=1}^{n} \eta_\alpha \frac{\partial}{\partial z_\alpha} \in H_p(M).$$

The Levi form is equal to zero at a point $p, L_\rho(p) = 0$, if $L_\rho(p)(\xi,\bar{\eta}) = 0$ for all $\xi, \eta \in H_p(M)$.

c) Let $(M,H(M)), (N,H(N))$ be two CR submanifolds of complex manifolds X, Y, respectively. We say that $f : M \longrightarrow N$ is a smooth CR mapping if f is smooth and $df(H(M)) \subset H(N)$. This also makes sense if M or N are complex manifolds.

A CR mapping f is by definition a CR isomorphism between $(M,H(M))$ and $(N,H(N))$ provided f is a smooth diffeomorphism such that $df(H(M)) = H(N)$.

Two CR submanifolds $(M,H(M))$, $(N,H(N))$ are said to be isomorphic if there exists a CR isomorphism between them.

If $(N,H(N))$ is in particular $(\mathbb{C},H(\mathbb{C}))$, then a smooth function $f : M \longrightarrow \mathbb{C}$ is a CR function if and only if

> for any smooth extension F of f onto a neighbourhood of M in X we have $\eta F(p) = 0$ for any $p \in M$ and any $\eta \in \overline{H_p(M)}$, i.e. F satisfies the Cauchy - Riemann equations at each holomorphic tangent space to M.

Notice that the last condition does not depend on the extension F of f.

Let $f : M \longrightarrow \mathbb{C}$ be a smooth CR function on a CR submanifold $M, M \subset X$. We say that the CR differential of f is not equal to zero at a point $p \in M$ if there exists $\xi \in H_p(M)$ such that $\xi f(p) \neq 0$. We write then $\partial_{CR} f(p) \neq 0$.

For convenience, we often write $\frac{\partial h(p)}{\partial z}$ instead of $\frac{\partial h(z)}{\partial z}\Big|_{z=p}$, as no ambiguity should arise by the context.

§2. Reducing the main theorem to a special case

a) In this section we reduce the main theorem (Theorem 0.1) to a special case, where the hypersurface M is embedded in $\mathbb{C}^n$ and the functions f,g have a special form.

The notation and assumptions are as in Theorem 0.1.

Denote

$$N = \{p \in M \, ; g(p) = 0\}.$$

Notice that N is a smooth submanifold of M. The function f/g is obviously a smooth CR function on $M-N$. If the quotient can be smoothly extended to M, then the extended function will also be CR on the whole hypersurface.

In order to prove that the function f/g extends to a smooth function on M, it suffices to establish this property in a neighbourhood of each point $p \in N$ in M.

Fix a point $p \in N$ and take a coordinate neighbourhood $(U;\zeta_1,\dots,\zeta_n)$ on the complex manifold X such that $p \in U \subset\subset X$ and p has coordinates $0=(0,\dots,0)$. Performing if necessary a $\mathbb{C}$-linear change of coordinates we can assume (see[5,§2]) that the hypersurface M in a neighbourhood of p is given by the equation

$$\text{(2.1)} \quad \operatorname{Im}\zeta_n = \varphi(\zeta_1,\dots,\zeta_{n-1}, \operatorname{Re}\zeta_n),$$

where φ is a smooth function and such that

$$\text{(2.2)} \quad \varphi(0) = 0, \ \frac{\partial\varphi(0)}{\partial\zeta_\alpha} = \frac{\partial\varphi(0)}{\partial\bar{\zeta}_\alpha} = \frac{\partial\varphi(0)}{\partial(\operatorname{Re}\zeta_n)} = 0, \qquad \alpha = 1,\dots,n-1.$$

Consequently we have

$$H_0(M) = \{a_1\frac{\partial}{\partial\zeta_1} + \dots + a_n\frac{\partial}{\partial\zeta_n} \in H_0(X);\ a_n = 0\},$$

$$\overline{H_0(M)} = \{b_1\frac{\partial}{\partial\bar{\zeta}_1} + \dots + b_n\frac{\partial}{\partial\bar{\zeta}_n} \in \overline{H_0(X)};\ b_n = 0\}.$$

Take the CR function g appearing in Theorem 0.1. We can smoothly extend it to a function (denoted also by g) on the neighbourhood U of p. We known that $\partial_{CR}g(p)\neq 0$ and that g satisfies the Cauchy-Riemann equations at p. Recalling the form of $H_0(M)$

and $\overline{H_o(M)}$, these conditions can be written in local coordinates

$$(2.3)\quad \begin{aligned} &\left(\frac{\partial g(0)}{\partial \zeta_1},\dots,\frac{\partial g(0)}{\partial \zeta_{n-1}}\right) \neq (0,\dots,0), \\ &\left(\frac{\partial g(0)}{\partial \bar{\zeta}_1},\dots,\frac{\partial g(0)}{\partial \bar{\zeta}_{n-1}}\right) = (0,\dots,0). \end{aligned}$$

If U is sufficiently small we can find by making use of (2.3) holomorphic functions $g_2,\dots,g_{n-1}$ in U, vanishing at the point p, and such that the mapping

$$(2.4)\quad U\cap M \ni \zeta = (\zeta_1,\dots,\zeta_n) \xrightarrow{\ \Phi\ } (g(\zeta), g_2(\zeta),\dots,g_{n-1}(\zeta),\zeta_n)$$

gives a CR isomorphism on a hypersurface $\Phi(M\cap U)$ lying in a neighbourhood of 0 in $\mathbb{C}^n$. In terms of the standard coordinates $(z_1,\dots,z_n)$ in $\mathbb{C}^n$ the inverse mapping Φ^{-1} has the form

$$\Phi^{-1}(z_1,\dots,z_n) = (\varphi_1(z_1,\dots,z_n),\dots,\varphi_{n-1}(z_1,\dots,z_n),z_n),$$
$$(z_1,\dots,z_n) \in \Phi(M\cap U).$$

Inserting Φ^{-1} into (2.1) we arrive at

$$(2.5)\quad \operatorname{Im} z_n = \varphi(\varphi_1(z_1,\dots,z_n),\dots,\varphi_{n-1}(z_1,\dots,z_n),\operatorname{Re} z_n).$$

Denote by $\varphi^* = \varphi^*(z_1,\dots,z_n)$ the function standing on the right-hand side of the above equation. According to (2.2), all partial dervatives of φ^* at the point $0 = (0,\dots,0)$ vanish

$$\varphi^*(0) = 0, \quad \frac{\partial \varphi^*(0)}{\partial z_\alpha} = \frac{\partial \varphi^*(0)}{\partial \bar{z}_\alpha} = 0, \quad \alpha = 1,\dots,n.$$

Therefore we can solve equation (2.5) with respect to $\operatorname{Im} z_n$ and find a smooth function ψ such that the hypersurface $\Phi(M\cap U)$ can be written in a neighbourhood of zero in the following way

$$\mathrm{Im} z_n = \psi(z_1, \ldots, z_{n-1}, \mathrm{Re} z_n),$$

where

$$\psi(0) = 0, \quad \frac{\partial\psi(0)}{\partial z_\alpha} = \frac{\partial\psi(0)}{\partial \bar{z}_\alpha} = \frac{\partial\psi(0)}{\partial(\mathrm{Re} z_n)} = 0, \quad \alpha = 1, \ldots, n-1.$$

Notice that to the function g there corresponds the function

$$g^*(z_1, \ldots, z_n) = z_1, \quad (z_1, \ldots, z_n) \in \Phi(M \cap U),$$

and to the function f there corresponds the function $f^* = f^*(z_1, \ldots, z_n)$ such that

$$f^*(0, z_2, \ldots, z_n) = 0 \quad \text{for} \quad (0, z_2, \ldots, z_n) \in \Phi(M \cap U).$$

b) Before formulating the reduced version of the main theorem we introduce some notation.

$$P = \{z = (z_1, \ldots, z_n) \in \mathbb{C}^n ; \ |x_\alpha| < 1, \ |y_\alpha| < 1, \ \alpha = 1, \ldots, n\},$$

$$\bar{P} = \{z = (z_1, \ldots, z_n) \in \mathbb{C}^n ; \ |x_\alpha| \leq 1, \ |y_\alpha| \leq 1, \ \alpha = 1, \ldots, n\},$$

where

$$z_\alpha = x_\alpha + iy_\alpha, \ \alpha = 1, \ldots, n.$$

$$\bar{Q} = \{z = (z_1, \ldots, z_n) \in \bar{P} \ ; \mathrm{Im}\ z_n = 0\}, \quad Q = P \cap \bar{Q}.$$

Points in $\bar{Q}$ will be denoted by $s = (z_1, \ldots, z_{n-1}, x_n)$ or by (z_1, t), where $t = (z_2, \ldots, z_{n-1}, x_n)$, depending on convenience.

The considerations performed in this section permit us to reduce the main theorem to the following

Theorem 2.1. *Let* M *be a smooth hypersurface in* $P \subset \mathbb{C}^n$, $n \geq 2$, *given by the equation*

$$y_n = \psi(z_1, \ldots, z_{n-1}, x_n), \quad (z_1, \ldots, z_{n-1}, x_n) \in Q,$$

where ψ *is a smooth function on* Q *admitting a smooth extension to a neighbourhood of* $\bar{Q}$, *and suppose that*

$$(2.6)\quad \psi(0)=0,\quad \frac{\partial\psi(0)}{\partial z_\alpha}=\frac{\partial\psi(0)}{\partial \bar z_\alpha}=\frac{\partial\psi(0)}{\partial x_n}=0,\quad \alpha=1,\dots,n-1.$$

Let f *be a smooth* CR *function on* M *which vanishes on*

$$N=M\cap\{(z_1,\dots,z_n)\in\mathbb{C}^n\ ;\ z_1=0\}$$

and can also be smoothly extended over the boundary of M *onto a hypersurface containing* M.
Then the function

$$\frac{f(z_1,\dots,z_n)}{z_1},\quad (z_1,\dots,z_n)\in M-N$$

can be smoothly extended to M.

§3. Proof of the reduced version of the main theorem

We retain all notation and assumptions of §2b and Theorem 2.1. Let us define the function

$$h(z_1,t)=h(z_1,\dots,z_{n-1},x_n)=$$

$$=f(z_1,\dots,z_{n-1},x_n+i\psi(z_1,\dots,z_{n-1},x_n))$$

$$\text{for } (z_1,t)=(z_1,\dots,z_{n-1},x_n)\in Q.$$

In order to show that the function $f(z_1,\dots,z_n)/_{z_1}$ can be extended to M it is sufficient to prove that the function $h(z_1,t)/_{z_1}$, $(z_1,t)\in Q$, $z_1\neq 0$, can be extended to Q. Notice that the last function is smooth on

$$Q-\{(z_1,t)\in Q\ ;\ z_1=0\}.$$

For our purpose it suffices to show that all partial derivatives of the function $h(z_1,t)/_{z_1}$ extend continuously to Q.

Denote by ∂_t and $\bar\partial_s$ the partial derivatives of the type

$$(3.1)\qquad \frac{\partial^{\alpha_2+\dots+\alpha_n}}{\partial z_2^{\alpha_2}\dots\partial z_{n-1}^{\alpha_{n-1}}\partial x_n^{\alpha_n}} \quad\text{and}\quad \frac{\partial^{\beta_1+\dots+\beta_n}}{\partial \bar z_1^{\beta_1}\dots\partial \bar z_{n-1}^{\beta_{n-1}}\partial x_n^{\beta_n}}$$

respectively, where $\alpha_2,\dots,\alpha_n$, $\beta_1,\dots,\beta_n$ are non-negative integers. Compute partial derivatives of the function $h(z_1,t)/_{z_1}$ for $z_1 \neq 0$.

$$\partial_t\,\bar\partial_s \frac{\partial^k}{\partial z_1^k}\Big(\frac{h(z_1,t)}{z_1}\Big) = \frac{\partial^k}{\partial z_1^k}\partial_t\,\bar\partial_s\Big(\frac{h(z_1,t)}{z_1}\Big) =$$

$$(3.2)\qquad = (-1)^k k!\,\frac{\sum_{\gamma=0}^{k}\frac{(-1)^\gamma}{\gamma!}\Big[\frac{\partial^\gamma}{\partial z_1^\gamma}\partial_t\,\bar\partial_s h(z_1,t)\Big]z_1^\gamma}{z_1^{k+1}}.$$

Write

$$H(z_1,t) = \sum_{\gamma=0}^{k}\frac{(-1)^\gamma}{\gamma!}\Big[\frac{\partial^\gamma}{\partial z_1^\gamma}\partial_t\,\bar\partial_s h(z_1,t)\Big]z_1^\gamma, \qquad (z_1,t)\in Q.$$

Assume that we have shown

$$(3.3)\qquad \frac{\partial^{\alpha+\beta}H(z_1,t)}{\partial z_1^\alpha\,\partial\bar z_1^\beta}\Big|_{z_1=0} = 0 \quad \text{for} \quad 0\le\alpha\le k,\ \beta\ge 0,\ (0,t)\in Q.$$

Therefore applying to the function $H(z_1,t)$ the Taylor formula with respect to z_1 at the center $(0,t)$, we obtain

$$H(z_1,t) = \frac{1}{(k+1)!}\Big(z_1\frac{\partial}{\partial z_1} + \bar z_1\frac{\partial}{\partial\bar z_1}\Big)^{k+1} H(z_1,t)\Big|_{z_1=0} + {}$$

$$+ z_1^{k+2}\,H_{k+2}(z_1,t),$$

$$(z_1,t)\in Q,$$

where $H_{k+2} = H_{k+2}(z_1,t)$ is a function on Q depending on the derivatives of H of order $k+2$. Moreover this function is bounded by a constant M_{k+2}

$$(3.4)\qquad |H_{k+2}(z_1,t)| \le M_{k+2}, \quad (z_1,t)\in Q.$$

Such a constant M_{k+2} exists because the function $H(z_1,t)$ can be smoothly extended to a neighbourhood of $\bar{Q}$.

According to (3.3), the first component of the above formula for H reduces to

$$\frac{1}{(k+1)!} \frac{\partial^{k+1} H(z_1,t)}{\partial z_1^{k+1}}\Big|_{z_1=0} z_1^{k+1}$$

and hence

$$H(z_1,t) = \frac{1}{(k+1)!} \frac{\partial^{k+1} H(z_1,t)}{\partial z_1^{k+1}}\Big|_{z_1=0} z_1^{k+1} + z_1^{k+2} H_{k+2}(z_1,t),$$

$$(z_1,t) \in Q.$$

Consequently, the partial derivatives in (3.2) have the form

$$\frac{\partial^k}{\partial z_1^k} \partial_t \bar{\partial}_s \left(\frac{h(z_1,t)}{z_1}\right) = \frac{(-1)^k}{k+1} \frac{\partial^{k+1} H(z_1,t)}{\partial z_1^{k+1}}\Big|_{z_1=0} +$$

$$+(-1)^k k!\, z_1 H_{k+2}(z_1,t),$$

$$(z_1,t) \in Q, \; z_1 \neq 0,$$

and the estimation (3.4) applied to the above equality yields

$$(3.5) \quad \left|\frac{\partial^k}{\partial z_1^k} \partial_t \bar{\partial}_s \left(\frac{h(z_1,t)}{z_1}\right) - \frac{(-1)^k}{k+1} \frac{\partial^{k+1} H(z_1,t)}{\partial z_1^{k+1}}\Big|_{z_1=0}\right| \leq$$

$$\leq k!\,|z_1| M_{k+2}$$

$$(z_1,t) \in Q, \; z_1 \neq 0.$$

Finally, it follows from the last inequality that the function

$$H^*(z_1,t) = \begin{cases} \dfrac{\partial^k}{\partial z_1^k} \partial_t \bar{\partial}_s \left(\dfrac{h(z_1,t)}{z_1}\right), & (z_1,t) \in Q, \quad z_1 \neq 0 \\[2ex] \dfrac{(-1)^k}{k+1} \dfrac{\partial^{k+1} H(z_1,t)}{\partial z_1^{k+1}}\Big|_{z_1=0}, & (0,t) \in Q \end{cases}$$

is continuous on Q. Recalling the considerations at the beginning of the proof, we have shown that the function $h(z_1,t)/_{z_1}$ can be extended to a smooth function on Q provided (3.3) is true. Thus to complete the proof of the theorem we need only show that the partial derivatives in (3.3) vanish. This is done in the following section.

§4. Proof of the vanishing of some partial derivatives

a) We preserve all the notation introduced in §2b and §3. In particular we recall that

$$h(z_1,t)=f(z_1,t,\psi(z_1,t)),\quad (z_1,t)\in Q,$$

where for simplicity we write $(z_1,t,\psi(z_1,t))$ instead of $(z_1,\dots,z_{n-1},$ $x_n+i\psi(z_1,\dots,z_{n-1},x_n))$.

$$H(z_1,t)=\sum_{\gamma=0}^{k}\frac{(-1)^\gamma}{\gamma!}\left[\frac{\partial^\gamma}{\partial z_1^\gamma}\partial_t\,\bar\partial_s h(z_1,t)\right]z_1^\gamma ,$$

with ∂_t, $\bar\partial_s$ defined in (3.1).

In this section we shall prove

$$(4.1)\qquad \frac{\partial^{\alpha+\beta}H(z_1,t)}{\partial z_1^\alpha\,\partial\bar z_1^\beta}\Big|_{z_1=0}=0\qquad \text{for } 0\le\alpha\le k,\beta\ge 0,\ (0,t)\in Q.$$

The main difficulty however in the proof of (4.1) consists in establishing that

$$(4.2)\qquad \partial_t\,\bar\partial_s h(z_1,t)\big|_{z_1=0}=0\quad \text{for}\quad (0,t)\in Q.$$

The proof of (4.2) is divided into two parts depending on the Levi form of M.
Denote by M_0 the set of those points of M at which the Levi form is non-vanishing and put $M_1=M-M_0$. Notice that M_0 is an open subset of M. Let $\bar M_0$ stand for the closure of M_0, and let Int M_1 be the interior of M_1 in M. Of course we have

$$M = \bar{M}_o \cup \operatorname{Int} M_1 .$$

At first we show (Subsection b) the property (4.2) for the points $(0,t) \in Q$ such that $(0,t,\psi(0,t) \in \bar{M}_o$, and then we proceed (Subsection c) to the points $(0,t) \in Q$ such that $(0,t,\psi(0,t)) \in \operatorname{Int} M_1$. Finally (Subsection d) we prove (4.1).

b) In this subsection we prove (4.2) for the points $(0,t) \in Q$ with $(0,t,\psi(0,t)) \in \bar{M}_o$. The function H being smooth, we need only to prove (4.2) for the points $(0,t) \in Q$ such that $(0,t,\psi(0,t)) \in M_o$.

Therefore fix a point $(0,t,\psi(0,t)) \in M_o$.
Using the Hans Lewy theorem [4, Th.2.6.13], (see also[6]) we can find a neighbourhood W of $(0,t,\psi(0,t))$ in P that the function $f|_{M\cap W}$ can be extended to a smooth function F on W which is holomorphic at least on

$$W^+ = \{z = (z_1,\ldots,z_n) \in W \,;\, \operatorname{Im} z_n > \psi(z_1,\ldots,z_{n-1},x_n)\}$$

or

$$W^- = \{z = (z_1,\ldots,z_n) \in W \,;\, \operatorname{Im} z_n < \psi(z_1,\ldots,z_{n-1},x_n)\}.$$

Because $f|_N$ vanishes, then it follows from the uniqueness theorem for holomorphic functions that

$$(4.3) \quad F(0,z_2,\ldots,z_n) = 0 \quad \text{for} \quad (0,z_2,\ldots,z_n) \in W^+ \quad (\text{or } W^-)$$

depending on where F is holomorphic.

Compute $\partial_t \bar{\partial}_s h(z_1,t)|_{z_1=0}$. We have

$$(4.4) \quad \partial_t \bar{\partial}_s h(z_1,t)|_{z_1=0} = \partial_t \bar{\partial}_s [F(z_1,t,\psi(z_1,t))]|_{z_1=0}.$$

The right-hand side of equality (4.4) is a finite sum of the products of partial derivatives of F at $(0,t,\psi(0,t))$ multiplied by partial derivatives of ψ at $(0,t)$.
Owing to the property (4.3) and the form of ∂_t, $\bar{\partial}_s$ (see(3.1)), all the terms vanish except possibly those which contain differentiations over $\bar{z}_1$. But the function F is holomorphic in W^+ or W^- and therefore such terms also vanish.

The following lemma results immediately from our considerations.

Lemma 4.1. *The function* $h = h(z_1,t)$, $(z_1,t) \in Q$ *has the property*

$$\partial_t \bar{\partial}_s h(z_1,t)|_{z_1=0} = 0 \quad \text{for } (0,t) \in Q \text{ such that } (0,t,\psi(0,t)) \in \bar{M}_0,$$

with ∂_t, $\bar{\partial}_s$ *defined in (3.1).*

c) In this subsection we prove (4.2) for the points $(0,t) \in Q$ such that $(0,t,\psi(0,t)) \in \text{Int } M_1$. This case is more complicated than in Subsection b.

Without loss of generality we can assume that the Levi form vanishes on the whole of M, i.e. $\text{Int } M_1 = M$. Using the theorem of Wells [10, Th.2.2] and the theorem of Rea [8,Prop,4.3] any sufficiently small neighbourhood of $(0,t,\psi(0,t))$ in M can be foliated by complex analytic hypersurfaces. We shall describe this more precisely.

Our hypersurface is of a special form, namely it is given by a function ψ having the properties (2.6). In view of the above mentioned theorems, our foliation of a neighbourhood of a point $(0,t,\psi(0,t))$, $t = (z_2,\dots,z_{n-1},x_n)$, also has a special form, namely it is given by a smooth mapping

$$\Lambda_t(\zeta_1,\dots,\zeta_{n-1},\sigma_n) = (\zeta_1,\dots,\zeta_{n-1},\lambda_t(\zeta_1,\dots,\zeta_{n-1},\sigma_n)),$$

$$(\zeta_1,\dots,\zeta_{n-1},\sigma_n) \in \Omega_t,$$

where

$$\Omega_t = \{(\zeta_1,\dots,\zeta_{n-1},\sigma_n) \in \mathbb{C}^{n-1} \times \mathbb{R}\,;\, |\zeta_1| < \varepsilon_t, |\sigma_n - x_n| < \varepsilon_t,$$
$$|\zeta_\gamma - z_\gamma| < \varepsilon_t,\ \gamma = 2,\dots,n-1\},$$

ε_t is a sufficiently small positive number.
Moreover the mapping Λ_t has the properties

(4.5) $$\frac{\partial(\text{Re } \lambda_t)(\zeta,\sigma_n)}{\partial\sigma_n} \neq 0, \qquad (\zeta;\sigma_n) \in \Omega_t;$$

(4.6) λ_t is holomorphic with respect to $\zeta = (\zeta_1,\dots,\zeta_{n-1})$ with σ_n fixed;

(4.7) Λ_t maps Ω_t onto a neighbourhood of $(0,t,\psi(0,t))$ in M and

$$\Lambda_t(0,t) = (0,t,\psi(0,t)).$$

This is described in detail in the papers by Rea [8,§3,§4] and Wells [10,§2].

Since f is a CR function on M, it follows that the function $f(\Lambda_t(\zeta,\sigma_n))$ is holomorphic with respect to ζ with σ_n fixed.

Recall that

$$h(z_1 t) = f(z_1, t, \psi(z_1, t)), \quad (z_1, t) \in Q.$$

It will be convenient for us to extend the function h in a natural way to the whole of P

$$h^*(z) = h^*(z_1,\dots,z_n) = h(z_1,\dots,z_{n-1}, \operatorname{Re} z_n),$$

$$z = (z_1,\dots,z_n) \in P.$$

Owing to the properties of the function f we have

$$(4.8) \quad h^*(0,z_2,\dots,z_n) = 0 \quad \text{for} \quad (0,z_2,\dots,z_n) \in P.$$

We shall now prove the following

<u>Lemma 4.2. Let the function</u> $h^* = h^*(z)$ <u>be as above. Then</u>

$$\frac{\partial^{\alpha+\beta} h^*(z)}{\partial \bar z_1^\alpha \, \partial \bar z_n^\beta}\Big|_{z_1=0} = 0, \quad z = (z_1,\dots,z_n) \in P,$$

$$\alpha, \beta = 0, 1, 2, \dots \,.$$

<u>Proof</u>. We prove this lemma by the induction on α. For $\alpha = 0$ we have

$$\frac{\partial^{\beta} h^*(z)}{\partial \bar z_n^\beta}\Big|_{z_1=0} = 0 \quad \text{because of (4.8).}$$

Assume that we have shown

$$(4.9) \quad \frac{\partial^{\alpha+\beta} h^*(z)}{\partial \bar z_1^\alpha \, \partial \bar z_n{}^\beta}\Big|_{z_1=0} \qquad \text{for} \quad 0 \le \alpha \le m-1$$

$$\beta \ge 0, \ (m-1 \ge 0).$$

Let $\alpha = m \geq 1$. We must verify that

$$(4.10)\quad \frac{\partial^{m+\beta} h^*(z)}{\partial \bar{z}_1^m \, \partial \bar{z}_n^\beta}\Big|_{z_1=0} = 0 \quad \text{for} \quad \beta = 0,1,2,\ldots .$$

We prove the property (4.10) by the induction with respect to β. First we notice that $h^* \circ \Lambda_t = f \circ \Lambda_t$ because for $(\zeta, \sigma_n) = (\zeta_1, \ldots, \zeta_{n-1}, \sigma_n) \in \Omega_t$ we have

$$h^*(\Lambda_t(\zeta,\sigma_n)) = h^*(\zeta, \lambda_t(\zeta,\sigma_n)) = h(\zeta, \mathrm{Re}\ \lambda_t(\zeta,\sigma_n)) =$$

$$= f(\zeta, \lambda_t(\zeta,\sigma_n)) = f(\Lambda_t(\zeta,\sigma_n)).$$

Consequently the function $h^* \circ \Lambda_t$ is holomorphic with respect to ζ with σ_n fixed, which implies that

$$\frac{\partial^m}{\partial \bar{\zeta}_1^m} [h^*(\zeta, \lambda_t(\zeta,\sigma_n))] \equiv 0 \quad \text{for} \quad (\zeta,\sigma_n) \in \Omega_t,$$

and in view of (4.9) and (4.6) we get

$$\frac{\partial^m h^*(z)}{\partial \bar{z}_1^m} = 0 \quad \text{for} \quad z = (0, \zeta_2, \ldots, \zeta_{n-1}, \lambda_t(0, \zeta_2, \ldots, \zeta_{n-1}, \sigma_n)).$$

Because t can be taken arbitrarily so that $(0,t) \in Q$ and h^* is independent of $\mathrm{Im}\ z_n$, we have

$$\frac{\partial^m h^*(z)}{\partial \bar{z}_1^m}\Big|_{z_1=0} = 0 \quad \text{for} \quad (0, z_2, \ldots, z_n) \in P.$$

The above assertion is equivalent to (4.10) for $\beta = 0$.

Assume that we have shown

$$(4.11)\quad \frac{\partial^{m+\beta} h^*(z)}{\partial \bar{z}_1^m \, \partial \bar{z}_n^\beta}\Big|_{z_1=0} = 0 \quad \text{for} \quad 0 \leq \beta \leq l-1, \quad (l-1 \geq 0).$$

Put $\beta=1$. Because the function $\zeta \longrightarrow h(\zeta,\lambda_t(\zeta,\sigma_n))$ is holomorphic for ζ such that $(\zeta,\sigma_n)\in\Omega_t$, then

$$(4.12)\quad \frac{\partial^{m+l}}{\partial\bar{\zeta}_1^m\,\partial\sigma_n^l}[h^*(\zeta,\lambda_t(\zeta,\sigma_n))]\equiv 0 \quad\text{for}\quad (\zeta,\sigma_n)\in\Omega_t.$$

Making use of (4.9) we obtain

$$(4.13)\quad \{\frac{\partial^{m+l}}{\partial\bar{\zeta}_1^m\,\partial\sigma_n^l}[h^*(\zeta,\lambda_t(\zeta,\sigma_n))]\}_{\zeta_1=0} = \frac{\partial^l}{\partial\sigma_n^l}[(\frac{\partial^m h^*}{\partial\bar{z}_1^m})(\zeta',\lambda_t(\zeta',\sigma_n))],$$

where $\zeta'=(0,\zeta_2,\ldots,\zeta_{n-1})$.

Now by the induction hypothesis (4.11) we see that the expression (4.13) is equal to

$$\sum_{\gamma=0}^{l}\binom{l}{\gamma}(\frac{\partial^{m+l}h^*}{\partial\bar{z}_1^m\,\partial z_n^\gamma\,\partial\bar{z}_n^{l-\gamma}})(\zeta',\lambda_t(\zeta',\sigma_n))(\frac{\partial\lambda_t(\zeta',\sigma_n)}{\partial\sigma_n})^\gamma(\frac{\partial\bar{\lambda}_t(\zeta',\sigma_n)}{\partial\sigma_n})^{l-\gamma}.$$

We know that the function h^* does not depend on $\operatorname{Im} z_n$ and hence the following equalities hold

$$(4.14)\quad \frac{\partial^{m+l}h^*(z)}{\partial\bar{z}_1^m\,\partial z_n^\gamma\,\partial\bar{z}_n^{l-\gamma}} = \frac{\partial^{m+l}h^*(z)}{\partial\bar{z}_1^m\,\partial\bar{z}_n^l},\quad z\in P,$$

Further, the last sum takes the form

$$(\frac{\partial^{m+l}h^*}{\partial\bar{z}_1^m\,\partial\bar{z}_n^l})(\zeta',\lambda_t(\zeta',\sigma_n))\sum_{\gamma=0}^{l}\binom{l}{\gamma}(\frac{\partial\lambda_t(\zeta',\sigma_n)}{\partial\sigma_n})^\gamma(\frac{\partial\bar{\lambda}_t(\zeta',\sigma_n)}{\partial\sigma_n})^{l-\gamma} =$$

$$=(\frac{\partial^{m+l}h^*}{\partial\bar{z}_1^m\,\partial\bar{z}_n^l})(\zeta',\lambda_t(\zeta',\sigma_n))[\frac{\partial\lambda_t(\zeta',\sigma_n)}{\partial\sigma_n}+\frac{\partial\bar{\lambda}_t(\zeta',\sigma_n)}{\partial\sigma_n}]^l =$$

$$=2^l(\frac{\partial^{m+l}h^*}{\partial\bar{z}_1^m\,\partial\bar{z}_n^l})(\zeta',\lambda_t(\zeta',\sigma_n))[\frac{\partial(\operatorname{Re}\lambda_t(\zeta',\sigma_n))}{\partial\sigma_n}]^l.$$

Gathering together (4.12),(4.13) and taking into account the last calculations we have in view of the property (4.5) of the function λ_t

$$(4.15)\quad \left(\frac{\partial^{m+1} h^*}{\partial \bar{z}_1^m\, \partial \bar{z}_n^1}\right)(\zeta',\lambda_t(\zeta',\sigma_n)) = 0 \quad \text{for} \quad (\zeta',\sigma_n) \in \Omega_t$$

Because t is an arbitrary point of Q and the function h^* does not depend on $\operatorname{Im} z_n$, (4.15) gives

$$(4.16)\quad \frac{\partial^{m+1} h^*(z)}{\partial \bar{z}_1^m\, \partial \bar{z}_n^1}\Big|_{z_1=0} = 0.$$

This ends the induction step and hence proves (4.10). But the proof of (4.10) ends the induction step for the proof of our lemma.

As an immediate consequence of the above lemma we have the following

Corollary 4.3. The function $h = h(z_1,t)$, $(z_1,t) \in Q$, satisfies the condition

$$\partial_t \bar{\partial}_s h(z_1,t)\big|_{z_1=0} = 0 \quad \text{for} \quad (0,t) \in Q \quad \text{such that}$$
$$(0,t,\psi(0,t)) \in \operatorname{Int} M_1,$$

where ∂_t, $\bar{\partial}_s$ were defined in (3.1).

d) If follows from Lemma 4.1 and Corollary 4.3 that

Lemma 4.4. The function $h = h(z_1,t)$, $(z_1,t) \in Q$, satisfies the condition

$$\partial_t \bar{\partial}_s h(z_1,t)\big|_{z_1=0} \quad \text{for} \quad (0,t) \in Q.$$

Finally we are in a position to prove the following

Proposition 4.5. The function H (see the beginning of this section) satisfies the following conditions

$$(4.17)\quad \frac{\partial^{\alpha+\beta} H(z_1,t)}{\partial z_1^\alpha\, \partial \bar{z}_1^\beta}\Big|_{z_1=0} = 0 \quad \text{for} \quad 0 \le \alpha \le k \quad \text{and} \quad \beta \ge 0, \quad (z_1,t) \in P.$$

Proof. We compute the derivatives occurring in (4.17).

$$\frac{\partial^{\alpha+\beta}H(z_1,t)}{\partial z_1^{\alpha}\,\partial\bar z_1^{\beta}}\Big|_{z_1=0} = \{\frac{\partial^{\alpha}}{\partial z_1^{\alpha}}[\sum_{\gamma=0}^{k}\frac{(-1)^{\gamma}}{\gamma!}\frac{\partial^{\gamma+\beta}\partial_t\,\bar\partial_s h(z_1,t)}{\partial z_1^{\gamma}\,\partial\bar z_1^{\beta}}z_1^{\gamma}]\}_{z_1=0} =$$

$$= \sum_{\gamma=0}^{k}\frac{(-1)^{\gamma}}{\gamma!}[\sum_{\sigma=0}^{\min(\alpha,\gamma)}\binom{\alpha}{\sigma}\frac{\partial^{\alpha+\beta+\gamma-\sigma}\partial_t\,\bar\partial_s h(z_1,t)}{\partial z_1^{\alpha+\gamma-\sigma}\partial\bar z_1^{\beta}}\gamma\ldots(\gamma-\sigma+1)z_1^{\gamma-\sigma}]_{z_1=0}$$

All terms in the above sum vanish except for those in which $\gamma = \sigma$, consequently the above expression is equal to

$$[\sum_{\gamma=0}^{\alpha}\frac{(-1)^{\gamma}}{\gamma!}\binom{\alpha}{\gamma}\frac{\partial^{\alpha+\beta}\partial_t\,\bar\partial_s h(z_1,t)}{\partial z_1^{\alpha}\,\partial\bar z_1^{\beta}}\gamma!]_{z_1=0} =$$

$$= \frac{\partial^{\alpha+\beta}\partial_t\,\bar\partial_s h(z_1,t)}{\partial z_1^{\alpha}\,\partial\bar z_1^{\beta}}\Big|_{z_1=0}\sum_{\gamma=0}^{\alpha}(-1)^{\gamma}\binom{\alpha}{\gamma} =$$

$$=(1-1)^{\alpha}\frac{\partial^{\alpha+\beta}\partial_t\,\bar\partial_s h(z_1,t)}{\partial z_1^{\alpha}\,\partial\bar z_1^{\beta}}\Big|_{z_1=0} .$$

Thus we conclude that the partial derivatives in (4.17) vanish for $0 < \alpha \le k$.

For $\alpha = 0$ we have

$$\frac{\partial^{\beta}H(z_1,t)}{\partial\bar z_1^{\beta}}\Big|_{z_1=0} = \frac{\partial^{\beta}\partial_t\,\bar\partial_s h(z_1,t)}{\partial\bar z_1^{\beta}}\Big|_{z_1=0},$$

which vanishes because of Lemma 4.4.
The proposition is proved.

§5. A generalization of the main theorem to systems of CR functions

a) In this section we generalize the main theorem, so that the generalization be useful for the explicit definition of the normal bundle to a CR submanifold given by the zero set of CR functions.

In general the set of zeros of CR functions on a CR manifold need not define a CR submanifold even if the set is a smooth

submanifold. This is shown in Example 5.3.
However in the case where the CR manifold is a hypersurface, this fact is true. This is the contents of the following lemma and corollary.

Lemma 5.1. *Let* M *be a smooth hypersurface in* $\mathbb{C}^n$ *given by a smooth function* $\rho : U \longrightarrow \mathbb{R}$, $U \subset \mathbb{C}^n, n \geq 2$,

$$M = \{z = (z_1, \ldots, z_n) \in U : \rho(z) = 0\}$$

such that $d\rho(z) \neq 0$ *for* $z \in M$. *Let* $f_1, \ldots, f_k$ *be smooth* CR *functions on* M *(the functions can be extended smoothly on* U*) and put*

$$N = \{p \in M \,;\, f_1(p) = \ldots = f_k(p) = 0\}.$$

Assume that

$$(5.1) \quad \operatorname{rank} \frac{\partial(f_1, \ldots, f_k, \bar{f}_1, \ldots, \bar{f}_k, \rho)}{\partial(z_1, \ldots, z_n, \bar{z}_1, \ldots, \bar{z}_n)}(p) = 2k + 1, \quad \text{for } p \in N.$$

Then N *is a smooth* CR *submanifold of* M.

Remark. The condition (5.1) does not depend on the choice of the smooth extension to U of the functions $f_1, \ldots, f_k$.

Proof. From the assumption (5.1) it follows that N is a smooth submanifold of M. In order to show that N is also a CR submanifold it suffices to show that

$$(5.2) \quad \operatorname{rank} \frac{\partial(f_1, \ldots, f_k, \rho)}{\partial(z_1, \ldots, z_n)}(p) = k + 1 \quad \text{for } p \in N.$$

Fix a point $p \in N$. Subject to a $\mathbb{C}$-linear change of complex coordinates in $\mathbb{C}^n$ we can assume that

$$(5.3) \quad \begin{aligned} &H_p(M) = \{a_1 \frac{\partial}{\partial z_1} + \ldots + a_n \frac{\partial}{\partial z_n} \in H_p(\mathbb{C}^n) \,;\, a_n = 0\} \\ &\frac{\partial \rho(p)}{\partial z_n} \neq 0, \ \frac{\partial \rho(p)}{\partial z_\alpha} = \frac{\partial \rho(p)}{\partial \bar{z}_\alpha} = 0 \quad \text{for } \alpha = 1, \ldots, n-1. \end{aligned}$$

$f_1,\dots,f_k$ being CR functions on M, it follows from the above that

$$(5.4)\qquad \frac{\partial f_\beta(p)}{\partial \bar{z}_\alpha} = 0, \quad \alpha = 1,\dots,n-1, \quad \beta = 1,\dots,k.$$

According to (5.1) and (5.3) we have

$$\operatorname{rank} \frac{\partial(f_1,\dots,f_k,\bar{f}_1,\dots,\bar{f}_k)}{\partial(z_1,\dots,z_{n-1},\bar{z}_1,\dots,\bar{z}_{n-1})}(p) = 2k-1$$

and hence by using (5.4) we find

$$(5.5)\qquad \operatorname{rank} \frac{\partial(f_1,\dots,f_k)}{\partial(z_1,\dots,z_{n-1})}(p) = k.$$

Using again (5.3) we arrive by (5.5) to (5.2).
This proves the lemma.

The lemma immediately implies

Corollary 5.2. Let X be a complex manifold and M a smooth hypersurface of X, $\dim_{\mathbb{C}} X = n$, $n \geq 2$, $\dim_{\mathbb{R}} M = 2n-1$. Let N be a subset of M given locally by the zero set of k smooth CR functions satisfying a condition of the type (5.1). Then N is a smooth CR submanifold of M.

Example 5.3. Let M be given in $\mathbb{C}^3$ by the system

$$\begin{cases} \rho_1(z_1,z_2,z_3) = \dfrac{z_1+\bar{z}_1}{2} + \dfrac{z_2+\bar{z}_2}{2} = 0 \\[2ex] \rho_2(z_1,z_2,z_3) = \dfrac{z_1+\bar{z}_1}{2} - \dfrac{z_2-\bar{z}_2}{2i} = 0 \end{cases}$$

It is easy to check that M is a smooth submanifold, $\dim_{\mathbb{R}} M = 4$ and $\dim_{CR} M = 1$. Take the holomorphic function $f(z_1,z_2,z_3) = z_1 + z_3^2$. The function $f|_M$ is a CR function. Let

$$N = \{z = (z_1,z_2,z_3) \in \mathbb{C}^3 \; ; \; \rho_1(z) = \rho_2(z) = f(z) = 0\}.$$

After simple calculations we get

$$\operatorname{rank} \frac{\partial(\rho_1,\rho_2,f,\bar{f})}{\partial(z_1,z_2,z_3,\bar{z}_1,\bar{z}_2,\bar{z}_3)}(0) = 4, \text{ where } 0 = (0,0,0),$$

$$\operatorname{rank} \frac{\partial(\rho_1,\rho_2, f)}{\partial(z_1,z_2,z_3)}(0) = 2, \quad \operatorname{rank} \frac{\partial(\rho_1,\rho_2, f)}{\partial(z_1,z_2,z_3)}(p) = 3$$

for $p \neq 0$ close to zero.

It follows that N is a smooth submanifold of M in a neighbourhood of zero but is not a CR submanifold of M.

b) Let M be a smooth hypersurface of $\mathbb{C}^n$, $\dim_{\mathbb{R}} M = 2n-1$, $n \geq 2$. Let CR functions $f_1,\dots,f_k$, $g_1,\dots,g_k$ be given on M such that

$$\{z \in M ; f_1(z) = \dots = f_k(z) = 0\} \supset \{z \in M ; g_1(z) = \dots = g_k(z) = 0\}.$$

Denote the smaller set by N. Furthermore assume that the system of functions $g_1,\dots,g_k$ satisfies a condition of the type (5.1). Put

$$N^\beta = \{q \in M ; g_\alpha(q) = 0,\ \alpha = 1,\dots,k, \alpha \neq \beta\}, \quad \beta = 1,\dots,k.$$

<u>Proposition 5.4</u>. <u>Under the above notation</u>, N^β <u>is a smooth CR submanifold of</u> M <u>and the function</u> f_α/g_β <u>can be smoothly extended to</u> N^β.

<u>Proof</u>. The first part of the proposition, i.e. the statement that N^β is a smooth CR submanifold of M follows by Corollary 5.2.

In order to show the second part, fix a point $p \in N^\beta$. Taking a sufficiently small neighbourhood V of p in $\mathbb{C}^n$ we can choose such holomorphic functions $h_1,\dots,h_{n-k}$ in V that the mapping

$$\Phi^\beta = (h_1,\dots,h_{n-k}, g_\beta, g_1,\dots,g_{\beta-1}, g_{\beta+1},\dots,g_k)$$

is a CR isomorphism of $M \cap V$ onto a hypersurface in $\mathbb{C}$. Notice that

$$\Phi^\beta|_{N^\beta\cap V} : N^\beta \cap V \longrightarrow \mathbb{C}^{n-k+1} \times \{0\},$$

and that $\Phi^\beta(N^\beta \cap V)$ is a smooth hypersurface in $\mathbb{C}^{n-k+1}$, where we identify $\mathbb{C}^{n-k+1} \times \{0\}$ with $\mathbb{C}^{n-k+1}$.

The function $g_\beta|_{N^\beta}$ is a smooth CR function on N^β whose CR differential is not singular at each point of N^β and moreover

$$\{q \in N^\beta ; f_\alpha(q) = 0\} \supset \{q \in N^\beta ; g_\beta(q) = 0\}.$$

Hence using the main theorem we conclude that the function

$$f_\alpha \circ (\Phi^\beta)^{-1} / g_\beta \circ (\Phi^\beta)^{-1} \quad \text{on} \quad \Phi^\beta(N^\beta \cap V - N)$$

extends to a smooth function on $\Phi^\beta(N^\beta \cap V)$, and consequently the function f_α/g_β extends to a smooth function on $N^\beta \cap V$. p being an arbitrary point of N^β, this implies the extendability of the function f_α/g_β to the whole of N^β. This proves the proposition.

§6. Application of the results to an explicit definition of the normal bundle to CR submanifolds

a) The normal bundle to a differential submanifold Y in a differential manifold Z can be defined as follows.

Choose a locally finite covering $\{U_i\}_{i\in I}$ of a neighbourhood of Y in Z by coordinate neighbourhoods with coordinates $x_i = (x_i^1, \ldots x_i^m)$ such that

$$Y \cap U_i = \{q \in U_i ; x_i^1(q) = \ldots = x_i^k(q) = 0\} , \quad k = \operatorname{codim}_Z Y.$$

If $f_{ij} = (f_{ij}^1, \ldots, f_{ij}^m)$, $i,j \in I$, are the transition functions between the coordinates x_i and x_j, then the normal bundle to Y is defined by the 1-cocycle

$$g_{ij}(x_j) = (g_{ij\beta}^\alpha(x_j))_{\alpha,\beta=1,\ldots,k},$$

$$g^{\alpha}_{ij\beta}(x_j) = \frac{\partial f^{\alpha}_{ij}(x_j)}{\partial x^{\beta}_j}, \text{ where } x_j = x_j(p),\ p \in Y \cap U_i \cap U_j.$$

Notice that if $p \in Y \cap U_i \cap U_j$ its coordinates have the form $(0,..,0, x^{k+1}_j(p),\dots,x^m_j(p))$.

The partial derivative $\frac{\partial f^{\alpha}_{ij}(x_j)}{\partial x^{\beta}_j}$ can be regarded as the limit

$$\lim_{q\to p} \frac{x^{\alpha}_i(q)}{x^{\beta}_j(q)}, \text{ where } p \in Y \cap U_i \cap U_j,$$

with q approaching p and belonging to the set

$$Y^{\beta}_{ij} = \{q \in U_i \cap U_j\ ;\ x^{\gamma}_j(q) = 0,\ \gamma = 1,\dots,k,\ \gamma \neq \beta\}.$$

We apply the same idea in the Cauchy-Riemann case.

b) Let X be a complex manifold of dimension n and let M be a smooth hypersurface of X, $\dim_{\mathbb{R}} M = 2n-1$, $n \geq 2$. Let N be a smooth CR submanifold of M which is locally given by the zero set of k smooth CR functions which satisfy a condition of the type (5.1).

Choose a sufficiently small covering of M, say $\{U_i\}_{i\in I}$ and CR functions $\varphi^1_j,\dots,\varphi^k_j$ on U_j satisfying a condition of the type (5.1) and such that

$$N \cap U_j = \{q \in U_j \subset M\ ;\ \varphi^1_j(q) = \dots = \varphi^k_j(q) = 0\}.$$

Put

$$N^{\beta}_j = \{q \in U_j\ ;\ \varphi^{\gamma}_j(q) = 0,\ \gamma = 1,\dots,k\ ,\ \gamma \neq \beta\}$$

In virtue of Proposition 5.4., the function $\varphi^{\alpha}_i/_{\varphi^{\beta}_j}$ which is well-defined on $U_i \cap (N^{\beta}_j - N)$ can be extended to a smooth function $\varphi^{\alpha}_{ij\beta}$ on $N^{\beta}_j \cap U_i$. Denote

$$\psi^{\alpha}_{ij\beta} = \varphi^{\alpha}_{ij\beta}\big|_{N^{\beta}_j \cap N \cap U_i}\ ,\quad \alpha,\beta = 1,\dots,k,$$

$$\psi_{ij} = (\psi^{\alpha}_{ij\beta})_{\alpha,\beta=1,\dots,k}.$$

It is easy to check that $\{\psi_{ij}\}_{i,j\in I}$ is a multiplicative 1-cocycle on N. The normal bundle to N can be defined by this cocycle,

Remark. The above definition of the normal bundle can be generalized to the following case:

X is a complex manifold of complex dimension n;
M is a smooth CR submanifold of X, $\dim_{\mathbb{R}} M = 2m-1$, $\dim_{CR} M = m-1$, $m \geq 2$;
N is a smooth CR submanifold of M given locally as the zero set of k CR functions on M satisfying a condition of the type (5.1).

In this case the normal bundle to N in M can be defined in exactly the same way as that described for hypersurfaces. This results from the fact (see[1,Prop.1.10]) that M is locally CR isomorphic to a hypersurface in $\mathbb{C}^m$.

References

[1] A.Andreotti, G.A.Fredricks, Embeddability of real analytic Cauchy--Riemann manifolds. Ann.Scuola Norm. Sup. Pisa, 6 (1979), 285-304.

[2] R.Dwilewicz, On the Hans Lewy theorem. To be submitted for publication in an italian journal.

[3] R.Dwilewicz, Embeddability of smooth Cauchy-Riemann manifolds. Preprint no.246, Inst. of Math. Polish Academy of Sciences, (1981), 1-86.

[4] L.Hörmander, An introduction to complex analysis in several variables. D. Van Nostrand Company, Inc., Princeton, New Jersey, 1966.

[5] L.R.Hunt, R.O,Wells, Jr., Holomorphic extension for nongeneric CR-submanifolds. In : Proceedings of Symposia in Pure Mathematics (Stanford 1973), vol.27, part II, 81-88. Amer. Math. Soc.,Providence, Rhode Island, 1975.

[6] H.Lewy, On the local character of the solutions of an atypical linear differential equation in three variables and a related theorem for regular functions of two complex variables. Ann. of Math., 64 (1956), 514-522.

[7] A.Newlander, L.Nirenberg, Complex analytic coordinates in almost complex manifolds. Ann. of Math., 65 (1957), 391-404.

[8] C.Rea, Levi-flat submanifolds and holomorphic extension of foliations. Ann. Scuola Norm. Sup. Pisa, 26 (1972), 665-681.

[9] F.Sommer, Komplex-analytische Blätterung reeler Mannigfaltigkeiten im $\mathbb{C}^n$. Math. Ann. 136 (1958), 111-133.

[10] R.O.Wells, Jr., Holomorphic hulls and holomorphic convexity of differentiable submanifolds. Trans. Amer. Math. Soc. 132 (1968), 245-262.

[11] R.O.Wells, Jr., Function theory on differentiable submanifolds. Contribution to Analysis, A collection of papers dedicated to Lipman Bers, Academic Press, New York, 1974, 407-441.

Institute of Mathematics
University of Warsaw
Pałac Kultury i Nauki, IXp.
Pl - 00 - 901 Warszawa, Poland

ON A BOUNDARY VALUE PROBLEM IN PSEUDOCONVEX DOMAINS

Adib A. Fadlalla (Cairo)

Contents page

Introduction

In [4] the Carathéodory limiting spherical shells in the Euclidean hyperball $K \subset \mathbb{C}^n$ are defined and studied. Their definition and existence in arbitrary domains in $\mathbb{C}^n$ is considered in [6], [7] and [8]; they are defined as the boundaries ∂B of the Carathéodory limiting balls B. In [4] it is proved that if $P \in \partial K$ and S_P is any Carathéodory limiting spherical shell having P as a point of contact, then $S_P \cap \partial K = \{P\}$.

In this article we are going to prove a similar result for strongly pseudoconvex domains, i.e., if $G \subset \mathbb{C}^n$ is a strongly pseudoconvex domain, $P \in \partial G$ and B is any Carathéodory limiting ball having P as a point of contact, then $\partial B \cap \partial G = \{P\}$. Such a result is incorrect if G is only pseudoconvex (cf. [10] and Theorems 5, 6 below). The proof is a consequence of the previous estimates of the Carathéodory distance obtained already by the author [6], together with the following theorem:

THEOREM 1. *Let $G \subset \mathbb{C}^n$ be a domain, $P, Q \in \partial G$, $P \neq Q$, be such that there exist peak functions on G at P and Q, respectively, then*

<u>there exists a function</u> f <u>holomorphic in a neighbourhood of</u> $\overline{G}$ (<u>the closure of</u> G) <u>such that</u>

(0.1) $|f(P)| = |f(Q)| = 1$, $f(Q) \neq f(P)$ and $|f(T)| < 1$,

<u>for all</u> $T \in \overline{G} - \{P,Q\}$.

A function f is called a <u>peak function</u> on G at $P \in \partial G$ if it is holomorphic in $\overline{G}$ and $|f(P)| > |f(T)|$ for all $T \in \overline{G} - \{P\}$. The existence of a peak function on a strongly pseudoconvex domain G at every boundary point $P \in \partial G$, was already proved in [5]. Thus Theorem 1 holds for any two boundary points $P,Q \in \partial G$ if G is a strongly pseudoconvex domain. If G is only pseudoconvex (not strongly), peak functions do not exist on G (cf. [9], Theorem 5). However, analogous results are also obtained here for pseudoconvex domains.

Theorem 1 will be used to get some estimates of the Carathéodory distance near the boundary of pseudoconvex domain.

1. Proof of the basic theorem (Theorem 1)

First we define the notations used in the proof. Hereafter $\mathfrak{z} = (z_1,\ldots,z_n) \in \mathbb{C}^n$ denotes a point, z is a point in $\mathbb{C}$ and $D_G(A,B)$ denotes the Carathéodory distance between two points $A,B \in G$, $G \subset \mathbb{C}^n$ is a domain [3].

Let x_1 and x_2 be two real numbers such that $x_1 < x_2$. Then x_1 and x_2 divide the real axis in $\overline{\mathbb{C}}$ into two intervals; in what follows $[x_1,x_2]$ and $[x_2,x_1]$ denote the finite and infinite closed intervals of the real axis, respectively.

Now let $h(\mathfrak{z})$ be a peak function on G at P, such that $h(P) = 1$ and $|h(T)| < 1$, $\forall T \in \overline{G} - \{P\}$; therefore $\zeta = h(\mathfrak{z})$ is a mapping of G into the unit disc Δ: $|\zeta| < 1$. Furthermore, let $g(\zeta)$ be a Riemann mapping of Δ onto the bounded domain D enclosed by the parabola $y = -kx^2$, $k > 0$ and the line $y = -\lambda$, $\lambda > 1$, such that $g(1) = 0$ and $g(h(Q)) = -i$ (k is a constant suitably chosen; in what follows we shall see that it is convenient to choose $k = 8$).

Now $z = h_1(\mathfrak{z}) = 1/g \circ h(\mathfrak{z})$ is a meromorphic function in a neighbourhood of $\overline{G}$, which maps G into the upper half plane $\operatorname{Im} z > 0$, such that $h_1(Q) = i$ and $h_1(P) = \infty$. Moreover, if $g \circ h(\mathfrak{z}) = u(\mathfrak{z}) - iv(\mathfrak{z})$ where $v(\mathfrak{z}) > 0$, then

$$h_1 = \frac{u + iv}{u^2 + v^2}, \quad \text{with} \quad \frac{v}{k} > u^2 \quad \text{in} \quad G,$$

which shows that if U_1 is a sufficiently small neighbourhood of P, then $I_1(\mathfrak{z}) = \operatorname{Im} h_1(\mathfrak{z}) > \frac{1}{2}k \geq 4$ (e.g. if $k \geq 8$) for all $\mathfrak{z} \in U_1 \cap G$.

Doing the same thing with Q, we get a function $h_2(\mathfrak{z})$, $\mathfrak{z} \in G$, which has values in the lower half plane $\operatorname{Im} z < 0$ and is such that $h_2(P) = -i$, and $I_2(\mathfrak{z}) = \operatorname{Im} h_2(\mathfrak{z}) < -4$ in $U_2 \cap G$, where U_2 is a sufficiently small neighbourhood of Q.

We can choose U_1 and U_2 sufficiently small such that $I_2(\mathfrak{z}) > -2$ in $U_1 \cap G$ and $I_1(\mathfrak{z}) < 2$ in $U_2 \cap G$. We consider the function $H(\mathfrak{z}) = h_1(\mathfrak{z}) + h_2(\mathfrak{z})$. It is meromorphic in a neighbourhood of $\overline{G}$ and there exists $x_1 > 0$ such that $H(\mathfrak{z})$ maps G into C' (the closed complex plane $\overline{C}$ cut along $[x_1, -x_1]$; this is because if $I(\mathfrak{z}) = \operatorname{Im} H(\mathfrak{z}) = I_1(\mathfrak{z}) + I_2(\mathfrak{z})$, then $I(\mathfrak{z}) > 2$ in $U_1 \cap G$, $I(\mathfrak{z}) < -2$ in $U_2 \cap G$ and $H(\mathfrak{z})$ is bounded on $\overline{G} - U_1 \cup U_2$. We can assume that $H(P) = \infty_u$ and $H(Q) = \infty_l$, where ∞_u and ∞_l are the points ∞ on the upper and lower edges of the cut $[x_1, -x_1]$ in C', respectively. To extend $H(\mathfrak{z})$ to neighbourhoods of P and Q, we match the lower edge of the cut of C' along $[x_2, -x_2]$, $x_2 > x_1$ with the upper edge of a similar cut in C'' and the upper edge of the same cut of C' to the lower edge of a similar cut of C''', respectively, where C'', C''' are copies of C'. The so constructed Riemann surface will be denoted by R. Thus $H(\mathfrak{z})$ is a meromorphic mapping of a neighbourhood of $\overline{G}$ into R.

We now consider the elliptic integral [1], [2]

$$w = \gamma(z) = \int_{x_2}^{z} \frac{dz}{\sqrt{(z^2 - x_1^2)(z^2 - x_2^2)}},$$

which maps the upper half plane of C' one to one onto a rectangular domain $OABC$, where 0, $a > 0$, $a + ib$ and ib $(b > 0)$ are the co-ordinates of O, A, B and C respectively, $\gamma(x_2) = 0$, $\gamma(-x_2) = A$, $\gamma(-x_1) = B$, $\gamma(x_1) = C$ and $\gamma(\infty_u) = E$, a point on the side OA different from O and A. Reflection in the interior of $[-x_1, x_1]$ extends γ to C' and $\gamma(C')$ will be the closed rectangular domain $OAA'D'$, where O', A' are the images of O, A in BC and $\gamma(\infty_l) = E'$, the image of E in BC. Further reflections in the upper and lower edges of the cut in C' across the interior of $[x_2,-x_2]$ extend γ to R.

Finally, let Θ be a Riemann mapping of the rectangular domain $OAA'O'$ onto the unit disc $|z|<1$; Θ has analytic continuations across the open sides of the rectangle $OAA'O'$. The function $f(\mathfrak{z}) = \Theta \circ \gamma \circ H(\mathfrak{z})$ posseses the required properties.

2. Applications

THEOREM 2. Let $G \subset \mathbb{C}^n$ be a strongly pseudoconvex domain, $P,Q \in \partial G$, $P \neq Q$. Furthermore, let $\{P_m\} \subset G$ be a conical sequence of points converging to P, $\{Q_m\} \subset G$ a sequence of points converging to Q and $A \in G$ a fixed point. We have

$$\lim_{m \to \infty} [D_G(P_m,Q_m) - D_G(A,P_m)] = +\infty$$

(for the Definition of a conical sequence of points see definition 1 in § 3 of this article).

Proof. Let f be as in Theorem 1. In [6] it is proved that

$$D_G(A,P_m) - D_E(f(A), f(P_m)) < k, \qquad 0 < k < +\infty$$

and

$$\lim_{m \to \infty} [D_E(f(P_m), f(Q_m)) - D_E(f(A), f(P_m))] = +\infty ,$$

where E is the unit disc. Since $D_G(P_m,Q_m) \geqslant D_E(f(P_m), f(Q_m))$, we get the result.

This result is not correct if G is pseudoconvex [9] (cf. Theorem 6 of this article).

THEOREM 3. Let G and P be as in Theorem 2. Then

$$\partial B \cap \partial G = \{P\},$$

where B is an arbitrary Carathéodory limiting ball (of any type) in G having P as a point of contact.

(In [6] and [7] different types of Carathéodory limiting balls were defined.)

Proof. The set of points $Q \in \partial G$ - for which a function f satisfying (0.1) - was denoted in [5], [6] and [7] by M_P and it is proved that $M_P \cap \partial B = \emptyset$. According to Theorem 1, $M_P = G - \{P\}$.

A similar result does not exist if G is pseudoconvex (cf. [10] and Theorem 6 of this article).

3. Some estimates of the Carathéodory distance in strongly pseudoconvex domains

Definition 1. Let $G \subset \mathbb{C}^n$ be a domain, $P \in \partial G$ and let ∂G in a neighbourhood of P be a C^2 - smooth hypersurface. Let $\{P_m\} \subset G$ be a sequence of points converging to P; $\{P_m\}$ is called a conical sequence of points converging to P, if there exists a cone K, with vertex P, axis: the 1-dimensional real interior normal to ∂G at P and whose semi-vertical angle $\alpha < \frac{1}{2}\pi$, such that $P_m \in K$ for all $m \geq m_o$ where m_o is a positive integer.

Definition 2. In Definition 1 let P be the origin; it is always possible to find an analytic rotation about P such that the axis of K will be the line $z_\mu = 0$ for $\mu = 2,\dots,n$, $x_1 = 0$, $y_1 > 0$. In this case ∂G in a neighbourhood U of P will be the hypersurface

(1) $$\psi(\mathfrak{z},\bar{\mathfrak{z}}) = i\,a\,(z_1 - \bar{z}_1) + \text{terms of degree} \geq 2, \quad a > 0,\dots$$

and

$$G \cap U = \{\mathfrak{z} : \mathfrak{z} \in U, \quad \psi(\mathfrak{z}) < 0\}.$$

The coordinate system $(z_1,\dots,z_n)$ in (1) defined in this manner is called a normal coordinate system of ∂G at P.

Now let $\{P_m\}$ be a conical sequence of points converging to $P \in \partial G$, $(z_1,\dots,z_n)$ be a normal coordinate system of ∂G at P and let $P_m = \mathfrak{z}_m = \{z_{1m},\dots,z_{nm}\}$; then there exists k such that

$$o < k \leq \frac{|z_{1m}|}{\|\mathfrak{z}_m\|} \leq 1 \quad \text{for all } m \geq m_o,$$

where $\|\mathfrak{z}_m\|$ is the euclidean length of the vector $\overrightarrow{PP_m}$.

If furthermore G is strongly pseudoconvex and $A \in G$, then Theorem 4 in [6] gives

(2) $$D_G(A,P_m) = k_m + \frac{1}{2}\log\frac{1}{r_m},$$

where $\{k_m\}$ is a bounded sequence of real constants and $r_m = \|\mathfrak{z}_m\|$.

THEOREM 4. Let G be a strongly pseudoconvex domain $P, Q \in \partial G$, and $\{P_m\} \subset G$ and $\{Q_m\} \subset G$ be two conical sequences of points converging to P and Q, respectively, then

$$D_G(P_m, Q_m) = k_m + \frac{1}{2} \log \frac{1}{r_m r_m'} , \tag{3}$$

where r_m and r_m' are the euclidean lengths of the vectors $\overrightarrow{PP_m}$ and QQ_m, respectively, and $\{k_m\}$ is a bounded sequence of real numbers.

P r o o f. (i) The proof of (3) in the case when G is the unit disc $E \subset \mathbb{C}$, P and Q are the points $z = 1$, $z = -1$ respectively, can be obtained by direct calculations.

(ii) Let $A \in G$ be a fixed point, we have by (2)

$$D_G(A, P_m) = \alpha_m + \frac{1}{2} \log \frac{1}{r_m}$$

$$D_G(A, Q_m) = \beta_m + \frac{1}{2} \log \frac{1}{r_m'} ,$$

where $\{\alpha_m\}$ and $\{\beta_m\}$ are bounded sequences of real numbers. By the triangle inequality we get

$$D_G(P_m, Q_m) \leqslant \alpha_m + \beta_m + \frac{1}{2} \log \frac{1}{r_m r_m'} . \tag{4}$$

(iii) From Theorem 1, let f be a function holomorphic in a neighbourhood of $\bar{G}$ such that $f(P) = 1$, $f(Q) = -1$ and $|f(\mathfrak{z})| < 1$ for all $\mathfrak{z} \in \bar{G} - \{P, Q\}$. Furthermore, let $f(P_m) = \xi_m$, $f(Q_m) = \eta_m$, $\gamma_m = z_1(P_m)$ with respect to a normal coordinate system $(z_1, \ldots, z_n)$ of ∂G at P and $\gamma_m' = w_1(Q_m)$ with respect to a normal coordinate system $(w_1, \ldots, w_n)$ of ∂G at Q. Finally, let $\xi_m' = 1 + i\gamma_m$, $\eta_m' = -1 - i\gamma_m'$. From the proof of Theorem 4 in [6] we have:

(a) $\xi_m', \eta_m' \in E$ for all $m \geqslant m_o$, where m_o is a sufficiently large positive integer;

(b) $\{\xi_m'\}$ and $\{\eta_m'\}$, where $m \geqslant m_o$, are conical sequences of points in E converging to 1 and -1, respectively;

(c) $D_E(\xi_m, \xi'_m)$ and $D_E(\eta_m, \eta'_m)$ are bounded.

Hence (i) gives

$$(5) \qquad D_E(f(P_m), f(Q_m)) = D_E(\xi_m, \eta_m) = k'_m + \frac{1}{2}\log\frac{1}{|\gamma_m|\ |\gamma'_m|}$$

$$= k''_m + \frac{1}{2}\log\frac{1}{r_m r'_m} ,$$

where $\{k'_m\}$, $\{k''_m\}$ are bounded sequences of real numbers. Since $D_G(P_m, Q_m) \geq D_E(f(P_m), f(Q_m))$, (3) follows from (4) and (5).

4. Estimates of the Carathéodory distance in pseudoconvex domains

Let $G \subset\subset \mathbb{C}^n$ be a pseudoconvex domain. In the neighbourhood of each boundary point (which is not strongly pseudoconvex), there exist analytic sets $A \subset \partial G$ [10]. Furthermore, if f is an analytic function in a neighbourhood of $\bar{G}$ and if $|f(P)| = \text{Max}\,|f(G)|$ for some ordinary point $P \in A$ then $f(z)$ will be constant and equal $f(P)$ for all $z \in A$ [9].

Now, we establish theorems for pseudoconvex domains analogous to Theorems 1-4 already proved for strongly pseudoconvex domains; we need the following definition:

Definition 3. Let $G \subset \mathbb{C}^n$ be a domain and $A \subset \partial G$ be an analytic set. A peak function on G at A is a function f analytic in a neighbourhood of $\bar{G}$ such that $f(z)$ is constant on A and equal 1 (say) and $|f(z)| < 1$ for all $z \in \bar{G} - A$.

THEOREM 5. *Let $G \subset\subset \mathbb{C}^n$ be a domain, $A, A' \subset \partial G$ be analytic sets, $A \cap A' = \emptyset$ and there exist peak functions on G at A and A', respectively, then there exists a function f analytic in a neighbourhood of $\bar{G}$ such that $f(z) = \exp i\theta_1$ (a constant) for all $z \in A$, $f(z) = \exp i\theta_2$ (a constant) for all $z \in A'$, $\theta_1 \neq \theta_2$, and $|f(z)| < 1$ for all $z \in \bar{G} - A \cup A'$.*

The proof is exactly the same as that of Theorem 1; we have only to replace P and Q by A and A', respectively, which is possible since peak functions at A (or A') are constants on A (or A') and are uniformly continuous on $\bar{G}$. The condition $A \cap A' = \emptyset$ is necessary to assure the existence of disjoint neighbourhoods of A and A'.

Now, if in Theorem 5 P, Q are ordinary points of A, then to every sequence of points $\{P_m\} \subset G$ converging to P, there exist sequences of points $\{Q_m\} \subset G$ converging to Q, such that $D_G(P_m,Q_m)$ is bounded [9]. Furthermore, if $B \subset G$ is any Carathéodory limiting ball of any type, having P as a point of contact, then $A \subset \partial B$ [10].

Now we have the following Theorem:

THEOREM 6. *In Theorem 5 we have:*

(i) *If* $\{P_m\} \subset G$ *is a conical sequence of points converging to* $P \in A$ *and* $\{Q_m\} \subset G$ *is any sequence of points converging to* $Q \in A'$ *and* $L \in G$ *is a fixed point*

$$\lim_{m \to \infty} [D_G(P_m,Q_m) - D_G(L,P_m)] = +\infty.$$

(ii) *If* $P \in A$ *and* $B \subset G$ *is any Carathéodory limiting ball of any type having* P *as a point of contact, then* $A' \cap \partial B = \emptyset$ *and* $A \subset \partial B$.

(iii) *Let* $\{P_m\}, \{Q_m\} \subset G$ *be two conical sequences of points converging to* $P \in A$ *and* $Q \in A'$*, respectively, then*

$$D_G(P_m,Q_m) = k_m + \frac{1}{2} \log \frac{1}{r_m r'_m} ,$$

where $\{k_m\}$ *is a bounded sequence of real numbers and* r_m *and* r'_m *are the euclidean lengths of the vectors* $\overrightarrow{PP_m}$ *and* $\overrightarrow{QQ_m}$*, respectively.*

The proofs of these statements are exactly the same as that of Theorems 2-4.

Now, we give the following illustrative example. Let $G' \subset \mathbb{C}^n$ be a domain and let $f_1,\dots,f_k$ be analytic functions on G'. If the set $G'' = \{\mathfrak{z} \in G', |f_j(\mathfrak{z})| < 1, j = 1,\dots,k\}$ is relatively compact in G', then G'' is called an *analytic polyhedron*. Let G be an open connected component of G''.

Let P be a smooth boundary point of G, then we can assume that ∂G in a neighbourhood of P is (say) the hypersurface $|f_1(\mathfrak{z})| = 1$. If $f_1(P) = \exp i\theta_0$ and

$$A'_{\theta_0} = \{\mathfrak{z} \in \partial G : f_1(\mathfrak{z}) = \exp i\theta_0\},$$

then A'_{θ_0} is an analytic set on ∂G. It is obvious that in the same way other analytic sets $A^j_\theta \subset \partial G$ can be found for $j = 1,2,\dots,k$

and for suitable values of Θ. However, the function $f(z) = \frac{1}{2}[f_j(z) + \exp i\Theta]$ is a peak function on G at A^j_Θ. Theorems 5 and 6 are applicable to G for $A^j_{\Theta_1}$ and $A^j_{\Theta_2}$, $\Theta_1 \neq \Theta_2$, i.e. for different analytic sets corresponding to the same function f_j, because in this case we are sure that $A^j_{\Theta_1} \cap A^j_{\Theta_2} = \emptyset$ and for analytic sets A^j_α, A^m_β belonging to different functions f_j, f_m only if we can prove that $A^j_\alpha \cap A^m_\beta = \emptyset$.

References

[1] AHLFORS, L.: Complex analysis, New York 1966.

[2] BIEBERBACH, L.: Conformal mapping, New York 1964.

[3] CARATHÉODORY, C.: Ueber das Schwarzsche Lemma bei zwei Komplexen Veränderlichen, Math. Ann. 97 (1927), 76-98.

[4] FADLALLA, A.A.: The Carathéodory limiting spherical shells in the Euclidean hypersphere, Mathematika 13 (1966), 69-75.

[5] ——: Global characterization of strongly pseudoconvex domains, Proc. Math. Phys. Soc. (U.A.R.) 31 (1969), 89-92.

[6] ——: Le principe du module maximum et la sphère limite de Carathéodory dans un domaine strictement pseudoconvexe I, Bull. Sci. Math. (2) 97 (1973), 193-205.

[7] ——: La boule limite de Carathéodory II, Bull. Sci. Math. (2) 97 (1973), 207-215.

[8] ——: La boule limite de Carathéodory III, Bull. Sci. Math. (2) 97 (1973), 217-224.

[9] ——: The Carathéodory metric near boundaries of pseudoconvex domains, Bull. U.M.I. (4) 8 (1973), 412-418.

[10] ——: Les boules limites de Carathéodory dans un domaine pseudoconvexe, Ann. Fac. Sci. Toulouse 1 (1979), 26-32.

Cairo University
Faculty of Science
Giza, Egypt

CARLEMAN APPROXIMATION ON UNBOUNDED SETS BY HARMONIC FUNCTIONS WITH NEWTONIAN SINGULARITIES

Paul M. Gauthier (Montréal)

In 1927 Carleman showed that for each function u continuous on the real axis and each positive continuous function ε, there is an entire function v such that

$$|u(x) - v(x)| < \varepsilon(x), \quad x \in \mathbb{R}.$$

Let D be a domain and F a subset of D, closed in the topology of D. Let $\mathbb{F}$ be a class of functions defined on D. We shall say that F is a set of Carleman approximation by function in the class $\mathbb{F}$ if for each pair of functions u and ε continuous on F, with ε positive, there exists a function $v \in \mathbb{F}$ satisfying

$$(1) \qquad |u(x) - v(x)| < \varepsilon(x), \quad x \in F.$$

The lecture of André Boivin at this conference treated the problem of Carleman approximation by meromorphic functions.

In this paper, we consider the analogous problem for harmonic approximation. The harmonic analogue of meromorphic functions which we will consider are the Newtonian functions, which we now define.

A function v is said to have a Newtonian singularity at a point $y \in \mathbb{R}^n$ if in some neighbourhood of y, v can be written the form

$$v(x) = \lambda K(x-y) + h(x),$$

where h is harmonic, λ is a real constant, and K is the Newtonian kernel

$$K(t) = \begin{cases} -\log|t|, & n = 2, \\ |t|^{2-n}, & n > 2. \end{cases}$$

A function v is said to be Newtonian on an open set D if it is harmonic on D except possibly for Newtonian singularities.

A set K is called a set of uniform approximation if for each function u continuous on K and harmonic on K^{o} and each positive constant ε, there is a function v, harmonic on (a neighbourhood of) K, such that

$$|u(x) - v(x)| < \varepsilon, \quad x \in K.$$

A set F is said to be locally a set of approximation if, for each $x \in F$, there is a neighbourhood $N(x)$ such that $\overline{N} \cap F$ is a set of uniform approximation.

THEOREM. *Let F be a relatively closed subset of a domain D in $\mathbb{R}^n$, $n \geq 2$. Then F is a set of Carleman approximation by functions Newtonian on D if and only if $F^{o} = \emptyset$ and F is locally a set of approximation.*

COROLLARY. *Let F and D be as in the Theorem. Then F is a set of Carleman approximation by functions Newtonian on D if and only if $\mathbb{R}^n \setminus F$ is thin at no point of F.*

Remark. Labrèche [3] has shown that F is a set of Carleman approximation by functions harmonic on F if and only if F satisfies the same conditions as in the above theorem. Thus, the necessity in our theorem is a consequence of the necessity in the Labrèche theorem, while the sufficiency in her result follows from the sufficiency in ours.

Proof of sufficiency. It will be helpful to state two lemmas. The first is the important theorem of Keldyš and Deny [1].

LEMMA A. *A compact set $K \subset \mathbb{R}^n$, $n \geq 2$, is a set of uniform approximation if and only if $\mathbb{R}^n \setminus K$ and $\mathbb{R}^n \setminus K^{o}$ are thin at the same points.*

The second lemma is due to Labrèche [3].

LEMMA B. *Suppose E_1 and E_2 are compact sets of uniform approximation in $\mathbb{R}^n$, $n \geq 2$, with $E_2^{o} = \emptyset$. Then $E_1 \cup E_2$ is also a set of uniform approximation.*

Now let

$$D = \bigcup_{1}^{\infty} D_j$$

be an exhaustion of D, where for each j, ∂D_j is smooth and $\bar{D}_j$ is compact in D_{j+1}. Also, the set D_0 is empty.

If F satisfies the conditions of the Theorem, we claim that for each j,

$$(2) \qquad F \cap (\bar{D}_{j+1} \setminus D_j)$$

is a set of uniform approximation. Since F is locally a set of approximation, there is a finite open cover of (2):

$$(3) \qquad F \cap (\bar{D}_{j+1} \setminus D_j) \subset \bigcup_{\alpha=1}^{m} N_\alpha$$

such that each

$$\bar{N}_\alpha \cap F$$

is a set of uniform approximation. From Lemma B and (3) it follows that (2) is a set of uniform approximation. Since ∂D_j is smooth, $\mathbb{R}^n \setminus \bar{D}_j$ is thin at no point of ∂D_j and so, by Lemma A, $\bar{D}_j$ is also a set of uniform approximation. Thus, by Lemma B,

$$(4) \qquad K_j = \bar{D}_j \cup [F \cap (\bar{D}_{j+1} \setminus D_j)]$$

is a set of uniform approximation, for each j.

Let ε_j, $j = 0,1,\dots$, be a sequence of positive numbers. We shall construct inductively a sequence p_j of Newtonian functions. Let u be continuous on F.

<u>Step 0</u>. Since $F \cap \bar{D}_1$ is a set of uniform approximation, there is a function q_0 harmonic on $F \cap \bar{D}_1$ such that

$$(5) \qquad |u(x) - q_0(x)| < \varepsilon_0/2, \qquad x \in F \cap \bar{D}_1.$$

Now by the harmonic Runge-type theorem [2, p.50] there is a Newtonian function p_0 on $\mathbb{R}^n$ such that

$$(6) \qquad |q_0(x) - p_0(x)| < \varepsilon_0/2, \qquad x \in F \cap \bar{D}_1,$$

and so, from (5) and (6),

$$|u(x) - p_0(x)| < \varepsilon_0, \qquad x \in F \cap \bar{D}_1.$$

<u>Step 1</u>. We introduce an auxiliary function u_1 on K_1. We set $u_1 = p_0$ on $\bar{D}_1$ and $u_1 = u$ on $F \cap \partial D_2$. Now extend u_1, by Tietze's

theorem, continuously to K_1 so that

$$|u_1(x) - u(x)| < \varepsilon_0, \quad x \in F \cap (\bar{D}_2 \setminus D_1).$$

Let s_1 be the principal part of u_1 on $\bar{D}_1$. That is,

$$s_1(x) = \sum_{k=1}^{m} \lambda_k \, K(x - y_k),$$

and $u_1 - s_1$ is continuous on K_1 and harmonic on K_1^o. As in Step 0, since K_1 is a set of uniform approximation and using the harmonic Runge theorem, there is a function r_1, Newtonian on $\mathbb{R}^n$, with

$$|r_1(x) - (u_1 - s_1)(x)| < \varepsilon_1, \quad x \in K_1.$$

Set $p_1 = r_1 + s_1$. Then p_1 is Newtonian on $\mathbb{R}^n$ and satisfies

$$|p_1(x) - p_0(x)| < \varepsilon_1 \quad \text{on } \bar{D}_1,$$

$$|p_1(x) - u(x)| < \varepsilon_0 + \varepsilon_1 \quad \text{on } K_1 \setminus D_1,$$

$$|p_1(x) - u(x)| < \varepsilon_1 \quad \text{on } \partial D_2 \cap F.$$

Step j. Now inductively we construct a sequence (p_j) of functions Newtonian on $\mathbb{R}^n$ and satisfying, for each j,

$$(7) \qquad |p_j(x) - p_{j-1}(x)| < \varepsilon_j \quad \text{on } \bar{D}_j,$$

$$(8) \qquad |p_j(x) - u(x)| < \varepsilon_{j-1} + \varepsilon_j \quad \text{on } K_j \setminus D_j,$$

$$|p_j(x) - u(x)| < \varepsilon_j \quad \text{on } \partial D_{j+1} \cap F.$$

It follows from (7) that (p_j) converges to a function v, Newtonian on D, if ε_j decrease rapidly. Now suppose that

$$x \in F \cap (\bar{D}_{j+1} \setminus D_j) = K_j \setminus D_j$$

and fix $m > j$. Then, from (7) and (8):

$$|p_m(x) - u(x)| \le |p_j(x) - u(x)| + \sum_{j+1}^{m} |p_k(x) - p_{k-1}(x)|$$

$$\le \varepsilon_{j-1} + \varepsilon_j + \sum_{j+1}^{m} \varepsilon_k.$$

Letting m tend to infinity, we have

$$(9) \qquad |u(x) - v(x)| \leq \sum_{j-1}^{\infty} \varepsilon_k, \qquad x \in F \cap (\bar{D}_{j+1} \setminus D_j).$$

Now let ε be a positive continuous function on F. Then, we can choose the sequence (ε_j) so small that (9) implies (1). Hence F is a set of Carleman approximation and the proof is complete.

The Corollary follows immediately from the Theorem and Lemma A.

References

[1] DENY, J.: Systèmes totaux de fonctions harmoniques, Ann. Inst. Fourier 1 (1949), 103-113.

[2] GAUTHIER, P.M. and W. HENGARTNER: Approximation uniforme qualitative sur des ensembles non bornés, Presses de l'Université de Montréal 1982.

[3] LABRÈCHE, M.: Thèse, Université de Montréal 1982.

Université de Montréal
Mathématiques
Montréal H3C 3J7, Canada

VALEURS FRONTIÈRES DES FONCTIONS HARMONIQUES OU HOLOMORPHES ET DE LEURS DÉRIVÉES

II. Cas de la boule

Bernard Gaveau (Paris)

Table des matières page

Introduction

Ce travail est la seconde partie de [11]. A l'origine, nous avions cherché à mettre en évidence des propriétés quantitatives des valeurs frontières spécifiques aux fonctions holomorphes, dans le bidisque ou la boule et qui ne soient pas vraies pour les fonctions harmoniques d'une théorie du potentiel convenable. Jusqu'à présent, cela est un échec;comme le montre les travaux de Debiard [3], [4] et de Putz [13], la théorie H^p et la théorie des intégrales d'aire des dérivées premières se généralise (assez difficilement d'ailleurs) des fonctions holomorphes de la boule aux fonctions harmoniques pour la métrique de Bergmann.

En utilisant les intégrales d'aire généralisées introduites par Malliavin [10] dans le bidisque et reprises dans une situation locale dans [8], nous avons mis en évidence dans [11] des propriétés globales des valeurs frontières de fonctions biharmoniques du bidisque et de leurs dérivées premières et secondes. Ces propriétés sont donc satisfaites par une classe plus générale que celle des fonctions holomorphes. Ici, nous développons des idées analogues dans la boule unité de $\mathbb{C}^n$ munie de la métrique de Bergmann en utilisant l'opérateur du quatrième-ordre Δ^2 (carré du laplacien de Bergmann). Le § 1 développe la théorie du potentiel pour Δ^2 ("théorie de l'élasticité"), aussi utilisée dans [7]. Le § 2 donne des identités de nature algébrique sur les dérivées des fonctions harmoniques de Bergmann. Le § 3 rappelle les estimées de dérivées premières dûes à Debiard [3], [4] et Putz [13]. Les § 4 à 7 donnent les estimées du type "intégrales d'aires" pour les dérivées secondes ou pour des fonctions non linéaires des dérivées premières; il est assez aisé de développer la théorie L^2; nous donnons les estimées L^∞-BMO par une méthode probabiliste et obtenons les estimées L^p par la théorème d'interpolation de Stroock. Enfin le § 8 donne une estimée de type "exponentielle intégrabilité" sur le volume de l'image de domaines admissibles par une application holomorphe de la boule dans $\mathbb{C}^2$ qui reste bornée.

1. Théorie de l'élasticité pour la métrique de Bergmann

a) le laplacien radial.

Soit ds^2 la métrique de Bergmann de la boule unité B de $\mathbb{C}^2$ (nous nous plaçons en 2 dimensions complexes pour simplifier la preuve, le cas n dimension se traitant pareillement exactement avec les changements ad hoc). Le laplacien est

(1) $$\Delta = (1-|Z|^2) \sum_{i,j=1}^{2} (\delta_{i\bar{j}}(Z_i \bar{Z}_j)) \frac{\partial^2}{\partial Z_i \partial \bar{Z}_j} .$$

Lemme 1 : Si $\varphi(|Z|^2)$ est une fonction C^2 ne dépendant que de $|Z|^2$, alors

(2) $$\Delta(\varphi(|Z|^2)) = (1-|Z|^2)((|Z|^2-|Z|^4)\varphi''(|Z|^2) + (2-|Z|^2)\varphi').$$

Preuve : on a

$$\Delta(\varphi(|Z|^2)) = (1-|Z|^2)\Big[(1-|Z_1|^2)(\varphi'+|Z_1|^2\varphi'') + (1-|Z_2|^2)(\varphi'+|Z_2|^2\varphi'') - 2\varphi''|Z_1|^2|Z_2|^2\Big]$$

$$=(1-|Z|^2)(\varphi''(|Z|^2-|Z|^4) + \varphi'(2-|Z|^2))$$

b) fonction de Green.

Lemme 2 : (voir aussi Debiard [3]) : la fonction de Green de Δ dans la boule B de $\mathbb{C}^2$ de pôle 0 est

(3) $$g(|Z|^2) = \frac{1}{|Z|^2} - 1 + \log|Z|^2.$$

Preuve : cela a la singularité ad. hoc $\frac{1}{|Z|^{2n-2}}$ à l'origine ; la condition du bord 0 sur ∂B et le lemme 1 montre que $\Delta g = 0$.

c) développement de l'élasticité dans une boule concentrique. Soit $R<1$ et $d\sigma$ l'aire euclidienne usuelle de $\partial B(0,R)$. La formule de la moyenne s'écrit

(4) $\int_{\partial B(0,R)} u(\sigma)d\sigma = u(0) + \int_{\partial B(0,R)} g_R(0,Z)\,\Delta u(Z)\,dv(Z)$

où g_R dénote la fonction de Green de $B(0,R)$ pour le ds^2 de Bergmann et $dv(Z) = \frac{1}{(1-|Z|^2)^3} dv_{eucl}(Z)$ est le volume de la métrique de Bergmann. D'après (3), il est clair que

(3) $$g_R(0,Z) = \log \frac{|Z|^2}{R^2} + \frac{1}{|Z|^2} - \frac{1}{R^2} .$$

Soit alors $\Phi_R(|Z|^2)$ une fonction telle que avec h de singularité plus faible, C_R constante

(5) $$\begin{cases} \Phi_R(|Z|^2) = C_R\, g_R(|Z|^2) + h(|Z|^2) \\ \Phi_R(R^2) = 0 \\ \dfrac{\partial \Phi_R}{\partial |Z|^2}(R^2) = 0 \\ \Delta \Phi_R = g_R \end{cases}$$

Appliquons la formule de Green à 2 fonctions α et β dans la boule $B(0,R)$ pour la métrique de Bergmann : notant dv le volume de Bergmann et $d\sigma$ l'élément d'aire de Bergmann sur $\partial B(0,R)$ et $\frac{\partial}{\partial n}$ la dérivée normale extérieure de Bergmann cette formule s'écrit :

(6) $$\int_{B(0,R)} (\alpha\Delta\beta - \beta\Delta\alpha)\, dv = \int_{\partial B(0,R)} \left(\alpha\frac{\partial\beta}{\partial n} - \beta\frac{\partial\alpha}{\partial n}\right) d\sigma$$

Appliquons cette formule avec $\beta = \Phi_R$ et $\alpha = \Delta u$. Le second membre disparait et il reste en tenant compte de (5)

(7) $$\int_{B(0,R)} (\Delta u\, g_R - \Phi_R\, \Delta^2 u)\, dv = C_R(\Delta u)(0)$$

ce que nous soustrayons de (4) pour obtenir

(8) $$\boxed{\int_{\partial B(0,R)} u\, d\sigma = u(0) + C_R(\Delta u)(0) + \int_{B(0,R)} \Phi_R(Z)\,(\Delta^2 u)(Z)\,dv(Z)}$$

ce qui sera notre formule fondamentale généralisant celles du disque et du bidisque.

d) calcul de Φ_R.

Nous devons trouver Φ_R satisfaisant (5) ; R étant fixé, nous poserons $\varphi = \Phi_R$ pour alléger l'écriture et $\psi = \varphi'$ (ici, nous écrivons $\varphi = \varphi(|Z|^2)$). Le calcul que nous allons faire s'apparente à un calcul de géométrie intégrale hyperbolique complexe fait dans [7] pour démontrer une formule d'inversion non locale pour la métrique de Bergmann. Nous avons à résoudre en posant $|Z|^2 = x$, en utilisant la condition (5) et les lemmes 1 et 2

$$(9) \quad (1-x)((x-x^2)\varphi'' + (2-x)\varphi') = \log \frac{x}{R^2} - \frac{1}{R^2} + \frac{1}{x} .$$

Posons $\varphi' = \psi$, l'équation sans second membre à la solution

$$(10) \quad \psi = C \frac{1-x}{x^2} \quad (C = \text{cste}).$$

La méthode de variation des constantes donne en reportant (10) dans (9)

$$C'(X) \frac{(1-x)^3}{x} = \log \frac{x}{R^2} - \frac{1}{R^2} + \frac{1}{x}$$

et par suite

$$C(X) = \int \frac{x \log x}{(1-x)^3} dx + \int \frac{dx}{(1-x)^3} - (\log R^2 + \frac{1}{R^2}) \int \frac{x\, dx}{(1-x)^3}$$

$$\int \frac{x\, dx}{(1-x)^3} = \frac{1}{2} \frac{1}{(1-x)^2} - \frac{1}{1-x}$$

$$\int \frac{dx}{(1-x)^3} = \frac{1}{2} \frac{1}{(1-x)^2}$$

$$\int \frac{x \log x\, dx}{(1-x)^3} = (\frac{1}{2} \frac{1}{(1-x)^2} - \frac{1}{1-x}) \log x - \int (\frac{1}{2} \frac{1}{(1-x)^2} - \frac{1}{(1-x)}) \frac{dx}{x}$$

$$= (\frac{1}{2} \frac{1}{(1-x)^2} - \frac{1}{1-x}) \log x + \log \frac{x}{1-x} - \frac{1}{2} \log \frac{x}{1-x} - \frac{1}{2} \frac{1}{(1-x)}$$

et par suite

$$C(t) = \frac{1}{2} \frac{1}{(1-t)^2} - \frac{1}{(1-t)} - \frac{1}{2} \log \frac{t}{1-t} - \frac{1}{2} \frac{1}{1-t} + \log \frac{t}{1-t}$$

$$(11) \quad + \frac{1}{2} \frac{1}{(1-t)^2} - (\log R^2 + \frac{1}{R^2})(\frac{1}{2} \frac{1}{(1-t)^2} - \frac{1}{(1-t)}) + K(R^2)$$

où $K(R^2)$ est tel que $C(R^2) = 0$ ce qui correspond à l'une des conditions (5) $\psi(R^2) = 0$) c'est-à-dire

$$- K(R^2) = \log R^2 \left(\frac{1}{2}\frac{1}{(1-R^2)^2} - \frac{1}{1-R^2}\right) + \frac{1}{2}\log\frac{R^2}{1-R^2} - \frac{1}{2}\frac{1}{1-R^2} +$$

$$+ \frac{1}{2}\frac{1}{(1-R^2)^2} - \left(\log R^2 + \frac{1}{R^2}\right)\left(\frac{1}{2}\frac{1}{(1-R^2)^2} - \frac{1}{(1-R^2)}\right)$$

d'où pour simplification

$$(12) \qquad -K(R^2) = \frac{1}{2}\left[\log\frac{R^2}{1-R^2} + \frac{1}{R^2}\right]$$

Utilisons (10), (11), (12) : alors φ satisfait

$$\varphi(x) = \int_x^{R^2} \varphi(t)\, dt$$

d'où

$$\varphi(x) = \int_x^{R^2} dt\left[\frac{1}{2}\frac{\log t}{1-t} - \frac{(1-t)\log(1-t)}{2t^2} + \frac{1}{2t} + \frac{1}{2(1-t)} - \right.$$

$$\left. - \left(\log R^2 + \frac{1}{R^2}\right)\left(\frac{1}{2(1-t)} + \frac{1}{2t} - \frac{1}{2t^2}\right) - \frac{1}{2}\left(\log\frac{R^2}{1-R^2} + \frac{1}{R^2}\right)\left(\frac{1-t}{t}\right)\right]$$

$$(13) \qquad = \frac{1}{2}\int_x^{R^2} dt\left[\frac{\log t}{1-t} - \frac{(1-t)\log(1-t)}{t^2} + \frac{1}{(1-t)}\left(1 - \log R^2 - \frac{1}{R^2}\right) + \frac{1}{t}\right]$$

$$+ \frac{1}{2}\int_x^{R^2} dt\left(\frac{1-t}{t^2}\right)\left[\log(1-R^2) - \frac{1}{R^2}\right]$$

Le second terme de (13) est exactement

$$\frac{1}{2}\left(\log(1-R^2) - \frac{1}{R^2}\right) g_R(|z|^2)$$

de suite

$$(14) \qquad \Phi_R(|z|^2) = \frac{1}{2}\left(\log(1-R^2) - \frac{1}{R^2}\right) g_R(|z|^2) + \frac{1}{2}\left(\log\frac{(1-R^2)}{(1-x)}\right)\left(1-\log R^2 - \frac{1}{R^2}\right)$$

$$+ \frac{1}{2}\log\frac{R^2}{x} + \frac{1}{2}\int_x^{R^2} dt\left(\frac{\log t}{1-t} - \frac{(1-t)\log(1-t)}{t^2}\right)$$

Revenant à (5) nous voyons bien que

$$(15) \qquad \Phi_R(|z|^2) = C_R\, g_R(|z|^2) + h_R(|z|^2)$$

où $C_R = \frac{1}{2}\left(\log(1-R^2) - \frac{1}{R^2}\right)$

et h_R est une fonction manifestement régulière près de $x = 0$ (sauf la partie $\frac{1}{2}\log\frac{R^2}{x}$ qui est de singularité plus faible que g_R).

e) passage à la limite si $R \longrightarrow 1$.

La différence avec la théorie de l'élasticité classique est que le passage à la limite dans la formule fondamentale (8) est légèrement plus compliqué parce que $C_R \longrightarrow \infty$ si $R \longrightarrow 1$.

Cependant fixons u de classe C^4 au voisinage de $\overline{B(0,1)}$ et utilisons (8) pour R fixé, que nous divisons membre à membre par C_R. En reportant la valeur de Φ_R donnée par (15), nous avons

$$\frac{1}{C_R}\int_{\partial B(0,R)} u d\sigma - \frac{1}{C_R} u(0) - \Delta u(0) = \frac{1}{C_R}\int_{B(0,R)} h_R(|Z|^2)(\Delta^2 u)(Z)\, dv(Z) +$$

$$(16) \qquad + \int_{B(0,R)} g_R(0,|Z|^2)(\Delta^2 u)(Z)\, dv(Z).$$

Maintenant considérons la première intégrale du 2nd membre de (16) ; si u est C^4, par (1),

$$\Delta^2 u = (1-|Z|^2)\, w$$

où w est une fonction régulière près de $\partial B(0,1)$

$$(17) \qquad \begin{aligned} dv &= \frac{1}{(1-|Z|^2)^3}\, dv_{eucl} \text{ , d'où} \\ \Delta^2 u\, dv &= \frac{1}{(1-|Z|^2)^2}\, dv_{eucl} \text{ .} \end{aligned}$$

Maintenant $h_R(|Z|^2)$ comprend le terme suivant

$$\log\frac{(1-R^2)}{(1-x)}\left(1-\log R^2 - \frac{1}{R^2}\right) \qquad \text{(voir (14))}$$

Par suite la contribution de ce terme est majorée par

$$(18) \qquad \frac{M}{C_R}\left(1 - \log R^2 - \frac{1}{R^2}\right)\left(\int_{B(0,R)} \log\frac{(1-R^2)}{(1-|Z|^2)}\, \frac{1}{(1-|Z|^2)^2}\, dv_{eucl}\right)$$

où M est majorant de $\|u\|_{C^4}$. Mais si $R \longrightarrow 1$, comme $1-\log R^2 - \frac{1}{R^2}$ s'annule à l'ordre 2 ce terme (18) tend vers 0. Toujours d'après (14), il y a aussi le terme suivant dans h_R.

$$\frac{1}{2}\log\frac{R^2}{x} + \frac{1}{2}\int_x^{R^2} dt\left(\frac{\log t}{1-t} - \frac{(1-t)\log(1-t)}{t^2}\right)$$

Or ce terme s'annule par définition si $R=1$ et $x \longrightarrow 1$ à l'ordre 1, et même à l'ordre 2 : en effet faisant $R=1$, la dérivée est

$-\frac{1}{2}\frac{1}{x} - \frac{1}{2}\frac{\log x}{1-x} \longrightarrow 0$ si $x \to 1$.

Par conséquent, ce terme de h_R donne une contribution à l'intégrale considérée qui se contrôle par

$$\frac{M}{C_R}\int_{B(0,1)} dv_{eucl}$$

qui tend vers 0 si $R \longrightarrow 1$, d'où

<u>Lemme 3</u> : <u>Pour</u> u <u>de classe</u> C^4 <u>près de la boule unité, on a</u>

$$(19) \qquad -(\Delta u)(0) = \int_{B(0,1)} g(0,|Z|^2)(\Delta^2 u)(Z)\, dv(Z).$$

<u>Remarque 1</u> : Formule que nous pouvons démontrer extrêmement facilement ainsi : appliquons la formule de Green pour la métrique de Bergmann (formule (6)) à $\alpha = \Delta u$, $\beta = g$, $R = 1$ et en tenant compte que $\Delta u = 0$ au bord de $B(0,1)$ si u est $C^4\ \overline{(B(0,1))}$. Alors

$$0 = \int_{\partial B(0,1)} \Delta u\, d\sigma = (\Delta u)(0) + \int_{B(0,1)} g(0,|Z|^2)(\Delta^2 u)(Z)\, dv\,(Z)$$

d'où le résultat.

<u>Remarque 2</u> : Nous allons appliquer la formule (19) dans la situation où u sera une fonction C^4 (et même C^∞) dans $B(0,1)$ seulement. Cependant nous ne pouvons pas en général appliquer directement (19) car les singularités de l'intégrale se contre balançent de façon compliquée.

<u>Nous procéderons ainsi</u> : pour $u \in C^4(B(0,1))$ et $\rho < 1$, appliquons la formule (19) à $u_\rho(Z) = u(\rho Z)$ qui est C^4 près de $\overline{B(0,1)}$. Clairement

$$(\Delta^2 u_\rho)(0) = \rho^2 (\Delta^2 u)(0).$$

Soit alors u une fonction et $u_n \longrightarrow u$ avec u_n régulière près de $\overline{B(0,1)}$. Nous déduisons alors que les suites $|\nabla^2 u_n|^2$ formeront une suite de Cauchy dans un espace L^2 à poids de la boule et cela nous permettra de conclure.

Convention : désormais on entendra par fonction harmonique réelle une fonction harmonique pour la métrique de Bergmann (1) à valeurs réelles.

2. Un calcul de dérivées secondes d'une fonction harmonique ou pluriharmonique

a) calcul de $\Delta(|f^2|)$.

Lemme 4 : Soit f harmonique réelle(pour la métrique de Bergmann) ou holomorphe. On a respectivement.

$$(20) \qquad \begin{aligned} \Delta(f^2) &= 2\|\nabla f\|^2 \\ \Delta|f|^2 &= 2\|\nabla f\|^2 \end{aligned}$$

où $\|\nabla f\|^2$ désigne le carré de la longueur en gradient de f pour la métrique de Bergmann

$$(21) \qquad (\nabla f|\nabla h) = \sum_{i,j} g^{ij} \frac{\partial f}{\partial Z_i} \left(\overline{\frac{\partial h}{\partial Z_j}}\right)$$

g^{ij} étant l'inverse de la métrique de Bergmann. De plus

$$\Delta(fh) = (\Delta f)h + (\Delta h)f + (\nabla f|\nabla h) + (\nabla h \mid \nabla f).$$

Preuve : évidente et classique.

Introdusons maintenant une notation fondamentale : posons

$$(22) \qquad \begin{cases} \nabla_n f = Z_1 \dfrac{\partial f}{\partial Z_1} + Z_2 \dfrac{\partial f}{\partial Z_2} \\ \nabla_t f = -\overline{Z}_2 \dfrac{\partial f}{\partial Z_1} + \overline{Z}_1 \dfrac{\partial f}{\partial Z_2} \end{cases}$$

∇_n est champ normal à la sphère concentrique à 0 et ∇_t est champ holomorphe tangent à cette sphère.

Lemme 5 : On a

$$(23)\quad (\nabla f \mid \nabla h) = \frac{1}{|z|^2}\left[(1-|z|^2)^2\, \nabla_n f\, \bar{\nabla}_n h + (1-|z|^2)\, \nabla_t f\, \bar{\nabla}_t h\right]$$

Preuve : de (22), on déduit

$$|z|^2 \frac{\partial f}{\partial z_1} = \bar{z}_1 \nabla_n f - z_2 \nabla_t f$$

$$|z|^2 \frac{\partial f}{\partial z_2} = \bar{z}_2 \nabla_n f + z_1 \nabla_t f$$

d'où, comme par polarisation, il suffit de voir (23) pour f=h, utilisant la définition

$$g^{ij} = (1-|z|^2)\begin{pmatrix} 1-|z_1|^2 & -z_1\bar{z}_2 \\ -\bar{z}_1 z_2 & 1-|z_2|^2 \end{pmatrix}$$

il vient

$$|z|^4 \|\nabla f\|^2 = (1-|z|^2)\Big[(1-|z_1|^2)(\bar{z}_1 \nabla_n f - z_2 \nabla_t f)(z_1 \overline{\nabla_n f} - \bar{z}_2 \overline{\nabla_t f}) +$$

$$+ (1-|z_2|^2)(\bar{z}_2 \nabla_n f + z_1 \nabla_t f)(z_2 \overline{\nabla_n f} + \bar{z}_1 \overline{\nabla_t f})$$

$$- z_1\bar{z}_2(\bar{z}_1 \nabla_1 f - z_2 \nabla_t f)(z_2 \overline{\nabla_1 f} + \bar{z}_1 \nabla_t f) - \bar{z}_1 z_2(z_1 \overline{\nabla_1 f} - \bar{z}_2 \overline{\nabla_t f})(\bar{z}_2 \nabla_1 f + z_1 \nabla_t f)\Big]$$

Le terme $|\nabla_n f|^2$ est $|z|^2(1-|z|^2)$

Le terme $|\nabla_t f|^2$ est $|z|^2$

Les termes croisés sont 0, d'où le résultat.

b) Calcul de $\Delta^2(f^2)$ ou $\Delta^2(|f|^2)$.

Lemme 6 : On a la formule pour f harmonique réelle ou holomorphe.

$$\Delta^2 f^2 = \sum_{i=1}^{6} K_i$$

avec

$$(24)\quad K_1 = |\nabla_n f|^2\, \Delta\left(\frac{(1-|z|^2)^2}{|z|^2}\right)$$

$$(25)\quad K_2 = \left(\nabla\left(\frac{(1-|z|^2)^2}{|z|^2}\right) \,\Big|\, \nabla(|\nabla_n f|^2)\right) + \text{conjugué}$$

(26) $$K_3 = \frac{(1-|z|^2)^2}{|z|^2} \Delta(|\nabla_n f|^2)$$

(27) $$K_4 = \Delta\left(\frac{(1-|z|^2}{|z|^2}\right) |\nabla_t f|^2$$

(28) $$K_5 = \left(\nabla\left(\frac{1-|z|^2}{|z|^2}\right) |\nabla(|\nabla_t f|^2)\right) + \text{conjugué}$$

(29) $$K_6 = \frac{1-|z|^2}{|z|^2} \Delta(|\nabla_t f|^2).$$

de même si f^2 est remplacé par $|f|^2$ lorsque f est holomorphe.

Preuve : évidente par les lemmes 5 et 4.

La singularité $|z|^2$ est apparente bien sûr. Ici tout ce qui nous intéresse est le comportement près de $|z|^2 = 1$. et les équivalentes que nous allons donner des 6 termes K_i seront relatifs au voisinage de $|z|^2 = 1$.

c) Estimée des termes K_i.

Nous noterons α une fonction qui est bornée et qui reste strictement bornée inférieurement par $\varepsilon > 0$ au voisinage de $|z|^2 = 1$ (bien sûr indépendante de f).

Lemme 7 : soit f harmonique réelle ou holomorphe : alors

(30) $$K_1 \simeq \alpha(1-|z|^2)^3 \, |\nabla_n f|^2$$

(31) $$K_2 \simeq \alpha(1-|z|^2)^3 \, (\overline{\nabla_n^2 f} \, \nabla_n f + \overline{\nabla_n f} \, \nabla_n \overline{\nabla_n f} + \text{conjugué})$$

(32) $$K_3 \simeq \alpha\left[(1-|z|^2)^3(\Delta_{eucl} f \, \overline{\nabla_n f} + \text{conjugué}) + (1-|z|^2)^4(|\nabla_n^2 f|^2) + \right.$$
$$\left. + (1-|z|^2)^3 \, |\nabla_t \nabla_n f|^2\right]$$

(33) $$K_4 \simeq \alpha(1-|z|^2) \, |\nabla_t f|^2$$

(34) $K_5 \simeq (1-|z|^2)^2 \left[(\nabla_n \nabla_t f)(\overline{\nabla_t f}) + (\nabla_n \overline{\nabla_t f})(\nabla_t f) + \text{conjugué}\right]$

(35) $K_6 \simeq \alpha \left[(1-|z|^2)^3(|\nabla_n \nabla_t f|^2 + |\nabla_n \overline{\nabla_t f}|^2) + (1-|z|^2)^2(|\nabla_t^2 f|^2 + |\nabla_t \overline{\nabla_t f}|^2) + \right.$

$\left. + (1-|z|^2)^2((\nabla_t \overline{\nabla_n f} + \nabla_n \nabla_t f)\overline{\nabla_t f} + (\bar\nabla_t \nabla_n f + \overline{\nabla_n \nabla_t f})(\nabla_t f)\right]$

Preuve : 1 calcul de K_1 :

$K_1 = |\nabla_n f|^2 \, \Delta\left(\frac{(1-|z|^2)^2}{|z|^2}\right)$. Utilisons la notion α_i et (23) en tenant compte de ce que le ∇_t des fonctions envisagées est 0

$$\Delta\left(\frac{(1-|z|^2)^2}{|z|^2}\right) = \alpha_1 (1-|z|^2)^3 + (1-|z|^2)^3 \alpha_2 + \frac{1}{|z|^2} \Delta((1-|z|^2)^2)$$

Mais utilisant le lemme 1

$$\Delta((1-|z|^2)^2) = (1-|z|^2)(-2(2-|z|^2)(1-|z|^2) + 2(|z|^2-|z|^4)) = 2(1-|z|^2)^3$$

d'où

$$\Delta\left(\frac{(1-|z|^2)^2}{|z|^2}\right) = (1-|z|^2)^3\alpha \quad \text{et donc (30)}$$

2 calcul de K_2

$$K_2 = \left(\nabla\left(\frac{(1-|z|^2)^2}{|z|^2}\right) \,\Big|\, \nabla(|\nabla_n f|^2)\right) + \text{conjugué}$$

Comme $\frac{(1-|z|^2)^2}{|z|^2}$ ne dépend que de la direction normale, dans la formule (23) il suffit de regarder le terme normal et

$$K_2 = \alpha(1-|z|^2)^2 \, \nabla_n\left(\frac{(1-|z|^2)^2}{|z|^2}\right) \overline{\nabla_n(|\nabla_n f|^2)} + \text{conjugué}$$

$$= \alpha(1-|z|^2)^3 \, (\overline{\nabla_n(|\nabla_n f|^2)} + \nabla_n(|\nabla_n f|^2))$$

$\nabla_n |\nabla_n f|^2 = \nabla_n^2 f \, \overline{\nabla_n f} + \nabla_n f \, \nabla_n(\overline{\nabla_n f})$ (que f soit harmonique réelle ou holomorphe).

$$K_2 = \alpha(1-|z|^2)^3 \, (\nabla_n^2 f \, \overline{\nabla_n f} + \nabla_n f \, \nabla_n \overline{\nabla_n f} + \overline{\nabla_n^2 f} \, \nabla_n f + \overline{\nabla_n f} \, \overline{\nabla_n \overline{\nabla_n f}})$$

et donc (31) est calculé.

3 calcul de K_3

Nous devons calculer (37) $\Delta(|\nabla_n f|^2) =$

$$= \Delta(\nabla_n f)\,\overline{\nabla_n f} + (\nabla(\nabla_n f)\,|\,\nabla(\overline{\nabla_n f})) + \text{conjugué} + \nabla_n f\,\Delta(\overline{\nabla_n f}).$$

Calculons $\Delta(\nabla_n f)$ = c'est

$$\Delta(\nabla_n f) = (1-|z|^2)\Big[(1-|z_1|^2)\frac{\partial^2}{\partial z_1\partial \bar z_1}\Big(z_1\frac{\partial f}{\partial z_1} + z_2\frac{\partial f}{\partial z_2}\Big) +$$

$$+ (1-|z_2|^2)^2\frac{\partial^2}{\partial z_2\partial \bar z_2}\Big(z_1\frac{\partial f}{\partial z_1} + z_2\frac{\partial f}{\partial z_2}\Big) - \bar z_1 z_2\frac{\partial^2}{\partial \bar z_1\partial z_2}\Big(z_1\frac{\partial f}{\partial z_1} + z_2\frac{\partial f}{\partial z_2}\Big) -$$

$$- z_1\bar z_2\frac{\partial^2}{\partial z_1\partial \bar z_2}\Big(z_1\frac{\partial f}{\partial z_1} + z_2\frac{\partial f}{\partial z_2}\Big)\Big].$$

Utilisons que $\Delta f = 0$

$$= (1-|z|^2)\Big[(1-|z_1|^2)\Big(\nabla_n\Big(\frac{\partial^2 f}{\partial z_1\partial \bar z_1}\Big)\Big) + (1-|z_2|^2)\Big(\nabla_n\Big(\frac{\partial^2 f}{\partial z_2\partial \bar z_2}\Big)\Big) -$$

$$- \bar z_1 z_2\,\nabla_n\Big(\frac{\partial^2 f}{\partial \bar z_1\partial z_2}\Big) - z_1\bar z_2\,\nabla_n\Big(\frac{\partial^2 f}{\partial z_1\partial \bar z_2}\Big)$$

$$(38)\quad \Delta(\nabla_n f) = (1-|z|^2)\Big[\frac{\partial^2 f}{\partial z_1\partial \bar z_1}\nabla_n|z_1|^2 + \frac{\partial^2 f}{\partial z_2\partial \bar z_2}\nabla_n|z_2|^2 + \frac{\partial^2 f}{\partial \bar z_1\partial z_2}\nabla_n(\bar z_1 z_2) +$$

$$+ \frac{\partial^2 f}{\partial z_1\partial \bar z_2}\nabla_n(z_1\bar z_2)\Big]$$

Or $\nabla_n \bar z_i z_j = \bar z_i z_j$, d'où

$$= (1-|z|^2)\Big(-\frac{\Delta f}{1-|z|^2} + \Delta_{\text{eucl}} f\Big) = (1-|z|^2)\,\Delta_{\text{eucl}} f .$$

Calculons les produits scalaires.

$$(39)\quad (\nabla(\nabla_n f)\,|\,\nabla(\overline{\nabla_n f})) = (1-|z|^2)^2(\nabla_n^2 f\,\overline{\nabla_n^2 f} + |\nabla_n\overline{\nabla_n f}|^2) + (1-|z|^2)(|\nabla_t\nabla_n f|^2) + |\nabla_t\overline{\nabla_n f}|^2)$$

d'où en utilisant (37), (38) et (39), on obtient (32).

4 <u>Calcul de</u> K_4

$$\Delta\Big(\frac{1-|z|^2}{|z|^2}\Big) = \Delta(1-|z|^2)\Big(\frac{1}{|z|^2}\Big) + \Big(\nabla\Big(\frac{1}{|z|^2}\Big)\Big|\,\nabla(1-|z|^2)\Big) + (1-|z|^2)\,\Delta\Big(\frac{1}{|z|^2}\Big)$$

d'où

$$= \alpha(1-|z|^2)$$

Par la définition (27) de K_4 nous déduisons (33).

5 Calcul de K_5

$$K_5 = \left(\nabla\left(\frac{1-|z|^2}{|z|^2}\right) \,\middle|\, \nabla(|\nabla_t f|^2)\right) + \text{conjugué}.$$

Dans la formule (23), il ne faut retenir que les termes en ∇_n, d'où

$$\left(\nabla\left(\frac{1-|z|^2}{|z|^2}\right) \,\middle|\, \nabla(|\nabla_t f|^2)\right) = (1-|z|^2)^2 \,\alpha\, \overline{\nabla_n(|\nabla_t f|^2)}$$

$$= (1-|z|^2)\,\alpha\, \overline{((\nabla_n \nabla_t f)\, \overline{\nabla_t f} + \nabla_n \overline{\nabla_t f}\, \nabla_t f)}$$

6 Calcul de K_6 : c'est le plus long

$$K_6 = \frac{(1-|z|^2)}{|z|^2}\, \Delta(|\nabla_t f|^2)$$

On a

$$\Delta(|\nabla_t f|^2) = \Delta(\nabla_t f)\, \overline{\nabla_t f} + (\nabla(\nabla_t f) \,|\, \nabla(\overline{\nabla_t f})) + \text{conjugué} + \Delta(\overline{\nabla_t f})\, \nabla_t f.$$

$$\equiv K_{6,1} + K_{6,2} + K_{6,3}\,.$$

calcul de $K_{6,1}$ et de $K_{6,3}$

Lemme 8 : si f est harmonique

$$\Delta(\nabla_t f) = (1-|z|^2)(+\nabla_t \overline{\nabla}_1 f + \nabla_1 \nabla_t f) \tag{40}$$

en effet

$$\begin{aligned}
\Delta(\nabla_t f) &= (1-|z|^2)\Big[(1-|z_1|^2)\frac{\partial^2}{\partial z_1 \partial \bar z_1}\left(\bar z_1 \frac{\partial f}{\partial z_2} - \bar z_2 \frac{\partial f}{\partial z_1}\right) + \\
&+ (1-|z_2|^2)\frac{\partial^2}{\partial z_2 \partial \bar z_2}\left(\bar z_1 \frac{\partial f}{\partial z_2} - \bar z_2 \frac{\partial f}{\partial z_1}\right) - z_1 \bar z_2 \frac{\partial^2}{\partial z_1 \partial \bar z_2}\left(-\bar z_2 \frac{\partial f}{\partial z_1} + \bar z_1 \frac{\partial f}{\partial z_2}\right) - \\
&- \bar z_1 z_2 \frac{\partial^2}{\partial \bar z_1 \partial z_2}\left(-\bar z_2 \frac{\partial f}{\partial z_1} + \bar z_1 \frac{\partial f}{\partial z_2}\right)\Big] = (1-|z|^2)\Big[(1-|z_1|^2)\nabla_t \frac{\partial^2 f}{\partial z_1 \partial \bar z_1} + \\
&+ (1-|z_2|^2)\nabla_t \frac{\partial^2 f}{\partial z_2 \partial \bar z_2} - z_1 \bar z_2 \nabla_t \frac{\partial^2 f}{\partial z_1 \partial \bar z_2} - z_2 \bar z_1 \nabla_t \left(\frac{\partial^2 f}{\partial \bar z_1 \partial z_2}\right)\Big] + \\
&+ (1-|z|^2)\, \nabla_1 \nabla_t f
\end{aligned} \tag{41}$$

puisque

$$(1-|z_1|^2)\frac{\partial^2 f}{\partial z_1 \partial z_2} - (1-|z_2|^2)\frac{\partial^2 f}{\partial z_1 \partial z_2} + z_1 \bar z_2 \frac{\partial^2 f}{\partial z_1^2} - \bar z_1 z_2 \frac{\partial^2 f}{\partial z_2^2} =$$

$$= (|z_2|^2 - |z_1|^2)\frac{\partial^2 f}{\partial z_1 \partial z_2} + z_1\bar{z}_2 \frac{\partial^2 f}{\partial z_1^2} - \bar{z}_1 z_2 \frac{\partial^2 f}{\partial z_2^2}$$

et que précisément.

$$\nabla_1 \nabla_t f = (z_1 \frac{\partial}{\partial z_1} + z_2 \frac{\partial}{\partial z_2})(-\bar{z}_2 \frac{\partial}{\partial z_1} + \bar{z}_1 \frac{\partial}{\partial z_2}) f$$

$$= -z_1\bar{z}_2 \frac{\partial^2 f}{\partial z_1^2} + z_2\bar{z}_1 \frac{\partial^2 f}{\partial z_2^2} + |z_1|^2 \frac{\partial^2 f}{\partial z_1 \partial z_2} - |z_2|^2 \frac{\partial^2 f}{\partial z_1 \partial z_2}$$

d'où en revenant à (41)

$$\Delta(\nabla_t f) = (1-|z|^2)\Big[\nabla_t(\frac{\Delta f}{1-|z|^2}) - \bar{z}_2 z_1 \frac{\partial^2 f}{\partial z_1 \partial \bar{z}_1} + \bar{z}_1\bar{z}_2 \frac{\partial^2 f}{\partial z_2 \partial \bar{z}_2} - \bar{z}_2^2 \frac{\partial^2 f}{\partial z_1 \partial \bar{z}_2} +$$

$$+ \bar{z}_1^2 \frac{\partial^2 f}{\partial \bar{z}_1 \partial z_2} + \nabla_1 \nabla_t f\Big]$$

Mais

$$\nabla_t \bar{\nabla}_1 f = (-\bar{z}_2 \frac{\partial}{\partial z_1} + \bar{z}_1 \frac{\partial}{\partial z_2})(\bar{z}_1 \frac{\partial}{\partial \bar{z}_1} + \bar{z}_2 \frac{\partial}{\partial \bar{z}_2}) f$$

$$= -\bar{z}_2\bar{z}_1 \frac{\partial^2 f}{\partial z_1 \partial \bar{z}_1} + \bar{z}_2\bar{z}_1 \frac{\partial^2 f}{\partial z_2 \partial \bar{z}_2} + \bar{z}_1^2 \frac{\partial^2 f}{\partial \bar{z}_1 \partial z_2} - \bar{z}_2^2 \frac{\partial^2 f}{\partial z_1 \partial \bar{z}_2}$$

d'où la formule (40)

$$\Delta(\nabla_t f) = (1-|z|^2)\left[\nabla_t \bar{\nabla}_1 f + \nabla_1 \nabla_t f\right]$$

Cela montre que

$$(42) \quad \begin{aligned} K_{6,1} &= (1-|z|^2)\left[\nabla_t \bar{\nabla}_1 f + \nabla_1 \nabla_t f\right] \overline{\nabla_t f} \\ K_{6,3} &= (1-|z|^2)\left[\overline{\nabla_t \bar{\nabla}_1 f + \nabla_1 \nabla_t f}\right](\nabla_t f) \end{aligned}$$

<u>calcul de</u> $K_{6,2}$

C'est par la formule (23)

$$(43) \quad \begin{aligned} K_{6,2} = \alpha\big(&(1-|z|^2)^2 |\nabla_n \nabla_t f|^2 + (1-|z|^2)|\nabla_t^2 f|^2 + (1-|z|^2)^2 |\nabla_n \overline{\nabla_t f}|^2 + \\ &+ (1-|z|^2)|\nabla_t \overline{\nabla_t f}|^2\big). \end{aligned}$$

Utilisons (42), (43), il vient le résultat (35)

3. Contrôle L^p des dérivées premières d'une fonction harmonique

Théorème 1 : Soit f une fonction harmonique réelle qui est $L^2(\partial B)$. Alors

$$(44) \qquad \int((1-|Z|^2)\,|\nabla_n f|^2 + |\nabla_t f|^2)\, dv_{eucl}(Z) \leq C\|f\|^2_{L^2(\partial B)}$$

Preuve : évident, car il suffit d'appliquer la formule de Green usuelle à f^2

$$\|f\|^2_{L^2} = \int_{\partial B(0,1)} f^2\, d\sigma = f^2(0) - \int_{B(0,1)} g(|Z|^2)\,\Delta f^2\, dv(Z)$$

et d'utiliser ensuite le lemme 2 de A. Debiard qui nous apprend que $g(|Z|^2) \simeq (1-|Z|^2)^2$ près de $|Z|^2 = 1$ et que $dv = \dfrac{1}{(1-|Z|^2)^3}\, dv_{eucl}$, et enfin la formule (23) du lemme 5.

Remarque 1 : Quelques secondes de réflexion permettent de démontrer (44), par une fonction pluriharmonique en utilisant l'analyse de Fourier sur SU(2) identifié à ∂B. Cependant, nous ne savons pas démontrer par cette technique (44) pour une fonction harmonique générale car le multiplicateur de Fourier correspond à la mesure harmonique de la métrique de Bergmann est assez compliqué.

Remarque 2 : Bien entendu le résultat (44) est beaucoup plus faible que les résultats de A. Debiard sur le sujet.

Nous pouvons introduire comme dans [3].

$$(45) \qquad g_t(f)^2(\sigma) = \int_0^1 |\nabla_t f|^2(r\sigma)\, dr$$

où $\sigma \in \partial B(0,1)$

$$(46) \qquad g_n(f)^2(\sigma) = \int_0^1 (1-|r|^2)\, |\nabla_n f|^2\, (r\sigma)\, dr$$

fonctions de Paley-Littlewood tangentielle complexe et radiale.

Alors on a d'après [3].

Théorème 2 : Si f est harmonique réelle et si $f \in L^p(\partial B)$ pour $2 \leq p < +\infty$, alors

(47)
$$\|g_t(f)\|_{L^p} \leq C\|f\|_{L^p(\partial B)}$$
$$\|g_n(f)\|_{L^p} \leq C\|f\|_{L^p(\partial B)}$$

(pour $p=2$, cela coïncide avec le théorème 1).

On introduit également la fonction d'aire brownienne ; plus précisément soit $Z_t(\omega)$ le mouvement brownien de la métrique de Bergmann supposé issu de 0 dont le générateur infinitésimal est $\frac{1}{2}\Delta$ avec Δ donné par (2). Alors posons pour f harmonique réelle

(41)
$$Y_t^2(f)(\omega) = \int_0^t \|\nabla f\|^2 (Z_\omega(s))ds$$

Là encore nous avons d'après [3].

Théorème 3 : Si f est harmonique réelle et $f \in L^p(\partial B)$, pour $2 \leq p < +\infty$, alors

(49)
$$E(Y_t^p) \leq C\|f\|_{L^p(\partial B)}$$

avec C indépendant de $t < +\infty$

(le temps de vie du processus $Z_\omega(t)$ est infini dans la boule $\|\nabla f\|^2$ désigne le carré de la longueur du gradient complet dans la métrique de Bergmann cf. (23)).

De même soit $A_\alpha^h(\sigma)$ le domaine admissible de Koranyi de sommet σ, de hauteur h et d'ouverture α (voir [ ] et [ ]) : la fonction d'aire classique est alors

(50)
$$S_\alpha^h(f)(\sigma) = \int_{A_\alpha^h(\sigma)} \|\nabla f\|^2 (Z)\, dv_{Berg}(Z).$$

On a alors (voir toujours [3])

Théorème 4 : Si f est harmonique réelle et $f \in L^p(\partial B)$ pour $2 \leq p < +\infty$, alors

(51) $\| S^h_\alpha(f)\|_{L^{p/2}} \leq C\|f\|_{L^p(\partial B)}$

De plus les normes de (50) et (49) sont équivalentes à la norme L^p.

Regardons le cas $f \in L^\infty$. On a

Lemme 9 : Si $f \in L^\infty$, est harmonique réelle, pour tout $t > s$.

(52) $E(Y_t^2 - Y_s^2 \mid \mathcal{B}_s) \leq C\|f\|^2_{L^\infty}$ (c indépendant de t,s)

En particulier il existe $\alpha, \beta > 0$ ne dépendant que de $\|f\|_{L^\infty}$ avec

(53) $E(\exp(\alpha Y_t^2)) \leq \beta$.

Preuve : On a en effet

$$
\begin{aligned}
E(Y_t^2 - Y_s^2 \mid \mathcal{B}_s) &= E_0 \left(\int_s^t \|\nabla f\|^2 (Z_\omega(u))\, du \mid \mathcal{B}_s\right) \\
&= E_{Z_s(\omega)} \left(\int_0^{t-s} \|\nabla f\|^2 (Z_\omega(u))du\right) \\
&\leq \int_{B(0,1)} g(Z_s(\omega),z) \|\nabla f\|^2(z)dv(z) .
\end{aligned}
$$

Mais ceci n'est autre que le potentiel de Green de Bergmann de $\|\nabla f\|^2 = \Delta f^2$ calculé en $Z_s(\omega)$ qui se majore donc par $2\|f\|^2_{L^\infty}$. Alors (52) découle de cela et implique que Y_t^2 est "BMO", d'où par le raisonnement usuel (cf. §. 1 et 2) l'estimée (53).

Du fait que pour tout $t < +\infty$ fixé, Y_t^2 est BMO découle que Y_t est une variable aléatoire BMO, i.e. on a

Lemme 10 : Soit $f \in L^\infty$, f harmonique réelle, t fixé. Alors

(54) $\| \sup_{s \leq t} E(|Y_t - E(Y_t \mid \mathcal{B}_s)| \mid \mathcal{B}_s)\|_{L^\infty} \leq C\|f\|_{L^\infty}$

où C est indépendant de $t > s$.

Preuve : Pour le voir il suffit de faire voir (54) avec Y_t remplacé par Y_t^2 (cf. lemme 8 du [11]).

Dans ce cas, cela se contrôle par

$$E(|Y_t^2 - E(Y_t^2|\mathcal{B}_s)|\,|\mathcal{B}_s) \leq E(|Y_t^2 - Y_s^2||\mathcal{B}_s) + E(|\int_s^t \ldots|\,|\mathcal{B}_s)$$

$$\leq 2\int g(Z_\omega(s),\zeta)\ \|\nabla f\|^2(\zeta)\ dv(\zeta)\ .$$

D'autre part, nous avons aussi

$$E(Y_t^2) = E(\int_0^t \|\nabla f\|^2(Z_s(\omega))ds)$$

$$\leq \int g(0,\zeta)\,\|\nabla f\|^2(\zeta)\ dv(\zeta) = \int_{\partial B} |f|^2\ d\sigma - |f|^2(0)$$

d'où (55) $E(Y_t^2) \leq \|f\|^2_{L^2(\partial B)}$

Par (55), (54) l'application

$$T_t : f \longmapsto Y_t(f)(\omega)$$

est un opérateur opérant de $L^2(\partial B)$ dans $L^2(\Omega)$ et de $L^\infty(\partial B)$ dans $BMO(\Omega)$. D'après C interpolation de Stroock, cela opère de $L^p(\partial B)$ dans $L^p(\Omega)$ avec constante indépendante de t vu (54) et (55) ce qui redémontre (49). Alors les estimées des normes L^p des intégrales d'aires et de Paley-Littlewood ((51) et (47)) s'obtiennent facilement. Du même coup, on obtient aussi le résultat.

Théorème 5 : Soit f harmonique réelle bornée. Il existe $\alpha, \beta > 0$ ne dépendant que de $\|f\|_{L^\infty(\partial B)}$

(56) $$\int_{\partial B} d\sigma\ \exp\left[\alpha(g_t(f)^2(\sigma) + g_n(f)^2(\sigma))\right] \leq \beta$$

et μ, $v > 0$ ne dépendant que de α, h et $\|f\|_{L^\infty(\partial B)}$

(57) $$\int_{\partial B} d\sigma \exp(u\ S_\alpha^h(f)(\sigma)) \leq v.$$

Preuve : partons de l'inégalité (53) et écrivons que

$$E_0 = \int_{\partial B} d\sigma\ E_0^\sigma$$

où E_0^σ est l'espérance pour le processus conditionné pour partir de 0 à $t = 0$ et arriver en $\sigma \in \partial B$ à $t = +\infty$. Notant $P_\sigma(z)$ le noyau de Poisson (de la métrique de Bergmann) de pôle $\sigma \in \partial B$ et calculé en $z \in B$, la fonction de Green du processus E_o est $g(0,z)\,\frac{P_\sigma(z)}{P_\sigma(0)} = g(0,z)\,P_\sigma(z)$ et donc par (53) et la convexité de exp, on a

$$(58)\qquad \int_{\partial B} d\sigma \exp \alpha\Big(\int_{B(0,1)} g(0,z)\,P_\sigma(z)\,\|\nabla f\|^2(z)\,dv(z)\Big) \leq \beta \quad .$$

L'intégrale sous l'exponentielle se minore par la même intégrale sur $A^h_\alpha(\sigma)$. La forme de $A^h_\alpha(\sigma)$ nous apprend que

$$(59)\qquad g(0,z)\,P_\sigma(z) \geqslant k_{\alpha,h} \quad \text{si} \quad z \in A^h_\alpha(\sigma) \ .$$

En effet, plaçons nous dans l'espace hermitien hyperbolique en version non compacte. D'après A. Debiard, [3]

$$(60)\qquad P_\sigma(z) = C\,\frac{h(Z)^2}{|p(\sigma,z)|^4}$$

et $A^h_\alpha(Z) = \{z \in D / h(z) \leq h,\ |p(\sigma,z)| \leq C'h(z)\}$

et $g(0,z) \sim h(Z)^2$, d'où le résultat.

Donc (58) et (59) nous apprennent que

$$\int_{\partial B} d\sigma \exp(\mu\, S^h_\alpha(f)(\sigma)) < \nu \quad .$$

c'est-à-dire (57). Le passage à (56) se fait par grossissement des fonctions de Paley-Littlewood de la façon utilisée de [11] par exemple.

4. Contrôle L^2 global des dérivées secondes

a) identités préliminaires.

Prenons f harmonique réelle définie sur $B(0,1)$ et se prolongeant C^4 à $\overline{B(0,1)}$. Appliquons à la fonction f^2 la formule (19) lemme 3 (①) qui nous dit que

$$(61)\qquad \int_{B(0,1)} g(|Z|^2)(\Delta^2 f^2)(Z)\,dv(Z) \leq C\|f\|^2_{L^2(\partial B)}$$

où on a controlé $\Delta(f^2)(0) = \|\nabla f\|^2(0)$ par la norme L^2 de f au bord.

Comme

$$g(|Z|^2)\, dv(Z) \simeq \frac{dv_{eucl}(Z)}{1-|Z|^2}$$

Les lemmes 6 et 7 nous disent que

$$\int_{B(0,1)} \Big[(1-|Z|^2)^2|\nabla_n f|^2+|\nabla_t f|^2+(1-|Z|^2)^3(|\nabla_n^2 f|^2+|\nabla_n \overline{\nabla}_n f|^2)+(1-|Z|^2)^2|\nabla_t \nabla_n f|^2 +$$

$$+(1-|Z|^2)^2(|\nabla_n \nabla_t f|^2+|\nabla_n \overline{\nabla_t f}|^2)+(1-|Z|^2)(|\nabla_t^2 f|^2+|\nabla_t \overline{\nabla_t f}|^2)\Big]\, dv_{eucl}(Z) \leq$$

$$(62)\quad \begin{aligned} &\leq C\|f\|^2_{L^2(\partial B)} + C'\Big|\int_{B(0,1)} \Big[(1-|Z|^2)^2(\overline{\nabla_n^2 f}\, \nabla_n f + \overline{\nabla_n f}(\overline{\nabla_n \overline{\nabla_n f}}) + \text{conjugué}) + \\ &+(1-|Z|^2)^2(\Delta_{eucl} f\, \overline{\nabla_n f} + \text{conjugué})+(1-|Z|^2)((\nabla_n \nabla_t f)(\overline{\nabla_t f}) + \\ &+ (\nabla_n \overline{\nabla_t f})(\nabla_t f) + \text{conjugué} + (\nabla_t \overline{\nabla_n f} + \nabla_n \nabla_t f)\, \overline{\nabla_t f} + \\ &+ (\overline{\nabla}_t \nabla_n f + \overline{\nabla_n \nabla_t f})(\nabla_t f))\Big]\, dv_{eucl}(z)\Big| \end{aligned}$$

où C' est constante indépendante de t. Clairement,

$$(63)\quad \Delta_{eucl} f \simeq (\nabla_n \overline{\nabla}_n f + \nabla_t \overline{\nabla}_t f + \nabla_n \overline{\nabla}_t f + \nabla_t \overline{\nabla}_n f)$$

Nous allons contrôle l'intégrale figurant au second membre de (62) par la technique $(\varepsilon, \frac{1}{\varepsilon})$; soit ε assez petit pour que $\varepsilon C' < 1$. Alors on a $a\bar{b} + b\bar{a} \leq \varepsilon|a|^2 + \frac{1}{\varepsilon}|b|^2$ car c'est $|\sqrt{\varepsilon}\, a + \sqrt{\frac{1}{\varepsilon}}\, b|^2 > 0$.

$$(64)\quad \begin{aligned} &(1-|Z|^2)^2(\overline{\nabla_n^2 f}\, \nabla_n f + \text{conjugué}) \leq \frac{1}{\varepsilon}(1-|Z|^2)|\nabla_n f|^2 + \varepsilon(1-|Z|^2)^3|\nabla_n^2 f|^2 \\ &(1-|Z|^2)^2(\overline{\nabla_n f}\, \overline{\nabla_n \overline{\nabla_n f}} + \text{conjugué}) \leq \frac{1}{\varepsilon}(1-|Z|^2)|\nabla_n f|^2 + \varepsilon(1-|Z|^2)^3|\nabla_n \overline{\nabla_n f}|^2 \end{aligned}$$

Cela règle le compte du premier terme en $(1-|Z|^2)^2$ de l'intégrale du 2nd membre de (62). On règle leur compte pareillement au 2nd terme en $(1-|Z|^2)^2$ en tenant compte de (63). Enfin, il y a les termes en $1-|Z|^2$: ils se contrôlent

$$(1-|Z|^2)((\nabla_n \nabla_t f)\overline{\nabla_t f} + \text{conjugué}) \leq \varepsilon(1-|Z|^2)^2|\nabla_n \nabla_t f|^2 + \frac{1}{\varepsilon}|\nabla_t f|^2$$

(65)

$$(1-|Z|^2)((\nabla_n \overline{\nabla_t f})(\nabla_t f) + \text{conjugué}) \leq \varepsilon(1-|Z|^2)^2|\nabla_n \overline{\nabla_t f}|^2 + \frac{1}{\varepsilon}|\nabla_t f|^2$$

Mais $\nabla_n \nabla_t = \nabla_t \nabla_n + \nabla_t$ par la formule (22)

$$\nabla_n \overline{\nabla}_t = \overline{\nabla}_t \nabla_n + \nabla_t$$

$$\overline{\nabla}_n \nabla_t = \nabla_t \overline{\nabla}_n + \nabla_t .$$

Par suite

$$(1-|Z|^2)((\nabla_t \overline{\nabla_n f})\overline{\nabla_t f} + \text{conjugué}) = (1-|Z|^2)\Big[((\overline{\nabla}_n \nabla_t f) \times \overline{\nabla_t f} - |\nabla_t f|^2) +$$

(66)

$$+ \text{conjugué}\Big] \leq \varepsilon(1-|Z|^2)^2|\nabla_n \overline{\nabla_t f}|^2 + \frac{1}{\varepsilon}|\nabla_t f|^2$$

b) contrôle L^2 des dérivées secondes.

Mettant ensemble (64), (65), (66) dans (62), on déduit alors le théorème suivant

Théorème 6 : Soit f harmonique réelle de $L^2(\partial B)$. On a alors

$$(67) \quad \int_{B(0,1)} (1-|Z|^2)^3(|\nabla_n^2 f|^2+|\nabla_n \overline{\nabla}_n f|^2)+(1-|Z|^2)^2(|\nabla_t \nabla_n f|^2+|\nabla_n \overline{\nabla_t f}|^2)+$$

$$+ (1-|Z|^2)(|\nabla_t^2 f|+|\nabla_t \overline{\nabla_t f}|^2)\Big] dv_{eucl} \leq C \|f\|^2_{L^2(\partial B)}$$

Preuve : Si f est C^4 jusqu'au bord, on applique (62), (64),(65), (66) et on estime alors les termes en $\frac{1}{\varepsilon}$ par l'estimées des dérivées $1^{\text{ère}}$ du théorème 1.

Si f est seulement L^2 du bord, soit $f_\rho(Z) = f(\rho Z)$ pour $\rho < 1$ et soit $\rho_n \longrightarrow 1$ et $f_n = f_{\rho_n}$. Alors f_n est C^4 et on peut appliquer l'estimée (67) à f_n au lieu de f et même avec $f_p - f_q$, pour déduire que $(f_p)_p$ est suite de Cauchy pour la norme L^2 à poids figurant au premier membre de (67). Comme $f_p \longrightarrow f$ en particulier au sens des distribution et que $f_p \longrightarrow g$ au sens de la norme figurant en ce premier membre de (67), on déduit la vérité de la formule pour f.

Remarque : ici encore, si f est pluriharmonique l'estimée du théorème 6 se fait en quelques secondes pour l'intégrale de Fourier sur $SU(2)$.

5. Fonction d'aire brownienne pour les dérivées d'ordre 2

Introduisons le processus

$$Y^2_{2,t}(f)(\omega) = \int_0^t \varphi(f)^2 \, (Z_\omega(s))ds \tag{68}$$

où on pose

$$\begin{aligned}\varphi(f)^2(z) = &(1-|Z|^2)^4(|\nabla_n^2 f|^2 + |\nabla_n \bar{\nabla}_n f|^2) + (1-|Z|^2)^3(|\nabla_n \nabla_t f|^2 + |\nabla_n \overline{\nabla_t f}|^2) + \\ &+(1-|Z|^2)^2(|\nabla_t^2 f|^2 + |\nabla_t \bar{\nabla}_t f|^2)\end{aligned} \tag{69}$$

Lemme 11 : On a pour f harmonique réelle avec $f \in L^2(\partial B)$ que

$$\|Y_{2,t}(f)\|_{L^2(\Omega)} \leq C\|f\|_{L^2(\partial B)} \tag{70}$$

où C est indépendant de f et t.

Preuve : $E(Y^2_{2,t}(f)) = E(Y^2_{2,\infty}(f)) = \int_{B(0,1)} g(|Z|^2)\varphi(f)^2(Z)\, dv(Z)$

et compte tenu du comportement de g au bord (cf formule $g(|Z|^2)\, dv = \frac{dv_{eucl}}{1-|Z|^2}$) cela est contrôlé par le 1er membre de la formule (67) du théorème 6 donc par $\|f\|^2_{L^2(\partial B)}$.

On a alors aussi

Lemme 12 : Sous l'hypothèse f harmonique réelle bornée, alors

$$E(Y^2_{2,t}(f) - Y^2_{2,s}(f) \mid \mathcal{B}_s) \leq C\|f\|^2_{L^\infty} \tag{71}$$

Preuve : le 1er membre de (71) est égal à

$$E\left(\int_s^t \varphi(f)^2(Z_u)du \,\big|\, \mathcal{B}_s\right) = E_{Z_s(\omega)}\left(\int_0^{t-s} \varphi(f)^2(Z_u)du\right)$$

(72)

$$= \int_{B(0,1)} g(Z_s(\omega),\zeta)\varphi(f)^2(\zeta)\, dv(\zeta)$$

Maintenant, essentiellement d'après (30) $\longrightarrow$ (35), et la formule $\Delta^2 f^2 = \sum_{i=1}^{6} K_i$, nous voyons que

$$\varphi(f)^2(\zeta) = (\Delta^2 f^2)(\zeta) + \alpha(1-|z|^2)^3\Big[|\nabla_n f|^2 + (|\nabla_n^2 f\, \nabla_n f + \overline{\nabla_n f}\, \nabla_n \overline{\nabla_n f} +$$

(73)

$$+ \text{conjugué}) + ((\Delta_{eucl} f)\,\overline{\nabla_n f} + \text{conjugué})\Big] + \alpha(1-|z|^2)^2\Big[(\nabla_n\nabla_t f)(\overline{\nabla_t f}) +$$

$$+ (\nabla_n \overline{\nabla_t f})(\nabla_t f) + (\nabla_t \overline{\nabla_n f})\,\overline{\nabla_t f} + \text{conjugué}\Big] + \alpha(1-|z|^2)|\nabla_t f|^2$$

Dans (73), majorons tous les termes en prenant le supremum M de α et en majorant la somme des doubles produits et de leur conjugué par la méthode $(\varepsilon, \frac{1}{\varepsilon})$ en prenant $\varepsilon M < 1$ et en mettant systématiquement le ε devant les dérivées secondes et en répartissant les puissances de poids convenablement, faisons ensuite repasser ses dérivées 2ndes du premier membre ; alors, il vient :

(74)

$$\varphi(f)^2(\zeta) \leq (\Delta^2 f^2)(\zeta) + M'((1-|z|^2)^3|\nabla_n f|^2 + (1-|z|^2)|\nabla_t f|^2 +$$
$$+ (1-|z|^2)^2|\nabla_n f|^2 + (1-|z|^2)|\nabla_t f|^2)$$

Reportons la majoration (74) dans (72) ; posant $z = Z_\omega(s)$

$$\int_{B(0,1)} g(z,\zeta)\varphi(f)^2(\zeta)dv(\zeta) \leq \lim_{R\to 1}\Big[\int_{B(0,R)} g_R(z,\zeta)(\Delta^2 f^2)(\zeta) +$$

$$+ M\int_{B(0,R)} g_R(z,\zeta)\,\|\nabla f\|^2\, dv(\zeta)\Big]$$

et comme $\Delta f^2 = \|\nabla f\|^2$

(75) $$\leq \lim_{R\to 1}\left(\int_{\partial B(0,R)} \|\nabla f\|^2(\zeta)d\sigma(\zeta) - \|\nabla f\|^2(0) + \int_{\partial B(0,R)} |f|^2(\zeta)d\sigma(\zeta) - |f|^2(0)\right)$$

Mais d'après le lemme de Putz (cf. [3] et [13]) si f est bornée harmonique

$$(1-|z|^2)\ |\nabla_n f| \leq C\|f\|_{L^\infty}$$
$$(1-|z|^2)^{1/2}\ |\nabla_t f| \leq C\|f\|_{L^\infty}\ .$$

(ce qui se voit d'ailleurs à la main), d'où

(76) $$\|\nabla f\|^2 \simeq (1-|z|^2)^2\ |\nabla_n f|^2 + (1-|z|^2)\ |\nabla_t f|^2 \leq C\|f\|^2_{L^\infty}$$

et d'après (76), (75) et (72), on conclut à l'affaire cherchée.

Le raisonnement usuel permet alors de conclure

Théorème 7 : 1 Soit $2 \leq p < +\infty$. Alors

(77) $$E(Y^p_{2,t}(f)) \leq C\|f\|^p_{L^p(\partial B)}$$

où f est harmonique réelle, C est indépendante de t et f.

2 Si $p = +\infty$ et f est donc bornée, il existe α, β ne dépendant que de $\|f\|_{L^\infty}$ avec

(78) $$E(\exp(\alpha\, Y_{2,\infty}(f))) \leq \beta$$

6. Passage aux fonctions d'aires et de Paley-Littlewood, pour les dérivées d'ordre 2

Pour f harmonique, introduisons

(79) $$g_2(f)^2(\sigma) = \int_0^1 \frac{\varphi(f)^2(r\sigma)}{1-r^2}\, dr$$

si $\sigma \in \partial B(0,1)$. De même posons

(80) $$S^h_{2,\alpha}(f)(\sigma) = \int_{A^h_\alpha(\sigma)} \varphi(f)^2(z)\ dv(z)$$

$\varphi(f)^2$ ayant été défini par la formule (69).

Ecrivons alors pour Φ fonction convexe

$$E_o(\Phi(Y^2_{2,t}(f))) \geq \int d\sigma\, \Phi(E^\sigma_o\ (Y^2_{2,t}(f))$$

Ensuite

$$E_o^{\sigma}(Y^2_{2,t}(f)) = \int g(0,z)\, P_{\sigma}(z)\, \varphi(f)^2(z)\, dv(z)$$

$$\geq \int_{A^h_{\alpha}(\sigma)} g(0,z)\, P_{\sigma}(z) \varphi(f)^2\, (z)\, dv(z)$$

$$\geq C_{\alpha,h} \int_{A^h_{\alpha}(\sigma)} \varphi(f)^2(z)\, dv(z) \qquad \text{par (61)}$$

$$= C_{\alpha,h}\, S^h_{Z,\alpha}(f)\,(\sigma)$$

Appliquons à $\Phi(X) = x^{p/2}$ $(p \geq 2)$ et $\Phi(X) = e^{\alpha x}$ $(\alpha > 0)$; il vient par (77), (78) et ce qui précède.

Théorème 8 : 1 Soit f harmonique réelle avec $f \in L^p(\partial B)$ $(p \geq 2)$. On a

$$(81) \qquad \int_{\partial B(0,1)} d\sigma\, S^h_{\alpha}(f)^{p/2}(\sigma) \leq C \|f\|^p_{L^p(\partial B)}$$

où C ne dépend que de α et h

2 si f est bornée, il existe $\mu,\ v > 0$ ne dépendant que de α, h et $\|f\|_{L^\infty}$ avec

$$(82) \qquad \int_{\partial B(0,1)} d\sigma \exp(\mu\, S^h_{\alpha}(f)(\sigma)) \leq v$$

Même résultat pour $g_2(f)$.

7. Fonctions d'aires non linéaires des dérivées premières

a) calcul de $\Delta^2 f^4$.

Calculons d'abord Δf^4 pour f harmonique réelle :

c'est $\Delta f^4 = 4\, f^2 \|\nabla f\|^2$. Calculons $\Delta^2 f^4 = 4 \Delta\, (f^2 \|\nabla f\|^2)$

c'est

$$\Delta\, (f^2 \|\nabla f\|^2) = \|\nabla f\|^4 + 2(\nabla f^2 | \nabla\, (\|\nabla f\|^2)) + f^2 \Delta (\|\nabla f\|^2)$$

$$(83) \qquad = \|\nabla f\|^4 + L + f^2 \sum_{i=1}^{6} K_i .$$

Calculons L : on a par (23) :

$$\frac{1}{4} L = \frac{1}{2} (\nabla f^2 | \nabla(\|\nabla f\|^2)) = f(\nabla f | \nabla(\|\nabla f\|^2))$$

(84)
$$= f(1-|Z|^2)^2 \nabla_n f \bar{\nabla}_n(\|\nabla f\|^2) +$$
$$f(1-|Z|^2) \nabla_t f \bar{\nabla}_t(\|\nabla f\|^2) + \text{conjugué}.$$

Calcul de $\bar{\nabla}_n(\|\nabla f\|^2)$

on a $\|\nabla f\|^2 = (1-|Z|^2)^2 |\nabla_n f|^2 + (1-|Z|^2)|\nabla_t f|^2$

$$\bar{\nabla}_n(\|\nabla f\|^2) = 2(1-|Z|^2)\alpha|\nabla_n f|^2 + (1-|Z|^2)^2 \bar{\nabla}_n(\|\nabla_n f\|^2) + \alpha\,(|\nabla_t f|^2)$$
$$+ (1-|Z|^2) \bar{\nabla}_n(|\nabla_t f|^2)$$

d'où

$$\bar{\nabla}_n(\|\nabla f\|^2) = (1-|Z|^2)^2\left[(\bar{\nabla}_n^2 f)(\nabla_n f) + (\bar{\nabla}_n \nabla_n f)\bar{\bar{\nabla}}_n f\right] + \alpha(1-|Z|^2)\left[\,|\nabla_n f|^2 +\right.$$

(85)
$$\left. + (\overline{\bar{\nabla}_n \nabla_t f})(\nabla_t f) + (\bar{\nabla}_n \nabla_t f)(\bar{\nabla}_t f)\right] + \alpha|\nabla_t f|^2.$$

Calcul de $\bar{\nabla}_t(f\|\nabla f\|^2)$

Comme ∇_t ne dérive pas $1-|Z|^2$; on a

$$\bar{\nabla}_t(\|\nabla f\|^2) = (1-|Z|^2)^2 \bar{\nabla}_t(|\nabla_n f|^2) + (1-|Z|^2)(\bar{\nabla}_t |\nabla_t f|^2)$$

d'où

$$\bar{\nabla}_t(\|\nabla f\|^2) = (1-|Z|^2)^2\left[\overline{\nabla_t \nabla_n f}\; \nabla_n f + (\bar{\nabla}_t \nabla_n f)(\bar{\nabla}_n f)\right] +$$

(86)
$$+ (1-|Z|^2)\left[(\overline{\nabla_t \nabla_t f})(\nabla_t f) + (\bar{\nabla}_t \nabla_t f)(\bar{\nabla}_t f)\right]$$

Posons alors

(87) $$U_{4,t}(f)^4(\omega) = \int_0^t \|\nabla f\|^4 (Z_\omega(s))\, ds.$$

b) Contrôle L^4.

<u>Lemme 13</u> : <u>Soit</u> f <u>harmonique réelle de classe</u> $L^4(\partial B)$. <u>Alors</u>

(88) $$E(U_{4,t}(f)^4) \leq C\|f\|^4_{L^4(\partial B)}$$

<u>où</u> C <u>est indépendant de</u> t <u>et</u> f.

Preuve : par (83), nous avons

$$\|\nabla f\|^4 = \Delta^2 f^4 - L - f^2 \ \Delta^2 f^2$$

Si nous reportons cela dans la définition de $U_{4,t}(f)$ (formule (87)) et que nous utilisons le lemme 3, (formule ((9)), nous obtenons un contrôle trivial de

$$E_o(U_{4,t}(f)^4) \leq E_o(U_{4,\infty}(f)^4)$$

(89) $$\leq \left| \int g(|Z|^2)\, \Delta^2 f^4 \, dv + \int g(|Z|^2)\, L\, dv + \right.$$

$$\left. + \int g(|Z|^2)\, f^2\, \Delta^2 f^2 \, dv \right| \equiv I_1 + I_2 + I_3$$

Le premier terme contrôle par (19) sans problème; étudiens

$I_3 = \left| \int g(|Z|^2)\, f^2\, \Delta^2 f^2 \, dv . \right|$

Soit $f^*(\sigma)$ la fonction maximale radiale

$$f^*(\sigma) = \sup_{0<r<1} |f(\sigma r)|$$

(90) $$I_3 \leq \int d\sigma\, (f^*)^2(\sigma) \left(\int_o^1 g(|Z|^2)\, (\Delta^2 f^2)(r\sigma) dv \right)$$

$$\leq A \|f^*\|_{L^4}^{1/2} \|g_2(f)\|_{L^4}^{1/2} \leq C \|f\|_{L^4}^4 .$$

(où on a contrôlé $\int_o^1 g(|Z|^2)(\Delta^2 f^2)(r\sigma) dv$ par la fonction de Paley Littlewood (79), les termes parasites (le double produit figurant dans $\Delta^2 f^2$ se contrôlant comme d'habiture)

étude de $I_2 = \left| \int g(|Z|^2)\, L\, dv \right|$

Ici L est donné par (84), (85) et (86) et I_2 se contrôle par

$$I_2 \leq \int d\sigma\, f^*(\sigma) \int_o^1 \frac{1}{1-r^2} \Big| (1-r^2)^2\, (\nabla_n f)(\bar{\nabla}_n (\|\nabla f\|^2)) +$$

$$+ (1-r^2)\, \nabla_t f\, (\bar{\nabla}_t \|\nabla f\|^2) + \text{conjugué} \Big| \, dr$$

puisque $g(|Z|^2)\, dv(Z) = \dfrac{dv_{eucl}(Z)}{1-|Z|^2}$

$$I_2 \leq \left| \int d\sigma f^*(\sigma) \int_0^1 \left[(1-r^2) \left((\nabla_n f)(\bar{\nabla}_n \|\nabla f\|^2) + \text{conjugué} \right) + (\nabla_t f \, \bar{\nabla}_t \|\nabla f\|^2 + \right.\right.$$

(91)

$$\left.\left. + \text{conjugué} \right] dr \right| $$

Nous majorons chacune des valeurs absolues : les termes normaux de (91) donnent :

$$(1-r^2) \left| (\nabla_n f)(\bar{\nabla}_n \|\nabla f\|^2) + \text{conjugué} \right| \leq (1-r^2)^3 \; (\nabla_n f)^2 (\bar{\nabla}_n^2 f) +$$

(92)

$$+ \left| \nabla_n f \right|^2 (\bar{\nabla}_n \nabla_n f) + \text{conjugué} \; \left| + (1-r^2) \right| (\nabla_n f)^3 + (\nabla_n f)(\overline{\nabla_n \nabla_t f})(\nabla_t f) +$$

$$+ (\nabla_n f)(\bar{\nabla}_n \nabla_t f)(\bar{\nabla}_t f) + \text{conjugué} \Big| + (1-r^2)(|\nabla_n f| |\nabla_t f|^2)$$

ce qui se majore par une technique (ε , $\frac{1}{\varepsilon}$) en prenant bien soin de mettre le ε devant les puissances quatrièmes des dérivées 1ères de façon à les faire repasser par la suite au 1er membre de (89) : on a

$$f^*(1-r^2)^3((\nabla_n f)^2(\bar{\nabla}_n^2 f) + \text{conjugué}) \leq \varepsilon (1-r^2)^3 |\nabla_n f|^4 + \frac{1}{\varepsilon} (1-r^2)^3 |\bar{\nabla}_n^2 f|^2 \; (f^*)^2$$

$$f^*(1-r^2)^3(|\nabla_n f|^2(\bar{\nabla}_n \nabla_n f) + \text{conjugué}) \leq \varepsilon (1-r^2)^3 |\nabla_n f|^4 + \frac{1}{\varepsilon} (1-r^2)^3 |\bar{\nabla}_n \nabla_n f|^2 (f^*)^2$$

$$f^*(1-r^2)^2((\nabla_n f)^3 + \text{conjugué}) \leq \varepsilon (1-r^2)^3 |\nabla_n f|^4 + \frac{1}{\varepsilon} (1-r^2) |\nabla_n f|^2 (f^*)^2$$

(93)

$$f^*(1-r^2)^2((\nabla_n f)\,\overline{(\nabla_n \nabla_t f)}\,(\nabla_t f) + \text{conjugué}) \leq \varepsilon (1-r^2)^2 |\nabla_n f|^2 |\nabla_t f|^2 + \frac{1}{\varepsilon} (1-r^2)^2 |\nabla_n \nabla_t f|^2 (f^*)^2$$

$$f^*(1-r^2)^2((\nabla_n f)(\bar{\nabla}_n \nabla_t f)\,\overline{(\nabla_t f)} + \text{conj.}) \leq \varepsilon (1-r^2)^2 |\nabla_n f|^2 |\nabla_t f|^2 + \frac{1}{\varepsilon} (1-r^2) |\bar{\nabla}_n \nabla_t f|^2 \quad (f^*)^2$$

$$f^*(1-r^2)(|\nabla_n f| |\nabla_t f|^2) \leq \varepsilon (1-r^2)^2 \; |\nabla_n f|^2 |\nabla_t f|^2 + \frac{1}{\varepsilon} \; |\nabla_t f|^2 (f^*)^2 \; .$$

Par suite dans (91) la 1ère intégrale au second membre se majore par

$$C \int d\sigma \int_0^1 \varepsilon \left[(1-r^2)^3 |\nabla_n f|^4 + (1-r^2)^2 |\nabla_n f|^2 \; |\nabla_t f|^2 \right] dr +$$

(94)

$$+ \frac{1}{\varepsilon} \int d\sigma (f^*(\sigma))^2 \int_0^1 \left[(1-r^2)^3 (|\bar{\nabla}_n^2 f|^2 + |\bar{\nabla}_n \nabla_n f|^2) + (1-r^2)^2 (|\nabla_n \nabla_t f|^2 + \right.$$

$$\left. + |\bar{\nabla}_n \nabla_t f|^2) + (1-r^2) |\nabla_n f|^2 + |\nabla_t f|^2 \right] dr$$

Pour ε assez petit, la première intégrale en ε de (94) peut repasser au premier membre de (89) puisque, le premier membre de (89) est

$$E_0(U_{4,\infty}(f)^4) = \int_B g(|Z|^2)\|\nabla f\|^4 \, dv(Z) = \int_B \|f\|^4(Z) \frac{dv_{eucl}(Z)}{1-|Z|^2}$$

(95) $$= \int_B \Big[(1-|Z|^2)^3 |\nabla_n f|^4 + 2(1-|Z|^2)^2 |\nabla_n f|^2 |\nabla_t f|^2 + (1-|Z|^2)|\nabla_t f|^4\Big] \, dv_{eucl}(Z)$$

La 2nd intégrale $\frac{1}{\varepsilon}$ de (94) se contrôle par Cauchy Schwarz, i.e. par $\frac{1}{\varepsilon}\|f^*\|^{1/2}_{L^4}(\|g_2(f)\|^{1/2}_{L^4} + \|g_1(f)\|^{1/2}_{L^4}) \leq C\|f\|^4_{L^4}$ en utilisant les fonctions de Paley-Littlewood $g_2(f)$ et $g_1(f)$ des dérivées 2ndes et 1ère respectivement (voir formules (79), (45) et (46) du ⑥ et du ③).

Ensuite, il faut majorer les termes tangentiels complexes de la 2nde valeur absolue dans (91). Utilisant (86), on a

(96) $$\Big|(\nabla_t f)\bar{\nabla}_t(\|\nabla f\|^2) + \text{conjugué}\Big| \leq (1-r^2)^2 \Big|(\nabla_t f)\overline{\nabla_t\nabla_n f}\nabla_n f + (\nabla_t f)(\bar{\nabla}_t\nabla_n f)(\overline{\nabla_n f}) + \text{conjugué}\Big| + (1-r^2)\Big|(\nabla_t f)^2\overline{\nabla_t^2 f} + |\nabla_t f|^2(\bar{\nabla}_t\nabla_t f) + \text{conjugué}\Big|$$

qu'on majore ensuite par la méthode $(\varepsilon \frac{1}{\varepsilon})$ selon

(97)
$$\left\{\begin{array}{l}
f^*(1-r^2)^2(\nabla_t f)(\nabla_n f)(\overline{\nabla_t\nabla_n f}) + \text{conjugué} \leq \varepsilon(1-r^2)^2|\nabla_t f|^2|\nabla_n f|^2 + \frac{1}{\varepsilon}(1-r^2)^2 f^{*2}|\nabla_t\nabla_n f|^2 \\
f^*(1-r^2)^2(\nabla_t f)\overline{\nabla_n f}(\bar{\nabla}_t\nabla_n f) + \text{conjugué} \leq \varepsilon(1-r^2)^2|\nabla_t f|^2|\nabla_n f|^2 + \frac{1}{\varepsilon}(1-r^2)^2 f^{*2}|\bar{\nabla}_t\nabla_n f|^2 \\
f^*(1-r^2)(\nabla_t f)^2(\overline{\nabla_t^2 f}) + \text{conjugué} \leq \varepsilon(1-r^2)\cdot|\nabla_t f|^4 + \frac{1}{\varepsilon} f^{*2}|\nabla_t^2 f|^2(1-r^2) \\
f^*(1-r^2)|\nabla_t f|^2(\bar{\nabla}_t\nabla_t f) + \text{conjugué} \leq \varepsilon(1-r^2)\cdot|\nabla_t f|^4 + \frac{1}{\varepsilon} f^{*2}(1-r^2)|\bar{\nabla}_t\nabla_t f|^2
\end{array}\right.$$

et donc dans (91) la seconde valeur absolue au second membre se majore par

(98) $$C\varepsilon\int d\sigma\int_0^1\Big[(1-r^2)^2|\nabla_t f|^2|\nabla_n f|^2 + (1-r^2)|\nabla_t f|^4\Big]dr + \frac{1}{\varepsilon}\int d\sigma\int_0^1 f^*(\sigma)^2\Big[(1-r^2)^2(|\nabla_t\nabla_n f|^2 + |\bar{\nabla}_t\nabla_n f|^2) + (1-r^2)(|\nabla_t^2 f|^2 + |\bar{\nabla}_t\nabla_t f|^2)$$

Dans (98), le terme en peut repasser au premier membre de (89) à cause de la relation (95). Le terme en $\frac{1}{\varepsilon}$ se contrôle par

$$\frac{1}{\varepsilon}\|f^*\|_{L^4}^{1/2}(\|g_2(f)\|_{L^4}^{1/2}) \leq C_\varepsilon \|f\|_{L^4}^4 .$$

Cela achève ainsi la démonstration du lemme 13.

c) contrôle BMO - L^∞

Lemme 14 : Soit f harmonique réelle bornée. On a alors pour $t \geq s$

$$(91) \quad \left\| \sup_{t \geq s} E(U_{4,t}(f)^4 - U_{4,s}(f)^4 \mid \mathcal{B}_s) \right\|_{L^\infty} \leq A \|f\|_{L^\infty}^4$$

où A ne dépend ni de t, ni de f

Preuve : l'espérance conditionnelle au 1er membre de (99) se majore par la méthode usuelle

$$(100) \quad E(U_{4,t}(f)^4 - U_{4,s}(f)^4 \mid \mathcal{B}_s) \leq \sup_{Z \in B} \int g(Z,\zeta) \|\nabla f\|^4(\zeta)\, dv(\zeta) .$$

Utilisons alors que

$$\|\nabla f\|^4 = \Delta^2 f^4 - L - f^2 \Delta^2 f^2$$

Maintenant, regardons chaque terme : on a

$$(101) \quad \int_{B(0,1)} g(Z,\zeta)\, \Delta^2 f^4\, dv(\zeta) = (\Delta f^4)(Z) = f^2 \|\nabla f\|^2 (Z)$$

à cause du lemme 3 formule (19). Mais

$$\|\nabla f\|^2(Z) = (1-|Z|^2)^2 |\nabla_n f|^2(Z) + (1-|Z|^2) |\nabla_t f|^2(Z)$$

et d'après le lemme de Putz ([ ] et [ ]) déjà utilisé au ⑤ formule (76), cela est contrôlé par $2\|f\|_{L^\infty}^2$, d'où

$$(102) \quad \int g(Z,\zeta)\, \Delta^2 f^4\, dv(Z) \leq 4\|f\|_{L^\infty}^4 .$$

Ensuite regardons, toujours par le lemme 3 formule (19)

$$\int g(Z,\zeta) f^2 \Delta^2 f^2\, dv(Z) \leq \|f\|_{L^\infty}^2 \int g(Z,\zeta)\, \Delta^2 f^2\, dv(Z)$$

$$(103) \quad \leq \|f\|_{L^\infty}^2 (\Delta f^2)(Z) \leq 2\|f\|_{L^\infty}^2 \|\nabla f\|^2(Z)$$

$$\leq 4\|f\|^4_{L^\infty}$$

Enfin vient le terme en L : par la définition (83) de L

(104) $\left|\int g(Z,\zeta)L(\zeta)dv(\zeta)\right| \leq \|f\|_{L^\infty}\int g(Z,\zeta)\left|(\nabla f \mid \nabla(\|\nabla f\|^2))\right| dv(\zeta).$

La encore, l'intégrale figurant au second membre de (104) se majore par la méthode (ε , $\frac{1}{\varepsilon}$) en mettant devant les dérivées lère à la puissance 4ème, et $\frac{1}{\varepsilon}$ devant les dérivées 2ndes au carré comme ela a été fait en (93) et (97) lemme 13 ; on refait alors passer les termes en $\varepsilon \times$(dérivées lère à la puissance 4) du côté du potentiel de Green (100) que l'on est en train de contrôler pour ε assez petit. Quant aux termes $\frac{1}{\varepsilon}\times$(dérivées 2nd au carré), ils sont majorés par $\|f\|^2_{L^\infty}$ exactement par la méthode du lemme 1.2

Cela étant, nous avons également le résultat suivant

Lemme 15 : pour tout t, on a

(105) $\left\| \sup_{s\leq t} E(\left|U_{4,t}(f) - E(U_{4,t}(f) \mid \mathcal{B}_s)\right| \mid \mathcal{B}_s)\right\|_{L^\infty(\Omega)} \leq A\|f\|_{L^\infty(B)}$

avec A indépendant de f et t

Ce qui se démontre à partir du lemme 14 formule (99) de la même façon que d'habitude (cf. disque [11]).

Les estimées (105) et (99), le théorème d'interpolation de Stroock [montrent alors

Théorème 9 : Soit f harmonique réelle.

① si f est $L^p(\partial B)$ pour $4 \leq p < +\infty$, alors

(106) $\|U_{4,t}(f)\|_{L^p(\Omega)} \leq A\|f\|_{L^p(\partial B)}$

où A est indépendant de t et f

② si $f \in L^\infty(\partial B)$, alors il existe α, β tels que

(107) $E(\exp(\alpha(U_{4,\infty}(f))^4)) \leq \beta$

où α, β ne dépendent que de $\|f\|_{L^\infty}$

Nous déduisons alors par la technique usuelle de conditionnement par le point d'arrivée et de minoration triviale que si on pose

$$S_{4,(\alpha)}^{(h)}(f)(\sigma) = \int_{A_\alpha^h(\sigma)} \|\nabla f\|^4(Z)\, dv(Z) \tag{108}$$

on a

Théorème 10 : Soit f harmonique réelle.

1 Si $f \in L^p(\partial B)$ pour $4 \leq p < +\infty$, alors

$$\|(S_{4,(\alpha)}^{(h)}(f))^{1/4}\|_{L^p(\partial B)} \leq A\|f\|_{L^p(\partial B)} \tag{109}$$

où A en dépend que de p et f, α et h.

2 Si $f \in L^\infty(\partial B)$, alors il existe μ, v ne dépendant que de $\|f\|_{L^\infty}$, α, h avec

$$\int_{\partial B} d\sigma \exp(\mu\, S_{4,(\alpha)}^{(h)}(f)) \leq v \tag{110}$$

Remarque : ici, l'opérateur $f \longrightarrow U_{4,t}(f)^{1/4}$ est bien sous linéaire sans problème contrairement au cas du bidisque.

8. Applications pseudoconformes de la boule dans $\mathbb{C}^2$

Nous appliquons ici les idées de [9] et nous donnons un début de réponse à la question posée dans ce texte pour le cas de la boule. Soit donc

$$F = (f,g) : B \longmapsto \mathbb{C}^2$$

une application pseudoconforme de la boule dans $\mathbb{C}^2$. Nous fixerons désormais une fois pour toute une ouverture α et une hauteur h pour un domaine admissible de sommet $\sigma \in \partial B$ et nous noterons pour abréger

$$(111)\qquad \begin{aligned} A(\sigma) &= A_{\alpha}^{h}(\sigma) \\ V(\sigma) &= \mathrm{Vol}(F(A(\sigma))) = \int_{A(\sigma)} |J(F)|^2 \, dv_{eucl} \end{aligned}$$

$V(\sigma)$ est donc le volume de l'image par F du domaine admissible $A(\sigma)$ compté avec la multiplicité de recouvrement de $J(F)$ est le Jacobien

$$|J(F)|^2 = \left| \text{dét} \begin{pmatrix} \frac{\partial f}{\partial Z_1} & \frac{\partial g}{\partial Z_1} \\ \frac{\partial f}{\partial Z_2} & \frac{\partial g}{\partial Z_2} \end{pmatrix} \right|^2 = \left| \text{dét} \begin{pmatrix} \frac{\partial u}{\partial Z_1} & \frac{\partial v}{\partial Z_1} \\ \frac{\partial u}{\partial Z_2} & \frac{\partial v}{\partial Z_2} \end{pmatrix} \right|^2$$

où on a posé $u = \mathcal{R}e\, f$, $g = \mathcal{R}e\, g$ et où on a tenu compte des équations de Cauchy-Riemann. Utilisons alors les formules (22) et leur inverse (voir lemme 5 du 2) : on a

$$|Z|^2 \frac{\partial u}{\partial Z_1} = \bar{Z}_1 \nabla_n u - Z_2 \nabla_t u$$

$$|Z|^2 \frac{\partial u}{\partial Z_2} = \bar{Z}_2 \nabla_n u + Z_1 \nabla_t u \;.$$

d'où

$$|Z|^4 \,\text{dét} \begin{pmatrix} \frac{\partial u}{\partial Z_1} & \frac{\partial v}{Z_1} \\ \frac{\partial u}{\partial Z_2} & \frac{\partial v}{\partial Z_2} \end{pmatrix} = (\bar{Z}_1 \nabla_n u - Z_2 \nabla_t u)(\bar{Z}_2 \nabla_n v + Z_1 \nabla_t v) - (\bar{Z}_1 \nabla_n v - Z_2 \nabla_t v) \times (\bar{Z}_2 \nabla_n u + Z_1 \nabla_t u)$$

$$= \bar{Z}_1 \bar{Z}_2 \nabla_n u \nabla_n v - Z_2 Z_1 \nabla_t u \nabla_t v + |Z_1|^2 \nabla_t v \nabla_n u - |Z_2|^2 \nabla_t u \nabla_n v - \bar{Z}_1 \bar{Z}_2 \nabla_n u \nabla_n v + Z_2 Z_1 \nabla_t u \nabla_t v +$$
$$+ |Z_2|^2 \nabla_t v \nabla_n u - |Z_1|^2 \nabla_n v \nabla_t u$$

$$= (|Z_1|^2 + |Z_2|^2) \nabla_t v \nabla_n u - (|Z_1|^2 + |Z_2|^2)(\nabla_t u \nabla_n v)$$

et par suite

$$V(\sigma) \leq \int_{A(\sigma)} |\nabla_t v \nabla_n u - \nabla_t u \nabla_n v|^2 \, dv_{eucl}$$

$$(112)\qquad \leq \int_{A(\sigma)} \left[|\nabla_t v|^2 |\nabla_n u|^2 + |\nabla_t u|^2 |\nabla_n v|^2 + |\nabla_t v \overline{\nabla_n v}\, \nabla_n u \overline{\nabla_t u}| + \overline{\nabla_t v}\, \overline{\nabla_n u} \nabla_t u \nabla_n v \right] dv_{eucl}$$

$$\leq C \int_{A(\sigma)} \left[|\nabla_t v|^2 |\nabla_n u|^2 + |\nabla_t u|^2 |\nabla_n v|^2 + |\nabla_t v|^2 |\nabla_n v|^2 + |\nabla_n u|^2 |\nabla_t u|^2 \right] dv_{eucl}$$

Maintenant, posons $w_\lambda = u + \lambda v$ (λ réel) et considérons

(113) $$B_{(\alpha,h)}(w_\lambda)(\sigma) = \int_{A(\sigma)} |\nabla_n w_\lambda|^2 |\nabla_t w_\lambda|^2 \, dv_{eucl}$$

D'après le théorème 10, formule (109), nous avons

(114) $$\|B_{(\alpha,h)}(w_\lambda)(\sigma)\|^k_{L^k(\partial B)} \leq C \|w_\lambda\|^{4k}_{L^{4k}(\partial B)} \leq C(\|u\|^{4k}_{L^{4k}(\partial B)} + \|v\|^{4k}_{L^{4k}(\partial B)})$$

si λ reste dans un borné de $\mathbb{R}$; cela résulte en effet que dans $S^{(h)}_{4,(\alpha)}(w_\lambda)(\sigma)$, les termes croisés dans $\|\nabla w_\lambda\|^4$ ont un poids $(1-|Z|^2)^3$ qui se compense exactement avec l'élément et volume de Bergmann. D'autre part posons

$$B_{(\alpha,h)}(w_\lambda)(\sigma) = \sum_{l=0}^{4} \lambda^l \, a_l(\sigma)$$

où ici en particulier le coefficient du terme de degré 4 est

(115) $$a_2(\sigma) = \int_{A(\sigma)} \left[|\nabla_n u|^2 |\nabla_t v|^2 + |\nabla_t u|^2 |\nabla_n v|^2 + \nabla_n u \overline{\nabla_t u} \, \overline{\nabla_n v} \nabla_t v + \dots \right] dv_{eucl}$$

et où de plus, par le lemme 9 de [11]

(116) $$|a_2(\sigma)| \leq C \max_{\lambda \in E} |B_{(\alpha,h)}(w_\lambda)(\sigma)|$$

où E est ensemble à 5 points et C est constante universelle indépendante de u, v et σ; α, h. Par suite de (115) et (116), il vient

(117) $$\int_{A(\sigma)} (|\nabla_n u|^2 |\nabla_t v|^2 + |\nabla_t u|^2 |\nabla_n v|^2) \, dv_{eucl} \leq a_2(\sigma) + \left| \int_{A(\sigma)} (\nabla_n u \overline{\nabla_t u} \, \overline{\nabla_n v} \nabla_t v + \dots) \, dv_{eucl} \right|$$

L'intégrale figurant au 2nd membre de (117) se majore par

(118) $$4 \int_{A(\sigma)} (|\nabla_n u|^2 |\nabla_t u|^2 + |\nabla_n v|^2 |\nabla_t v|^2) \, dv_{eucl} = 4 \, (B_{(\alpha,h)}(u)(\sigma) + B_{(\alpha,h)}(v)(\sigma))$$

Reportons alors (118) dans (117), puis (117) dans (112)

$$V(\sigma) \leq a_2(\sigma) + C(B_{(\alpha,h)}(u)(\sigma) + B_{(\alpha,h)}(v)(\sigma))$$

par (116)

$$(119) \quad V(\sigma) \leq C \max_{\lambda \in E} (B_{(\alpha,h)}(w_\lambda)(\sigma)) + C(B_{(\alpha,h)}(u)(\sigma) + B_{(\alpha,h)}(v)(\sigma))$$

Par conséquent l'estimée (114) donne

$$(120) \quad \|V(\sigma)\|^k_{L^k(\partial B)} \leq C(\|u\|^{4k}_{L^{4k}(\partial B)} + \|v\|^{4k}_{L^{4k}(\partial B)})$$

puisque toutes les fonctions de σ figurant au 2nd membre de (119) sont de puissance k^i intégrables si u et v sont de puissance $4k^i$ intégrables.

De même par (119)

$$\exp(\mu\, V(\sigma)) \leq \exp\left[C'\mu\, (B_{(\alpha,h)}(u)(\sigma) + B_{(\alpha,h)}(v)(\sigma))\right]$$

Supposons alors u et v bornées. Utilisons le théorème 10 (formule (110)) et l'inégalité de Cauchy Schwarz ainsi que le fait que

$$B_{(\alpha,h)}(u)(\sigma) \leq S^{(h)}_{4(\alpha)}(u)(\sigma),$$

on conclut qu'il existe μ', v' ne dépendant que de $\|u\|_{L^\infty}$, $\|v\|_{L^\infty}$ et α, h avec

$$(121) \quad \int_{\partial B} d\sigma \exp(\mu' V(\sigma)) \leq v'$$

Nous résumons (120) et (121) dans le théorème.

Théorème 11 : Soit $F = (f,g)$ une application pseudoconforme de la boule B de $\mathbb{C}^2$ à valeurs $\mathbb{C}^2$ et posons $u = \mathcal{R}e\, f$, $v = \mathcal{R}e\, g$, $V(\sigma) = \mathrm{Vol}\, F(A_{(\alpha,h)}(\sigma))$.

① Supposons $\mu, v \in L^p(\partial B)$ pour $4 \leq p < +\infty$. Alors

$$\|V(\sigma)\|_{L^{p/4}(\partial B)} \leq C(\|u\|_{L^p(\partial B)} + \|v\|_{L^p(\partial B)})$$

où C ne dépend que de α et h.

② Supposons u, $v \in L^\infty(\partial B)$. Alors il existe μ, $v > 0$ ne dépendant que de $\|u\|_{L^\infty}$, $\|v\|_{L^\infty}$, α et h tels que

$$\int_{\partial B} d\sigma \exp(\mu V(\sigma)) \leq v.$$

References

[1] BERTHIER,A.M., B. GAVEAU : Critere de convergence des fonctionnelles de Kac et applications en mecanique quantique et en geometrie, Journal of Functional Analysis 29 (1978), 416-424.

[2] COURANT,R., D. HILBERT : Methods of Mathematical Physics II, Interscience, New York 1965.

[3] DEBIARD,A. : Espaces H^p au dessus de l'espace hermitien hyperbolique, Bull. Sci. Math. 2 103 (1979), 305-331.

[4] ——: Integrales d'aire de Lusin-Calderon dans l'espace hermitien hyperbolique, C.R.A.S. Paris 281 (1975), 123-126 et 282 (1976), 1231-1234.

[5] ——, B. GAVEAU : Potentiel fin et algebres de fonctions III, Journal of Functional Analysis 21 (1976), 448-468.

[6] GARSIA, A.: Martingale inequalities, Benjamin, Reading 1973.

[7] GAVEAU, B.: Geometrie integrale dans l'espace hermitien hyperbolique, a paraitre.

[8] MALLIAVIN, M.P., P. MALLIAVIN : Integrales de Lusin-Calderon pour les fonctions biharmoniques, Bull. Sci. Math. 101 (1977), 357-384.

[9] ——, ——: Integrales de volume pour les applications pseudoconformes du bidisque, C.R.A.S. Paris 286 (1978), 617-619.

[10] MALLIAVIN, P.: Processus a temps bidimensionnel et integrale d'aire dans le bidisque, C.R.A.S. Paris 285 (1977), 221-224.

[11] GAVEAU, B.: Valeurs frontieres des harmoniques ou holomorphes et de leurs derivees I. Cas du disque et du bidisque, Bull. Sci. Math. 2 104 (1980), 233-272.

[12] ——: Valeurs frontieres des fonctions harmoniques ou holomorphes et de leurs derivees. Cas de la boule, C.R.A.S. Paris 288 (1979), 403-406.

[13] PUTZ, N.: A generalized area theorem for harmonic functions on hyperbolic spaces, Trans. Am. Math. Soc. 168 (1972), 243-258.

[14] STEIN, E.: Singular integrals and differentiability properties of functions, Princeton Univ. Press, Princeton 1970.

[15] ——: Topics in harmonic analysis related to the Littlewood-Paley theory, Annals of Math. studies, Princeton Univ. Press, Princeton 1970.

[16] ——: Boundary behaviour of holomorphic functions of several complex variables, Princeton Univ. Press, Princeton 1972.

[17] STROOCK, D.: A Fefferman-Stein type interpolation theorem and applications to probability and analysis, Comm. Pure Appl. Math. 26 (1973), 477-495.

[18] ZYGMUND, A.: Trigonometric series II, Cambridge Univ. Press, Cambridge 1959.

[19] DEBIARD, A., B. GAVEAU : Valeurs frontières des fonctions holomorphes d'intégrale de Dirichlet finie et capacité sous elliptique, Bull. Soc. Sci. Lettres Łódź 31, 1 (1981), 10 pp.

Université Pierre et Marie Curie
Mathématiques
4, Place Jussieu
F-75230 Paris Cedex 05, France

ON CAUCHY-RIEMANN DERIVATIVES IN SEVERAL REAL VARIABLES

Tadeusz Iwaniec (Warszawa)

Contents

Summary

The differential operators ∂ and $\overline{\partial}$ are constructed in the spaces of several real variables. They generalize the usual (two-dimensional) Cauchy-Riemann operators, which play an important role in the theories of complex functions, differential equations, and quasicoformal mappings.

On Cauchy-Riemann Derivatives in Several Real Variables

The Beltrami type differential equation

$$\partial\bar{f} = Q\,\partial f$$

in the space $\mathbb{R}^n$ is investigated and its connections with quasiconformality of f is studied. As an application, L^p-estimates of derivatives of quasiconformal mappings in $\mathbb{R}^n$ are given.

Introduction

Let $f = f(z)$ be a C^1 class function of the complex variable $z = x + iy$. The Cauchy-Riemann complex derivatives of f are defined by

$$\text{(1)} \qquad \frac{\partial f}{\partial \bar{z}} = \frac{1}{2}\left(\frac{\partial f}{\partial x} + i\frac{\partial f}{\partial y}\right), \quad \frac{\partial f}{\partial z} = \frac{1}{2}\left(\frac{\partial f}{\partial x} - i\frac{\partial f}{\partial y}\right).$$

For $f = u + iv$ they take the following form

$$\text{(2)} \qquad \frac{\partial f}{\partial \bar{z}} = \frac{1}{2}(u_x - v_y) + \frac{i}{2}(u_y + v_x), \quad \frac{\partial f}{\partial z} = \frac{1}{2}(u_x + v_y) + \frac{i}{2}(v_x - u_y).$$

The operators $\frac{\partial}{\partial \bar{z}}$ and $\frac{\partial}{\partial z}$ are very important for the theory of analytic functions, differential equations and quasiconformal mappings in the complex plane. Some new interesting and promising generalizations of those in n real variables have been suggested by L. Ahlfors, [1].

Let $f : \mathbb{R}^n \longrightarrow \mathbb{R}^n$ be a C^1 mapping. We denote by $Df(x)$ Jacobi matrix of f at the point x, by $D^*f(x)$ the matrix transposed to $Df(x)$ and by I the unit $n \times n$-matrix. Then the operators of Ahlfors are given by

$$\text{(3)} \qquad Sf = \frac{1}{2}(Df + D^*f) - \frac{1}{n}(\mathrm{Tr}Df)I, \quad Af = \frac{1}{2}(Df - D^*f) + \frac{1}{n}(\mathrm{Tr}Df)I.$$

For $n = 2$ we write

$$\text{(4)} \qquad Sf = \frac{1}{2}\begin{pmatrix} u_x - v_y, & u_y + v_x \\ u_y + v_x, & v_y - u_x \end{pmatrix} \quad \text{and} \quad Af = \frac{1}{2}\begin{pmatrix} u_x + v_y, & u_y - v_x \\ v_x - u_y, & u_x + v_y \end{pmatrix}.$$

This shows that S and A are generalizations of $\frac{\partial}{\partial \bar{z}}$ and $\frac{\partial}{\partial z}$.

Let us recall that S takes its origin from an infinitesimal deformation of a system of partial differential equations describing conformal mappings in $\mathbb{R}^n$.

Let Ω be a domain in $\mathbb{R}^n$ and let $F:\Omega \longrightarrow \mathbb{R}^n$ be an orientation preserving diffeomorphism. We say that F is conformal if the tangent map $F'(x):T_x\Omega \longrightarrow \mathbb{R}^n$ is a similarity transformation of the tangent space $T_x\Omega$ into $\mathbb{R}^n$. This condition expresses the fact that Jacobi matrix $DF(x)$ is proportional to an orthogonal one, namely

$$(5)\qquad D^*F(x)DF(x) = J(x,F)^{2/n}I,$$

where $J(x,F) = \det DF(x)$ stands for Jacobian of F.

Assume that a Riemannian matric tensor is defined on the domain $\Omega \subset \mathbb{R}^n$. We identify it with a matrix-valued function $G = G(x)$, symmetric and positive at every point $x \in \Omega$. We accept, without changing the conformal structure on Ω, the following normalization $\det G(x) \equiv 1$. Then the conformality of a diffeomorphism $F:\Omega \longrightarrow \mathbb{R}^n$ means that

$$(6)\qquad D^*F(x)DF(x) = J(x,F)^{2/n}G(x),$$

F is called quasiconformal diffeomorphism. Clearly the identity map $F_o(x) = x$ satisfies (5). Consider one-parameter deformation families $F_t(x) = x + tf(x) + o(|t|)$ and $G_t(x) = I + t\nu(x) + O(|t|)$, where f is a mapping and $\nu(x)$ is a symmetric matrics with zero trace. We investigate the equation

$$(7)\qquad D^*F_t(x)DF_t(x) = J(x,F_t)^{2/n}G_t(x)$$

which implies some integrability conditions on $G_t(x)$ (see theorem of Weyl-Schouten from Riemannian geometry). We will be intersted in a first approximation of (7), namely in the following asymt-otic expansion

$$D^*F_t(x)DF_t(x) = J(x,F_t)^{2/n}G_t(x) + o(|t|)$$

This leads us to the study of Ahlfors equation

$$(8)\qquad Sf = \tfrac{1}{2}\nu$$

In other words Sf measures the quasiconformality of the infinitesimal deformation of the identity map F_o.

We are not going to detailed discussion of (8) for which we refer to the papers of L. Ahlfors [1], J.Sarvas [11] and B. Bojarski and T. Iwaniec [5_1]. We should have to say that in general the solutions of the homogeneous equation Sf = O are not conformal mappings. There is no linear theory of conformal mappings in space at all. However S has many remarkable applications in the theory of quasiconformal mappings in several variables.

In this paper we concentrate attention on non-linear equations and we suqqest other analogues of Cauchy-Riemann derivatives for purposes of quasiconformal theory. The idea has been developped in connection to Beltrami equation studies

$$(9) \qquad \frac{\partial f}{\partial \bar{z}} = q(z) \frac{\partial f}{\partial z},$$

where $q(z)$ is a complex measurable function satisfying the ellipticity condition $|q(z)| \le q_o < 1$. Roughly speaking the theory of quasiconformal mappings in the complex plane reduces itself to the study of equation (9). Some analytic and functional properties of operators $\frac{\partial}{\partial \bar{z}}$ and $\frac{\partial}{\partial z}$ are of principle importance in these investigations, see [3], [4], [12]. As we shall see later certain ideas of those methods may be directly carried forward quasiconformal mappings in space. We shall construct Cauchy-Riemann derivatives $\bar{\partial} f$ and ∂f of f which acts between Riemannian manifolds and we shall express the quasiconformality by Beltrami's type equation

$$(10) \qquad \bar{\partial} f = Q \partial f$$

In contrast to the two-dimensional case our operators $\bar{\partial}$ and ∂ will be non-linear. An application of (10) will be illustrated on the problem of L^P-estimation of derivatives of quasiconformal mappings. We shall use a similar trick as it was originally applied by Bojarski in [4_2]. The main result of this paper is an L^P estimation of $\bar{\partial} f$ and ∂f, see Theorem 1. Finally we should say that in the construction of $\bar{\partial} f$ and ∂f we were a little bit inspired by non-linear elastostatics which seems to be allied with quasiconformal mappings.

§ 1. Cauchy-Riemann operators for linear mappings

Let E and F be n-dimensional oriented vector spaces with the scalar products $<\,,\,>_E$ and $<\,,\,>_F$ respectively. We denote by Hom (E,F) the space of all linear mappings from E into F. For every $A \in \text{Hom}(E,F)$ the adjoint map $A^* \in \text{Hom}(F,E)$ is defined by $<Ax,y>_F = =<x,A^*y>_E$ for $x \in E$ and $y \in F$. We introduce a scalar product in Hom(E,F) by setting $<A,B> = \text{Tr}A^*B$. The norm in Hom(E,F) is then equal to $\|A\| = <A,A>^{1/2}$.

Let us denote by detA the determinant of A, thus we have a function

$$(11) \qquad \det : \text{Hom}(E,F) \longrightarrow \mathbb{R},$$

which may be viewed as a homogeneous polynomial of degree n with respect to $A \in \text{Hom}(E,F)$. We are looking for the differential of that function. At every point $A \in \text{Hom}(E,F)$ this will be a functional on Hom(E,F) which we denote by det'A. Since Hom(E,F) is Hilbert space we may identify def'A with an element of Hom(E,F) such that

$$(12) \qquad \det(A+X) = \det A + <\det'A,\ x> + o(\|X\|)$$

for each $X \in \text{Hom}(E,F)$.
Let us introduce the adjugate mapping by $\text{Adj}A = (\det'A)^* \in \text{Hom}(F,E)$ this means that

$$\det(A+X) = \det A + \text{Tr}(\text{Adj}A)X + o(\|X\|),$$

where

$$(13) \qquad \text{Adj} : \text{Hom}(E,F) \longrightarrow \text{Hom}(F,E)$$

is a homogeneous polynomial of degree (n-1) with respect to $A \in \text{Hom}(E,F)$ taking values in Hom(F,E). For $n = 2$ Adj is simply a linear transformation. In any case

$$(14) \qquad \text{Adj}A = \det A\ A^{-1}$$

for invertable $A \in \text{Hom}(E,F)$.
Finally, we consider the following (n-1)-homogeneous form

(15) $\quad \| \ \|^{n-2}\mathrm{id}^* : \mathrm{Hom}(E,F) \longrightarrow \mathrm{Hom}(F,E)$

i.e. any $A \in \mathrm{Hom}(E,F)$ is transformed into $\|A\|^{n-2}A^*$.
Now, we define Cauchy-Riemann operators

$$\bar{\partial}, \partial : \mathrm{Hom}(E,F) \longrightarrow \mathrm{Hom}(F,E)$$

as follows

$$\bar{\partial}A = \|A\|^{n-2}A^* - n^{\frac{n-2}{n}}\ \mathrm{Adj}A$$

(16)

$$\partial A = \|A\|^{n-2}A^* + n^{\frac{n-2}{n}}\ \mathrm{Adj}A$$

Another way to arrive at $\bar{\partial}$ and ∂ is the study of the following n-forms

$$\phi = \| \ \|^n - n^{\frac{n}{2}}\det : \mathrm{Hom}(E,F) \longrightarrow \mathbb{R},$$

$$\Psi = \| \ \|^n + n^{\frac{n}{2}}\det : \mathrm{Hom}(E,F) \longrightarrow \mathbb{R},$$

whose differentials are equal to $n\bar{\partial}^*$ and $n\partial^*$, i.e. one has formulas

$$\phi(A+X) = \phi(A) + n\mathrm{Tr}\bar{\partial}\ AX + o(\|X\|)$$

$$\Psi(A+X) = \Psi(A) + n\mathrm{Tr}\partial AX + o(\|X\|)$$

for every $A, X \in \mathrm{Hom}(E,F)$.

On the other hand by Hadamard's inequality we see that $\phi(A)$ and $\Psi(A)$ are non-negative for each $A \in \mathrm{Hom}\ (E,F)$. Furthermore ϕ vanishes on the orientation preserving homothetic mappings $A : E \longrightarrow F$, i.e., on those A with $\langle Ax, Ay\rangle_F = \lambda\langle x, y\rangle_E$ for $x, y \in E$. Similarly Ψ vanishes on the orientation changing homothetic mappings. Since the minima of ϕ and Ψ are zeros of their differentials one has the following

<u>Proposition 1 Non-trivial solutions of equation</u> $\bar{\partial}A = 0$, $(\partial A = 0)$ <u>are conformal mappings from</u> E <u>into</u> F <u>which preserve</u> (change) <u>the orientation</u>.

We end this paragraph by noting the following identity

$$(17)\qquad \|\partial A\|^2 - \|\bar{\partial} A\|^2 = 4n^{\frac{n}{2}} \|A\|^{n-2} \det A$$

which immediately follows from the definition of $\bar{\partial}A$ and ∂A. This will be used in the study of L^P-estimations of $\bar{\partial}$ and ∂.

§ 2. Definition of Cauchy-Riemann derivatives for mappings

Suppose that (M,g) and (N,h) are oriented Riemannian manifolds of the same dimension n. We denote by TM and TN their tangent boundles. Any C^1 map $f: M \longrightarrow N$ induces a morphism Df from TM into TN over the base map f.

$$\begin{array}{ccc} TM & \xrightarrow{\ Df\ } & TN \\ \downarrow & & \downarrow \\ M & \xrightarrow{\ f\ } & N \end{array} \qquad\qquad Df \in \mathrm{Mor}_f(TM, TN)$$

For every $x \in M$ $Df(x)$ is a linear map between tangent spaces T_xM and $T_{f(x)}N$ in which some orientations and scalar products are naturally induced by tensors g and h respectively. According to the considerations of the previous paragraph one can define $\bar{\partial}$ and ∂ derivatives on $\mathrm{Hom}(T_xM, T_{f(x)}N)$. For notational simplicity we write $\bar{\partial}f(x)$ and $\partial f(x)$ instead of $\bar{\partial}Df(x)$ and $\partial Df(x)$. We shall call $\bar{\partial}f$ and ∂f Cauchy-Riemann derivatives of $f: M \longrightarrow N$. In local coordinates they may by written as follows

$$(18)\qquad \bar{\partial}f(x) = \|Df(x)\|^{n-2} D^*f(x) - n^{\frac{n-2}{2}} \mathrm{Adj} f(x),$$

$$\partial f(x) = \|Df(x)\|^{n-2} D^*f(x) + n^{\frac{n-2}{2}} \mathrm{Adj} f(x),$$

where the norm $\| \ \|$, the adjoint map D^* and the adjugate operation Adj should be considered with respect to the corresponding scalar structure in $\mathrm{Hom}(T_xM, T_{f(x)}N)$.

Suppose for a moment that $f: M \longrightarrow N$ is a diffeomorphism such that $\bar{\partial}f = 0$. By Proposition 1 we see that for every $x \in M$ the tangent map $Df(x): T_xM \longrightarrow T_{f(x)}N$ is a homothetic transformation with positive determinant. This means that f is a conformal equivalence between manifolds M and N. Similarly diffeomorphic solutions of $\partial f = 0$ are conformal mappings with negative Jacobian. This allows us to interpret all solutions of $\bar{\partial} f = 0$ as a generalization of holomorphic functions. We shall call them regular mappings while the solutions of $\partial f = 0$ will be called antiregular mappings in correspondence to antiholomorphic functions.

Let us examine the situation when M and N are domains in $\mathbb{R}^n$ with standard metrics. Identifying Df with Jacobi matrix $\left(\frac{\partial f^\alpha}{\partial x_\beta}\right)$ formulas (18) take the forms

$$(\bar{\partial} f)_{ij} = |Df|^{n-2} D^* f - n^{\frac{n-2}{2}} \operatorname{Adj} f =$$

$$= \left| \sum_{\alpha\beta} \left(\frac{\partial f^\alpha}{\partial x\beta}\right)^2 \right|^{\frac{n-2}{2}} \frac{\partial f^j}{\partial x_i} - (-1)^{i+j} n^{\frac{n-2}{2}} \frac{\partial (f^1, \ldots, \widehat{f^j}, \ldots, f^n)}{\partial (x_1, \ldots, \widehat{x}_i, \ldots, x_n)}$$

(19)

$$(\partial f)_{ij} = |Df|^{n-2} D^* f + n^{\frac{n-2}{2}} \operatorname{Adj} f =$$

$$= \left| \sum_{\alpha\beta} \left(\frac{\partial f^\alpha}{\partial x_\beta}\right)^2 \right|^{\frac{n-2}{2}} \frac{\partial f^j}{\partial x_i} + (-1)^{i+j} n^{\frac{n-2}{2}} \frac{\partial (f^1, \ldots, \widehat{f^j}, \ldots, f^n)}{\partial (x_1, \ldots, \widehat{x}_i, \ldots, x_n)}$$

where $\wedge$ means that corresponding variables are deleted from the formula.

Now we would like to point out some relations with non-linear elasticity theory.

Suppose that $f : \Omega \longrightarrow \mathbb{R}^n$ is a mapping of the class $W_n^1(\Omega)$. We consider a material body whose particles occupy Ω and we interpret f as a deformation of the above configuration of particles. Then $Df(x)$ is considered as a stress tensor with total energy being defined by the integral

$$(20) \qquad \int_\Omega |Df(x)|^n dx$$

We fix a boundary value of f, this means that we are given a map $\varphi \in W_n^1(\Omega)$ such that $f - \varphi \in \overset{\circ}{W}{}_n^1(\Omega)$ – the completion of $C_o^\infty(\Omega)$ mappings in the norm of $W_n^1(\Omega)$. By the interpretation of f this mapping must minimize energy integral (20). From the general theory of calculus of variations we know that such minimizer exists, is unique and satisfies Lagrange-Euler equation

$$(21)\qquad \operatorname{div}|Df|^{n-2}D^*f = 0$$

which should be understood in the sense of distributions, i.e. for every test mapping $\eta : \Omega \longrightarrow \mathbb{R}^n$, $\eta \in \overset{\circ}{W}{}^1_n(\Omega)$ it holds

$$\int_\Omega |Df|^{n-2} D\eta D^* f = 0$$

On the other hand it is know that the columns of any adjugate matrix Adjf are free divergence.
This means that

$$(22)\qquad \int_\Omega D\eta \operatorname{Adj} f = 0 \qquad \text{for any} \quad \eta \in \overset{\circ}{W}{}^1_n(\Omega) .$$

This property of Adjf is a particular one of so called Null-Lagranians, see [2]. Now, equation (21) takes the form

$$(23)\qquad \operatorname{div} \bar{\partial} f = 0$$

which may be treated as an n-dimensional analogue of Laplace equation acting on mappings $f : \Omega \longrightarrow \mathbb{R}^n$. Conformal mappings are particular solutions of (23). Since Dirchlet problem for (23) is always solvable then (23) gives a useful tool for dealing with conformal mappings in space.

§ 3. Quasiregular mappings

Let Ω be a domain in $\mathbb{R}^n$ and let $f : \Omega \longrightarrow \mathbb{R}^n$ belong to Sobolev space $W^1_{n,loc}(\Omega)$

Definition. The map f is said to be quasiregular if and only if

$$(24)\qquad |Df(x)|^n \le n^{\frac{n}{2}} \mathcal{K} J(x,f), \qquad \text{for almost all} \quad x \in \Omega$$

where $\mathcal{K}$ is a constant independent of x, $\mathcal{K} \ge 1$. Hereafter we denote by $|A|$ norm of the matrix A such that $|A|^2 = \operatorname{Tr} A^* A = \Sigma A_{ij}^2$.

If $\mathcal{K} = 1$ the map f becomes a generalized conformal transformation

(1 - quasiregular). We see that Jacobian of a quasiregular map is non-negative almost everywhere and it may vanish only with all partial derivatives of f. If $J(x,f) \neq 0$ we define a matrix $G(x)$ such that

$$D^{*}f(x)Df(x) = J(x,f)^{2/n}G(x) \tag{25}$$

One can simply extend the validity of (25) for almost every $x \in \Omega$ by defining $G(x)$ to be the unit matrix I when $J(x,f) = 0$. In view of (24) one has the following properties of G:

- $G(x)$ is symmetric and positive
- G is a bounded matric-valued function
- $\det G(x) = 1$.

These allow us to interpret G as a Riemannian tensor defined on Ω, In general G is not smooth even not continuous but only a measurable matrix-valued function. Nevertheless we may correctly define, as above, the Cauchy derivatives $\bar{\partial}_G f$ and $\partial_G f$ for an arbitrary map $f : \Omega \longrightarrow \mathbb{R}^n$ of the class $W^1_{p,loc}(\Omega)$. Equation (25) means that $\bar{\partial}_G f(x) = 0$ for almost every $x \in \Omega$. We shall express this condition by means of $\bar{\partial}f$ and ∂f - Cauchy derivatives in Euclidean matric. To this end we observe that (25) implies $|Df|^2 = J(x,f)^{2/n}\mathrm{Tr}G$. Therefore

$$|Df|^{n-2}D^{*}fDf = J(x,f)\,(\mathrm{Tr}G)^{\frac{n-2}{2}}G$$

or equivalently

$$|Df|^{n-2}D^{*}f = (\mathrm{Tr}G)^{\frac{n-2}{2}}\,G\,\mathrm{Adj}f$$

By the definition of operators $\bar{\partial}$ and ∂ we have

$$|Df|^{n-2}D^{*}f = \tfrac{1}{2}(\partial f + \bar{\partial}f), \quad \mathrm{Adj}f = \tfrac{1}{2}\, n^{\frac{2-n}{2}}(\partial f - \bar{\partial}f)$$

Hence we get

$$\partial f + \bar{\partial}f = (\tfrac{1}{n}\mathrm{Tr}G)^{\frac{n-2}{2}}\,G(\partial f - \bar{\partial}f)$$

$$[I + (\tfrac{1}{2}\mathrm{Tr}G)^{\frac{n-2}{2}}\,G]\bar{\partial}f = [(\tfrac{1}{n}\mathrm{Tr}G)^{\frac{n-2}{2}}\,G - I]\partial f$$

Since the matrix $I+(\frac{1}{2}\mathrm{Tr}G)^{\frac{n-2}{2}}G$ is positive, in particular non-singular, then we may write

(26) $\bar{\partial}f(x)=Q(x)\,\partial f(x)$

where

(27) $$Q=\frac{(\frac{1}{n}\mathrm{Tr}G)^{\frac{n-2}{2}}G-I}{(\frac{1}{n}\mathrm{Tr}G)^{\frac{n-2}{2}}G+I}$$

In this way system (25) yields a linear relation between $\bar{\partial}f$ and ∂f. We interpret (26) as a generalization of Beltrami's system in n variables.

Here are some properties of $Q(x)$.

- $Q(x)$ is symmetric matrix for almost every $x\in\Omega$.
- the eigenvalues $q_1(x)\le q_2(x)\le\ldots,\le q_n(x)$ are equal to

(28) $$q_i=\frac{(\frac{1}{n}\Sigma\lambda_\alpha)^{\frac{n-2}{2}}\lambda_i-1}{(\frac{1}{n}\Sigma\lambda_\alpha)^{\frac{n-2}{2}}\lambda_i+1}\ , \qquad i=1,2,\ldots,n,$$

where $0<\lambda_1(x)\le\lambda_2(x)\le\ldots\le\lambda_n(x)$ are eigenvalues of $G(x)$.

Let us denote by $K=\operatorname{ess\,sup}\limits_{x\in\Omega}\frac{\lambda_n(x)}{\lambda_1(x)}$ - the dilatation of the map f.

It is easy to prove that

- $$\operatorname{ess\,sup}_{x\in\Omega} q_n(x)=Q_o\le\frac{\sqrt{k^{n-1}}-1}{\sqrt{k^{n-1}}+1}<1$$

In that case a relevant norm of $Q(x)$ will be defined by $|Q(x)|=\sup\limits_{|h|=1}|Q(x)h|=q_n(x)$. Therefore we have

- $|Q(x)|\le Q_o<1$

This will be called the ellipticity condition of the system (26). Gathering together the above remarks we infer the following.

Proposition 2. Let $f:\Omega \longrightarrow \mathbb{R}^n$ be a mapping of the class $W^1_{n,loc}(\Omega)$. Then f is quasiregular if and only if

(29) $\bar{\partial} f(x) = Q(x)\partial f(x)$,

(30) $\|Q(x)\| \leq Q_o < 1$

Proposition 2 reduces the theory of quasiregular mappings to the study of Beltrami system (29). In this case we shall say that f is Q_o quasiregular mapping.

Let us observe that for arbitrary matrices A,B one has $|AB| \leq \|A\| \cdot |B|$. Hence

Corollary. Any Q_o quasiregular map satisfies

(31) $|\bar{\partial} f(x)| \leq Q_o |\partial f(x)|$

for almost all $x \in \Omega$.

Remark. The idea of Q_o-quasiregularity may be extended for mappings $f: M \longrightarrow N$ acting between Riemannian manifolds. We recall that a volume element dx on M is well definite via tensor g. In particular the integral

$$\int_M \| Df(x) \|^p dx$$

makes a sense and we may say that $f \in W^1_{n,loc}(M)$ provided

$$\int_{M'} \| Df(x) \|^n dx \quad < \infty$$

for any compact $M' \subset M$,

A map $f: M \longrightarrow N$ is said to be Q_o-quasiregular, $0 \leq Q_o < 1$, if $f \in W^1_{n,loc}(M)$ and $\| \bar{\partial} f(x) \| \leq Q_o \| \partial f(x) \|$ for almost every $x \in M$. Here the operators $\bar{\partial}, \partial$ and the norm $\| \ \|$ are considered with respect to the metrics on M and N.

From now on M and N be exclusively domains of $\mathbb{R}^n$ with Euclidean matrics.

§ 4. L^P - estimates

Mojority properties of solutions of Beltrami system (9) have been proved owing to their apriori estimates in Sobolev spaces $W^1_{p,loc}(\Omega)$, $1 < p < \infty$. The essential part of those studies is the inequality

$$(32) \qquad \left\| \frac{\partial w}{\partial z} \right\|_{L^P(\mathbb{R}^2)} \le A_p \left\| \frac{\partial w}{\partial \bar{z}} \right\|_{L^P(\mathbb{R}^2)}, \quad 1 < p < \infty.$$

which holds for any $w \in W^1_p(\mathbb{R}^2)$ with a constant A_p independent of w. This was successfully used in the study of general linear elliptic systems on the plane with discontinuous coefficients, (see [4],[12] and others). The readers familiar with those methods easily observe that the identity

$$(33) \qquad \left\| \frac{\partial w}{\partial z} \right\|_{L^2(\mathbb{R}^2)} = \left\| \frac{\partial w}{\partial \bar{z}} \right\|_{L^2(\mathbb{R}^2)}$$

is equally important.

Our aim is to generalize (32) and (33) for operators $\bar{\partial}$ and ∂.

Theorem 1. Let $2 \le p < \infty$, then for every $f : \mathbb{R}^n \longrightarrow \mathbb{R}^n$ from the class $W^1_{\frac{pn}{2}}(\mathbb{R}^n)$ we have

$$(34) \qquad \left\| |Df|^{\frac{2-n}{2}} \partial f \right\|_{L^P} \le A_p \left\| |Df|^{\frac{2-n}{2}} \bar{\partial} f \right\|_{L^P}$$

$$(35) \qquad \left\| |Df|^{\frac{2-n}{2}} \bar{\partial} f \right\|_{L^P} \le A_p \left\| |Df|^{\frac{2-n}{2}} \partial f \right\|_{L^P}$$

$$(36) \qquad \left\| |Df|^{\frac{2-n}{2}} \partial f \right\|_{L^2} = \left\| |Df|^{\frac{2-n}{2}} \bar{\partial} f \right\|_{L^2}$$

where $A_p = A_p(n)$ does not depend of f. For $n = 2$ this theorem is reduced to inequality (32) which can be rather simply derived from L^P - estimates of so called Hilbert transform. If $n > 2$ we cannot appeal to the theory of singular integrals because of non-linearity of $\bar{\partial}$ and ∂. That is why the proof will be more complicated. We shall use some results on regularity and stability of conformal mappings.

The basic fact we need will be an inequality on BMO norms due to Fefferman and Stein.

The property (36) arises from the following observation. For every map $f:\mathbb{R}^n \longrightarrow \mathbb{R}^n$ of the class $W_n^1(\mathbb{R}^n)$ the integral $\int J(x,f)dx$ vanishes. In fact. If $f \in C_0^\infty(\mathbb{R}^n)$, then

$$\int J(x,f)dx = \int df^1 \wedge df^2 \wedge \ldots \wedge df^n = \int d(f^1 df^2 \wedge \ldots \wedge df^n) = 0$$

by Stokes' theorem. The general case may be obtained by using an approximation argument.
Therefore (17) implies

$$\int |Df(x)|^{2-n} |\partial f(x)|^2 dx = \int |Df(x)|^{2-n} |\bar{\partial} f(x)|^2 dx$$

proving (36).
Now we pass to the proof of (34) and (35).

We may assume without loss of generality that f has compact support.
First we shall prove the following stability lemma.
Lemma 1. For every $\sigma > 0$ there exists $C_\sigma = C_\sigma(n) > 0$ such that: If $f : B(x_o,R) \longrightarrow \mathbb{R}^n$ is a map of the class $W_n^1(B(x_o,R))$, where $B = B(x_o,R)$ denotes the ball with centre x_o and radius R, then there axist a 1-quasiregular map $g \in W_n^1(B(x_o,R))$ such that

$$(37) \qquad \int_{B(x_o,\frac{1}{2}R)} \big| |Df(x)|^n - |Dg(x)|^n \big| dx \le \sigma^{n+1} \int_{B(x_o,R)} |Df(x)|^n dx +$$

$$+ C_\sigma \int_{B(x_o,R)} |Df(x)|^{2-n} |\bar{\partial} f(x)|^2 dx$$

Proof. The proof will be ineffective and we will not be able to give any uper bound of C_σ.
Since (37) is invariant under shifts and homothetic transformations of variables x and f we may restrict our considerations to the case when B is the unit ball and we may assume that

$$\int_{B(0,1)} |Df(x)|^n dx = 1 \quad , \quad \int_{B(0,1)} f(x)dx = 0$$

Contradicting the lemma we suppose that there are $\sigma > 0$ and a sequence $\{f_j\}$ of mappings such that

$$(38) \qquad \int_{B(0,1)} |Df_j(x)|^n dx = 1 \quad , \quad \int_{B(0,1)} f_j(x)dx = 0, \quad j = 1,2,\ldots$$

and

$$(39) \qquad \int_{B(0,\frac{1}{2})} ||Df_j(x)|^n - |Dg(x)|^n| dx > \sigma^{n+1} \int_{B(0,1)} |Df_j(x)|^n dx +$$

$$+ j \int_{B(0,1)} |Df_j(x)|^{2-n} |\bar{\partial} f_j(x)|^2 dx$$

The last being fulfiled with arbitrary 1-quasiregular map $g \in W_n^1(B)$. In particular we see that

$$(40) \qquad \lim_{j\to\infty} \int_B |Df_j|^{2-n} |\bar{\partial} f_j|^2 = 0$$

In view of (38) one can choose a subsequence $f_{j_\alpha}(x)$, $\alpha = 1,2,\ldots$ weakly converging to a map $g \in W_n^1(B)$. We shall prove that g is a 1-quasiregular mapping. For this it suffices to show that $|Dg(x)|^n \leq n^{\frac{n}{2}} J(x,g)$ for almost all $x \in B$. Let us consider an arbitrary compact subset $F \subset B(0,1)$. We state without proof the following property of Jacobian

$$(41) \qquad \int_F J(x,g)dx = \lim_{\alpha\to\infty} \int_F J(x,f_{j_\alpha})dx$$

which is well known fact in quasiconformal theory, see for example [9₁].

Obviously we have

$$\int_F |Dg|^n \leq \lim_{\alpha\to\infty} \int_F |Df_{j_\alpha}|^n$$

Therefore one can write

$$\operatorname{Tr}\int_F \bar\partial g Dg = \int_F |Dg|^n - n^{\frac{n}{2}} \int_F J(x,g) \leq \lim_{\alpha\to\infty} \int_F [|Df_{j_\alpha}|^n - n^{\frac{n}{2}} J(x,f_{j_\alpha})] =$$

$$(42) \qquad = \lim_{\alpha\to\infty} \operatorname{Tr}\int_F \bar\partial f_{j_\alpha} Df_{j_\alpha} = 0$$

The last equality holds because of the inequality

$$(43) \qquad |\operatorname{Tr}\int_F \bar\partial f_{j_\alpha} Df_{j_\alpha}| \leq \int_F |\bar\partial f_{j_\alpha}||Df_{j_\alpha}| \leq \Big(\int_F |Df_{j_\alpha}|^{2-n}|\bar\partial f_{j_\alpha}|^2\Big)^{1/2}\Big(\int_F |Df_{j_\alpha}|^n\Big)^{1/2}$$

where the right hand side goes to zero in account of (38) and (40). From (42) we conclude

$$\int_F |Dg|^n \leq n^{\frac{n}{2}} \int_F J(x,g)$$

for every compact $F \subset B(0,1)$. This implies that $|Dg(x)|^n \leq n^{\frac{n}{2}} J(x,g)$ for almost all $x \in B(0,1)$, i.e. g is a 1-quasiregular mapping. On the other hand by (41) and (42) we get

$$\int_{B(0,\frac{1}{2})} |Dg|^n = \operatorname{Tr}\int_{B(0,\frac{1}{2})} \bar\partial g Dg + n^{\frac{n}{2}} \cdot \int_{B(0,\frac{1}{2})} J(x,g) = n^{\frac{n}{2}} \cdot \int_{B(0,\frac{1}{2})} J(x,g) =$$

$$= \lim_{\alpha\to\infty}[\operatorname{Tr}\int_{B(0,\frac{1}{2})} \bar\partial f_{j_\alpha} Df_{j_\alpha} + n^{\frac{n}{2}} \cdot \int_{B(0,\frac{1}{2})} J(x,f_{j_\alpha})] = \lim_{\alpha\to\infty} \int_{B(0,\frac{1}{2})} |Df_{j_\alpha}|^n$$

which proves that f_{j_α} goes to g in $W_n^1(B(0,\frac{1}{2}))$.
Here we have used the property of uniform convexity of the space $W_n^1(B(0,\frac{1}{2}))$. Thus we have reached a contradiction with (38) (39).
The proof of the lemma is complete.
The next lemma concerns a regularity property of Jacobians of 1-quasiregular mappings.
Lemma 2. *Let $B(x_0,R)$ be a ball in $\mathbb{R}^n$ and let $g \in W_n^1(B(x_0,R))$ be a 1-quasiregular map. Then for every $\sigma \in (0,\frac{1}{2}]$ it holds*

$$(44) \qquad ⨍_{B(x_0,\sigma R)} |J(x,g) - ⨍_{B(x_0,\sigma R)} J(y,g)dy|dx \leq C(n)\,\sigma ⨍_{B(x_0,\frac{1}{2}R)} J(y,g)dy.$$

Hereafter we use the notation $\fint_B \varphi$ for the integral mean value of a function φ, $\fint_B \varphi = \frac{1}{|B|}\int_B \varphi$.

The proof of this lemma is based on two facts. The first says that the function $v(x) = J(x,y)^{\frac{n-2}{2n}}$ is harmonic. An elementary proof of this fact may be found in $[5_2]$.

Secondly, given harmonic function $v \in L^P(B(x_o,R))$, $2 \le p \le \frac{2n}{n-2}$, then

$$\fint_{B(x_o,\sigma R)} \left| |v|^P - \fint_{B(x_o,\sigma R)} |v|^P \right| \le C(n,p)\sigma \fint_{B(x_o,\frac{1}{2}R)} |v|^P \quad \text{for } o < \sigma \le \tfrac{1}{2}.$$

For the details we refer to $[9_1]$.

Now, observe that $J(x,g) = n^{-\frac{n}{2}}|Dg(x)|^n$, so (44) reads

$$(45)\qquad \fint_{B(x_o,\sigma R)} \left| |Dg|^n - \fint_{B(x_o,\sigma R)} |Dg|^n \right| \le C(n)\sigma \fint_{B(x_o,\frac{1}{2}R)} |Dg|^n \quad \text{for } 0 < \sigma \le \tfrac{1}{2}.$$

On the other hand by (37) we have.

$$(46)\qquad \fint_{B(x_o,\sigma R)} \left| |Df|^n - |Dg|^n \right| \le \sigma \fint_{B(x_o,R)} |Df|^n + \sigma^{-n} C_\sigma \fint_{B(x_o,R)} |\partial f|^2 |Df|^{2-n}$$

$$\text{for } 0 < \sigma \le \tfrac{1}{2}.$$

Adding (45) and (46) side by side one gets

$$(47)\qquad \fint_{B(x_o,\sigma R)} \left| |Df|^n - \fint_{B(x_o,\sigma R)} |Dg|^n \right| \le C(n)\sigma \fint_{B(x_o,\frac{1}{2}R)} |Dg|^n + \sigma \fint_{B(x_o,R)} |Df|^n +$$

$$+ \sigma^{-n} C_\sigma \fint_{B(x_o,R)} |\partial f|^2 |Df|^{2-n}$$

We eliminate unconvienient term $\fint_{B(x_o,\frac{1}{2}R)} |Dg|^n$ in the right hand side of (47) by using once again inequality (46) with $\sigma = \frac{1}{2}$ getting

$$\fint_{B(x_o,\,\frac{1}{2}R)} |Dg|^n \le \fint_{B(x_o,\frac{1}{2}R)} |Df|^n + \frac{1}{2}\fint_{B(x_o,R)} |Df|^n + 2^n C_{1/2} \fint_{B(x_o,R)} |\partial f|^2 |Df|^{2-n}$$

Therefore (47) implies

$$\fint_{B(x_o,\sigma R)} \Big| |Df|^n - \fint_{B(x_o,\sigma R)} |Dg|^n \Big| \le C(n)\sigma \fint_{B(x_o,R)} |Df|^n + C(n,\sigma) \fint_{B(x_o,R)} |Df|^{2-n} |\bar\partial f|^2$$

And by simple observation that

$$\fint_{B(x_o,\sigma R)} \Big| |Df|^n - \fint_{B(x_o,\sigma R)} |Df|^n \Big| = \fint_{B(x_o,\sigma R)} \Big| \Big(|Df|^n - \fint_{B(x_o,\sigma R)} |Dg|^n\Big) - \fint_{B(x_o,\sigma R)}\Big(|Df|^n - \fint_{B(x_o,\sigma R)} |Dg|^n\Big) \Big| \le$$

$$\le 2 \fint_{B(x_o,\sigma R)} \Big| |Df|^n - \int_{B(x_o,\sigma R)} |Dg|^n \Big|$$

we are led to the following

$$\text{(48)} \qquad \fint_{B(x_o,\sigma R)} \Big| |Df|^n - \fint_{B(x_o,\sigma R)} |Df|^n \Big| \le C(n)\sigma \fint_{B(x_o,R)} |Df|^n + C(n,\sigma) \fint_{B(x_o,R)} |Df|^{2-n} |\bar\partial f|^2$$

This is the main inequality from which we shall derive L^p - estimation of $\bar\partial f$ and ∂f. Recall that (48) is valid for any ball $B(x_o,R) \subset \mathbb{R}^n$ and any parameter σ such that $0 < \sigma \le \frac{1}{2}$.
We introduce the following auxiliary functions

$$\text{(49)} \qquad u(x) = |Df(x)|^n, \quad v(x) = |Df(x)|^{2-n} |\bar\partial f(x)|^2$$

According to our hypothesis on f we have; $u, v \in L^{p/2}(\mathbb{R}^n)$, $\frac{p}{2} \ge 1$.
We express (48) by means of the following maximal functions

$$(Mv)(x) = \sup_{R>0} \fint_{B(x,R)} v(y)\,dy$$

$$u^{\sharp}(x) = \sup_{R>0} ⨍_{B(x,R)} \left| u(z) - ⨍_{B(x,R)} u(y)dy \right| dz$$

Therefore (48) reads as follows

(50) $\quad u^{\sharp}(x) \leq C(n)\sigma(Mu)(x) + C(n,\sigma)(Mv)(x)$ for any $x \in \mathbb{R}^n$ and $0 < \sigma \leq \frac{1}{2}$.

Now we are ready to apply the inequality of Fefferman and Stein [7] (see also $[9_1]$) which says that: there is a constant $B_q = B(n,q)$, $1 \leq q < \infty$ that for every $u \in L^q(\mathbb{R}^n)$ it holds

(51) $\quad \|Mu\|_{L^q} \leq B_q \|u^{\sharp}\|_{L^q}$

This and (50) imply

$$\|Mu\|_{L^q} \leq \sigma B_q \|Mu\|_{L^q} + C(n,q,\sigma)\|Mv\|_{L^q}, \quad q = \frac{p}{2} \geq 1.$$

Choosing σ small enough to satisfy $\sigma B_q \leq \frac{1}{2}$ we get

(52) $\quad \|Mu\|_{L^q} \leq C_q \|Mv\|_{L^q}$

On the other hand we have

(53) $\quad \|u\|_{L^1} \leq \|v\|_{L^1}$

In fact, By the obvious identity

$$\mathrm{Tr} \int \bar{\partial} f D f = \int |Df|^n - n^{\frac{n}{2}} \int J(x,f) = \int |Df|^n$$

and by Hölder's inequality we infer

$$\int |Df|^n \leq \int |\bar{\partial} f||Df| \leq \left(\int |Df|^n\right)^{1/2} \left(\int |Df|^{2-n} |\bar{\partial} f|^2\right)^{1/2}.$$

Hence

$$\|u\|_{L^1} = \int |Df|^n \leq \int |Df|^{2-n} |\bar{\partial} f|^2 = \|v\|_{L^1}.$$

Assume for a moment that $q > 1$. In this case the norms $\|Mu\|_{L^q}$ and $\|u\|_{L^q}$ are equivalent because of Hardy-Littlewood inequality

$$\|u\|_{L^q} \leq \|Mu\|_{L^q} \leq \frac{qC(n)}{q-1} \|u\|_{L^q}.$$

Therefore (52) implies

$$(54) \qquad \|u\|_{L^{p/2}} \leq N_p \|v\|_{L^{p/2}}$$

Let us observe that the above arguments have led us to (54) with a factor N_p which is unbounded when p approaches 2.
This result is unexpected in account of (53). But we may improve that by using some interpolation methods getting a constant N_p bounded in every finite interval $2 \leq p \leq T$. For details we refer to $[9_2]$. Now returning to the previous notations we write (54) as follows

$$\left[\int |Df|^{\frac{np}{2}}\right]^{2/p} \leq N_p \left[\int \left(|Df|^{\frac{2-n}{2}} |\bar{\partial} f|^2\right)^p\right]^{2/p}$$

which is stronger then (34). Finally if change the orientation of $\mathbb{R}^n$ then one would immediately get (35).

§ 5. On L^p-estimates of derivatives of a quasiregular map

We know, by the definition, that any quasiregular map $f : \Omega \to \mathbb{R}^n$ belongs to $W^1_{n,loc}(\Omega)$. A fundamental result of quasiconformal theory says that f actually belongs to $W^1_{p,loc}(\Omega)$ for some $p > n$. This fact was first discovered by B. Bojarski [4] in dimension 2. Later F. Gehring [8] and then Meyers and Elcrat [6] have proved it for all dimensions. In 1976 J. Reshetniak [10] obtained an asympt-otic estimation $p(n,k) \geq \frac{C(n)}{k-1}$ if K goes to 1. The exact value of the best possible exponent $p(n,k)$ is unknown. We shall show some connection of this problem with L^p estimates of Cauchy-derivatives.

According to the result of Bojarski we have: any solution of equation (9) belongs to $W^1_{p,\ loc}(\Omega)$ for each p satisfying $q_o A_p < 1$, where A_p is the norm from inequality (32).

Our aim is to prove a similar result concerning solutions of n-dimensional Beltrami system (29). Since we do not know whether A_p from Theorem 1 increases with respect to p we then introduce the following increasing function

(55) $$\mathcal{A}_p = \sup_{2 \leq s \leq p} A_s$$

Obviously $\mathcal{A}_2 = A_2 = 1$.

<u>Theorem 2. Let</u> $f : \Omega \longrightarrow \mathbb{R}^n$ <u>be a solution of Beltrami system</u>

(56) $$\bar{\partial} f(x) = Q(x) \partial f(x), \quad |Q(x)| \leq Q_0 < 1.$$

<u>Then</u> $f \in W^1_{\frac{np}{2},loc}(\Omega)$ <u>for every</u> $p \geq 2$ <u>such that</u>

(57) $$Q_0 \mathcal{A}_p < 1.$$

In particular we have the following

<u>Corollary:</u>

(58) $$\lim_{k \to 1} p(n,K) = \infty$$

In fact, if K goes to 1, then $Q_0 = Q_0(n,k) \leq \frac{\sqrt{K^{n-1}} - 1}{\sqrt{K^{n-1}} - 1}$ tends to zero, so the condition $Q_0 \mathcal{A}_p < 1$ is satisfied with arbitrarily large p whenever K is sufficiently close to 1. Hence the property (58) follows.

<u>Lemma 3. Let</u> $f : \Omega \longrightarrow \mathbb{R}^n$ <u>be a mapping from Sobolev space</u> $W^1_{q,loc}(\Omega)$, $2 \leq q \leq T$. <u>Suppose that</u> f <u>satisfies the following weak inverse Poincaré's inequality</u>

(59) $$\left[⨍_{B(x_0,R)} |Df|^q \right]^{1/q} \leq \frac{C}{R} \left[⨍_{B(x_0,2R)} \left| f - ⨍_{B(x_0,2R)} f \right|^q \right]^{1/q}$$

<u>for every ball</u> $B(x_0,2R) \subset \Omega$, <u>where</u> C <u>does not depend on the ball</u> $B(x_0,2R)$.

Then there exist $q' = q+\nu$, $\nu = \nu(n,C,T) > 0$ and a constant $C' = C'(n,q,C)$ such that: $f \in W^1_{q',loc}(\Omega)$ and

$$(60)\qquad \Big[\fint_{B(x_o,R)} |Df|^{q'} \Big]^{1/q'} \le \frac{C'}{R} \Big[\fint_{B(x_o,2R)} |f - \fint_{B(x_o,2R)} f|^{q'} \Big]^{1/q'}$$

for any ball $B(x_o,2R) \subset \Omega$.

This lemma may be simply derived from the well known lemma of Gehring concerning Hölder's inverse inequalities [8],[9_2]

The proof of Theorem 2. Let s be an arbitrary number from the interval $[2,p]$ such that $f \in W^1_{\frac{ns}{2},loc}(\Omega)$. Clearly the number $s=2$ satisfies those conditions . We are going to prove that

$$(61)\qquad \Big[\fint_{B(x_o,R)} |Df|^{\frac{ns}{2}} \Big]^{2/ns} \le \frac{C(n) \mathcal{A}_p}{(1-Q_o \mathcal{A}_p) R} \Big[\fint_{B(x_o,2R)} |f - \fint_{B(x_o,2R)} f|^{\frac{ns}{2}} \Big]^{2/ns}$$

for every ball $B(x_o 2R) \subset \Omega$.

For this we take a function $\varphi \in C^\infty_o(B(x_o,2R))$ such that

$$\varphi(x) \equiv 1 \quad \text{on} \quad B(x_o,R) \quad , \quad |\nabla\varphi(x)| \le \frac{C(n)}{R} \ ,$$

where $C(n)$ is a constant independent of R.

We multiply both sides of (56) by $|\varphi|^{n-2}\varphi$ getting

$$(62)\qquad \bar\partial[\varphi Df] = Q\partial[\varphi Df]$$

For notational simplicity we introduce the matrix $A = \varphi Df = D\varphi f - f \otimes \nabla\varphi$, where the symbol $f \otimes \nabla\varphi$ means tensor product of f and $\nabla\varphi$, ie $(f \otimes \nabla\varphi)_{ij} = f^i \frac{\partial\varphi}{\partial x_j}$, $i,j = 1,2,\ldots,n$.

$$(63)\qquad |f \otimes \nabla\varphi| \le |f||\nabla\varphi|$$

Equation (62) reads $\bar\partial A = Q\partial A$, and in view of (30) we have

$$(64)\qquad \big| |A|^{\frac{2-n}{2}} \bar\partial A \big| \le Q_o \big| |A|^{\frac{2-n}{2}} \partial A \big|$$

In connection with this inequality we consider the following $\frac{n}{2}$-homogeneous forms

$$S(A) = |A|^{\frac{2-n}{2}} \bar{\partial} A \qquad \text{and} \qquad R(A) = |A|^{\frac{2-n}{2}} \partial A$$

Thus (64) implies

$$(65) \qquad |S(D\varphi f - f \otimes \nabla\varphi)| \le Q_o |R(D\varphi f - f \otimes \nabla\varphi)|$$

For S and R we have the following estimates

$$(66) \qquad |A|^{\frac{n}{2}} \le \tfrac{1}{2}(|S(A)| + |R(A)|), \text{ (because } 2|A|^{\frac{n-2}{2}} A^* = S(A) + R(A)),$$

for each $\varepsilon \in (0,1)$ and for every matrices A,B

$$(67) \qquad |S(A)| \le |S(A-B)| + \varepsilon |A|^{\frac{n}{2}} + C(n)\varepsilon^{\frac{2-n}{2}} |B|^{\frac{n}{2}},$$

$$(68) \qquad |R(A-B)| \le |R(A)| + \varepsilon |A|^{\frac{n}{2}} + C(n)\varepsilon^{\frac{2-n}{2}} |B|^{\frac{n}{2}}.$$

This is the general fact valid for arbitrary $\frac{n}{2}$-homogeneous forms which are smooth at $A \neq 0$.
Now (65) and (67),(68) imply

$$(69) \qquad |SD\varphi f| \le Q_o |RD\varphi f| + \delta |D\varphi f|^{\frac{n}{2}} + C(n)\delta^{\frac{2-n}{2}} |f \otimes \nabla\varphi|^{\frac{n}{2}}$$

for arbitrary $\delta \in (0,1)$.
Obviously $\varphi f \in W^1_{\frac{ns}{2}}(\mathbb{R}^n)$ so, we may apply inequality (34) getting

$$\| R D\varphi f\|_s \le A_s \| SD\varphi f\|_s \le Q_o A_s \| R D\varphi f\|_s + \delta A_s \| |D\varphi f)|^{\frac{n}{2}} \|_s +$$
$$+ C(n)\sigma^{\frac{2-n}{2}} A_s \| |f|^{\frac{n}{2}} |\nabla\varphi|^{\frac{n}{2}} \|_s$$

By the assumption on s we know that $Q_o A_s \le Q_o A_p < 1$.

Therefore

(70) $$\| \mathcal{R} D\varphi f\|_s \le \frac{\delta \mathcal{A}_p}{1-\Omega_o \mathcal{A}_p} \| |D\varphi f|^{\frac{n}{2}}\|_s + \frac{C(n)\delta^{\frac{2-n}{2}} \mathcal{A}_p}{1-\Omega_o \mathcal{A}_p} \| |f|^{\frac{n}{2}} |\nabla\varphi|^{\frac{n}{2}}\|_s$$

Using (69) once again we get

(71) $$\| SD\varphi f\|_s \le \frac{\delta}{1-\Omega_o \mathcal{A}_p} \| |D\varphi f|^{\frac{n}{2}}\|_s + \frac{C(n)\delta^{\frac{2-n}{2}}}{1-\Omega_o \mathcal{A}_p} \| |f|^{\frac{n}{2}} |\nabla\varphi|^{\frac{n}{2}}\|_s$$

Adding (70) and (71) side by side and applying (66)

$$\| |D\varphi f|^{\frac{n}{2}}\|_s \le \frac{\delta(1+\mathcal{A}_p)}{2(1-\Omega_o \mathcal{A}_p)} \| |D\varphi f|^{\frac{n}{2}}\|_s + \frac{C(n)\delta^{\frac{2-n}{2}}(1+\mathcal{A}_p)}{2(1-\Omega_o \mathcal{A}_p)} \| |f|^{\frac{n}{2}} |\nabla\varphi|^{\frac{n}{2}}\|_s$$

Now, we put $\delta = \frac{1-\Omega_o \mathcal{A}_p}{1+\Omega_o \mathcal{A}_p}$ getting

$$\| |D\varphi f|^{\frac{n}{2}}\|_s \le C(n) \left[\frac{1+\mathcal{A}_p}{1-\Omega_o \mathcal{A}_p}\right]^{\frac{n}{2}} \| |f|^{\frac{n}{2}} |\nabla\varphi|^{\frac{n}{2}}\|_s .$$

In other words

$$\left[\int |D\varphi f|^{\frac{ns}{2}}\right]^{2/ns} \le C(n)^{2/n} \frac{1+\mathcal{A}_p}{1-\Omega_o \mathcal{A}_p} \left[\int |f|^{\frac{ns}{2}} |\nabla\varphi|^{\frac{ns}{2}}\right]^{2/ns} .$$

On account of the properties of φ we conclude

(72) $$\left[⨍_{B(x_o,R)} |Df|^{\frac{ns}{2}}\right]^{2/ns} \le \frac{C(n)\mathcal{A}_p}{1-\Omega_o \mathcal{A}_p} \left[⨍_{B(x_o,2R)} |f|^{\frac{ns}{2}}\right]^{2/ns} .$$

Finally (61) follows from (72) if we replace f by $f - ⨍_{B(x_o,R)} f$.

In order to complete the proof of Theorem 2 we observe that the factor in front of the integral in the right hand side of (61) does not depend on s. On the basis of Lemma 3 one can find a number $s' = s+\mu, \mu = \mu(n,p) > 0$ that f belongs to $W^1_{\frac{ns'}{2},loc}(\Omega)$. According to the previous considerations (61) remains valid for s'. In other words we enlarged s of validity of (61) by a term μ independent of s.

This procedure may be continued as long as s will not exceed s p . Obviously, after a finite number of such steps we may reach p. This completes the proof of Theorem 2.

Institute of Mathematics of the
Polish Academy of Sciences (Śniadeckich 8)
PL-00-950 Warszawa, P.O.Box 137, Poland

References

[1] Ahlfors, L.:"Conditions for quasiconformal deformations in several variables", contributions to analysis (a collection of papers dedicated to Lipman Bers), Academic Press, New York 1974, pp. 19-25.

[2] Ball, J.M., Currie, J.C . and Olver, P.J.:"Null lagrangians, weak continuity, and variational problems of arbitrary order", J. Functional Analysis. Vol.41, No 2, 1981, pp.135-174.

[3] Bers, L. and Nirenberg, L.:"On a representation theorem for linear elliptic system with discontinuous coefficients and its applications, Convegno Internazionale sulle Equazioni Lineari alle derivate parziali, agosto 1951, Ed. Gremonese, Roma (1955), pp. 111-140.

[4] Bojarski, B.V.:"Homeomorphic solutions of Beltrami systems, (Russian), Dokl. Akad. Nauk SSSR 102 (1955) pp. 661-664. "Generalized solutions of first order elliptic equations with discontinuous coefficients",(Russian), Math. Sborn., vol.43 (85): 4, 1957, pp. 451-503

[5] Bojarski, B.V. and Iwaniec, T."Topics in quasiconformal theory in several variables", Proceedings of The First Finnish-Polish Summer School in Complex Analysis, Podlesice 1977 pp.21-44. "Another approach to Liouvill'e theorem, Mathematische Nachrichten

[6] Elcrat, A. and Meyers, N.G.:"Some results on regularity for solutions of non-linear elliptic systems and quasiregular functions", Duke Math. J. 42, 1975, pp. 121-136.

[7] Fefferman, C. and Stein, E.M.:"H^p spaces of several variables" Acta Math. 129, 1972 pp. 137-193.

[8] Gehring, F.W.:"The L^p-integrability of the partial derivatives of a quasiconformal mapping", Acta Math. 130, 1973 pp. 265-277

[9] Iwaniec, T.:"On L^p-integrability in PDE's and quasiregular mappings for large exponents" to appear in Annales Academiae Scientiarum Fennicae. "Note on the maximal function inequalities". to appear

[10] Reshetniak, J.G.:"Estimates for stability in the Liouvill'e theorem and L^p-integrability of derivatives of quasiconformal mappings", (Russian), Sib. Math. J., 17, 1976, pp.868-896.

[11] Sarvas, J.:"Ahlfors' trivial deformations and Liouville's theorem in $\mathbb{R}^n$", Colloquium talk in Joensuu, August 1978,

[12] Vekua, I.W.:"Generalized analytic functions", Pergamon Press, 1962.

THE DECOMPOSITION THEOREMS IN THE BIDISC

Piotr Jakóbczak (Kraków)

1. Introduction. In 1970 Henkin proved in [2] that if D is a strictly pseudoconvex domain in $\mathbb{C}^n$ with C^2 boundary, and s is a fixed point of D, then every function $f \in A(D)$ which vanishes at s can be written in the form

$$(1) \qquad f(z) = \Sigma_{i=1}^{n} (z_i - s_i) f_i(z)$$

for some functions $f_i \in A(D)$, and the mapping

$$f \longmapsto (f_1, \ldots, f_n) \in (A(D))^n$$

is linear and continuous. (We set $A(D)$ to be the algebra of all functions holomorphic in D and continuous in $\overline{D}$). Øvrelid showed in [7] that the decomposition (1) holds with $z_1 - s_1, \ldots, z_n - s_n$ replaced by any finite family $g_1, \ldots, g_N \in A(D)$ such that $g_i(s) = 0$, $i = 1, \ldots, N$, the germs of the functions g_i at s generate the ideal of germs at s of holomorphic functions which vanish at s, and the only common zero of $g_1, \ldots, g_N$ in $\overline{D}$ is the point s. On the other hand Ahern and Schneider proved in [1] that if $f \in A(D)$, then there exist functions $f_i(z,s)$ holomorphic in $D \times D$ and continuous in the set $(\overline{D} \times \overline{D}) \setminus \{(z,z): : z \in \partial D\}$, such that

$$f(z) - f(s) = \Sigma_{i=1}^{n} (z_i - s_i) f_i(z,s), \quad z,s \in D.$$

By a result of [5], a similar decomposition holds also for every function $f \in A(D \times D)$ such that $f|_{\Delta} \equiv 0$, where $\Delta = \{(z,z): z \in \overline{D}\}$. Therefore it is natural to ask whether one can prove the following result: Given a family of functions $g_1(z,s), \ldots, g_N(z,s) \in A(D \times D)$ which satisfy the appropriate conditions, and an arbitrary function $f \in A(D \times D)$ such that

$f|_\Delta \equiv 0$, find functions $f_1(z,s),\dots,f_N(z,s)$ holomorphic in $D\times D$ and continuous in $(\bar{D}\times\bar{D})\setminus\{(z,z): z\in\partial D\}$, such that

(2) $\quad f(z,s) = \Sigma_{i=1}^N g_i(z,s)f_i(z,s),\ z,s\in D.$

We describe now a generalization of the decomposition problem (2). First we fix some notation which will be used in the sequel. For every positive integers N and s, let $L(N,s)$ be the set of all sequences of the form $I=(i_1,\dots,i_s)$, with $1\le i_1,\dots,i_s\le N$. Set $C^{N,s}=\{f = =(f_I)_{I\in L(N,s)}$: for every I, $f_I\in C(D\times D)$, and $(f_I)_I$ is skew-symmetric with respect to $I\}$. Let also $C^{N,0}=C(D\times D)$. (Note that $C^{N,s}=0$ for $s>N$). Let $g=(g_1,\dots,g_N)$ be a finite family of functions continuous in $D\times D$. For every $s=1,2,\dots$, define the operator $P=P_g=P_{g,s}$: $: C^{N,s}\longrightarrow C^{N,s-1}$ by the formula

(3) $\quad (Pf)_J = \Sigma_{j=1}^N g_j f_{Jj},$

where $f=(f_I)_I\in C^{N,s}$, and for $J=(j_1,\dots,j_{s-1})$, we set $Jj=(j_1,\dots \dots,j_{s-1},j)$. Note that $P\circ P=0$. Define also $P_{g,0}$ on $C^{N,0}$ as the zero operator. Set

$$A^{N,s}=\{f=(f_I)_I\in C^{N,s}:\ \text{for every } I,\ f_I\in A(D\times D)\},$$

where $s=1,2,\dots$, and let $A^{N,0}=A(D\times D)$. The generalization of the decomposition problem, mentioned above, is the following: Let $g = =(g_1,\dots,g_N)$ be a finite family of functions from $A(D\times D)$. Under which conditions on g, for every $s=0,1,\dots,N-1$, and for every $f\in A^{N,s}$ with $f|_\Delta\equiv 0$ and $P_g f=0$, there exists $u=(u_J)_J \in C^{N,s+1}$ such that for each J, u_J is holomorphic in $D\times D$ and continuous in $(\bar{D}\times\bar{D})\setminus\{(z,z): z\in\partial D\}$, and $P_g u=f$?

We shall prove a result in this direction, in the most simple situation when D is the unit disc in $\mathbb{C}$ centered at 0, and under some rather restrictive assumptions on the functions $g_j(z,s)$. On the other hand we do not require that the functions g_j and the decomposed system $f=(f_I)_I$ be continuous on the diagonal of the boundary.

Since now let U denote the unit disc in $\mathbb{C}$ with centre at 0, and let $Q = (\bar{U} \times \bar{U}) \setminus \{(z,z) : z \in \partial U\}$. Let also $\Delta = \{(z,z) : z \in U\}$. Define $A_Q = \mathcal{O}(U^2) \cap C(Q)$, and set

$$A_Q^{N,s} = \{f = (f_I)_I \in C^{N,s} : f_I \in A_Q\},$$

$s = 1,2,\ldots$, and let $A_Q^{N,O} = A_Q$. We equipe $A_Q^{N,s}$ with the topology of componentwise uniform convergence on compact subsets of Q. Denote by $(A_Q^{N,s})_O$ ths subspace of $A_Q^{N,s}$ consisting of those systems which vanish on Δ. Set also

$$\mathcal{O}^{N,s} = \{f = (f_I)_I \in C^{N,s} : f_I \in \mathcal{O}(U^2)\}.$$

Let $g_1,\ldots,g_N \in A_Q$ be such that:

(a) $\Delta = \{(z,s) \in Q : g_1(z,s) = \ldots = g_N(z,s) = O\}$.

(b) For every $z \in U$, the germs of the functions $g_1,\ldots,g_N$ at the point (z,z) generate the ideal of germs at (z,z) of holomorphic functions which vanish on Δ.

(c) There exists a neighbohood V of $\partial(U^2)$ in $\bar{U}^2$ such that $g_1,\ldots,g_N$ have no zeros in $V \cap (Q \setminus \Delta)$.

<u>THEOREM 1</u>. <u>Suppose that</u> $g_1,\ldots,g_N$ <u>satisfy the conditions</u> (a), (b) <u>and</u> (c). <u>Let</u> s <u>be an integer</u>, $O \leq s \leq N-1$. <u>Then for every</u> $f \in (A_Q^{N,s})_O$ <u>such that</u> $P_g f = O$ <u>there exists</u> $u \in A_Q^{N,s+1}$ <u>such that</u>

$$P_g u = f. \tag{4}$$

In particular, for $s = O$, we obtain the decomposition (2). In this case, we can state even more:

<u>THEOREM 2</u>. <u>If</u> $g_1,\ldots,g_N \in A_Q$ <u>satisfy the conditions</u> (a), (b) <u>and</u> (c), <u>then for every</u> $f \in A_Q$ <u>such that</u> $f|_\Delta \equiv O$ <u>there exist functions</u> $f_1,\ldots,f_N \in A_Q$ <u>such that</u>

$$f(z,s) = \Sigma_{i=1}^N g_i(z,s) f_i(z,s), \quad (z,s) \in Q.$$

<u>Moreover, the mapping</u>

$$(A_Q)_O \ni f \longrightarrow (f_1,\ldots,f_N) \in A_Q^{N,1} \tag{5}$$

is linear and continuous.

Since the natural inclusion mapping $A(U^2) \longrightarrow A_Q$ is continuous, we can state the result which includes in this particular case the Øvrelid theorem. Set $A_o(U^2) = \{f \in A(U^2) : f|_\Delta \equiv 0\}$.

COROLLARY. *For every* $f \in A_o(U^2)$ *there exist functions* $f_1,\dots,f_N \in A_Q$ *which depend linearly on* f, *such that the decomposition* (2) *holds in* Q, *and for every compact subset* $K \subset Q$, *the mapping*

$$A_o(U^2) \ni f \longrightarrow (f_1|_K,\dots,f_N|_K) \in (A(K))^N$$

is continuous.

We want to stress that it is natural to assume that the functions $g_1,\dots,g_N$ satisfy the conditions (a) and (b), while the condition (c) seems to be superfluous. However we were not able to prove the result without the assumption (c). (It will be clear from the proof that this assumption can be replaced by some less restrictive conditions, but which are not essentially different from (c)).

I am very much indebted to Dr. M. Jarnicki for his helpful suggestions and discussions on this subject.

2. P r o o f of Theorem 1. In the proof we apply the techniques of Øvrelid [7] and the results on the solution of the $\bar\partial$-problem in pseudoconvex polyhedra [3](cf. also [4]), applied to the particular case, when the considered domain is the bidisc.

Let $g_1,\dots,g_N \in A_Q$ satisfy the assumptions of the theorem, and let $f = (f_I)_I \in (A_Q^{N,s})_o \cap \ker P_g$. Since $g_1,\dots,g_N$ and all functions f_I vanish on Δ, there exist (uniquely determined) functions $G_1,\dots,G_N \in A_Q$ and $F = (F_I)_I \in A_Q^{N,s}$, such that $g_i(z,s) = (z-s)G_i(z,s)$ and

$$f_I(z,s) = (z-s)F_I(z,s) \tag{6}$$

for every $(z,s) \in Q$. Set $G = (G_1,\dots,G_N)$. Note that in order to prove (4) it is sufficient to show that there exists $u \in A_Q^{N,s+1}$ such that $P_G u = F$.

It follows from (b) that the functions $G_1,\dots,G_N$ have no common zeros in Q. Therefore by the methods of [6] we can prove the following result:

LEMMA 1. *For every* $h \in \mathcal{O}^{N,s}$ *with* $P_G h = 0$, $s = 0,\dots,N-1$, *there exists* $v \in \mathcal{O}^{N,s+1}$ *such that* $P_G v = h$.

Since $P_g f = 0$, we have also $P_G F = 0$. Therefore, by the above lemma, there exists $v = (v_J)_J \in \mathcal{O}^{N,s+1}$ such that

$$(7) \qquad P_G v = F.$$

By the assumption (b) there exists an index i, $1 \le i \le N$, such that $G_i|_\Delta$ is not identically zero. We may assume that G_1 is such a function. It follows from (c) that $G_1|_{Q\cap V}$ does not vanish at any point.

Following [7] let $N_i = \{(z,s) \in Q \setminus \Delta : g_i(z,s) = 0\}$. The sets N_i are closed in $\mathbb{C}^2 \setminus \Delta(\overline{U})$; therefore there exist functions $\tilde{\varphi}_i \in C^\infty(\mathbb{C}^2 \setminus \Delta(\overline{U}))$, such that $0 \le \tilde{\varphi}_i \le 1$, $\Sigma_{i=1}^N \tilde{\varphi}_i \equiv 1$, and each $\tilde{\varphi}_i$ is zero in a neighborhood of N_i in $C^2 \setminus \Delta(\overline{U})$. Moreover, because of the condition (c), we may assume (shrinking V if necessary) that $\tilde{\varphi}_1|_{V\setminus\Delta(\overline{U})} \equiv 1$, while $\tilde{\varphi}_i|_{V\setminus\Delta(\overline{U})} \equiv 0$, $i = 2,\dots,N$. Choose a real number p such that $1 < p < 2$, and let q be given by the equation $1/p + 1/q = 1$. Since $G_{1|Q\cap V}$ does not vanish, it is not difficult to see that there exists a neighborhood W of Δ in U^2 such that for every $J = (j_1,\dots,j_{s+1})$

(a) $(v_J/G_1)|_{W\cap V}$ is in $L^q(W\cap V)$ (with respect to four-dimensional Lebesgue measure in C^2),

(8) (b) $\|(v_J/G_1)|_{(\{z\}\times U)\cap W}\|_q \longrightarrow 0$ $(\|(v_J/G_1)|_{(U\times\{s\})\cap W}\|_q \longrightarrow 0)$

as $z \longrightarrow \partial U$ $(s \longrightarrow \partial U)$,

where $\| \ \|_q$ denotes the L^q-norm with respect to the two-dimensional Lebesgue measure on $\{z\} \times U$ and on $U \times \{s\}$. There exists a function $\varphi_0 \in C^\infty(\mathbb{C}^2 \setminus \Delta(\partial U))$ such that $0 \le \varphi_0 \le 1$, $\operatorname{supp}\varphi_0 \subset W$, $\varphi_0 \equiv 1$ in some neighborhood W_1 of Δ in U^2, and such that:

(a) The coefficients of $\overline{\partial}\varphi_0$ are in $L^p(U^2)$.

(b) There exists a constant $c > 0$ such that for every $z, s \in U$,

(9) $$\|\bar{\partial}\varphi_o|_{(\{z\}\times U)\cap W}\|_p \leq c \quad \text{and} \quad \|\bar{\partial}\varphi_o|_{(U\times\{s\})\cap W}\|_p < c,$$

where $\| \ \|_p$ denotes the sum of the L^p-norms of the coefficients of $\bar{\partial}\varphi_o$ with respect to the two-dimensional Lebesgue measure on $\{z\} \times U$ and on $U \times \{s\}$.

For the convenience of the reader, we describe the construction of the function φ_o. Let W be a given neighborhood of Δ in U^2. It is useful to consider the image of the bidisc $\bar{U}^2$ by the transformation $H : \mathbb{C}^2 \ni (z,s) \longmapsto (\frac{1}{2}(z+s), \frac{1}{2}(z-s)) \in \mathbb{C}^2$. We will call the new coordinates again by (z,s). Then $H(\bar{U}^2) = \{(z,s) : |z+s| \leq 1, |z-s| \leq 1$, and $H(\Delta(U)) = \bar{U} \times \{0\}$. The image of the set W in $H(\bar{U}^2)$ is a neighborhood of $U \times \{0\}$ in $H(U^2)$; it will be called again by W. Choose a decreasing function $\chi \in C^\infty([0,1))$ such that $\chi(x) > 0$ and $|\chi'(x)| \leq 1$ for every $x \in [0,1)$, $\chi(x) \longrightarrow$ as $x \longrightarrow 1$ and such that the set $V_1 = \{(z,s) : z \in U, |s| \leq \chi(|z|)\}$ is contained in W. Choose also a decreasing function $\eta \in C^\infty([0,1])$, such that $\eta \equiv 1$ in a neighborhood of 0, $\eta \equiv 0$ in a neighborhood of 1, and $|\eta'(x)| \leq 2$ for all $x \in [0,1]$. Denote $z = x+iy$, $s = u+iv$, and identify a point $(z,s) \in \mathbb{C}^2$ with a point $(x,y,u,v) \in \mathbb{R}^4$. Set

$$\varphi(x,y,u,v) = \eta(\sqrt{u^2+v^2}/\chi(\sqrt{x^2+y^2})).$$

Then $\varphi \in C^\infty(V_1)$, $\operatorname{supp} \varphi \subset V_1$, and $\varphi \equiv 1$ in some neighborhood of $U \times \{0\}$ in $H(U^2)$. We can assume that φ is smooth in all of $H(Q)$, by setting $\varphi \equiv 0$ in $H(Q) \setminus V_1$. Set $\varphi_o = \varphi \circ H$.

Calcutating the partial derivative φ_x, estimating the integral of $|\varphi_x|^p$ over $H(U^2)$, and making a change of variables u,v into polar coordinates, we conclude that

$$\int_{H(U^2)} |\varphi_x|^p dxdydudv \leq c \int_U \chi^{2-p}(\sqrt{x^2+y^2})dxdy.$$

Since $1 < p < 2$, the last integral is finite. Similarly we can check that φ_y, φ_u and φ_v are in $L^p(H(U^2))$. Therefore, φ_o satisfies the condition (9)(a).

Now fix $z \in U$. Let $H_z : \{z\} \times U \longrightarrow U$ be the identification map, $H_z(z,\xi) = \xi$. The set $U_z = H_z((\{z\} \times U) \cap \operatorname{supp}\varphi_0)$ is a neighbourhood of z in U. Consider any $w \in C$ with $|w| = 1$. Let $I_{z,w} = U_z \cap \{z+tw : t \geq 0\}$; then $I_{z,w}$ is an interval, with the length not exceeding 1. Having chosen the function φ as above, we can prove that there exists a constant $c > 0$, independent of z and w, such that for every point $z + tw \in I_{z,w}$,

$$|\varphi_x(H \circ H_z^{-1}(z+tw))| \leq c/t,$$

and that the same estimate holds for φ_y, φ_u and φ_v instead of φ_x. This gives the first part of the condition (9)(b). The case of the slices $U \times \{s\}$ can be treated similarly.

Let $\mathcal{E}_{(o,r)}(Q)$ ($\mathcal{E}^\infty_{(o,r)}(U^2)$) be the space of differential forms of order $(0,r)$ with coefficients in Q (with coefficients of the class $\mathcal{E}^\infty(U^2)$). Set

$$L_r = \{u \in \mathcal{E}^\infty_{(o,r)}(U^2) : u \in \mathcal{E}_{(o,r)}(Q),\ \bar{\partial}u \in \mathcal{E}_{(o,r+1)}(Q)\},$$

$r = 0,1,2$. Let L_r^s be defined similarly as $C^{N,s}$ with the only change that now $u_I \in L_r$ for every $I \in L(N,s)$. Denote also $M_r^s = \{u \in L_r^s : u|_{W_1} \equiv 0\}$. Define the operator P_G on L_r^s and on M_r^s in a similar way as in (3). Choose a function $\varphi \in \mathcal{E}^\infty(\mathbb{C}^2 \setminus \Delta(\partial U))$ so that $0 \leq \varphi \leq 1$, $\varphi \equiv 1$ outside W_1, and $\varphi \equiv 0$ in some neighbourhood W_2 of Δ in U^2. Let $k = (\varphi\tilde{\varphi}_1/G_1, \ldots, \varphi\tilde{\varphi}_N/G_N)$. Define the operators $\Psi = \Psi_r^s : M_r^s \cap \ker P_G \to M_r^{s+1}$ by the formula

$$(\Psi u)_J = \Sigma_{l=1}^{s+1} (-1)^{s+l-1} (k)_{j_l} u_{J \setminus j_l},$$

where, for $J = (j_1, \ldots, j_{s+1})$, we set $I \setminus j_l = (j_1, \ldots, j_l, \ldots, j_{s+1})$ (j_l is omitted). Note that

(1) $$P_G \Psi = \mathrm{Id}|_{M_r^s \cap \ker P_G},$$

and that $\Psi_r^s = 0$ for $s \geq N$, since $M_r^{s+1} = 0$ in this case. Set

$$Rf = \varphi_o v + \Psi_r^s((1-\varphi_o)F),$$

where v and F are defined in (7) and (6). (Note that this definition makes sense, since $(1-\varphi_o)F|_{W_1} \equiv 0$). For $n = 1,2,\ldots,$ set $U_n = \{z \in \mathbb{C} : |z| < 1-1/n\}$, and let $D_n = U \times U_n$. Let $R_n f = Rf|_{D_n}$. Put

$$(L_n)_r = \{u \in \mathcal{C}^\infty_{(o,r)}(D_n) : u \in \mathcal{C}_{(o,r)}(\bar{D}_n),\ \bar\partial u \in \mathcal{C}_{(o,r+1)}(\bar{D}_n)\},$$

and define $(L_n)_r^s$ and $(M_n)_r^s$ similarly as L_r^s and M_r^s. Also, define in an obvious way the operators $P_G = (P_G^n)_r^s : (L_n)_r^s \longrightarrow (L_n)_r^{s-1}$, and $\Psi = (\Psi_n)_r^s : (M_n)_r^s \cap \ker(P_G^n)_r^s \longrightarrow (M_n)_r^{s+1}$. Let the operators $\bar\partial : (L_n)_r^s \longrightarrow (L_n)_{r+1}^s$ be given by $\bar\partial((u_I)_I) = (\bar\partial u_I)_I$. Note that

(11) $\quad P_G P_G = 0, \quad \bar\partial P_G = P_G\bar\partial$, and $\quad P_G R_n f = F|_{D_n}$.

The following lemma is a special case of [3, Theorem](see also [4, Theorem 2]):

LEMMA 2. *Let $r = 1,2$ and $n = 1,2,\ldots,$ be fixed. Then there exists a linear operator $(\chi_n)_r : (L_n)_r \cap \ker\bar\partial \longrightarrow (L_{n-1})_r$, such that for every $u \in (L_n)_r \cap \ker\bar\partial$,*

$$\bar\partial(\chi_n)_r u = u,$$

and

$$||(\chi_n)_r u||_{D_n} \le c||u||_{D_n}$$

for some c independent of u.

We will use later the explicit form of the operator $(\chi_n)_r$ which is given in [3].

Now define an operator $(\chi_n)_r^s : (L_n)_r^s \cap \ker\bar\partial \longrightarrow (L_n)_{r-1}^s$ by setting $(\chi_n)_r^s((u_I)_I) = ((\chi_n)_r u_I)_I$, and let

(12) $\quad T_n f = R_n f - P_G(\chi_n)_1^{s+2}(\Psi\bar\partial R_n f - P_G(\chi_n)_2^{s+3}\Psi\bar\partial\Psi\bar\partial R_n f).$

(In the case $s = N-1$ or $s = N-2$, $T_n f$ has simpler form, since some operators in its definition are then trivial). $T_n f$ is well-defined, since $\mathrm{dom}(\chi_n)_2^{s+3} = (L_n)_2^{s+3}$, the form on which the operator $(\chi_n)_1^{s+2}$ acts is $\bar\partial$-closed, in virtue of (10) and (11), and all the operators Ψ act on the forms which vanish in W_1. Note also that $Tf \in (L_n)_0^{s+1}$, that each component $(T_n f)_J$ is holomorphic in D_n (by (10) and (11)), and that $P_G(T_n f) = F$. We will show that for each $J \in L(N,s+1)$ there exists a function $u_J \in A_Q$, such that $(T_n f)_J \longrightarrow u_J$ pointwise in Q; then $u = (u_J)_J \in A_Q^{N,s+1}$, and $P_G u = F$, which will finish the proof.

We recall that according to [3], the operator $(\chi_n)_1$ acting on a $\bar\partial$-closed form $\omega \in (L_n)_1$ can be expressed as

$$(\chi_n)_1\omega(z,s) = \frac{-1}{(2\pi i)^2}\int_{D_n}\frac{(\bar\zeta_1-\bar z)d\bar\zeta_2 - (\bar\zeta_2-\bar s)d\bar\zeta_1}{(|\zeta_1-z|^2+|\zeta_2-s|^2)^2}\wedge d\zeta_1\wedge d\zeta_2\wedge\omega(\zeta)$$

$$+\frac{1}{2\pi i}\int_{|\zeta_2|<1-1/n}\frac{d\zeta_2}{\zeta_2-s}\wedge\omega(z,\zeta_2)$$

$$+\frac{1}{(2\pi i)^2}\int_{D_n}\bar\partial_\zeta\left(\frac{\bar\zeta_2-\bar s}{|\zeta_1-z|^2+|\zeta_2-s|^2}\right)\wedge d\zeta_1\wedge d\zeta_2\,\frac{\omega(\zeta)}{\zeta_1-z}$$

$$(13)\qquad +\frac{1}{(2\pi i)^2}\int_{U\times\partial U_n}\frac{(\bar\zeta_2-\bar s)d\zeta_1\wedge d\zeta_2\wedge\omega(\zeta)}{(\zeta_1-z)(|\zeta_1-z|^2+|\zeta_2-s|^2)}+\frac{1}{2\pi i}\int_{|\zeta_1|<1}\frac{d\zeta_1}{\zeta_1-z}\wedge\omega(\zeta_1,s)$$

$$+\frac{1}{(2\pi i)^2}\int_{D_n}\bar\partial_\zeta\left(\frac{\bar\zeta_1-\bar z}{|\zeta_1-z|^2+|\zeta_2-s|^2}\right)\wedge d\zeta_1\wedge d\zeta_2\wedge\frac{\omega(\zeta)}{\zeta_2-s}$$

$$+\frac{1}{(2\pi i)^2}\int_{\partial U\times U_n}\frac{(\bar\zeta_1-\bar z)d\zeta_1\wedge d\zeta_2\wedge\omega(\zeta)}{(\zeta_2-s)(|\zeta_1-z|^2+|\zeta_2-s|^2)} = \Sigma_{k=1}^{7}\chi_n^{(k)}\omega(z,s).$$

One can also check, using the formula (1.7) of [4], that

$$(14)\qquad (\chi_n)_2\omega(z,s) = \frac{-1}{(2\pi i)^2}\int_{D_n}\frac{(\bar\zeta_2-\bar s)d\bar z - (\bar\zeta_1-\bar z)d\bar s}{(|\zeta_1-z|^2+|\zeta_2-s|^2)^2}\wedge d\zeta_1\wedge d\zeta_2\wedge\omega(\zeta),$$

where $\omega \in (L_n)_2$. By a direct computation we obtain that for $K = (k_1,\dots,k_{s+2})$, $(\Psi \bar\partial R_n f)_K$ is a linear combination of the terms

$$(15) \qquad \frac{\tilde\varphi_i V_{K\setminus i}}{G_i}\,\bar\partial\varphi_0 \quad\text{and}\quad \frac{(1-\varphi_0)\tilde\varphi_i\,\bar\partial\,\tilde\varphi_j F_{K\setminus(i,j)}}{G_i G_j}\ ; \quad i,j = k_1,\dots,k_{s+2},$$

where v and F are defined in (7) and (6), respectively. Similarly, for $L = (l_1,\dots,l_{s+3})$, $(\Psi\bar\partial\Psi\bar\partial R_n f)_L$ is a linear combination of the terms

$$(16) \qquad \frac{V_{L\setminus(i,j)}\tilde\varphi_i\bar\partial\tilde\varphi_j}{G_i G_j}\wedge\bar\partial\tilde\varphi_0,\ \frac{(1-\varphi_0)F_{L\setminus(i,j,k)}\tilde\varphi_i\bar\partial\tilde\varphi_j\wedge\bar\partial\tilde\varphi_k}{G_i G_j G_k}\ \text{, and}$$

$$\frac{F_{L\setminus(i,j,k)}\tilde\varphi_i\tilde\varphi_j\bar\partial\tilde\varphi_k}{G_i G_j G_k}\wedge\bar\partial\varphi_0,\quad i,j,k = l_1,\dots,l_{s+3}.$$

It follows from (14),(16) and our choice of the functions $\tilde\varphi_i$, that each $(\Psi\bar\partial\Psi\bar\partial R_n f)_L$ is a restriction to D_n of a (0,2)-form $(\Psi\bar\partial\Psi\bar\partial Rf)_L$ which is $\mathcal{C}^\infty$ in U^2 and continuous in $\bar U^2$, that the coefficients of $(\chi_n)_2^{s+3}\Psi\bar\partial\Psi\bar\partial R_n f$ converge boundedly in U^2 and uniformly on compact subsets of U^2, and that the limit system $(w_L)_L$ has the form

$$w_L = \frac{-1}{(2\pi i)^2}\int_{U\times U}\frac{(\bar\zeta_2-\bar s)d\bar z-(\bar\zeta_1-\bar z)d\bar s}{(|\zeta_1-z|^2+|\zeta_2-s|^2)^2}\wedge d\bar\zeta_1\wedge d\bar\zeta_2\wedge(\Psi\bar\partial\Psi\bar\partial\ Rf)_L(\zeta),$$

and is therefore smooth in U^2 and continuous in $\bar U^2$.Hence the system $P_G(\chi_n)_2^{s+3}\Psi\bar\partial\Psi\bar\partial R_n f$ is also convergent uniformly on compact subsets of U^2 and boundedly on U^2 to a system $(t_K)_K$ of differential forms of class $\mathcal{C}^\infty$ in U^2 and continuous in $\bar U^2$.

It follows from the choice of $\tilde\varphi_i$ that the only terms in (15) which need not vanish in V, are of the form $(\tilde\varphi_i v_{K\setminus i}/G_i)\bar\partial\varphi_0$. We egzamine now the behaviour of the integral operators in (13), acting on $(\Psi\bar\partial R_n f)_K - (P_G(\chi_n)_2^{s+3}\Psi\bar\partial\Psi\bar\partial R_n f)_K$. It is simple to show that if a system $(t_K^{(n)})_K$ of forms which are defined and continuous in $\bar D_n$ and smooth in D_n converges boundedly in U^2 and uniformly on compact subsets of U^2 to $(t_K)_K$ (where $(t_K)_K$ is defined above), then

$(\chi_n)_1 t_k^{(n)} \longrightarrow \chi_1 t_K$ uniformly on conpact subsets of U^2 and the limit form $\chi_1 t_K$ has the coefficients which are smooth in U^2 and continuous in $\bar{U}^2$. (Here $\chi_1 t_K$ denotes the operator of the solution of the $\bar\partial$-problem in U^2 from [3]; the expression for χ_1 is similar to that of (13), with U_n and D_n replaced by U and U^2). Therefore $(\chi_n)_1(P_G(\chi_n)_2^{s+3}\Psi\bar\partial\Psi\bar\partial R_n f)_K \longrightarrow \chi_1 t_K$ uniformly on compact subsets of U^2.

Similarly, for every form ω from (15) which vanishes on V, $(\chi_n)_1(\omega|_{D_n}) \longrightarrow \chi_1\omega$ uniformly on compact subsets of U^2, and $\chi_1\omega$ is continuous in $\bar{U}^2$ and smooth in U^2.

It remains to verify the behaviour of the integral operators in (13), acting on the forms $\omega = (\tilde\varphi_i v_{K\setminus i}/G_i)\bar\partial\varphi_o$. Note first that since $\bar\partial\varphi_o \equiv 0$ on $\partial(U^2)\setminus\Delta(\partial U)$, then for every ω, $\chi_n^{(7)}\omega(z,s) \equiv 0$. Moreover, given $(z,s) \in U^2$, the only singularity of the coefficients of the forms occuring in the integrals defining $\chi_n^{(1)}\omega(z,s)$, $\chi_n^{(3)}\omega(z,s)$ and $\chi_n^{(6)}\omega(z,s)$, is at $\zeta = (z,s)$, and we integrate over the set $\operatorname{supp}\bar\partial\varphi_o$. Therefore, by (8)(a),(9)(a), and by the use of the Lebesgue dominated convergence theorem and Hölder inequality, we conclude that $\chi_n^{(i)}\omega(z,s) \longrightarrow \chi^{(i)}\omega(z,s)$ uniformly on compact subsets of U^2, and that the functions $\chi^{(i)}\omega$ are smooth in U^2 and extend continuosly to Q, $i = 1,3,6$. It is simple to see that there exists a function $\chi^{(2)}$ which is $\mathcal{C}^\infty$ in U^2 and such that $\chi_n^{(2)}\omega(z,s) \longrightarrow \chi^{(2)}\omega(z,s)$ uniformly on compact subsets of U^2. Moreover, by (8)(b),(9)(b) and Hölder inequality we verify that $\chi^{(2)}\omega$ extends continuously to Q. Similarly, the function $\chi_n^{(5)}\omega(z,s) = \chi^{(5)}\omega(z,s)$ (it does not depend on n) extends continuously to Q. The similar considerations can be made for the operators $\chi_n^{(4)}\omega(z,s)$.

We have thus shown that there exists a system of functions $(t_K)_K$ $K \in L(n,s+2)$, such that each $t_K \in C^\infty(U^2) \cap C(Q)$, and such that each component $((\chi_n)_1^{s+2}(\Psi\bar\partial R_n f - P_G(\chi_n)_2^{s+3}\Psi\bar\partial\Psi\bar\partial R_n f))_K = t_K^{(n)}$ tends to t_K uniformly on compact subsets of U^2. Therefore the same is true for $P_G t^{(n)}$, and hence we conclude from (12) that there exist functions $u_J \in C^\infty(U^2) \cap C(Q)$, $J \in L(N,s+1)$, such that $(T_n f)_J \longrightarrow u_J$ uniformly on compact subsets of U^2. Since each $(T_n f)_J$ is holomorphic in D_n, the limit functions u_J are holomorphic in U^2. This completes the proof.

3. Proof of Theorem 2. For every $f \in A_Q$ with $f|_\Delta \equiv 0$, there exists a uniquely determined function $F \in A_Q$ such that

$$f(z,s) = (z-s)F(z,s), \quad z,s \in D.$$

It is easy to see that the mapping $f \longrightarrow F$ is linear and continuous. By Theorem 1, there exist functions $h_i \in A_Q$, $i = 1,\dots,N$, such that

$$z - s = \Sigma_{i=1}^N g_i(z,s)h_i(z,s).$$

Therefore

$$f(z,s) = \Sigma_{i=1}^N g_i(z,s)f_i(z,s)$$

with $f_i = h_i F$, and it is clear that the mapping (5) is linear and continuous. This ends the proof.

References

[1] AHERN, P. and R.SCHNEIDER: The boundary behavior of Henkin's kernel, Pacific J. Math. 66 (1976), 9-14.

[2] HENKIN, G.M.: Approximation of functions in strictly pseudoconvex domains and a theorem of Z. L. Lejbenzon, Bull. Acad. Polon. Sci. Sér. Sci. Math. Astronom. Phys. 19 (1971), 37-42 [in Russian].

[3] —— : Uniform estimates of the solution of the $\bar{\partial}$-problem in Weil domains, Uspehi Mat. Nauk 26 (1971), 211 - 212 [in Russian].

[4] —— and A.G.SERGEEV: Uniform estimates of the solutions of the $\bar{\partial}$-equation in pseudoconvex polyhedra, Mat. Sbornik 112 (1980) 522-565 [in Russian].

[5] JAKÓBCZAK, P.: Extension and decomposition operators in products of strictly pseudoconvex sets, Ann. Polon. Math., to appear.

[6] KELLEHER, J. and B.TAYLOR: Finitely generated ideals in rings of analytic functions, Math. Ann. 193 (1971), 225-237

[7] ØVRELID, N.: Generators of the maximal ldeals of A(D), Pacific J.Math. 39 (1971), 219-223.

Institute of Mathematics
Jagiellomian University
Reymonta 4, PL-30-059 Kraków, Poland

THE GROWTH OF COMPOSITIONS OF A PLURISUBHARMONIC FUNCTION WITH ENTIRE MAPPINGS

Christer O. Kiselman (Uppsala)

Resumo. La kresko de kunligoj de plursubharmona funkcio kun entjeraj mapoj. Oni pridiskutas la kreskon de la kunligo $z \mapsto v(x(z))$ de plursubharmona funkcio v difinita sur $\mathbb{C}^n$ kun holomorfa mapo x de $\mathbb{C}^m$ al $\mathbb{C}^n$. La funkcio v estas fiksita sed x kaj z varias; do oni estas kondukita al la studado de plursubharmonaj funkcioj sur spaco de nefinia dimensio. Montriĝas ke la kresko de $v \circ x$ estas maksimuma krom kiam x apartenas al tre malgranda, pli precize polusa, aro en la spaco de ĉiuj mapoj.

Background

The purpose of my talk is to discuss the growth of the composition $v \circ x$ of a plurisubharmonic function v defined on $\mathbb{C}^n$ with an entire mapping x from $\mathbb{C}^m$ to $\mathbb{C}^n$, i.e., the behavior at infinity of $v(x(z))$, $z \in \mathbb{C}^m$, as a function of both z, a variable in $\mathbb{C}^m$, and x, a variable in a suitable space of mappings. Thus $v(x(z))$ is considered, for a fixed v, as the value of a function of (z,x) varying in an infinite-dimensional space. This is because we want to distinguish the exceptionally slow growth of $v \circ x$ from its typical growth. Although we start with functions defined only on $\mathbb{C}^m$ and $\mathbb{C}^n$, we are led to study plurisubharmonic functions on an infinite-dimensional space, viz. the space of mappings.

The result is that $v \circ x$ grows exceptionally slowly, in a precise sense, only when x belongs to a polar set in the space of all mappings considered. This result has a dual character. On the one hand, it says that those entire mappings whose growth is smaller than maximal form a very small set. As an example, the mappings of order less than ρ is polar in a Banach space of mappings of order at most ρ. From this point of view the function v is only used to measure the growth of the mappings x. On the other hand, the result says that the restriction $v|X$ of the plurisubharmonic function v to the subset $X = x(\mathbb{C}^m)$ of $\mathbb{C}^n$ has a growth at infinity which is almost as fast as the global growth of v, except for a small class of sets X and the smallness of this exceptional class is expressed by the polarity of the set of corresponding mappings x. Thus, from this point of view, the structure of v is of interest, and the mappings x are only used to parametrize the sets $X = x(\mathbb{C}^m)$. In brief, the growth of $v \circ x$ can be slow for two different reasons: either x grows slowly or v grows slowly on the range of x.

When the mapping x is linear or affine, the problem becomes completely finite-dimensional, and this case has been treated before, see Kiselman [1981, Theorems 4.1 and 5.1]. In another paper, Kiselman [1982], I have discussed the growth of $v(x(z))$ when z tends to the origin: this means that we consider the density (or Lelong number) of v at the point $x(0)$, for the behavior of v at a finite point is essentially described by its density. Thus, in that case, the underlying space is also of infinite dimension, for we consider all holomorphic mappings satisfying a relation like $x(0) = 0$, but the classification of the behavior is much less complicated. This is due to the following simple fact. The behavior at $-\infty$ of an increasing convex function on the real line is well described by the limit of its slope (and this limit is nothing but the Lelong number); at least it gives a good first term in its asymptotic development. But at $+\infty$ the behavior cannot be parametrized by a real number, it is much richer, and even if we majorize the growth, it gives rise to all the curious kinds of scales like order and type which have been studied over the centuries. Thus, there cannot exist any simple classification like density $= \alpha$ and density $< \alpha$.

Exceptional sets

In finite-dimensional real analysis the most frequently occurring exceptional sets are of course the Lebesgue null sets. In potential theory the sets of Newtonian capacity zero appear, and for many problems it is known that they form the correct class, i.e. a set can appear as an exceptional set for a certain property if and only if it is of capacity zero. The polar sets in $\mathbb{C}^n$, i.e. the sets where a plurisubharmonic function which is not the constant $-\infty$ assumes that value, have Newtonian capacity zero and hence Lebesgue measure zero. Until recently it was not known whether the exceptional sets that appear e.g. in the lim-sup-star theorem (called negligible sets) are polar, but this has been proved by Bedford and Taylor [1981]. This, of course, reinforced the interest of this class of sets, but it should be noted that the various results of the author showing that certain exceptional sets are polar [1981, 1982] are independent of any capacity theory in that the plurisubharmonic functions needed are obtained by rather constructive methods.

In infinite dimensions measure theory and potential theory become completely non-trivial, and the most popular kind of exceptional set is the class of meager sets (Baire first category). However, the polar sets can be defined as easily as in finite dimensions, and many of their properties remain the same. Also they take into account the complex structure of the space which the meager sets do not. Since, finally, it is known that they form the correct class for many problems in finite dimensions (see e.g. Kiselman [1981, Theorem 2.5]) there is no reason why they should not be at least as well suited as the meager sets for problems in infinite-dimensional complex analysis. We proceed to define them.

If X is a subset of a vector space E over $\mathbb{C}$ and $u: X \to [-\infty,+\infty[$ is a numerical function not taking the value $+\infty$, we say that u is *plurisubharmonic* if, for every linear map $f: \mathbb{C}^n \to E$, the composition $u \circ f$ is defined in an open set in $\mathbb{C}^n$ and plurisubharmonic in the usual sense there. In particular, X must be *finitely open*, i.e. it must cut every finite-dimensional subspace of E in an open set, and u must be upper semicontinuous in X for this finite topology. For such semicontinuous functions, the property of being plurisubharmonic just means that they satisfy the mean value inequality

$$u(x) \leq \int_0^1 u(x + e^{2\pi i s}y)ds$$

for all $x \in X$ and all $y \in E$ such that the disk

$$x + Dy = \{x + ty \in E;\ t \in \mathbb{C},\ |t| \leq 1\}$$

is contained in X. Now, a set Y is called *polar* in X if there is a plurisubharmonic function in X which is $-\infty$ on Y and which is not identically $-\infty$ in any component of X, these components being defined by the finite topology. If this function can, moreover, be chosen so that it is upper semicontinuous with respect to a certain topology, then we say that Y is *polar in* X *for this topology*. However, for the definition of the components of X we still have to use the finite topology.

Polar sets also appear in other problems than those mentioned above; see for example Lelong [1970, 1977]. For interesting results on spaces such that all compact sets are polar, see Lelong [1980] and Dineen, Meise and Vogt [1982].

Results

Let $E_h(\mathbb{C}^m,\mathbb{C}^n)$ or just E_h denote the space of all holomorphic mappings $x: \mathbb{C}^m \to \mathbb{C}^n$ such that

$$\|x\|_h = \sup_{\substack{z\in\mathbb{C}^m \\ z\neq 0}} |x(z)|e^{-h(\log|z|)} < +\infty ,$$

where (e.g.)

$$|x(z)| = (\sum_1^n |x_j(z)|^2)^{1/2} , \quad z \in \mathbb{C}^m .$$

Here h is any real-valued function on the real line. The global growth of a plurisubharmonic function v will be measured by

$$\hat{v}(t) = \sup_{|z|\leq e^t} v(z) , \quad t \in \mathbb{R},$$

so that we always have

$$v(z) \leq \hat{v}(\log|z|) , \quad z \in \mathbb{C}^n , \quad \text{and}$$

$$v(x(z)) \leq \hat{v}(\log \|x\|_h + h(\log|z|)) , \quad (x,z) \in E_h \times \mathbb{C}^m .$$

For any function $h:\mathbb{R} \to [-\infty,+\infty]$ we let H denote the largest minorant of h whose graph is a polygon with slopes which are non-negative integers, to be precise,

$$H(t) = \sup_{j\in\mathbb{N}} (jt - \tilde{h}(j)) , \quad t \in \mathbb{R},$$

where $\tilde{h}(j)$ is chosen so that the affine function $t \mapsto jt - \tilde{h}(j)$ becomes a tangent to the graph of h, i.e.

$$\tilde{h}(j) = \sup_{t\in\mathbb{R}} (jt - h(t)) , \quad j \in \mathbb{N} .$$

If two convex increasing functions are given we say that k is *small in comparison with* h if there are positive numbers $x_j < y_j < z_j$, $j \in \mathbb{N}$, such that

$$\frac{y_j - x_j}{z_j - x_j} \to 0 \quad \text{and}$$

$$H(y_j) - \sup(h(x_j), k(z_j)) \to +\infty \quad \text{as} \quad j \to +\infty .$$

We can now state the main result.

Theorem. Let $h, k : \mathbb{R}\to\mathbb{R}$ be two convex increasing functions. We suppose that $k \leq h$, that h grows faster than any linear function and that k is small in comparison with h. Let v be plurisubharmonic in $\mathbb{C}^n$ and not constant, and let $f: E_h(\mathbb{C}^m,\mathbb{C}^n) \to \mathbb{R}$ be any function. Then the set

$$P = \{x \in E_h(\mathbb{C}^m, \mathbb{C}^n);\ v(x(z)) \le \hat{v}(f(x) + k(\log|z|)) \ \text{for all}\ \ z \in \mathbb{C}^m\}$$

is polar in $E_h(\mathbb{C}^m, \mathbb{C}^n)$.

As a simple case of the theorem we state:

Corollary. Let ρ be a positive number. The set of all entire mappings $\mathbb{C}^m \to \mathbb{C}^n$ of order less than ρ is polar in the Banach space of all entire mappings satisfying an estimate

$$|x(z)| \le C e^{|z|^\rho}, \quad z \in \mathbb{C}^m.$$

To prove the corollary, let

$$h(t) = e^{\rho t},$$

$$k(t) = e^{\rho t - 2\rho\sqrt{t}}, \quad t \ge 1,$$

both functions being continued as constants for $t \le 1$. It can then easily be checked that k is small in comparison with h. Let

$$v(z) = \log|z|, \quad z \in \mathbb{C}^n.$$

Then the definition of P means that $x \in P$ if and only if

$$\log|x(z)| \le f(x) + \exp(\rho \log|z| - 2\rho(\log|z|)^{1/2}), \quad z \in \mathbb{C}^m, \quad |z| \ge e.$$

Now, if x is of order less than ρ, there is an $\varepsilon > 0$ and a constant C such that

$$|x(z)| \le C e^{|z|^{\rho-\varepsilon}},$$

whence

$$\log|x(z)| \le \log C + \exp((\rho - \varepsilon)\log|z|), \quad z \in \mathbb{C}^m.$$

It is therefore clear that $x \in P$ if we define f as

$$f(x) = \sup_{t \ge 1} (\log C + \exp((\rho - \varepsilon)t) - \exp(\rho t - 2\rho\sqrt{t})), \quad x \in E_h.$$

Now E_h is the space of all entire mappings satisfying an estimate

$$|x(z)| \le C e^{|z|^\rho}, \quad z \in \mathbb{C}^m,$$

(in fact with $C = \|x\|_h$), and all elements of E_h has order at most ρ. This shows that the corollary follows from the theorem.

Methods

The proof of the theorem is rather long. The methods of proof are not very different from those of Kiselman [1981], which can be described as convex analysis applied to complex analysis. The Legendre transformation and infimal convolution are fundamental tools, as well as the minimum principle established in Kiselman [1978]. However, the fact that the plurisubharmonic functions are defined on infinite-dimensional spaces introduces some difficulties; in particular it becomes a non-trivial statement to say that a function is not identically minus infinity. More precisely, to prove that the function constructed on E_h is not identically $-\infty$ means that we have to find an element in E_h, i.e. an entire mapping, satisfying precise growth conditions. This boils down to solving an interpolation problem:

Proposition. Let $h:\mathbb{R}\to\mathbb{R}$ be convex and increasing, and suppose that it increases faster than any linear function. Given any sequence (a_j) of complex numbers without accumulation point and a number $\varepsilon > 0$, any sufficiently fastgrowing sequence (j_p) has the following property: for every sequence (b_p) of complex numbers there exists an entire function x such that $x(a_{j_p}) = b_p$, $p \in \mathbb{N}$, and

$$\|x\|_h \leq \|x\|_H \leq (1+\varepsilon)\sum_p |b_p| \exp(-H(\log|a_{j_p}|)) .$$

Thus, the function x in the proposition belongs to E_h if the right-hand side of the estimate is finite, but the numbers b_p can be chosen so that the plurisubharmonic function on E_h is finite at x. In general, it is not possible to replace H by h in the right-hand side of this estimate, so the difficulties caused by the appearance of the minorant H to h are intrinsic to the interpolation problem.

References

Bedford, E., and Taylor, B.A. 1981. Some potential theoretic properties of plurisubharmonic functions. Manuscript.

Dineen, S., Meise, R., and Vogt, D. 1982. Compact polar sets in Fréchet spaces. Manuscript.

Kiselman, C.O. 1978. The partial Legendre transformation for plurisubharmonic functions. Invent. Math. 49, 137-148.

Kiselman, C.O. 1981. The growth of restrictions of plurisubharmonic functions. Mathematical Analysis and Applications, 435-454. Advances in Math. Suppl. Studies, vol. 7B, ed. L. Nachbin. Academic Press.

Kiselman, C.O. 1982. Stabilité du nombre de Lelong par restriction à une sous-variété. Séminaire P. Lelong - H. Skoda 1980/81 et Colloque de Wimereux 1981, 324-336. Lecture Notes in Mathematics 919. Springer-Verlag.

Lelong, P. 1970. Fonctions plurisousharmoniques et ensembles polaires sur une algèbre de fonctions holomorphes. Séminaire P. Lelong 1969, 1-20. Lecture Notes in Mathematics 116. Springer-Verlag.

Lelong, P. 1977. Sur l'application exponentielle dans l'espace des fonctions entières. Infinite Dimensional Holomorphy and Applications, 297-311. North-Holland Mathematics studies 12, ed. M.C. Matos.

Lelong, P. 1980. A class of Fréchet complex spaces in which the bounded sets are $\mathbb{C}$-polar sets. Manuscript, 13 p.

Institute of Mathematics
Uppsala University
Thunbergsvägen 3, S-752 38 Uppsala, Sweden

THE ROOTS OF UNITY AND THE m-MEROMORPHIC EXTENSION OF FUNCTIONS

Ralitza Krumova Kovačeva (Sofia)

Contents

Summary

For a function f analytic on the closed unit disk and a positive integer m, denote by $R_{n,m} = R_{n,m}(f)$ the rational function $p_{n,m}/q_{n,m}$, where $q_{n,m} \neq 0$, $\deg p_{n,m} \leq n$, $\deg q_{n,m} \leq m$, and the function

(1) $\varphi_n(z) = (f q_{n,m} - p_{n,m})(z)/(z^{n+m+1} - 1), \quad |z| \leq 1,$

is analytic. Let $R_{n,m} = P_{n,m}/Q_{n,m}$, where $P_{n,m}$ and $Q_{n,m}$ have no common divisor and the polynomial $Q_{n,m}$ is monic. Denote by $\| \ \|$ the norm of the $(m+1)$-dimensional space of polynomial coefficients. As a further generalization of a generalized theorem of Montessus de Ballore (1902), due to E. B. Saff (1972), the author proves that if there exists a polynomial

(2) $Q(z) = (z - a_1)\dots(z - a_m), \quad 1 < |a_1| \leq \dots \leq |a_m|,$

such that

(3) $\limsup_{n\to\infty} \|Q_{n,m} - Q\|^{1/n} \leq q < 1,$

then f is m-meromorphic in a disk $D_R = \{z, |z| < R\}$, where R satisfies the condition

(4) $R q \geq |a_m|$

and all the zeros of Q (including their multiplicities) are poles of f in D_R. As a consequence the author obtains a criterion for the m-meromorphic extensibility of f onto D_R with $R>1$.

Introduction

Let f be a function analytic on the closed unit disk $\bar{D}=\{z, |z| \leq 1\}$ (we write: $f\in\mathcal{H}(\bar{D})$) and m – a fixed positive integer (we write: $m\in\mathbb{N}$). For each $n\in\mathbb{N}$ we denote by $R_{n,m}=R_{n,m}(f)$ the rational function of the form $R_{n,m}=p_{n,m}/q_{n,m}$, where $q_{n,m}\not\equiv 0$, $\deg p_{n,m}\leq n$, $\deg q_{n,m}\leq m$, such that the function (1) is analytic. It is well known that $R_{n,m}$ always exists and is unique whenever $p_{n,m}$ and $q_{n,m}$ are as specified (cf. [6]).

For each $n\in\mathbb{N}$ we set $R_{n,m}=P_{n,m}/Q_{n,m}$, where $P_{n,m}$ and $Q_{n,m}$ have no common divisor and the polynomial $Q_{n,m}$ is monic. If $\deg Q_{n,m}=m$, then $R_{n,m}$ interpolates f in all the points $\exp[2\pi ik/(n+m+1)]$, $k=0,1,\ldots,n+m$, and we have the representation

$$(f-R_{n,m})(z)=(z^{n+m+1}-1)\phi_n(z), \text{ where } \phi_n\in\mathcal{H}(D).$$

Next, let us denote by $\mathcal{M}_m(D_R)$, $R>1$, $D_R=\{z, |z|<R\}$, the class of all functions m-meromorphic in D_R, i. e. the class of all functions which are meromorphic in D_R and have precisely m poles there (including their multiplicities).

The main result of this paper is

THEOREM 1. *Let $f\in\mathcal{H}(\bar{D})$ and $m\in\mathbb{N}$ be fixed. Denote by $\|\ \|$ the norm of the $(m+1)$-dimensional space of polynomial coefficients. If there exists a polynomial (2) of degree m such that, for some q, the estimate (3) holds, then f is m-meromorphic in a disk D_R, where R satisfies the condition (4) and $a_1,\ldots,a_m$ are the poles of f in D_R.*

The following result (Theorem A), due to Saff [6], generalizes the classical theorem of Montessus de Ballore [5] with respect to interpolating rational functions.

THEOREM A. *Let $f\in\mathcal{M}(D_R)$, $R>1$, and $a_1,\ldots,a_m$ be the poles of f in D_R; $1<|a_1|\leq\ldots<|a_m|$. Then for all sufficiently large $n\in\mathbb{N}$ the function $R_{n,m}=R_{n,m}(f)$ has exactly m $(m<\infty)$ poles and the sequence $(R_{n,m})$ converges to f as $n\to\infty$, uniformly and geometrically on $D_R'=D_R-\{a_j, j=1,\ldots,m\}$. If K is a compact subset of D_R', then*

$$\limsup_{n\to\infty}\|R_{n,m}-f\|_K^{1/n}\leq(1/R)\|id\|_K,$$

where $\|\ \|_K$ denotes the uniform norm on K and $id(z)\equiv z$.

Theorem A yields (cf. [8]) the estimate (3) with Q given by (2) and any q satisfying (3). The above result and Theorem 1 give a criterion for the m-meromorphic extensibility of the function f onto D_R with $R>1$, namely:

THEOREM 2. Let $f\in\mathcal{H}(\bar{D})$ and $m\in\mathbb{N}$ be fixed. Then $f\in\mathcal{M}_m(D_R)$, $R>1$, if and only if there exists a polynomial Q of degree exactly m, different from zero in $\bar{D}$, and such that

$$\limsup_{n\to\infty}\|Q_{n,m}-Q\|^{1/n}\le(1/R)\max\{|a_j|,\ j=1,\dots,m\},$$

where a_j, $j=1,\dots,m$, are all the zeros of Q (including their multiplicities) and, at the same time, the poles of f in D_R.

Preliminary statements

Suppose the function $f(z)=f_0+f_1z+f_2z^2+\dots$ is analytic on $\bar{D}_\rho$, $\rho>1$. Let, for $n=0,1,2,\dots$,

$$I_n(f)=(1/2\pi i)\int_{|z|=\rho}(z^{n+1}-1)^{-1}f(z)\,dz=f_n+f_{2n+1}+f_{3n+2}+\dots$$

In order to prove Theorem 1 we need a few lemmas.

LEMMA 1. For each $n=0,1,2,\dots$ there exists a unique polynomial P_n of degree n, such that

$$I_k(P_n)=\delta_{k,n},\quad k=0,1,2,\dots$$

The lemma is obvious [1]. The polynomials P_n can be generated by the recurrence formula

$$P_0(z)=1,\quad P_n(z)=z^n-\Sigma_{k=0}^{n-1}P_k(z)I_k(z^n),\quad\text{where } \Sigma_{k=0}^{n-1}=\sum_{k=0}^{n-1}.$$

To introduce the polynomials P_n, $n=0,1,2,\dots$, in the solution of inverse problems of the m-meromorphic extension, we follow an idea of Saff and Karlsson [7].

LEMMA 2. The polynomials P_n, $n=0,1,2,\dots$, satisfy the condition

$$\lim_{n\to\infty}\|P_n\|_{\bar{D}}^{1/n}=1. \tag{5}$$

Proof. For each $n\in\mathbb{N}$ we set

$$P_n(z)=a_{n,n}z^n+\dots+a_{n,k}z^k+\dots+a_{n,0},\quad a_{n,n}=1.$$

Hence

$$I_k(P_n) = a_{n,k} + \ldots + a_{n,lk+l-1}, \quad lk + l - 1 \le n, \quad k = 0, 1, 2, \ldots$$

We are going to show that each coefficient $a_{n,k}$, $k = 0, \ldots, n-1$, is equal to $1, -1$, or 0, and this suffices to conclude (5); cf. [2].

We notice first that, for all $n \ge 5$,

$$\text{(6)} \quad \text{if} \quad \begin{matrix} n = 2\rho - 1, \\ n = 2\rho, \end{matrix} \quad \text{then} \quad \begin{matrix} a_{n,n-1} = \ldots = a_{n,\rho} = 0, \; a_{n,\rho-1} = -1, \\ a_{n,n-1} = \ldots = a_{n,\rho-1} = a_{n,\rho-2} = 0, \end{matrix} \quad \text{respectively.}$$

We obtain $P_1(z) = z - 1$, $P_2(z) = z^2 - 1$, $P_3(z) = z^3 - z$, $P_4(z) = z^4 - 1$.

The coefficient $a_{n,0}$ can be calculated as follows. Let μ be the Möbius function:

$$\mu(l) = \begin{cases} 1 \text{ if } l = 1; \\ (-1)^k \text{ if } l = p_1, \ldots, p_k, \text{ where } p_1, \ldots, p_k \ge 2 \text{ are distinct primes;} \\ 0 \text{ if } p^2 | l \text{ for some } p \ge 2, \text{ where } p \text{ is a prime} \end{cases}$$

($x|l$ means that x is a divisor of l). Let us recall that (cf. [3]), for each $n \in \mathbb{N}$,

$$\Sigma_{l|n} \mu(l) = 1 \text{ if } n = 1, \text{ and } = 0 \text{ otherwise.}$$

Suppose $n = 2\rho - 1$. We have

$$0 = \Sigma_{l=1}^{\rho-1} \mu(l) I_{l-1}(P_n) = \Sigma_{l=1}^{\rho-1} \Sigma_{l|m+1} a_{n,m},$$

where we set $a_{n,m} = 0$ whenever $m > n$. This gives

$$\Sigma_{m=0}^{2\rho-1} a_{n,m} \Sigma_{l=1,\, l|m+1}^{\min(m+1,\rho-1)} \mu(l) = 0.$$

By the above mentioned property of the Möbius function and by (6), we get

$$\text{(7)} \quad a_{n,0} = \Sigma_{l=1,\, l|\rho}^{\rho-1} \mu(l) - \Sigma_{l=1,\, l|2\rho}^{\rho-1} \mu(l) = \mu(2\rho) = \mu(n+1).$$

Suppose next $n = 2\rho$. Applying the same method, we obtain

$$\text{(8)} \quad a_{n,0} = \mu(2\rho + 1) = \mu(n+1).$$

Let us notice now that, if $n \in \mathbb{N}$ is an integer such that $n+1$ is a prime, then

$$\text{(9)} \quad P_n(z) = z^n - 1.$$

Indeed, for each pair (k,n) of nonnegative integers, we have

$$I_k(z^n) = 1 \ \text{if} \ k+1 \mid n+1, \ \text{and} \ = 0 \ \text{otherwise}.$$

Using the above relations, we can bring the recurrence formula for P_n, $n \in \mathbb{N}$, to the the form

$$(10) \qquad P_n(z) = z^n - \Sigma^n_{k=1,\, k|n+1} P_{k-1}(z).$$

We are going to distinguish three cases. First, let $n+1 = p^l$, $l \in \mathbb{N}$, $l \geq 2$ (hereafter $p, p_k \geq 2$, where $k = 1, 2, \ldots,$ are primes). In this case we have

$$(11) \qquad P_n(z) = P_{p^l-1}(z) = z^{p^l-1} - z^{p^{l-1}-1}.$$

Next, let $n+1 = p_1 \ldots p_i$, where $p_k \neq p_j$ for $k \neq j$; $j, k = 1, \ldots, i$. Formula (10) gives

$$P_{p_1\ldots p_i-1}(z) = z^{p_1\ldots p_i-1} - \Sigma_{1\leq k_1<\ldots<k_{i-1}\leq i} P_{p_{k_1}\ldots p_{k_{i-1}}-1}(z)$$

$$- \ldots - \Sigma_{1\leq k_1\leq i} P_{p_{k_1}-1}(z) - 1.$$

If $i = 2$, then, by (9), we obtain

$$P_{p_1p_2-1}(z) = z^{p_1p_2-1} - z^{p_1-1} - z^{p_2-1} + 1.$$

If $i > 2$, we apply (10) to the sum $\Sigma(1)$, where

$$\Sigma(j) = \Sigma_{1\leq k_1<\ldots<k_{i-j}\leq i} P_{p_{k_1}\ldots p_{k_{i-j}}-1}(z), \quad j = 1, \ldots, i.$$

Hence

$$P_{p_1\ldots p_i-1}(z) = z^{p_1\ldots p_i-1} - \Sigma_{1\leq k_1<\ldots<k_{i-1}\leq i} z^{p_{k_1}\ldots p_{k_{i-1}}-1}$$

$$+ \Sigma_{1\leq k_1<\ldots\leq k_{i-2}\leq i} P_{p_{k_1}\ldots p_{k_{i-2}}-1}(z) + \ldots$$

$$+ (i-2)\, \Sigma_{1\leq k_1\leq i} P_{p_{k_1}-1}(z) + i - 1.$$

Applying successively (10) to each of the sums $\Sigma(j)$, $j = 2, \ldots, i-1$, and making the corresponding calculations, we arrive, by (7) and (8), at the formula

(12) $P_n(z) = P_{p_1\dots p_i-1}(z) = z^{p_1\dots p_{i-1}-1}$

$$+\Sigma_{j=1}^{i-1}(-1)^j\,\Sigma_{1\le k_1<\dots<k_{i-j}\le i}\, z^{p_{k_1}\dots p_{k_{i-j}}-1} + (-1)^i.$$

Therefore in the special cases: $n+1=p$; $n+1=p^l$, $l\in\mathbb{N}$, $l\ge 2$; $n+1=p_1\dots p_i$, where $p_1,\dots,p_i\ge 2$ are distinct primes, our statement on the coefficients $a_{n,0},\dots,a_{n,n-1}$ holds true.

Finally, consider the general case $n+1=p_1^{l_1}\dots p_i^{l_i}$, $l_k\in\mathbb{N}$, $l_k\ge 2$, $k=1,\dots,i$, where $p_k\ne p_j$ for $k\ne j$; $j,k=1,\dots,i$. We set $l=l_1+\dots+l_i$. Using the same argument as in the previous part, we obtain

$$P_n(z) = z^{p_1^{l_1}\dots p_i^{l_i}-1} + \Sigma_{j=1}^{l+1-i}(-1)^j\,\Sigma_{\substack{k_1+\dots+k_i=l-j\\ 0\le k_s\le l_s,\ s=1,\dots,i}}\, z^{p_{k_1}\dots p_{k_i}-1}$$

$$+(-1)^i z^{p_1^{l_1-1}\dots p_i^{l_i-1}}.$$

Formulae (10) - (12) and the last one can be written in a compact way:

(13) $P_n(z) = \Sigma_{k|n+1}\,\mu\left(\frac{n+1}{k}\right) z^{k-1}$.

Now we are in a position to prove (5). Formula (13) gives

(14) $\|P_n\|_{\overline{D}} \le 1+[\tfrac{1}{2}n]$.

For each $n\in\mathbb{N}$ we set $m_n=\inf\|p_n\|_{\overline{D}}$, where the infimum is taken over all polynomials of the form $p_n(z)=z^n+\dots$ Obviously,

$$m_n \le \|P_n\|_{\overline{D}}.$$

The above estimate and (14) give (4) (we recall that $m_n\to 1$ as $n\to\infty$; cf. [2]).

Let $R>1$ be a fixed arbitrary number. We set $\Gamma_R=\partial D_R$. For each sufficiently large $n\in\mathbb{N}$ we obtain from (13) the estimate

(15) $\min\{|P_n(z)|,\ z\in\Gamma_R\}\ge R^n-[\tfrac{1}{2}n]R^{[\frac{1}{2}n]}>0$.

Applying the lemma of Walsh - Bernstein (cf. [9]) to the polynomials P_n, $n\in\mathbb{N}$, we obtain, by (14), that for every $z\in\Gamma_R$:

(16) $|P_n(z)|\le(1+[\tfrac{1}{2}n])R^n$.

The estimates (15) and (16) yield

$$R^n - [\tfrac{1}{2}n]R^{[\frac{1}{2}n]} \le \min\{|P_n(z)|,\ z\in\Gamma_R\} \le |P_n(z)| \le \|P_n\|_{\Gamma_R} \le (1+[\tfrac{1}{2}n])R^n.$$

This means that

$$(17) \qquad \lim_{n\to\infty} |P_n(z)|^{1/n} = R \quad \text{for} \quad z\in\Gamma_R,$$

and that the convergence is uniform.

Remark. A more detailed description of the above proof, including the calculations left here for a reader, will appear in [4].

LEMMA 3. *Suppose the function* $g(z) = g_0 + g_1 z + g_2 z^2 + \ldots$ *is analytic on* $\overline{D}$. *If* $I_n(g) = 0$ *for each* $n = 0, 1, 2, \ldots,$ *then* $g \equiv 0$.

Proof. We begin with proving that g is an entire function. Assume that this is not the case, i.e. that

$$\limsup_{n\to\infty} |g_n|^{1/n} = 1/\rho > 0.$$

Obviously $\rho > 1$. Let ϑ be an arbitrary number such that $\exp(3\vartheta) < \rho$. For all sufficiently large $n\in\mathbb{N}$, $n > N_1$, we have $|g_n| \le [\exp(n\vartheta)]\,\rho^{-n}$. Moreover, there exists a subsequence Λ of $\mathbb{N}$ such that $|g_n|^{1/n} \to \rho^{-1}$ as n tends to ∞ within Λ. Thus for all $n_k \subset \Lambda$, $n_k > N_2 \ge N_1$, we get

$$(18) \qquad [\exp(-n_k\vartheta)]\rho^{-n_k} \le |g_{n_k}| \le [\exp(n_k\vartheta)]\rho^{-n_k}.$$

Let $n_k \subset \Lambda$, $n_k \ge N_2$. By the hypotheses of Lemma 3 we have

$$-g_{n_k} = \Sigma_{l\ge 2}\, g_{ln_k+l-1}.$$

Hence

$$|g_{n_k}| \le C_1[\exp(2n_k\vartheta)]\rho^{-2n_k}$$

(hereafter C_j, $j\in\mathbb{N}$, are positive constants independent of n). From the last estimate we obtain (cf. (18)): $1 \le C_1[\exp(3\vartheta n_k)]\,\rho^{-n_k}$, what is impossible in the case when $n_k\in\Lambda$ is sufficiently large because $\exp(3\vartheta) < \rho$. Therefore the function g is entire, i.e. $|g_n|^{1/n} \to 0$ as $n\to\infty$.

Now we shall prove that g is a polynomial. Suppose the contrary and let ε, $0 < \varepsilon < \frac{1}{2}$, be an arbitrary number. For all $n\in\mathbb{N}$, $n > N_3$, we have $|g_n| < \varepsilon^n$. Because of the assumption that g is not a polynomial, we can assert that there exists a number n_0, $n_0\in\mathbb{N}$, $n_0 > N_3$, such that $|g_{n_0}| > 0$ and $|g_n|^{1/n} < |g_{n_0}|^{1/n_0}$ for all $n > n_0$. The condition $I_{n_0}(g) = 0$ gives

$$0<|g_{n_o}|\le\Sigma_{l\ge 2}|g_{n_o}|^{1+(l-1)/n_o}=|g_{n_o}|^{2+1/n_o}(1-|g_{n_o}|^{1+1/n_o})^{-1}.$$

This means that

$$1<2|g_{n_o}|^{1+1/n_o}\le 2\varepsilon^{1+1/n_o},$$

and the obtained inequality contradicts the choice of ε. Consequently g is a polynomial and we finally conclude that $g\equiv 0$.

LEMMA 4. Suppose the function $f(z)=f_o+f_1z+f_2z^2+\dots$ is analytic in D_ρ, $\rho>1$. Then

(19) $\quad f(z)=\Sigma_{n\ge 0}I_n(f)P_n(z)$

and the convergence is uniform on each compact subset of D_ρ. If $\rho_o=\sup\{\rho,\ f\in\mathcal{H}(D_\rho)\}$, then

(20) $\quad \limsup_{n\to\infty}|I_n(f)|^{1/n}=1/\rho_o.$

Proof. The lemma follows from Lemmas 2 and 3. Indeed, obviously we have

(21) $\quad \limsup_{n\to\infty}|I_n(f)|^{1/n}\le 1/\rho_o.$

Let R, $1<R<\rho_o$, be an arbitrary number. Lemma 2 (cf. (16) and (17)) and the inequality (21) imply that the series $F(z)=\Sigma_{n\ge 0}I_n(f)P_n(z)$ converges uniformly on Γ_R, i.e. the function F is analytic (on $\overline{D}_R$). On the other hand, the function $\Phi=f-F$ satisfies the hypotheses of Lemma 3. Consequently $\Phi\equiv 0$ and (19) holds true.

Suppose now that $\limsup_{n\to\infty}|I_n(f)|^{1/n}<1/\rho_o$. Let ρ_1 be a number such that $\rho_1>\rho_o$ and

$$\limsup_{n\to\infty}|I_n(f)|^{1/n}<1/\rho_1<1/\rho_o.$$

Then the function F has to be analytic on $\overline{D}_{\rho_1}$ and the previous part gives $F=f$. This contradicts the definition of ρ_o, thus implying (20).

In this way we have proved all the lemmas needed. It is worthwhile to mention that, under the assumption that Lemma 4 is valid, Saff and Karlsson [7] have proved the following

THEOREM B. If the hypotheses of Theorem 1 are satisfied, then either (a) the function f is meromorphic in $\overline{\mathbb{C}}$ with at most $m-1$ poles or (b) f is m-meromorphic in D_R, where R satisfies the condition (4) and $a_1,\dots,a_m$ are the (finite) poles of f.

Thus our Theorem 1 is a more precise form of Theorem B.

Proof of the main result

The proof of Theorem 1 consists of two parts and both of them have an independent meaning.

We show first that in the conditions of the theorem the function $F = fQ$ can be analytically extended to the disk D_R, where $R = |a_m|/q$ (we recall that $|a_m| = \max\{|a_j|$, where $Q(a_j) = 0$, $j = 1,\dots,m\}$.

Suppose $R_0 = \sup\{R,\ F \in \mathcal{H}(D_R)\} \leq |a_m|$. Lemma 4 gives

$$(22)\qquad \limsup_{n\to\infty} |I_n(F)|^{1/n} = 1/R_0 .$$

Let $a_1,\dots,a_{m_1}$ be the zeros of Q which lie in D_{R_0}, $0 \leq m_1 < m$. The function $F_1(z) = f(z)(z-a_1)\dots(z-a_{m_1})$ is analytic in D_{R_0} and, by Lemma 4,

$$\limsup_{n\to\infty} |I_n(F_1)|^{1/n} = 1/R_0 .$$

By the definition of $R_{n,m}$, for all sufficienly large $n \in \mathbb{N}$ it follows the equality

$$(23)\qquad I_{n+m}(F_1(Q_{n,m} - Q)) = I_{n+m}(F_1 Q).$$

This gives

$$\limsup_{n\to\infty} |I_n(F_1 Q)|^{1/n} \leq q/R_0 ,$$

so $F_1 Q$ is analytic in the disk $D_{R_0/q}$. Yet

$$(F_1 Q)(z) = f(z)\prod_{j=1}^{m_1}(z-a_j)\prod_{j=1}^{m}(z-a_j) = F(z)\prod_{j=1}^{m_1}(z-a_j),$$

and hence $F \in \mathcal{H}(D_{R_0/q})$. This result contradicts our assumption that F is analytic only in the disk D_{R_0}; cf. (22). Thus we conclude that $F \in \mathcal{H}(\overline{D}_{|a_m|})$.

Let now ε be an arbitrary number such that $0 < \varepsilon < |a_m| - 1$. We can prove as before (cf. (23)) that $F \in \mathcal{H}(D_{R_\varepsilon})$, where $R_\varepsilon = (|a_m| - \varepsilon)/q$. Because of the arbitrariness of ε we conclude that $F \in \mathcal{H}(D_R)$.

The second part of the proof aims at showing that the function f has a pole at each point $a_1,\dots,a_m$. The relation $F \in \mathcal{H}(D_R)$, where $R = |a_m|/q$, means that the function f is meromorphic in the disk D_R

with possible poles at the points $a_1,\dots,a_m$.

Suppose that $f\in\mathcal{M}_{m_1}(D_R)$, where $m_1<m$. Denote by ω the polynomial whose zeros are the poles of f (in D_R), and set $\Phi=f\omega$. Obvoiusly, we have the inequality

(24) $$\limsup_{n\to\infty}|I_n(\Phi)|^{1/n}\le 1/R.$$

In the same way as in the previous part (cf. (23)), we obtain that Φ can be analytically extended to the disk $D_{R/q}$, i.e. that we can replace in the estimate (24) the constant $1/R$ with q/R. Let $N\in\mathbb{N}$, $N\ge 2$. By the same considerations, the constant $1/R$ in (24) can be further replaced by q^N/R. Hence $|I_n(\Phi)|^{1/n}\to 0$ as $n\to\infty$, i.e. the function Φ is entire. In such a case the function f is meromorphic in $\overline{\mathbb{C}}$ and has less than m poles which coincide with zeros of the polynomial Q. We shall prove that this is impossible.

Indeed, let $\Phi(z)=\Phi_0+\Phi_1 z+\Phi_2 z^2+\dots$ It follows from the hypotheses of Theorem 1 that, for all sufficiently large $n\in\mathbb{N}$, $n>N_4$, we have $\deg Q_{n,m}=m$. Set

$$Q_{n,m}(z)=c_{n,0}\,z^m+\dots+c_{n,m},\quad Q(z)=c_0z^m+\dots+c_m,\quad c_{n,0}=c_0=1.$$

Set further $C_2=1+\max\{|c_k|,\ k=0,\dots,m\}$. For all n sufficiently large, $n>N_5\ge N_4$, we have the estimates

(25) $$|c_{n,k}|\le C_2,\quad k=0,\dots,m.$$

Suppose $n>N_5$. The definition of $R_{n,m}$ gives $\Sigma_{\nu=0}^{\infty}\Phi_\nu I_{n+m}(z^\nu Q_{n,m}(z))=0$. A routine calculation shows that

(26) $$-\Phi_n=\Sigma_{k\ge 1}\,\Phi_{n+k}\,I_{n+m}(z^{n+k}\,Q_{n,m}(z)).$$

We obtain, further, from (25):

(27) $$|I_{n+m}(z^k\,Q_{n,m}(z))|\le C_2(m+1)=C_3.$$

Now, let ε be an arbitrary number such that $0<\varepsilon<(1+C_3)^{-1}$. There is an integer N_6 with the property:

(28) $$|\Phi_n|<\varepsilon^n\ \text{ for } n>N_6\ge N_5.$$

Let $n>N_6$. By (26) and (27), we obtain

(29) $$|\Phi_n|\le C_3(|\Phi_{n+1}|+\Sigma_{k\ge 2}|\Phi_{n+k}|).$$

An application of the same estimate to Φ_{n+1} gives

$$(30)\qquad |\Phi_n| \le C_3(1+C_3)(|\Phi_{n+2}| + \Sigma_{k\ge 3}|\Phi_{n+k}|).$$

Let $N \in \mathbb{N}$, $N \ge 3$. Estimating $\Phi_{n+2}, \ldots, \Phi_{n+N}$ in the same way as in (29) and inserting the result into (30), we obtain

$$|\Phi_n| \le C_3(1+C_3)^N \Sigma_{k\ge 1}|\Phi_{n+N+k}|.$$

Further, this implies (cf. (28)):

$$|\Phi_n| \le C_4(1+C_3)^N \varepsilon^N.$$

The last estimate is valid for each $N \ge 3$ and because of the choice of ε (we recall that $\varepsilon < (1+C_3)^{-1}$) we conclude that $\Phi_n = 0$. Consequently $\Phi_n = 0$ for $n > N_6$. Thus the function Φ is a polynomial. Yet, if this were the case, the function f would be rational with a denominator ω of degree less than m. Then, for all $n > \deg\omega$, the polynomials $P_{n,m}$ and $Q_{n,m}$ should be equal to Φ and ω, respectively. Yet, this would contradict the conditions of the theorem (we have to remember that Q is a polynomial of degree exactly m). Thus the proof is completed.

The author is very much obliged to Prof. A. A. Gončar and to S. Suetin for useful discussions and critical remarks.

References

[1] GELFOND, A. O.: Die Differenzenrechnung, VEB Deutscher Verlag der Wissenschaften, Berlin 1958.

[2] ГОЛУЗИН, Г.М.: Геометрическая теория функций комплексного переменного, изд. 2-е, Изд. "Наука", Москва 1966.

[3] HARDY, H, and E. M. WRIGHT: An introduction to the theory of numbers, 5th ed., Clarendon Press, Oxford 1979.

[4] KOVACHEVA, R. K. [КОВАЧЕВА, Р.К., KOVAČEVA, R. K.]: Die Einheitswurzeln und die ihnen entsprechenden Basispolynome, Serdica, to appear.

[5] MONTESSUS de BALLORE, R. de: Sur les fractions continues algebriques, Bull. Soc. Math. France 30 (1902), 28-36.

[6] SAFF, E. B.: An extension of Montessus de Ballore's theorem on the convergence of interpolating rational functions, J. Appr. Theory 6 (1972), 63-67.

[7] —— and J. KARLSSON: Singularities of functions determined by the poles of Padé approximants, in: Padé approximation and its applications, Amsterdam 1980, Proceedings (Lecture Notes in Math. 888), Springer-Verlag, Berlin - Heidelberg - New York 1981, pp. 238-254.

[8] ——, R. S. VARGA and A. SHARMA: An extension to rational functions of a theorem of J. L. Walsh on differences of interpolating polynomials, International Conference on Constructive Function Theory, Varna 1981, Proceedings, to appear.

[9] WALSH, J. L.: Interpolation and approximation by rational functions in the complex domain, 5th ed. (Colloq. Publ. 20), American Math. Soc., Providence, RI 1969; Russian translation: Дж.Л. УОЛШ: Интерполяция и аппроксимация рациональными функциями в комплексной области, Изд. Иностранной Литературы, Москва 1966.

Institute of Mathematics of the
Bulgarian Academy of Sciences
BG-1090 Sofia, P.O. Box 373, Bulgaria

ON BIHOLOMORPHIC INVARIANTS RELATED TO HOMOLOGY GROUPS

Tadeusz Krasiński (Łódź)

C o n t e n t s

Summary

In this paper examples and applications of the semi-norms on homology groups of complex manifolds, introduced in [4] , are given.

Introduction

Many properties of mappings between manifolds are related to homology groups. In the first part [4] of this paper we defined semi-norms on homology groups in the sense of de Rham [7] , which are a modification of those introduced by Chern, Levine and Nirenberg in [2] . Each of them was associated with a form of the class C^∞ . They were applied to investigating holomorphic mappings between complex manifolds. Necessary conditions for the existence of a holomorphic mapping which is homotopic to the given mapping of the class C^∞ were given in terms of some topologies on homology groups, induced by the semi-norms.

In this part, we give some examples and applications of the semi-norms. Namely, in the case of Riemann surfaces the semi-norms are equivalent to the extremal length of the homology classes. The examples give a partial answer to the question how various semi-norms and topologies on homology groups can be generated by C^∞ - forms.

1. The notion of semi-norms

First, we shall recall definitions and theorems given in [4] .

Let M be a complex manifold, χ a C^∞ - form on M , F a family of real C^∞ - functions (including constant functions) bounded by 1 . Then the function

$$N_{\chi,F}(\gamma) := \sup_{u \in F} \inf_{T \in \gamma} |T(d^c u \wedge \chi)| ,$$

where $\gamma \in H^c(M,\mathbb{C})$ (compact homology group of M in the sense of de Rham [7]), $d^c := i(\bar{\partial} - \partial)$ and T is a current belonging to γ, is a semi-norm on $H^c(M,\mathbb{C})$ (see [4]) (in fact, in [4] F is the full family F_M of C^∞ - functions bounded by 1 , but the proof is the same in this case). When $F = F_M$, we put $N_\chi := N_{\chi,F_M}$.

R e m a r k. *If* χ *is a real form, then* $N_{\chi,F}$ *can be defined on* $H^c(M,\mathbb{R})$.

The following theorems (all appear in [4]) play a fundamental role in applications and calculations of the semi-norms.

THEOREM 1. *If* $d\chi = 0$ *and* F_χ *is the family of functions* u *of* F , *for which* $dd^c u \wedge \chi = 0$, *then, for each* $\gamma \in H^c(M,\mathbb{C})$,

$$N_{\chi,F}(\gamma) = \sup_{u \in F_\chi} \inf_{T \in \gamma} |T(d^c u \wedge \chi)| .$$

COROLLARY. *If* T_0 *is an arbitrary cycle belonging to* γ , *then*

$$N_{\chi,F}(\gamma) = \sup_{u \in F_\chi} |T_0(d^c u \wedge \chi)| .$$

THEOREM 2. *If* M,N *are complex manifolds,* $f: M \to N$ *is a holomorphic mapping and* χ *is a closed* C^∞ - *form on* N , *then, for each* $\gamma \in H^c$ M,$\mathbb{C}$) , *the inequality*

$$N_\chi(f_*\gamma) \le N_{f^*\chi}(\gamma)$$

holds, where f_* *is the linear map on homologies, induced by* f .

Each set $\mathfrak{X}$ of pairs (χ,F) , where χ is a C^∞ - form on M and F is a family of real C^∞ - functions on M bounded by 1 , generates a family of semi-norms on $H^c(M,\mathbb{C})$. Next, this family defines a locally convex topology on $H^c(M,\mathbb{C})$ (not necessarily Hausdorff).

THEOREM 3. If M,N are complex manifolds, $f: M \to N$ is a holomorphic mapping and Φ is a family of closed C^∞ - forms on N, then the function $f_*: H^c(M,\mathbb{C}) \to H^c(N,\mathbb{C})$ is continuous with respect to the topologies generated on $H^c(M,\mathbb{C})$ and $H^c(N,\mathbb{C})$ by the families of the semi-norms $\{N_{f^*\chi}, \chi \in \Phi\}$ and $\{N_\chi, \chi \in \Phi\}$, respectively.

COROLLARY. Let Φ_M, Φ_N be the sets of all closed C^∞-forms on N and M, respectively, and $f: M \to N$ - a diffeomorphism. If the linear mapping $f_*: H^c(M,\mathbb{C}) \to H^c(N,\mathbb{C})$ is not continuous with respect to the topologies generated by the families $\{N_\chi, \chi \in \Phi_M\}$ and $\{N_\chi, \chi \in \Phi_N\}$, then the homotopy class of f does not contain any holomorphic mappings.

2. Examples

Now, we give some examples which show that various topologies on the homology groups can be introduced by the semi-norms.

EXAMPLE 1. Let M be a ring in $\mathbb{C}^n$, i.e. $M: = \{z \in \mathbb{C}^n: a < \|z\| < b,\ a,b \in R,\ 0 < a < 1 < b\}$. Since M is homotopic to the $(2n-1)$ - dimensional sphere $S: = \{z \in \mathbb{C}^n : \|z\| = 1\}$, then $H^c_{2n-1}(M,\mathbb{R}) \cong \mathbb{R}$. Let us define

$$\chi: = (dd^c \log \|z\|)^{n-1} = \underbrace{dd^c \log \|z\| \wedge \ldots \wedge dd^c \log \|z\|}_{(n-1)\ -\ \text{times}} .$$

The form χ is a real closed form of type $(n-1,n-1)$. We shall prove that the semi-norm N_χ is a norm on $H^c_{2n-1}(M,\mathbb{R})$.

Let γ_0 be the homology class belonging to $H^c_{2n-1}(M,\mathbb{R})$, represented by the cycle T_0:

$$T_0(\varphi): = \int_S \varphi$$

where φ is a real C^∞ - form on M and the orientation of S is induced by the canonical orientation of the ball $B: = \{z \in C^n : \|z\| < 1\}$. It is easy to see that the function

$$u(z): = \frac{\log\|z\| - \log a}{\log b - \log a}$$

belongs to F_M . Moreover, $u \in (F_M)_\chi$ since $dd^c u \wedge \chi = 0$.
Hence, from the corollary of theorem 1 we have

$$N_\chi(\gamma_0) = \sup_{\tilde{u} \in (F_M)_\chi} \left| \int_S d^c \tilde{u} \wedge \chi \right| \geq \left| \int_S d^c u \wedge \chi \right| .$$

But

$$\int_S d^c u \wedge \chi = \frac{1}{\log b/a} \int_S d^c \log \|z\| \wedge (dd^c \log \|z\|)^{n-1} =$$

$$= \frac{1}{2^n \log b/a} \int_S d^c \log \|z\|^2 \wedge (dd^c \log \|z\|^2)^{n-1} =$$

$$= \frac{1}{2^n \log b/a} \int_S \frac{1}{\|z\|^2} d^c \|z\|^2 \wedge \left(d \frac{1}{\|z\|^2} \wedge d^c \|z\|^2 + \frac{1}{\|z\|^2} dd^c \|z\|^2\right)^{n-1} =$$

$$= \frac{1}{2^n \log b/a} \int_S d^c \|z\|^2 \wedge (dd^c \|z\|^2)^{n-1} = \frac{1}{2^n \log b/a} \int_B (dd^c \log\|z\|^2)^n =$$

$$= \frac{(-1)^{\frac{n(n-1)}{2}}}{\log b/a} \int_B \left(\frac{i}{2}\right)^n dz_1 \wedge d\bar{z}_1 \wedge \ldots \wedge dz_n \wedge d\bar{z}_n = \frac{(-1)^{\frac{n(n-1)}{2}}}{\log b/a} \operatorname{vol} B \neq 0 .$$

Hence $N_\chi(\gamma_0) > 0$ and, in consequence, N_χ is a norm on $H^C_{2n-1}(M,\mathbb{R})$. The topology induced by the norm N_χ is the canonical topology of $\mathbb{R}$.

<u>EXAMPLE 2</u>. Let M be the subset of $\mathbb{C}^n$ defined as follows:

$M: = K-\bar{B}$ where $K: = \{z \in \mathbb{C}^n : \|z\| < 1\}$ and $B: = \bigcup_{k=1}^{\infty} K_k$,

$K_k: = \{z \in \mathbb{C}^n: \|z-w_k\| < \frac{1}{4(k+1)(k+2)}$, $w_k: = (0,\ldots,0,\frac{k}{k+1})\}$,

$k = 1,2,\ldots$.

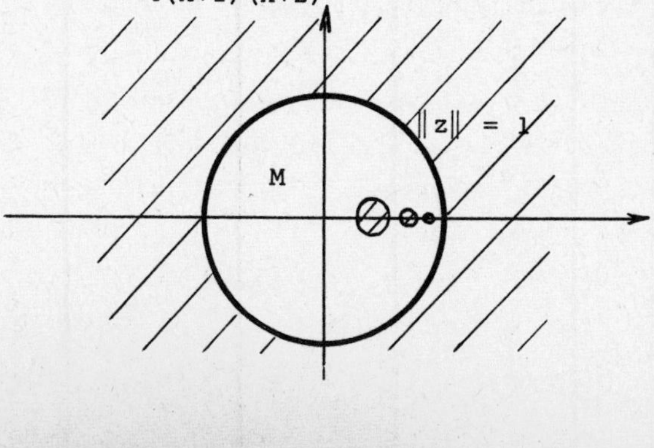

The homology group $H^c_{2n-1}(M,\mathbb{R})$ is isomorphic to $\bigoplus_{n=1}^{\infty} \mathbb{R}$ and the algebraic base of $H^c_{2n-1}(M,\mathbb{R})$ are, for instance, the homology classes γ_i generated by the cycles T_i :

$$T_i(\varphi) := \int_{S_i} \varphi \ , \quad S_i := \{z \in C^n : \|z-w_i\| = \frac{1}{2(i+1)(i+1)}\} .$$

Let us define $\chi_i(z) := (dd^c \log \|z-w_i\|)^{n-1}$, $F_i := \{u_i\}$,

$$u_i(z) := \frac{\log \|z-w_i\| - \log \frac{1}{4(i+1)(i+2)}}{\log \frac{2i+1}{i+1} - \log \frac{1}{4(i+1)(i+2)}} .$$ Similarly as in example 1,

we can show that $N_{\chi_i, F_i}(\gamma_i) =: a_i \neq 0$, $i = 1,2,\ldots$.

Since each form χ_i extended to $\mathbb{C}^n$ has the only singularity at the point w_i , then $\int_{S_j} d^c u_i \wedge \chi_i = 0$ for $i \neq j$.

Hence we have obtained a sequence of semi-norms with the property

$$N_{\chi_i, F_i}(\gamma_j) = \begin{cases} a_i \neq 0 & \text{for} \quad i = j \\ 0 & \text{for} \quad i \neq j \end{cases} .$$

The topology on $H^c_{2n-1}(M,\mathbb{R}) \cong \bigoplus_{n=1}^{\infty} \mathbb{R}$ generated by this sequence is the topology induced on $\bigoplus_{n=1}^{\infty} R$ by that of the cartesian product $\prod_{n=1}^{\infty} \mathbb{R}$ (we have $\bigoplus_{n=1}^{\infty} \mathbb{R} \subset \prod_{n=1}^{\infty} \mathbb{R}$) .

<u>EXAMPLE 3</u>. Let χ be a closed positive form (see [5], [3]) of type $(n-1,n-1)$ on a compact complex manifold M . Then, the semi-norm N_χ on $H^c_{2n-1}(M,\mathbb{R})$ is trivial. In fact, the family F_χ contains functions u satisfying the equation $dd^c u \wedge \chi = 0$. In local coordinates, we receive a partial differential equation of the elliptic type. Since the solutions of the equation of this type satisfy the maximum principle (see [6]) , therefore they are constants. Hence, for each $\gamma \in H^c_{2n-1}(M,\mathbb{R})$ and $T_o \in \gamma$,

$$N_\chi(\gamma) = \sup_{u \in F} \inf_{T \in \gamma} |T(d^c u \wedge \chi)| = \sup_{u \in F_\chi} |T_o(d^c u \wedge \chi)| = 0 .$$

3. The semi-norms and the extremal length

In this section by a slight modification we shall define semi--norms which, in the case of compact Riemann surfaces, are equivalent to the extremal length of the homology classes.

Let $\mathfrak{U} = \{U_i\}$ be a locally finite open covering of a complex manifold M and $F(\mathfrak{U})$ - a family of systems $u = \{u_i\}$ of real C^∞ - functions on M, such that

(i) $u_i : U_i \to \mathbb{R}$

(ii) $\sup_{x,y \in U_i} |u_i(x) - u_i(y)| \le 1$

(iii) $du_i = du_j$ in $U_i \cap U_j$.

From (iii) we obtain that each $u \in F(\mathfrak{U})$ generates a global form on M, denoted by du and, similarly, forms $d^c u$ and $dd^c u$. Let χ be a form on M. We define

$$N_{\chi, F(\mathfrak{U})}(\gamma) := \sup_{u \in F(\mathfrak{U})} \inf_{T \in \gamma} |T(d^c u \wedge \chi)| , \quad \gamma \in H^c(M, \mathbb{C}) .$$

<u>THEOREM 4</u>. <u>The function</u> $N_{\chi, F(\mathfrak{U})}$ <u>is a semi-norm on</u> $H^c(M,\mathbb{C})$.

<u>P r o o f.</u> The proof is the same as that of theorem 1 in [4] .

The following theorem, analogous to theorem 1, is also true.

<u>THEOREM 5</u>. <u>If</u> $d\chi = 0$ <u>and</u> $F_\chi(\mathfrak{U})$ <u>is the family of those elements</u> u <u>from</u> $F(\mathfrak{U})$ <u>for which</u> $dd^c u \wedge \chi = 0$, <u>then, for each</u> $\gamma \in H^c(M,\mathbb{C})$,

$$N_{\chi, F(\mathfrak{U})}(\gamma) = \sup_{u \in F_\chi(\mathfrak{U})} \inf_{T \in \gamma} |T(d^c u \wedge \chi)| .$$

It is easy to prove

THEOREM 6. *If* $\mathfrak{U}$ *and* $\mathfrak{B}$ *are open finite and simply connected coverings of a complex manifold* M , *then the semi-norms* $N_{\chi,F(\mathfrak{U})}$ *and* $N_{\chi,F(\mathfrak{B})}$ *are equivalent*.

Now, let M be a compact Riemann surface. Then $H_1^c(M,\mathbb{R}) \cong H_1(M,\mathbb{R}) \cong H_1^{sing}(M,\mathbb{R})$. For $\gamma \in H_1^{sing}(M,\mathbb{R})$, the extremal length of γ (see [1], [2], [8]) is defined by

$$\lambda(\gamma) := \sup_{\rho} \frac{\left(\inf_{C \in \gamma} \int_C \rho\right)^2}{\iint_M \rho^2}$$

where ρ ranges over nonnegative lower semicontinuous densities which are not identically zero. Then $\lambda^{\frac{1}{2}}$ is a semi-norm on $H_1^{sing}(M,\mathbb{R})$. Since we have the canonical isomorphism between $H_1(M,\mathbb{R})$ and $H_1^{sing}(M,\mathbb{R})$, then we can compare semi-norms on these vector spaces.

THEOREM 7. *Let* $\mathfrak{U}$ *be an open finite simply connected covering of* M . *Then, for* $\chi = \text{const.} \neq 0$, *the semi-norms* $N_{\chi,F(\mathfrak{U})}$ *and* $\lambda^{\frac{1}{2}}$ *are equivalent on* $H_1^{sing}(M,\mathbb{R}) \cong H_1(M,\mathbb{R})$.

Proof. For $\chi = \text{const.} \neq 0$, we have

$$N_{\chi,F(\mathfrak{U})}(\gamma) = \sup_{u \in F(\mathfrak{U})} \inf_{T \in \gamma} |T(d^c u \wedge \chi)| =$$

$$= |\chi| \sup_{u \in F_\chi(\mathfrak{U})} \inf_{T \in \gamma} |T(d^c u)| .$$

If we take $\chi := 1$ for simplicity, then $u = \{u_i\}$ belongs to $F_\chi(\mathfrak{U})$ if and only if $dd^c u = 0$. This means that the functions u_i are harmonic. Hence $N_{\chi,F(\mathfrak{U})}(\gamma) = N(\gamma,\mathfrak{U})$ where $N(\cdot,\mathfrak{U})$ is the semi-norm defined by S.S. Chern, H.I. Levine and L. Nirenberg in [2], p. 126. This semi-norm was proved to be equivalent to the extremal length (see [2], p. 129). So, $N_{\chi,F(\mathfrak{U})}$ is equivalent to $\lambda^{\frac{1}{2}}$. This concludes the proof.

We shall now give an example of a semi-norm which depends on a covering.

EXAMPLE 4. Let T be a torus $T := \mathbb{C}/\Gamma$ where $\Gamma := \{ni + m : n,m \in \mathbb{Z}\}$ is a net in $\mathbb{C}$. Let $\Pi: \mathbb{C} \to T$ be the canonical projection. It is known that $H_1^c(T,\mathbb{R})$ is isomorphic to $\mathbb{R}^2$ and a base of $H_1^c(T,\mathbb{R})$ are curves γ_1, γ_2

$$\gamma_1: [0,1] \ni t \to \Pi(ti) \in T$$

$$\gamma_2: [0,1] \ni t \to \Pi(t) \in T .$$

Let W_i, $i = 1,2,\ldots,k$, be a covering of the square $K := \{z = x + iy \in \mathbb{C} : 0 \le x \le 1,\ 0 \le y \le 1\}$ by open unit squares. Then $\Pi(W_i) =: U_i$ are open in T and the system $\mathfrak{U} := \{U_i,\ i = 1,2,\ldots,k\}$ is a covering of T. We shall prove that, for $\chi = 1$,

$$N_{\chi,F(\mathfrak{U})}(\gamma) = \max(|a_1|, |a_2|)$$

where $\gamma = a_1\gamma_1 + a_2\gamma_2$ for $a_1, a_2 \in \mathbb{R}$. From theorem 5 we have $N_{\chi,F(\mathfrak{U})}(\gamma) = N_{\chi,F_\chi(\mathfrak{U})}(\gamma)$ where $F_\chi(\mathfrak{U}) \subset F(\mathfrak{U})$ is the family of those systems $u = \{u_i,\ i = 1,2,\ldots,k\}$ for which u_i are harmonic functions. From property (iii) in the definition of $N_{\chi,F(\mathfrak{U})}$ it follows that, for each $u = \{u_i\} \in F_\chi(\mathfrak{U})$ there exists a harmonic function $\tilde{u}$ on $\mathbb{C}$ such that $\Pi^*u_i - \tilde{u}$ are constant on each component of the set $\Pi^{-1}(U_i) =: V_i$. Since $\frac{\partial\tilde{u}}{\partial x}$, $\frac{\partial\tilde{u}}{\partial y}$ are bounded on K and $\frac{\partial\tilde{u}}{\partial x} = \frac{\partial v_i}{\partial x}$, $\frac{\partial\tilde{u}}{\partial y} = \frac{\partial v_i}{\partial y}$ in every V_i, where $v_i := \Pi^*u_i$, therefore $\frac{\partial\tilde{u}}{\partial y}$ and $\frac{\partial\tilde{u}}{\partial y}$ are constants. Hence $\tilde{u}(x,y) = ax + by + c$ for some $a,b,c \in \mathbb{R}$. Since $\gamma = a_1\gamma_1 + a_2\gamma_2$, we have

$$N_{\chi,F_\chi(\mathfrak{U})}(\gamma) = \sup_{u \in F_\chi(\mathfrak{U})} \left|a_1\int_{\gamma_1} d^c u + a_2\int_{\gamma_2} d^c u\right| =$$

$$= \sup_{u \in F_\chi(\mathfrak{U})} \left|a_1\int_{\tilde{\gamma}_1} \Pi^* d^c u + a_2\int_{\tilde{\gamma}_2} \Pi^* d^c u\right| =$$

$$= \sup_{u \in F_\chi(\mathcal{U})} |a_1 \int_{\tilde{\gamma}_1} d^c\tilde{u} + a_2 \int_{\tilde{\gamma}_2} d^c\tilde{u}|$$

where $\tilde{\gamma}_1 : [0,1] \ni t \to ti \in \mathbb{C}$, $\tilde{\gamma}_2 : [0,1] \ni t \to t \in \mathbb{C}$, and $\tilde{u} = a_u x + b_u x + c_u$ is a harmonic function on $\mathbb{C}$ associated with u. Hence

$$N_{\chi, F_\chi(\mathcal{U})}(\gamma) = \sup_{u \in F_\chi(\mathcal{U})} |a_1 a_u - a_2 b_u| .$$

Since, for each $u_i \in u$, we have $\sup_{x,y \in U_i} |u_i(x) - u_i(y)| \le 1$, therefore $|a_u| + |b_u| \le 1$. This implies

$$N_{\chi, F_\chi(\mathcal{U})}(\gamma) = \sup_{|a|+|b|\le 1} |a_1 a - a_2 b| = \max(|a_1|, |a_2|),$$

which completes the proof.

References

[1] ACCOLA, R.D.M.: Differentials and extremal length on Riemann surfaces, Proc. Nat. Acad. Sci. USA 46 (1960), pp. 540-543.

[2] CHERN, S.S., H.I. LEVINE and L. NIRENBERG: Intrinsic norms on complex manifolds, Global Analysis, Papers in honor of K.Kodaira, ed.by D.C. Spencer and S. Iynaga, Univ. of Tokyo Press and Princeton Univ. Press, Tokyo 1969, pp. 119-139.

[3] KRASIŃSKI, T.: Stoll semi-norms and biholomorphic invariants, Thesis, 1980 (in Polish).

[4] KRASIŃSKI, T.: Semi-norms on homology groups of complex manifolds, Proc. of the International Conference of Complex Analysis and Applications, Varna 1981 (in print).

[5] LELONG, P.: Fonctions plurisousharmoniques et formes différentielles positives, Gordon and Breach, Paris-Londres-New York 1968.

[6] MIRANDA, C.: Equazioni alle derivate parziali di tipo ellitico, Springer 1955.

[7] RHAM, G.de: Variétés différentiables, Hermann, Paris 1955.

[8] RODIN, B and L. SARIO: Principal Functions, D.van Nostrand Company, Inc., 1968.

Institute of Mathematics
University of Łódź
Banacha 22, PL-90-258 Łódź, Poland

BIHOLOMORPHIC INVARIANTS ON RELATIVE HOMOLOGY GROUPS

Wiesław Królikowski (Łódź)

Contents

Summary

The first part of the paper gives an extension of Krasiński's results on biholomorphic invariants related to homology groups [4,5], in the sense of applying instead of the usual homology groups the relative homology groups. Besides we give some biholomorphic invariants on complex manifolds which in particular cases reduce to the semi-norms of Chern, Levine, Nirenberg [3] and of Krasiński [4,5].

Introduction

Let M be a complex manifold. We consider currents in the sense of de Rham on M. One can construct (in a standard way) the homology group $H(M)$. We are interested in the properties and applications of biholomorphic invariants on the space $H(M)$.

Certain biholomorphic invariants on $H(M)$ have been introduced and investigated by Chern, Levine and Nirenberg [3]. Another approach to biholomorphic invariants on complex manifolds has been proposed by Krasiński [4,5] in connection with some results of Stoll [8].

Wiesław Królikowski

We propose to define certain semi-norms on relative homology groups of a complex manifold. Such an idea is due to B. Gaveau and J.Ławrynowicz. Our definition depends on open subsets of the manifold. For the empty set it reduces to that introduced by Krasiński on the (usual) homology groups [4].

One of the most interesting properties of our semi-norms is that they generate biholomorphic invariants immediately on manifolds. Owing to their properties we call them "quasi-capacities". We give some examples and applications of the quasi-capacities. It is quite easy to calculate or estimate them. If the complex dimension of the manifold is one, we can choose the quasi-capacities so that, for the empty set, they always coincide with the semi-norms investigated by Chern, Levine and Nirenberg [3].

On the other hand [6] our semi-norms generate the topologies on relative homology groups. In terms of these topologies we can find the sufficient conditions for existence of holomorphic mappings within the class of homotopic mappings between complex manifolds. These results correspond to those obtained by Krasiński for (usual) homology groups [4].

Finally we give the relations between our semi-norms and the extremal length of the relative homology classes. In the case of compact Riemann surface, one of the form of the Chern, Levine, Nirenberg semi-norms is equivalent to the extremal length of the homology classes. This result is basic in our considerations.

1. Definition and basic properties of the function $N(A,\chi,F)$

Let M be a complex manifold of complex dimension n. We denote by $D_i^c(M)$ the complex vector space of homogeneous currents (in the sense of de Rham) of degree i with compact support.

Let A be a subset of M such that $\operatorname{int} A \neq \emptyset$. In $D_i^c(M)$ we introduce the following relation $\sim$:

$$(T_1 \sim T_2) \Leftrightarrow [\operatorname{supp}(T_1 - T_2) \subset A], \quad T_1, T_2 \in D_i^c(M).$$

It is clear that this is an equivalence relation. Any equivalence class of currenst $T \in D_i^c(M)$ will be called a <u>relative current modulo</u> A and denoted by $[T]_s^A$ or $[T]_s$.

It is easy to check

LEMMA 1. The set of all relative currents modulo A is a complex vector space.

The above space will be denoted by $D_i^c(M,A)$.

It is easy to show that if $T_1 \sim T_2$ then $bT_1 \sim bT_2$, where b is the usual boundary operator for currents. According to the above property we can define the boundary operator b for the relative currents modulo A by the formula: $b[T]_s = [bT]_s$.

The relative currents satisfy the theorem of de Rham [7] on regularization. Thus we have

THEOREM 1. Let A be an open subset of M. Let $[T]_s \in D_i^c(M,A)$. Then there exist linear operators $[T]_s \longrightarrow R[T]_s$, $[T]_s \longrightarrow V[T]_s$ satisfying the following conditions:

(i) $R[T]_s \in D_i^c(M,A)$, $V[T]_s \in D_{i+1}^c(M,A)$,

(ii) $R[T]_s = [T]_s + bV[T]_s + Vb[T]_s$.

Proof. Let $[T]_s \in D_i^c(M,A)$. By the definition of $[T]_s$ we have

$$[T]_s = \{T' \in D_i^c(M);\ \mathrm{supp}(T'-T) \subset A\}.$$

By the theorem of de Rham, there exist linear operators $(T'-T) \longrightarrow R(T'-T)$, $(T'-T) \longrightarrow V(T'-T)$ which satisfy the following conditions:

(i) $R(T'-T) = (T'-T) + bV(T'-T) + Vb(T'-T)$,

(ii) $\mathrm{supp}\ R(T'-T) \subset A$ and $\mathrm{supp}\ V(T'-T) \subset A$.

By the linearity of the operators R and V we have

$$\mathrm{supp}(RT'-RT) \subset A,\ \mathrm{supp}(VT'-VT) \subset A.$$

Consequently, we obtain

$$RT' \sim RT,\ VT' \sim VT \quad \text{for any } T' \in [T]_s.$$

According to the above property we can define the linear operators $[T]_s \longrightarrow R[T]_s$, $[T]_s \longrightarrow V[T]_s$ by the formulae

(1) $R[T]_s \overset{\text{def}}{=} [RT]_s, \quad V[T]_s \overset{\text{def}}{=} [VT]_s,$

respectively.

By the identity of currents RT and $T+bVT+VbT$, we have

$$[RT]_s = [T+bVT+VbT]_s = [T]_s + b[VT]_s + [VbT]_s.$$

Hence, by (1), we get

$$[RT]_s = [T]_s + bV[T]_s + Vb[T]_s.$$

The proof of Theorem 1 is thus completed.

Introducing, in a standard way, the notion of a relative cycle modulo A and the relative boundary modulo A, we can define the relative homology space modulo A, $H_i^c(M,A)$, as the quotient space $Z_i^c(M,A)/B_i^c(M,A)$, where $Z_i^c(M,A)$ is the space of all relative cycles modulo A and $B_i^c(M,A)$ is the space of all relative boundaries modulo A.

Then, let χ be a C^∞-form of degree $(k,\ )$, $0 \le k \le n$, $0 \le \ell \le n$, $k+\ell < 2n$ on M and let F be the family of C^∞-functions u on M satisfying the condition $-1 \le u \le 1$.

<u>D e f i n i t i o n 1</u>. Let $N(A,\chi,F) : H^c_{2n-(k+\ell+1)}(M,A) \longrightarrow \mathbb{R}_+ \cup \{0\}$, where $\mathbb{R}_+$ denotes the real half-line, be the function given by the formulae

$$N(A,\chi,F)(\Gamma) \overset{\text{def}}{=} \sup_{u\in F} \inf_{[T]_s\in\Gamma} [T]_s(d^c u\wedge\chi), \quad \Gamma \in H^c_{2n-(k+\ell+1)}(M,A),$$

where

$$[T]_s(d^c u\wedge\chi) \overset{\text{def}}{=} \inf_{T'\in[T]_s} |T'(d^c u\wedge\chi)|.$$

We denote by $WH_i^c(M,A)$ the subspace of $H_i^c(M,A)$ consisting of those elements Γ which are generated by the relative currents $[T]_s$, where T is a cycle. Then we have

<u>THEOREM 2</u>. <u>The function</u> $N(A,\chi,F)$, <u>defined above, is a semi-norm</u>

<u>on</u> $WH^{c}_{2n-(k+\ell+1)}(M,A)$ <u>for any open subset</u> $A\subset M$.

<u>Proof</u>. According to the definition of a semi-norm we have to show that

(i) $N(A,\chi,F)(\Gamma)\geq 0$ for any $\Gamma\in WH^{c}_{2n-(k+\ell+1)}(M,A)$,

(ii) $N(A,\chi,F)(a\Gamma)=|a|\ N(A,\chi,F)(\Gamma)$ for any $a\in\mathbb{C}$ and $\Gamma\in WH^{c}_{2n-(k+\ell+1)}(M,A)$,

(iii) $N(A,\chi,F)(\Gamma_1+\Gamma_2)\leq N(A,\chi,F)(\Gamma_1)+N(A,\chi,F)(\Gamma_2)$ for any pair $\Gamma_1,\Gamma_2\in WH^{c}_{2n-(k+\ell+1)}(M,A)$,

(iv) $N(A,\chi,F)(\Gamma)<+\infty$ for any $\Gamma\in WH^{c}_{2n-(k+\ell+1)}(M,A)$

for any open subset $A\subset M$.

The condition (i) is obvious.

To prove the condition (ii), let $a\in\mathbb{C}$ and $\Gamma\in WH^{c}_{2n-(k+\ell+1)}(M.A)$. Then we have

$$N(A,\chi,F)(a\Gamma)\overset{\text{def}}{=}\sup_{u\in F}\ \inf_{[T]_s\in a\Gamma}\ \inf_{\tilde{T}\in[T]_s}|\tilde{T}(d^cu\wedge\chi)|$$

$$=\sup_{u\in F}\ \inf_{[S]_s\in\Gamma}\ \inf_{R\in a[S]_s}|R(d^cu\wedge\chi)|=\sup_{u\in F}\ \inf_{[S]_s\in\Gamma}\ \inf_{\tilde{S}\in[S]_s}|(a\tilde{S})(d^cu\wedge\chi)|$$

$$=|a|\sup_{u\in F}\ \inf_{[S]_s\in\Gamma}\ \inf_{\tilde{S}\in[S]_s}|\tilde{S}(d^cu\wedge\chi)|=|a|\ N(A,\chi,F)(\Gamma).$$

The condition (iii) is satisfied as well. Indeed, let $\Gamma_1,\Gamma_2\in WH^{c}_{2n-(k+\ell+1)}(M,A)$. Then we have

$$N(A,\chi,F)(\Gamma_1+\Gamma_2)\overset{\text{def}}{=}\sup_{u\in F}\ \inf_{[T]_s\in\Gamma_1+\Gamma_2}\ \inf_{\tilde{T}\in[T]_s}|\tilde{T}(d^cu\wedge\chi)|$$

$$\leq\sup_{u\in F}\ \inf_{[T_1]_s\in\Gamma_1}\ \inf_{[T_2]_s\in\Gamma_2}\ \inf_{T\in[T_1]_s+[T_2]_s}|T(d^cu\wedge\chi)|$$

$$\leq \sup_{u\in F} \inf_{[T_1]_s\in\Gamma_1} \inf_{[T_2]_s\in\Gamma_2} \inf_{\tilde{T}_1\in[T_1]_s} \inf_{\tilde{T}_2\in[T_2]_s} |(\tilde{T}_1+\tilde{T}_2)(d^c u\wedge\chi)|$$

$$\leq \sup_{u\in F} \inf_{[T_1]_s\in\Gamma_1} \inf_{[T_2]_s\in\Gamma_2} \inf_{\tilde{T}_1\in[T_1]_s} \inf_{\tilde{T}_2\in[T_2]_s} (|\tilde{T}_1(d^c u\wedge\chi)| + |\tilde{T}_2(d^c u\wedge\chi)|)$$

$$\leq \sup_{u\in F} \inf_{[T_1]_s\in\Gamma_1} \inf_{\tilde{T}_1\in[T_1]_s} |\tilde{T}_1(d^c u\wedge\chi)| + \sup_{u\in F} \inf_{[T_2]_s\in\Gamma_2} \inf_{\tilde{T}_2\in[T_2]_s} |\tilde{T}_2(d^c u\wedge\chi)|$$

$$= N(A,\chi,F)(\Gamma_1) + N(A,\chi,F)(\Gamma_2)$$

To prove the inequality $N(A,\chi,F)(\Gamma) < +\infty$ for $\Gamma \in WH^c_{2n-(k+\ell+1)}(M,A)$, let $[T]_s$ be an arbitrary relative current modulo A with $\Gamma \in WH^c_{2n-(k+\ell+1)}(M,A)$ such that T is a cycle. By the theorem of de Rham [7], there exist linear operators $T \longrightarrow RT$ and $T \longrightarrow VT$ which satisfy the condition

$$RT = T + bVT + VbT = T + bVT.$$

Then, by Theorem 1, we have

$$R[T]_s - [T]_s = b[VT]_s, \quad bR[T]_s = 0.$$

In consequence, we get $R[T]_s \in \Gamma$. Hence

$$N(A,\chi,F)(\Gamma) \leq \sup_{u\in F} [RT]_s(d^c u\wedge\chi).$$

By the theorem of de Rham [7] again, there exists a closed C^∞-form ψ with compact support such that

$$RT(d^c u\wedge\chi) = \psi(d^c u\wedge\chi) = \int_M \psi\wedge d^c u\wedge\chi.$$

Then we have

$$N(A,\chi,F)(\Gamma) \leq \sup_{u\in F} |\int_M \psi\wedge d^c u\wedge\chi|.$$

We are going to show that

$$\sup_{u\in F}\left|\int_M \psi\wedge d^c u\wedge\chi\right| < +\infty .$$

Indeed, by the theorem of Stokes we get

$$\sup_{u\in F}\left|\int_M \psi\wedge d^c u\wedge\chi\right| \leq \sup_{u\in F}\left|\int_M d^c(u\psi\wedge\chi)\right| + \sup_{u\in F}\left|\int_M u d^c\psi\wedge\chi\right| + \sup_{u\in F}\left|\int_M u\psi\wedge d^c\chi\right|$$

Since ψ is a fixed C^∞-form with compact support and the functions $u \in F$ satisfy the condition $-1 \leq u \leq 1$, we have

$$\sup_{u\in F}\left|\int_M u d^c\psi \wedge \chi\right| + \sup_{u\in F}\left|\int_M u\psi\wedge d^c\chi\right| < +\infty .$$

The form $u\psi\wedge\chi$, as a form of degree $2n-1$, can be written as

$$u\psi\wedge\chi = \varphi^{(n-1,n)} + \varphi^{(n,n-1)},$$

where $\varphi^{(n-1,n)}$, $\varphi^{(n,n-1)}$ denote the forms of degree $(n-1,n)$ and $(n,n-1)$, respectively, defined on M. If we take

$$(u\psi\wedge\chi)^c \overset{\text{def}}{=} i(\varphi^{(n,n-1)} - \varphi^{(n-1,n)}),$$

then

$$d^c(u\psi\wedge\chi) = i(\bar{\partial}-\partial)(\varphi^{(n-1,n)} + \varphi^{(n,n-1)}) = i(\bar{\partial}\varphi^{(n,n-1)} - \partial\varphi^{(n-1,n)}),$$

$$d[(u\psi\wedge\chi)^c] = (\partial+\bar{\partial})(i\varphi^{(n,n-1)} - i\varphi^{(n-1,n)}) = i(\bar{\partial}\varphi^{(n,n-1)} - \partial\varphi^{(n-1,n)}).$$

Therefore

$$d^c(u\psi\wedge\chi) = d[(u\psi\wedge\chi)^c].$$

Finally, by the theorem of Stokes,

$$\int_M d^c(u\psi\wedge\chi) = \int_M d[(u\psi\wedge\chi)^c] = 0$$

for any function $u \in F$.

In this way the proof of Theorem 2 is completed.

2. The function $N(A,\chi,F)$ as a biholomorphic invariant

Let M and P be complex manifolds of the same complex dimension n, and $f: M \longrightarrow P$ a holomorphic mapping. If $T \in D_i^c(M)$, we define f_*T on P by the formulae

$$(f_*T)(\phi) = T(f^*\phi),$$

where ϕ is an arbitrary C^∞-form on P of degree $(2n-i)$. Moreover,

$$\text{(2)} \qquad \operatorname{supp}(f_*T) \subset f(\operatorname{supp} T).$$

Let A be an arbitrary open subset of M and $[T]_s^A \in D_i^c(M.A)$. By the definition of a relative current modulo A, We have $\operatorname{supp}(\tilde{T}-T) \subset A$ for any $\tilde{T} \in [T]_s^A$. By (2) we get

$$\operatorname{supp}(f_*\tilde{T} - f_*T) = \operatorname{supp}[f_*(\tilde{T}-T)] \subset f[\operatorname{supp}(\tilde{T}-T)] \subset f(A) \subset B,$$

where B is an arbitrary open subset of P such that $f(A) \subset B$. According to the above property we can define the linear mapping $\tilde{f}_*: D_i^c(M,A) \longrightarrow D_i^c(P,B)$, induced by f, by the formulae

$$\tilde{f}_*([T]_s^A) \overset{\text{def}}{=} [f_*T]_s^B, \quad [T]_s^A \in D_i^c(M,A).$$

Then, by means of $\tilde{f}_*$, we define the linear mapping $f_{**}: H_i^c(M,A) \longrightarrow H_i^c(P,B)$ by the formulae

$$f_{**}([[T]_s^A]) \overset{\text{def}}{=} [\tilde{f}_*([T]_s^A)], \quad [[T]_s^A] \in H_i^c(M,A).$$

THEOREM 3. *Let M and P be complex manifolds of the same complex dimension n. If $f: M \longrightarrow P$ is a holomorphic mapping then, for any open subset $A \subset M$ and for any $\Gamma \in H^c_{2n-(k+\ell+1)}(M,A)$, we have*

$$N(B,\chi,F_P)(f_{**}\Gamma) \le N(A,f^*\chi,f^*F_P)(\Gamma) \le N(A,f^*\chi,F_M)(\Gamma),$$

where B is an arbitrary open subset of P such that $f(A) \subset B$, f_{**} is the induced homomorphism on relative homology spaces, and the semi-norms are taken in M and P, respectivelly.

Proof. Let $\Gamma \in H^{c}_{2n-(k+\ell+1)}(M,A)$. By the definitions, we have

$$N(B,\chi,F_P)(f_{**}\Gamma) \overset{\text{def}}{=} \sup_{u\in F_P} \inf_{[S]^B_s \in f_{**}\Gamma} \inf_{\tilde{S}\in[S]^B_s} |\tilde{S}(d^c u\wedge\chi)|$$

$$\sup_{u\in F_P} \inf_{[T]^A_s\in\Gamma} \inf_{R\in\tilde{f}_*([T]^A_s)} |R(d^c u\wedge\chi)| \leq \sup_{u\in F_P} \inf_{[T]^A_s\in\Gamma} \inf_{\tilde{T}\in[T]^A_s} |f_*\hat{T}(d^c u\wedge\chi)|$$

$$= \sup_{u\in F_P} \inf_{[T]^A_s\in\Gamma} \inf_{\tilde{T}\in[T]^A_s} |\tilde{T}[f^*(d^c u\wedge\chi)]| = \sup_{u\in F_P} \inf_{[T]^A_s\in\Gamma} \inf_{\tilde{T}\in[T]^A_s} |\tilde{T}(f^* d^c u\wedge f^*\chi)|$$

$$= \sup_{u\in F_P} \inf_{[T]^A_s\in\Gamma} \inf_{\tilde{T}\in[T]^A_s} |\tilde{T}(d^c f^* u\wedge f^*\chi)| \leq \sup_{v\in F_M} \inf_{[T]^A_s\in\Gamma} \inf_{\tilde{T}\in[T]^A_s} |\tilde{T}(d^c v\wedge f^*\chi)|$$

$$= N(A,f^*\chi,F_M)(\Gamma)$$

Thus the function $N(A,\chi,F)$ is a biholomorphic invariant in the following sense:

COROLLARY 1. Let M and P be complex manifolds of the same complex dimension n. If $f: M \longrightarrow P$ is biholomorphic then, for any open subset $A \subset M$ and for any element $\Gamma \in H^{c}_{2n-(k+\ell+1)}(M,A)$, we have

$$N(f(A),\chi,F_P)(f_{**}\Gamma) = N(A,f^*\chi,F_M)(\Gamma),$$

where f_{**} is the induced homomorphism on relative homology spaces, and the semi-norms are taken in M and P, respectively.

3. Semi-norms $N(A,\chi,F)$ as a generalization of Chern-Levine-Nirenberg's, Krasiński's and Stoll's semi-norms

Suppose $A=\emptyset$. Let $\Gamma=[[T]^{\emptyset}_s]$ be an arbitrary element of $WH^c_i(M,\emptyset)$. By the definition of Γ we have $b[\tilde{T}]^{\emptyset}_s=[b\tilde{T}]^{\emptyset}_s=0$, for any relative current $[\tilde{T}]^{\emptyset}_s$ in Γ. Then, by the definition of a relative current, we have

$$[T]_s^{\emptyset} = \{T' \in D_i^c(M);\ \mathrm{supp}(T'-T) \subset \emptyset \ \text{ and } \mathrm{supp}\ \ bT' \subset \emptyset\}.$$

Therefore $Z_i^c(M,\emptyset) = Z_i^c(M)$. However, by the definition of Γ, we have

$$\Gamma = [[T]_s^{\emptyset}] \overset{\mathrm{def}}{=} \{[T']_s^{\emptyset} \in Z_i^c(M,\emptyset); [T']_s^{\emptyset} - [T]_s^{\emptyset} = b[S]_s^{\emptyset},$$

where $[S]_s^{\emptyset}$ is an element of $D_{i+s}^c(M,\emptyset)\}$.

Hence $\Gamma = [[T]_s^{\emptyset}] = \gamma_T$, where γ_T is a usual homology class of the current T, so $WH_i^c(M,\emptyset) = H_i^c(M)$. Thus it is clear that

$$N(\emptyset,\chi,F)(\Gamma) \overset{\mathrm{def}}{=} \sup_{u\in F} \inf_{[T']_s^{\emptyset}\in\Gamma} \inf_{R\in[T']_s^{\emptyset}} |R(d^c u\wedge\chi)|$$

$$= \sup_{u\in F} \inf_{\{T'\}\in\gamma_T} \inf_{R\in\{T'\}} |R(d^c u\wedge\chi)| = \sup_{u\in F} \inf_{T'\in\gamma_T} |T'(d^c u\wedge\chi)| \overset{\mathrm{def}}{=} N_{\chi,F}(\gamma_T),$$

where $N_{\chi,F}$ are the semi-norms defined by Krasiński on the usual homology groups (cf.Krasiński [4]). The relations between Krasiński's semi-norms and semi-norms introduced by Chern, Levine, Nirenberg and by Stoll are contained in [5].

4. Quasi-capacities on complex manifolds induced by the semi-norms $N(A,\chi,F)$

Let $T \in D_{2n-(k+L+1)}^c(M)$ be a cycle ($bT = 0$), satisfying the condition

$$(3) \qquad 0 < \sup_{u\in F} |T(d^c u\wedge\chi)| < +\infty,$$

where χ and F are as in Definition 1. As an example of such a current may serve a current represented by any closed form not identically zero.

Let $O(M)$ denote a family of all open subsets of manifold M.

<u>Definition 2</u>. Let $C(\chi,T,F) : O(M) \longrightarrow \mathbb{R}_+ \cup \{0\}$ be the function given by the formulae

$$C(\chi,T,F)(A) \overset{\text{def}}{=} N(A,\chi,F)([[T]_s^A]),\ A \in O(M).$$

The number $C(\chi,T,F)(A)$ will be called the quasi-capacity of A. We are going to derive basic proporties of the quasi-capacity.

LEMMA 1. *For any open subset* A *of* M *we have*

$$C(\chi,T,F)(A) < +\infty.$$

P r o o f. By the assumption (3), it is abvious.

LEMMA 2.

$$C(\chi,T,F)(M) = 0$$

P r o o f. This is obvious the current being identically zero belongs to $[T]_s^M$.

LEMMA 3. *We have*

$$C(\chi,T,F)(\emptyset) = \sup_{u\in F} \inf_{T\in\gamma_T} |T(d^c u \wedge \chi)|,$$

where γ_T *denotes a usual homology class of the current* T.

P r o o f. By considerations of Section 3, it is obvious as well.

LEMMA 4. *For any open subsets* A *and* B *of* M *such that* $A \subset B$, *we have*

$$C(\chi,T,F)(B) \leq C(\chi,T,F)(A).$$

P r o o f. By the inclusion

$$[T]_s^B \supset [T]_s^A \quad \text{for } A \subset B,\ A,B \in O(M)$$

we have

$$[T]_s^B(d^c u \wedge \chi) \leq [T]_s^A(d^c u \wedge \chi)$$

for any $u \in F$. Hence

$$\inf_{[\tilde{T}]^B_s \in [[T]^B_s]} [\tilde{T}]^B_s(d^c u \wedge \chi) \leq \inf_{[T']^A_s \in [[T]^A_s]} [T']^A_s(d^c u \wedge \chi)$$

for any $u \in F$. Finally

$$C(\chi,T,F)(B) \overset{\text{def}}{=} \sup_{u \in F} \inf_{[\tilde{T}]^B_s \in [[T]^B_s]} [\tilde{T}]^B_s(d^c u \wedge \chi)$$

$$\leq \sup_{u \in F} \inf_{[T']^A_s \in [[T]^A_s]} [T']^A_s(d^c u \wedge \chi) \overset{\text{def}}{=} C(\chi,T,F)(A),$$

as desired.

COROLLARY 2. *For any open subsets* A *and* B *of* M, *we have*

$$C(\chi,T,F)(A \cup B) \leq C(\chi,T,F)(A) + C(\chi,T,F)(B).$$

COROLLARY 3. *For any sequence* $A_1, A_2, \ldots$ *of open subsets of* M *we have*

$$C(\chi,T,F)(A_1 \cup A_2 \cup \ldots) \leq C(\chi,T,F)(A_1) + C(\chi,T,F)(A_2) + \ldots,$$

$$C(\chi,T,F)(A_1 \cup A_2 \cup \ldots) \leq \inf\{c(\chi,T,F)(A_n),\ n = 1,2,\ldots\}$$

$$\leq C(\chi,T,F)(A_1 \cap A_2 \cap \ldots).$$

LEMMA 5. *For any open subset* A *of* M *such that* $\operatorname{supp} T \subset A$, *we have*

$$C(\chi,T,F)(A) = 0$$

Proof. This is obvious because in our case we have $[T]^A_s = [0]^A_s$.

THEOREM 4. *Let* M *and* P *be complex manifolds of the same complex dimension* n. *If* $f: M \longrightarrow P$ *is biholomorphic then for any open subset* A *of* M *we have*

$$C(\chi, f_* T, F)[f(A)] = C(f^*\chi, T, f^*F)(A).$$

P r o o f. This theorem is an immediate consequence of Corollary 1.

The properties of the quasi-capacities (generated by the semi-norms $N(A,\chi,F)$), defined by

$$C_1(\chi,T,F)(A) \overset{\text{def}}{=} \sup_{u\in F} [T]_s^A(d^c u \wedge \chi)$$

and

$$C_2(\chi,T,F)(A) \overset{\text{def}}{=} \sup_{u\in F} \{T\}_s^A(d^c u \wedge \chi) = \sup_{u\in F} \inf_{\tilde{T}\in\{T\}_s^A} |\tilde{T}(d^c u \wedge \chi)|,$$

are quite analogous if we suppose that A is an arbitrary open subset of M and

$$\{T\}_s^A = \{\tilde{T} \in [T]_s^A;\ \tilde{T} = T + T_A, \text{ where } T_A \text{ is a current with connected compact support in } A\}.$$

In further sections we give some examples and applications of the quasi-capacities.

5. The case of complex tori

In this section we give an application of the introduced quasi-capacity, where the currents in question are represented by curves. (By a curve we mean a finite sum of Jordan arcs).

We consider the complex torus $\Pi = \mathbb{C}/\Gamma_0$ with the complex structure induced from $\mathbb{C}$, where $\Gamma_0 \overset{\text{def}}{=} \{ni + m; n,m \in \mathbb{Z}\}$. Let π denote the canonical mapping $\mathbb{C} \longrightarrow \mathbb{C}/\Gamma_0 = \Pi$ and K be defined as

$$K \overset{\text{def}}{=} \{z \in \mathbb{C}; z = x + iy,\ 0 \le x \le 1,\ 0 \le y \le 1\}.$$

We take a finite open covering of K consisting of unit open squares. We denote these squares by K_i, $i = 1,2,\dots,k$. Then the sets $\pi(K_i) = U_i$, $i = 1,2,\dots,k$, are open in Π and form an open covering of Π. Let

$\mathfrak{U} = \{U_i\}_{i=1,\dots,k}$. We define the family $F(\mathfrak{U})$ of collections $u = \{u_i; i = 1,2,\dots,k\}$, where u_i, $i = 1,2,\dots,k$ is a C^∞-function satisfying the following conditions:

1) $u_i : U_i \longrightarrow \mathbb{R}$,

2) $du_i = du_j$ in $U_i \cap U_j \neq 0$

3) $\sup\limits_{p,q \in U_i} |u_i(p) - u_i(q)| \leq 1$

4) $dd^c u_i = 0$

The functions u_i, $i = 1,\dots,k$, are harmonic. We shall show that there is a linear function $\tilde{u}(x,y) = ax + by + c$, $(x,y) \in \mathbb{C}$, such that

$$\pi^* u_i - \tilde{u} = \text{const}, \quad i = 1,\dots,k$$

on each component of $\pi^{-1}(U_i)$.

Indeed, the open sets $V_i = \pi^{-1}(U_i)$, $i = 1,\dots,k$, form a covering of the plane $\mathbb{C}$. The functions $v_i = \pi^* u_i$ are harmonic on V_i and satisfy the condition: on each non-empty component of $\pi^{-1}(U_i) \cap \pi^{-1}(U_j)$ we have $v_i - v_j = \text{const}_{ij}$. There exists a harmonic function $\tilde{u}$ on $\mathbb{C}$ such that $\tilde{u} - v_i = \text{const}$ on each component of V_i.

Now, we consider the partial derivatives $\tilde{u}_x$ and $\tilde{u}_y$. In a square K, $\tilde{u}_x$ and $\tilde{u}_y$ are harmonic and bounded. By the equalities

$$\tilde{u}_x = (v_i)_x, \quad \tilde{u}_y = (v_i)_y \quad \text{in } V_i, \quad i = 1,2,\dots,k.$$

they are bounded in $\mathbb{C}$. Since $\tilde{u}_x$ and $\tilde{u}_y$ are harmonic, they are real constants: $\tilde{u}_x = a$, $\tilde{u}_y = b$. In consequence, we obtain $\tilde{u}(x,y) = ax + by + c$, $c \in \mathbb{R}$. On each component of $\pi^{-1}(U_i)$ the oscillation of $\tilde{u}$ is less than one (because $\tilde{u} - v_i = \text{const}$ and $v_i = \pi^* u_i$). Owing to the fact that $\tilde{u}$ is linear we shall get the conditions for a and b considering $\tilde{u}$ in an arbitrary square on $\mathbb{C}$, for example in $K = \{(x,y) : 0 < x < 1, 0 < y < 1\}$. It is easy to show that the condition

$$|u(x_1,y_1) - u(x_2,y_2)| \leq 1 \quad \text{for} \quad (x_i,y_i) \in K, \quad i = 1,2$$

is equivalent to $|a| + |b| \leq 1$.

Now, we consider a cycle C_o on Π, represented by a curve

$$\tilde{C}_o \overset{\text{def}}{=} [0,1) \ni t \longrightarrow \pi(t + \tfrac{1}{2}i) \in \Pi.$$

We shall calculate the quasi-capacities $C_2(1,C_o,F(\mathfrak{v}))$ on Π. By definition

$$C_2(1,C_o,F(\mathfrak{v}))(A) = \sup_{u \in F(\mathfrak{v})} \{C_o\}_s^A (d^c u)$$

for any open subset A of Π. In our case we have

$$\{C_o\}_s^A = \{C;\ C = C_o + C_A, \text{ where } C_A \text{ is a current represented by closed arc lying in } A\}.$$

Let $\tilde{K}$ be a set consisting at the open square $\{(x,y): 0 < x < 1, 0 < y < 1\}$ and the segments $y = 0,\ 0 \leq x < 1$ and $x = 0,\ 0 \leq y < 1$. Let A be an arbitrary open subset of Π. It is obvious that there exists a unique subset $\tilde{A}$ of $\tilde{K}$ such that $A = \pi(\tilde{A})$. If we identify the following points

$$(0,y) \text{ and } (1,y) \quad \text{for} \quad 0 \leq y < 1,$$

$$(x,0) \text{ and } (x,1) \quad \text{for} \quad 0 \leq x < 1,$$

then we may treat $\tilde{A}$ as a connected set in $\tilde{K}$. It is easy to see that for any $C \in \{C_o\}_s^A$ there exists a curve $\tilde{C} \subset \tilde{K}$ such that $C = \pi(\tilde{C})$. Moreover, we have

$$\int_C d^c u = \int_{\tilde{C}} \pi^* d^c u$$

for any $u \in F(\mathfrak{v})$.

Since there is a one-to-one correspondence between $\{C_o\}_s^A$ and $\{\tilde{C}_o\}_s^{\tilde{A}}$, we have the following equalities

$$C_2(1,C_0,F(\mathfrak{V}))(A) \overset{\text{def}}{=} \sup_{u\in F(\mathfrak{V})} \{C_0\}_s^A(d^c u)$$

$$= \sup_{u\in F(\mathfrak{V})} \inf_{C\in\{C_0\}_s^A} \left|\int_C d^c u\right| = \sup_{u\in F(\mathfrak{V})} \inf_{\tilde{C}\in\{\tilde{C}_0\}_s^{\tilde{A}}} \left|\int_{\tilde{C}} \pi^* d^c u\right|$$

$$= \sup_{u\in F(\mathfrak{V})} \inf_{\tilde{C}\in\{\tilde{C}_0\}_s^{\tilde{A}}} \left|\int_{\tilde{C}} d^c \tilde{u}\right|,$$

where $\tilde{u} = a_u x + b_u y + c$ is a harmonic function in $\mathbb{C}$ corresponding to $u \in F(\mathfrak{V})$. Since

$$d^c\tilde{u} = i(\bar{\partial} - \partial)(a_u x + b_u y + c) = a_u dy - b_u dx,$$

then

$$C_2(1,C_0,F(\mathfrak{V}))(A) = \sup_{u\in F(\mathfrak{V})} \inf_{\tilde{C}\in\{\tilde{C}_0\}_s^{\tilde{A}}} \left|a_u \int_C dy - b_u \int_C dx\right|.$$

For the clearness of calculations we need the following

<u>D e f i n i t i o n 3.</u> Let A be an arbitrary domain in Π. We put

$$\alpha(A) = \begin{cases} 1 & \text{if for any } 0 \le a < 1 \text{ we have } \{(x,y): x=a,\ 0\le y<1\} \cap \tilde{A} \ne \emptyset, \\ 1 - \max\{\Sigma(b_n - a_n)\}, & \text{where maximum is taken over all } \sum_n (b_n - a_n) \\ & \text{so that } \{(x,y): a_1 \le x \le b_1,\ a_2 \le x \le b_2, \ldots, 0 \le y < 1, \\ & 0 \le a_1 \le b_1 < a_2 \le b_2 < \ldots \le 1\} \cap \tilde{A} = \emptyset, \text{ if there} \\ & \text{is some } 0 \le a < 1 \text{ for which } \{(x,y): x = a,\ 0 \le y < 1\} \cap \tilde{A} = \emptyset. \end{cases}$$

(Geometrically, $\alpha(A)$ is the diameter of A along the x-axis)

<u>D e f i n i t i o n 4</u>. Similarly we define the number $\beta(A)$ (the diameter of A along the y-axis), where A is a domain in Π. We put

$$\beta(A)=\begin{cases} 1 & \text{if for any } 0 \le c \le 1 \text{ we have } \{(x,y): 0 \le x \le 1, y=c\} \cap \tilde{A} \neq \emptyset, \\ 1-\max\{\sum_k (d_k-c_k)\}, & \text{where maximum is taken over all } \sum_k (d_k-c_k) \\ & \text{so that } \{(x,y): c_1 \le y \le d_1, c_2 \le y \le d_2, \ldots, 0 \le x < 1, \\ & 0 \le c_1 \le d_1 < c_2 \le d_2 < \ldots < 1\} \cap \tilde{A} = \emptyset, \text{ if there is some} \\ & 0 \le c < 1 \text{ for which } \{(x,y): 0 \le x < 1, y=c\} \cap \tilde{A} = \emptyset. \end{cases}$$

Now, let $u \in F(\mathfrak{V})$. We distinguish six cases:

C a s e $\underline{1}$. $a_u = 0$. In this case we have

$$\inf_{\tilde{C} \in \{\tilde{C}_o\}_s^{\tilde{A}}} |-b_u \int_{\tilde{C}} dx| = |b_u| \inf |\int_{\tilde{C}_o} dx + \int_{\tilde{C}_{\tilde{A}}} dx| = |b_u| \inf_{\tilde{C}_{\tilde{A}}} |1 + \int_{\tilde{C}_{\tilde{A}}} dx|$$

$$= |b_u| \; [1-\alpha(A)].$$

In consequence we have

$$\sup_{-1 \le b_u \le 1} \; \inf_{\tilde{C} \in \{\tilde{C}_o\}_s^{\tilde{A}}} |-b_u \int_{\tilde{C}} dx| = 1 - \alpha(A).$$

C a s e $\underline{2}$. $b_u = 0$. We have

$$\inf_{\tilde{C} \in \{\tilde{C}_o\}_s^{\tilde{A}}} |a_u \int_{\tilde{C}} dy| \le |a_u| \, |\int_{\tilde{C}_o} dy| = |a_u| \; 0 = 0.$$

C a s e $\underline{3}$. $0 < a_u < 1$, $-1 < b_u < 0$. In this case

$$\inf_{\tilde{C} \in \{\tilde{C}_o\}_s^{\tilde{A}}} |a_u \int_{\tilde{C}} dy - b_u \int_{\tilde{C}} dx| = \inf_{\tilde{C}_{\tilde{A}}} |a_u \int_{\tilde{C}_{\tilde{A}}} dy - b_u \int_{\tilde{C}_{\tilde{A}}} dx - b_u|.$$

By Definitions 3 and 4, we have

$$(4) \qquad -\alpha(A) \le \int_{\tilde{C}_{\tilde{A}}} dx \le \alpha(A),$$

(5) $\quad -\beta(A) \le \int_{\tilde{C}_{\tilde{A}}} dy \le \beta(A).$

Therefore

$$-a_u\beta(A) + b_u\alpha(A) - b_u \le a_u \int_{\tilde{C}_{\tilde{A}}} dy - b_u \int_{\tilde{C}_{\tilde{A}}} dx - b_u \le a_u\beta(A) - b_u\alpha(A) - b_u.$$

We observe that $a_u\beta(A) - b_u\alpha(A) - b_u \ge 0$. Then we have

$$\inf_{\tilde{C}_{\tilde{A}}} \left| a_u \int_{\tilde{C}_{\tilde{A}}} dy - b_u \int_{\tilde{C}_{\tilde{A}}} dx - b_u \right| = \begin{cases} 0, \text{ if } -a_u\beta(A) + b_u\alpha(A) - b_u \le 0, \\ -a_u\beta(A) + b_u[\alpha(A)-1], \text{ if} \\ \qquad a_u\beta(A) + b_u\alpha(A) - b_u > 0. \end{cases}$$

Now, consider a function $f(a_u, b_u) \overset{\text{def}}{=} -a_u\beta(A) + b_u[\alpha(A) - 1]$ for $0 < a_u < 1$ and $-1 < b_u < 0$. It is easy to see that

$$\sup\{-a_u\beta(A) + b_u[\alpha(A) - 1] : 0 < a_u < 1, -1 < b_u < 0\} = 1 - \alpha(A).$$

In consequence we have

$$\sup_{\substack{0<a_u<1 \\ -1<b_u<0}} \inf_{\tilde{C}\in\{\tilde{C}_0\}_s^{\tilde{A}}} \left| a_u \int_{\tilde{C}} dy - b_u \int_{\tilde{C}} dx \right| = \sup_{\substack{0<a_u<1 \\ -1<b_u<0}} \inf_{\tilde{C}_{\tilde{A}}} \left| a_u \int_{\tilde{C}_{\tilde{A}}} dy - b_u \int_{\tilde{C}_{\tilde{A}}} dx - b_u \right|$$

$$= \sup\{-a_u\beta(A) + b_u[\alpha(A) - 1] : 0 < a_u < 1, -1 < b_u < 0\} = 1 - \alpha(A).$$

C a s e <u>4.</u> $-1 < a_u < 0$, $0 < b_u < 1$ - analogous to the case 3.

C a s e <u>5.</u> $0 < a_u < 1$, $0 < b_u < 1$. We have

$$-a_u\beta(A) - b_u\alpha(A) - b_u \le a_u \int_{\tilde{C}_{\tilde{A}}} dy - b_u \int_{\tilde{C}_{\tilde{A}}} dx - b_u \le a_u\beta(A) + b_u\alpha(A) - b_u.$$

We observe that $-a_u\beta(A) - b_u\alpha(A) - b_u < 0$. Therefore

$$\inf_{\tilde{C}_{\tilde{A}}} \left| a_u \int_{\tilde{C}_{\tilde{A}}} dy - b_u \int_{\tilde{C}_{\tilde{A}}} dx - b_u \right| = \begin{cases} 0, \text{ if } a_u\beta(A) + b_u\alpha(A) - b_u \geq 0 \\ b_u[1-\alpha(A)] - a_u\beta(A), \text{ if} \\ \qquad a_u\beta(A) + b_u\alpha(A) - b_u < 0. \end{cases}$$

Now we consider a function $g(a_u,b_u) \overset{\text{def}}{=} b_u[1-\alpha(A)] - a_u\beta(A)$ for $0 < a_u < 1$ and $0 < b_u < 1$. It is easy to see that

$$\sup\{b_u[1-\alpha(A)] - a_u\beta(A) : 0 < a_u < 1,\ 0 < b_u < 1\} = 1 - \alpha(A).$$

Then

$$\sup_{\substack{0<a_u<1\\0<b_u<1}} \inf_{\tilde{C}\in\{\tilde{C}_o\}_s^{\tilde{A}}} \left| a_u \int_{\tilde{C}} dy - b_u \int_{\tilde{C}} dx \right| = \sup_{\substack{0<a_u<1\\0<b_u<1}} \inf_{\tilde{C}_{\tilde{A}}} \left| a_u \int_{\tilde{C}_{\tilde{A}}} dy - b_u \int_{\tilde{C}_{\tilde{A}}} dx - b_u \right|$$

$$= \sup\{b_u[1-\alpha(A)] - a_u\beta(A) : 0 < a_u < 1,\ 0 < b_u < 1\} = 1 - \alpha(A).$$

C a s e <u>6.</u> $-1 < a_u < 0$, $-1 < b_u < 0$ - analogous to case 5.

Finally we obtain

$$C_2(1,C_o,F(\mathfrak{W}))(A) = 1 - \alpha(A).$$

If we take a cycle C_1 on Π represented by a curve $\tilde{C}_1$:

$$\tilde{C}_1 \overset{\text{def}}{=} [0,1) \ni t \longrightarrow \pi(\tfrac{1}{2} + it) \in \Pi,$$

then, by analogous considerations, we get

$$C_2(1,C_1,F(\mathfrak{W}))(A) = 1 - \beta(A).$$

<u>COROLLARY 4</u>. <u>Let</u> A <u>be an open subset of</u> Π. <u>By the definition of</u> $C(1,C_o,F(\mathfrak{W}))(A)$, $C_1(1,C_o,F(\mathfrak{W}))(A)$ <u>and</u> $C_2(1,C_o,F(\mathfrak{W}))(A)$ <u>we have</u>

$$N(A,1,F(\mathfrak{W}))([[C_o]_s^A]) = C_1(1,C_o,F(\mathfrak{W}))(A) \leq C_2(1,C_o,F(\mathfrak{W}))(A)$$

$$= 1 - \alpha(A).$$

In analogy

$$N(A,1,F(\mathfrak{V}))([[C_1]_s^A]) = C_1(1,C_1,F(\mathfrak{V}))(A) \le C_2(1,C_1,F(\mathfrak{V}))(A)$$

$$=1-\beta(A).$$

The first equality in above expressions is a direct consequence of the following

THEOREM 5. Let χ and F be as in Definition 1. Let F_χ denote the subfamily of F consisting of functions u satisfying the differential equation

$$dd^c u \wedge \chi = 0.$$

Suppose χ is closed. Then, for any open subset $A \subset M$ and for any element $\Gamma \in H^c_{2n-(k+\ell+1)}(M,A)$, we have

$$N(A,\chi,F_\chi)(\Gamma) = \sup_{u \in F_\chi} [T]_s^A(d^c u \wedge \chi), \quad [T]_s^A \in \Gamma.$$

P r o o f. Let $\Gamma \in H^c_{2n(k+\ +1)}(M,A)$. If $\Gamma = 0$ then $N(A,\chi,F)(\Gamma) = 0$. On the other hand, $[0]_s^A \in \Gamma$, where 0 is a current being identically zero. Then $\sup\{[0]_s^A(d^c u\ \chi) : u \in F_\chi\} = 0$.

Suppose next $\Gamma \neq 0$. Let $[T]_s^A$ be an arbitrary element of Γ. For any $[\tilde{T}]_s^A \in \Gamma$ there exists an element $[\tilde{S}]_s^A \in D^c_{2n-(k+\ell)}(M,A)$ such that $[\tilde{T}]_s^A = [T]_s^A + b[\tilde{S}]_s^A$. Let $\varphi = d^c u \wedge \chi$, $u \in F_\chi$. We have

$$[\tilde{T}]_s^A(\varphi) \overset{\text{def}}{=} \inf_{R \in [\tilde{T}]_s^A} |R(\varphi)| = \inf_{R \in [T]_s^A + b[\tilde{S}]_s^A} |R(\varphi)|$$

$$\le \inf_{T' \in [T]_s^A} \inf_{S' \in [\tilde{S}]_s^A} |(T' + bS')(\varphi)| = \inf_{T' \in [T]_s^A} \inf_{S' \in [\tilde{S}]_s^A} |T'(\varphi) + S'(d\varphi)|$$

$$= \inf_{T' \in [T]_s^A} |T'(\varphi)| = [T]_s^A(\varphi).$$

In a simillar way we obtain $[T]_s^A(\varphi) \leq [\tilde{T}]_s^A(\varphi)$. The proof of Theorem 5 is thus completed.

6. The case of an annulus in the plane

We consider an annulus $P(a) \overset{\text{def}}{=} \{z \in \mathbb{C};\ 1 < |z| < a,\ a \in \mathbb{R},\ a > 1\}$. Let $\chi \equiv 1$ and F_1 denote the family of harmonic functions u on $P(a)$ satisfying the condition $0 < u < 1$.

Let A be an arbitrary domain of $P(a)$. We shall estimate the quasi-capacities $C_2(1, C_o, F_1)(A)$, where C_o is a cycle on $P(a)$ represented by a curve

$$C_o \overset{\text{def}}{=} [0, 2\pi) \ni t \longrightarrow re^{it},\ 1 < r < a.$$

By definition

$$C_2(1, C_o, F_1)(A) = \sup_{u \in F_1} \inf_{C \in \{C_o\}_s^A} \left|\int_C d^c u\right|.$$

Now, we need the following

D e f i n i t i o n 5. Let A be a domain in $P(a)$. Let $I \overset{\text{def}}{=} \{(x,y) : y = 0, x > 0\} \cap P(a)$. Then we put

$$\Theta(A) = \begin{cases} \sup\{|\operatorname{Arg} p - \operatorname{Arg} q| : p, q \in A\}, \text{ if } A \cap I = \emptyset; \\ 2\pi, \text{ if } A \cap I \neq \emptyset \text{ and for any } \Theta,\ 0 \leq \Theta < 2\pi, \text{ we have} \\ \{z = re^{i\Theta};\ 1 < r < a\} \cap A \neq \emptyset; \\ 2\pi - \max(\Theta_2 - \Theta_1), \text{ where maximum is taken over all } \Theta_2 - \Theta_1 \geq 0 \\ \quad \text{so that } \{z = re^{i\Theta};\ 1 < r < a,\ \Theta_1 \leq \Theta \leq \Theta_2\} \cap A = \emptyset, \\ \quad \text{if } A \cap I \neq \emptyset \text{ and there is some } \Theta \text{ for which} \\ \quad \{z = re^{i\Theta};\ 1 < r < a\} \cap A = \emptyset. \end{cases}$$

Now, let us take a function $u_o \overset{\text{def}}{=} \log r / \log a$. It is clear that $u_o \in F_1$ and

$$C_2(1,C_o,F_1)(A) \geq \inf_{C\in\{C_o\}_s^A} |\int_C d^c u_o|.$$

Since

$$\inf_{C\in\{C_o\}_s^A} |\int_C d^c u_o| = \frac{1}{\log a} \inf_{C\in\{C_o\}_s^A} |\int_C d\theta| = \frac{1}{\log a} \inf_{C_A} |\int_{C_o} d\theta + \int_{C_A} d\theta|$$

$$= \frac{1}{\log a} \inf_{C_A} |2\pi - \int_{C_A} d\theta| = \frac{2 - \Theta(A)}{\log a},$$

then

$$C_2(1,C_o,F_1)(A) \geq [2\pi - \Theta(A)]/\log a.$$

On the other hand we have

$$C_2(1,C_o,F_1)(A) \leq \sup_{u\in F_1} |\int_{C_o} d^c u| = 2\pi/\log a.$$

The latter equality has been proved by Chem, Levine and Nirenberg [3]. Finally we get

$$[2\pi - \Theta(A)]/\log a \leq C_2(1,C_o,F_1)(A) \leq 2\pi/\log a.$$

7. The case of (nonrelative homology groups

Let us consider a particular case of the situation described in Sections 5 and 6 : $A = \emptyset$. By considerations of Section 3, we have $[[T]_s^{\emptyset}] = \gamma_T$, where γ_T denotes the (usual) homology class of cycle T,

$$N(\emptyset,\chi,F)([[T]_s^{\emptyset}]) = N_{\chi,F}(\gamma_T)$$

and $N_{\chi,F}$ are the semi-norms of Krasiński defined on the (usual) homology group [4].

Hence, for complex tori we obtain

$$1 = 1-\alpha(\emptyset) = C_2(1,C_o,F(\mathfrak{v}))(\emptyset) = C_1(1,C_o,F(\mathfrak{v}))(\emptyset) = N(\emptyset,1,F(\mathfrak{v}))([[C_o]_s^{\emptyset}])$$

$$= N_{1,F(\mathfrak{v})}(\gamma_{C_o})$$

and, analogously

$$1 = 1 - \beta(\emptyset) = C_2(1, C_1, F(\mathfrak{U}))(\emptyset) = C_1(1, C_1, F(\mathfrak{U}))(\emptyset) =$$

$$= N(\emptyset, 1, F(\mathfrak{U}))([[C_1]_s^{\emptyset}]) = N_{1,F(\mathfrak{U})}(\gamma_{C_1}).$$

Thus we arrive at a result of Krasiński [5]:

$$N_{1,F(\mathfrak{U})}(\gamma_{C_0}) = N_{1,F(\mathfrak{U})}(\gamma_{C_1}) = 1$$

In the case of an annulus in the plane we have $C_2(1, C_0, F_1)(\emptyset) = 2\pi/\log a$. In the other hand

$$C_2(1, C_0, F_1)(\emptyset) = C_1(1, C_0, F_1)(\emptyset) = N(\emptyset, 1, F_1)([[C_0]_s^{\emptyset}])$$

$$= N_{1,F_1}(\gamma_{C_0}) = N(\gamma_{C_0}),$$

where N is a semi-norm defined by Chern, Levine and Nirenberg on $H_1(\mathbf{P}, \mathbf{Z})$, [3]. Thus we obtain a result of Chern, Levine and Nirenberg [3]: $N(\gamma_{C_0}) = 2\pi/\log a$.

8. The semi-norms and the (relative) extremal length

In this section by a slight modification of Definition 1 we shall define semi-norms of the form $N(A, \chi, F)$ which in the case of compact Riemann surfaces, have some relationship with the (relative) extremal length of the relative homology classes.

Definition 6. Let $\mathfrak{U} = \{U_i\}_{i \in I}$ be a locally finite open covering of a complex manifold M and $F(\mathfrak{U})$-a family of systems $u = \{u_i\}_{i \in I}$ of real C^{∞}-functions on M, such that

(i) $u_i : U_i \longrightarrow \mathbf{R}$,

(ii) $\sup \{|u_i(x) - u_i(y)| : x, y \in U_i\} < +\infty$,

(iii) $du_i = du_j$ in $U_i \cap U_j \neq \emptyset$.

(From (iii) we deduce that every $u \in F(\mathfrak{U})$ generates a global form on M, denoted by du and, similarly, the forms $d^c u$ and $dd^c u$). Let χ be a C^∞-form on M of degree (k,ℓ), $0 \le k \le n$, $0 \le \ell \le n$, $k+\ell < 2n$. We define the function $N(A,\chi,F(\mathfrak{U})): H^c_{2n-(k+\ell+1)}(M,A) \longrightarrow \mathbb{R}_+ \cup \{0\}$, by the formulae

$$N(A,\chi,F(\mathfrak{U}))(\Gamma) \overset{def}{=} \sup_{u \in F(\mathfrak{U})} \inf_{[T]^A_s \in \Gamma} [T]^A_s (d^c u \wedge \chi), \Gamma \in H^c_{2n-(k+\ell+1)}(M,A),$$

where A is an arbitrary open subset of M.

Now, let R be a compact Riemann surface. Then

$$H^c_1(R) \cong H_1(R) \cong H^{sing}_1(R),$$

where $H^{sing}_1(R)$ is the singular homology group, and

$$WH^c_1(R,A) \cong WH_1(R,A) \cong WH^{sing}_1(R,A)$$

for any open subset of R. Since there is a canonical isomorphism between $WH^c_1(R,A)$ and $WH^{sing}_1(R,A)$, then we can compare functions defined on $WH^c_1(R,A)$ with those on $WH^{sing}_1(R,A)$, respectively.

We are going to compare the semi-norms of the form $N(A,\chi,F(\mathfrak{U}))$, defined on $WH^c(R,A)$, where R is a Riemann surface, with the (relative) extremal length of the relative homology classes. This (relative) extremal length we define in the following way.

<u>Definition 7</u>. Let R be a Riemann surface and A-an arbitrary open subset of R. We define the function $\Lambda(A): H^{sing}_1(R,A) \longrightarrow \mathbb{R}_+ \cup \{0\}$ by the formulae

$$\Lambda^{\frac{1}{2}}(A)(\Gamma) \overset{def}{=} \sup_{Q} \inf_{[T]^A_s \in \Gamma} \Big[\inf_{\tilde{T} \in [T]^A_s} \int_{\tilde{T}} Q|dz| / \Big(\iint_R Q^2 dxdy\Big)^{\frac{1}{2}} \Big],$$

where $Q \ge 0$ ranges over all lower semicontinuous linear densities which are not identically zero [1].

Remark 1. By considerations of Section 3, for $A=\emptyset$ we have $\Gamma=[[T]_s^{\emptyset}]=\gamma_T$ and

$$\Lambda^{\frac{1}{2}}(\emptyset)(\Gamma)\overset{\text{def}}{=}\sup_Q \inf_{T'\in\gamma_T}\left[\int_{T'}Q|dz|/\left(\iint_R Q^2dxdy\right)^{\frac{1}{2}}\right]=\mathcal{L}^{\frac{1}{2}}(\gamma_T).$$

THEOREM 6. *Suppose, R is a compact Riemann surface without boundary. Then* $\Lambda^{1/2}(A)(\Gamma)$, $\Gamma\in WH_1^{sing}(R,A)$ *defines a semi-norm on the real vector space* $WH_1^{sing}(R,A)$, i.e. *we have*

(i) $0\le\Lambda^{1/2}(A)(\Gamma)<+\infty$ *for any* $\Gamma\in WH_1^{sing}(R,A)$,

(ii) $\Lambda^{1/2}(A)(a\Gamma)=|a|\Lambda^{1/2}(A)(\Gamma)$ *for any* $a\in\mathbb{R}$ *and* $\Gamma\in WH_1^{sing}(R,A)$,

(iii) $\Lambda^{1/2}(A)(\Gamma_1+\Gamma_2)\le\Lambda^{1/2}(A)(\Gamma_1)+\Lambda^{1/2}(A)(\Gamma_2)$ *for any pair* $\Gamma_1,\Gamma_2\in WH_1^{sing}(R,A)$,

for any open subset A *of* R.

Proof. Let A be an open subset of R and $\Gamma=[[T_o]_s^A]\in WH_1^{sing}(R,A)$. By the definition of $WH_1^{sing}(R,A)$ we may assume that T_o is a cycle. The inequality $\Lambda^{1/2}(A)(\Gamma)\ge 0$ is obvious. We shall prove that $\Lambda^{1/2}(A)(\Gamma)<+\infty$. Indeed, we have

$$\Lambda^{1/2}(A)(\Gamma)\overset{\text{def}}{=}\sup_Q \inf_{[T]_s^A\in\Gamma}\left[\inf_{\tilde T\in[T]_s^A}\int_{\tilde T}Q|dz|/\left(\iint_R Q^2dxdy\right)^{1/2}\right]$$

$$\le\sup_Q \inf_{[T_c]_s^A\in\Gamma}\left[\inf_{\tilde T\in[T_c]_s^A\cap\gamma_{T_c}}\int_{\tilde T}Q|dz|/\left(\iint_R Q^2dxdy\right)^{1/2}\right],$$

where the current T_c is cycle and γ_{T_c} is a (usual) homology class of T_c. Then, by the equality $\gamma_{T_c}=\gamma_{T_o}$ for $T_c\in\gamma_{T_o}$

$$\sup_Q \inf_{[T_c]_s^A\in\Gamma}\frac{\inf_{\tilde T\in[T_c]_s^A\cap\gamma_{T_c}}\int_{\tilde T}Q|dz|}{\left(\iint_R Q^2dxdy\right)^{1/2}}\le\sup_Q \inf_{[T_c]_s^A\in\Gamma\ T_c\in\gamma_{T_o}}\frac{\inf_{\tilde T\in[T_c]_s^A\cap\gamma_{T_o}}\int_{\tilde T}Q|dz|}{\left(\iint_R Q^2dxdy\right)^{1/2}}$$

$$\leq \sup_Q \inf_{[T_c]_s^A \in \Gamma,\ T_c \in \gamma_{T_0}} \frac{\int_{T_c} Q|dz|}{(\iint_R Q^2 dxdy)^{1/2}} .$$

Now, we need the following

<u>LEMMA 6. Let</u> $T \in Z_i^c(M)$ <u>and</u> $[T]_s^A \in Z_i^c(M,A)$, <u>where</u> A <u>is an open subset of</u> M. <u>Then for any</u> $T' \in \gamma_T$ <u>we have</u>

$$[T']_s^A \in [[T]_s^A]] .$$

<u>P r o o f</u>. Let $T' \in \gamma_T$. By the definition of the homology class γ_T there exists a current $S' \in D_{i+1}^{sing}(M)$ such that $T' - T = bS'$. Since $[T' - T - bS']_s^A = 0$, we have $[T']_s^A - [T]_s^A = b[S']_s^A$. Thus $b[T']_s^A = 0$ and, in consequence, $[T']_s^A \in [[T]_s^A]]$.

By LEMMA 6 we have

$$\sup_Q \inf_{[T_c]_s^A \in \Gamma_1 T_c \in \gamma_{T_0}} \frac{\int_{T_c} Q|dz|}{(\iint_R Q^2 dxdy)^{1/2}} = \sup_Q \inf_{T_c \in \gamma_{T_0}} \frac{\int_{T_c} Q|dz|}{(\iint_R Q^2 dxdy)^{1/2}} \stackrel{1}{=}$$

$$= \mathcal{L}^{\frac{1}{2}}(\gamma_{T_0})$$

Finally we get

$$(6) \qquad \Lambda^{1/2}(A)(\Gamma) = \Lambda^{1/2}(A)([[T_0]_s^A]) \leq \mathcal{L}^{1/2}(\gamma_{T_0}) < +\infty ,$$

because the extremal length $\mathcal{L}^{1/2}$ is a semi-norm on the one-dimensional homology group $H_1^{sing}(R)$.

To prove a condition (ii), let $a \in \mathbf{R}$ and $\Gamma \in WH_1^{sing}(R,A)$. Then we have

$$\Lambda(A)(a\Gamma) \overset{\text{def}}{=} \sup_Q \inf_{[T]_s^A \in a\Gamma} \frac{\left(\inf_{\tilde{T}\in[T]_s^A} \int_{\tilde{T}} Q|dz|\right)^2}{\iint_R Q^2 dxdy} = \sup_Q \inf_{[S]_s^A \in \Gamma} \frac{\left(\inf_{\tilde{T}\in a[S]_s^A} \int_{\tilde{T}} Q|dz|\right)^2}{\iint_R Q^2 dxdy}$$

$$= \sup_Q \inf_{[S]_s^A \in \Gamma} \frac{\left(\inf_{\tilde{S}\in[S]_s^A} \int_{a\tilde{S}} Q|dz|\right)^2}{\iint_R Q^2 dxdy} = \sup_Q \inf_{[S]_s^A \in \Gamma} \frac{\left(\inf_{\tilde{S}\in[S]_s^A} a\int_{\tilde{S}} Q|dz|\right)^2}{\iint_R Q^2 dxdy} = a^2\Lambda(A)(\Gamma).$$

The condition (iii) is satisfied as well. Indeed, let Γ_1, Γ_2 $WH_1^{sing}(R,A)$. Then we have

$$\Lambda^{1/2}(A)(\Gamma_1+\Gamma_2) \overset{\text{def}}{=} \sup_Q \inf_{[T]_s^A \in \Gamma_1+\Gamma_2} \frac{\inf_{\tilde{T}\in[T]_s^A} \int_{\tilde{T}} Q|dz|}{\left(\iint_R Q^2 dxdy\right)^{1/2}}$$

$$\leq \sup_Q \inf_{[T_1]_s^A \in \Gamma_1} \inf_{[T_2]_s^A \in \Gamma_2} \frac{\inf_{\tilde{T}\in[T_1]_s^A+[T_2]_s^A} \int_{\tilde{T}} Q|dz|}{\left(\iint_R Q^2 dxdy\right)^{1/2}} \leq \sup_Q \inf_{[T_1]_s^A \in \Gamma_1} \inf_{[T_2]_s^A \in \Gamma_2}$$

$$\frac{\inf_{\tilde{T}_1\in[T_1]_s^A} \inf_{\tilde{T}_2\in[T_2]_s^A} \int_{\tilde{T}_1+\tilde{T}_2} Q|dz|}{\left(\iint_R Q^2 dxdy\right)^{1/2}} = \sup_Q \Bigg[\inf_{[T_1]_s^A \in \Gamma_1} \frac{\inf_{\tilde{T}_1\in[T_1]_s^A} \int_{\tilde{T}_1} Q|dz|}{\left(\iint_R Q^2 dxdy\right)^{1/2}} +$$

$$+ \inf_{[T_2]_s^A \in \Gamma_2} \frac{\inf_{\tilde{T}_2\in[T_2]_s^A} \int_{T_2} Q|dz|}{\left(\iint_R Q^2 dxdy\right)^{1/2}} \Bigg] \leq \Lambda^{1/2}(A)(\Gamma_1) + \Lambda^{1/2}(A)(\Gamma_2).$$

Now, let $F_o(\mathfrak{V})$ denote the subfamily of $F(\mathfrak{V})$ consisting of systems $u=\{u_i\}_{i\in I}$ satisfying the following condition:

(iv) there exists a constant $C_F > 0$ such that for any $i \in I$ we have

$$\iint_{z_i(U_i)} \left|\frac{du_i}{dz_i}\right| dx_i dy_i \leq C_F,$$

where $z_i = x_i + iy_i$ is a local coordinate on U_i.

THEOREM 7. *Let* R *be a compact Riemann surface without boundary. Then there exist a finite open covering* $\mathfrak{U} = \{U_i, i = 1, \dots, k\}$ *of* R *consisting of simply connected sets and a constant* $C > 0$ *such that*

$$N(A, \chi, F_o(\mathfrak{U}))(\Gamma) \leq C \cdot \Lambda^{1/2}(A)(\Gamma), \quad \Gamma \in H_1^c(R, A) \cong H_1^{sing}(R, A)$$

for any open subset A *of* R.

P r o o f. We choose the covering $\mathfrak{U} = \{U_i; i = 1, \dots, k\}$ in the following way. Suppose that each U_i of $\mathfrak{U}$ is the domain of definition of some map. A local coordinate of this map is denoted by z_i, $i = 1, \dots, k$.

Now, let $\Gamma = [[T]_s^A] \in H_1^c(R, A) \cong H_1^{sing}(R, A)$, let C be a curve of $[T]_s^A \in \Gamma$ and let $u \in F_o(\mathfrak{U})$. We define a linear density g^u on R in the following way:

$$g^u \stackrel{\text{def}}{=} \{g_i^u; \ i = 1, \dots, k\}, \quad \text{where} \quad g_i^u = \left|\frac{du_i}{dz_i}\right| \quad \text{in } U_i.$$

It is easy to see that

$$g_i^u = \left|\frac{du_i}{dz_i}\right| = \left|\frac{du_i}{dz_j} \frac{dz_j}{dz_i}\right| = \left|\frac{du_j}{dz_j}\right| \left|\frac{dz_j}{dz_i}\right| = g_j^u \left|\frac{dz_j}{dz_i}\right|.$$

For $\mathfrak{U}$ and g^u defined above we have

$$(7) \qquad \int_C |d^c u| \leq \int_C g^u.$$

Indeed, the curve C can be written as $C_1 + \dots + C_p$, where C_s, $s = 1, \dots, p$, denotes a curve in some $U_{i(s)}$, with the parametrization

$$[0,1) \ni t \longrightarrow z_{i(s)}(t) = x_{i(s)}(t) + iy_{i(s)}(t) \in U_{i(s)},$$

and the functions $x_{i(s)}$, $y_{i(s)}$ have their derivatives almost every where, which are integrable. Then we have

$$\int_C |d^c u| = \sum_{s=1}^{p} \int_{C_s} |d^c u_{i(s)}| = \sum_{s=1}^{p} \int_0^1 \left| -\frac{\partial u_{i(s)}}{\partial y_{i(s)}} x'_{i(s)} + \frac{\partial u_{i(s)}}{\partial x_{i(s)}} y'_{i(s)} \right| dt$$

$$\leq \sum_{s=1}^{p} \int_0^1 \left| \frac{\partial u_{i(s)}}{\partial x_{i(s)}} x'_{i(s)} + \frac{\partial u_{i(s)}}{\partial y_{i(s)}} y'_{i(s)} + i\left(- \frac{\partial u_{i(s)}}{\partial y_{i(s)}} x'_{i(s)} + \frac{\partial u_{i(s)}}{\partial x_{i(s)}} y'_{i(s)}\right)\right| dt$$

$$\leq \sum_{s=1}^{p} \int_0^1 \left|\frac{du_{i(s)}}{dz_{i(s)}}\right| |z'_{i(s)}| dt = \sum_{s=1}^{p} \int_{C_s} g^u_{i(s)} |dz_{i(s)}| = \int_C g^u$$

For any $u \in F_o(\mathfrak{V})$ this implies

$$\left|\int_C d^c u \wedge \chi\right| \leq \int_C |d^c u| \; |\chi| \leq \max_{p \in R} |\chi(p)| \int_C |d^c u| \leq C_f \int_C g^u$$

where $C_f = \max_{p \in R} |\chi(p)|$.

Now, let $[T]^A_s \in \Gamma$. By (7) and by the definition of $F_o(\mathfrak{V})$ we get

$$\inf_{\tilde{T} \in [T]^A_s} \left|\int_{\tilde{T}} d^c u \wedge \chi\right| \leq C_f \inf_{\tilde{T} \in [T]^A_s} \int_{\tilde{T}} g^u \leq C_f \frac{\inf_{\tilde{T} \in [T]^A_s} \int_{\tilde{T}} g^u}{(\iint_R (g^u)^2 dxdy)^{1/2}} (\iint_R (g^u)^2 dxdy)^{1/2}$$

$$\leq C_f C_F \left[\inf_{\tilde{T} \in [T]^A_s} \int_{\tilde{T}} g^u / (\iint_R (g^u)^2 dxdy)^{1/2}\right]$$

Thus

$$\inf_{[T]^A_s \in \Gamma} \inf_{\tilde{T} \in [T]^A_s} \left|\int_{\tilde{T}} d^c u \wedge \chi\right| \leq \inf_{[T]^A_s \in \Gamma} \inf_{\tilde{T} \in [T]^A_s} \int_{\tilde{T}} |d^c u| \cdot |\chi|$$

$$\leq C_f C_F \inf_{[T]^A_s \in \Gamma} \left[\inf_{\tilde{T} \in [T]^A_s} \int_{\tilde{T}} g^u / (\iint_R Q^2 dxdy)^{1/2}\right] \leq C_f C_F \sup_Q \inf_{[T]^A_s \in \Gamma}$$

$$\times \left[\inf_{\tilde{T} \in [T]^A_s} \int_{\tilde{T}} Q / (\iint_R Q^2 dxdy)^{1/2}\right] = C \, \Lambda^{1/2}(A)(\Gamma),$$

where $C = C_f C_F$. The above inequalities are fulfiled for any $u \in F_o(\mathfrak{V})$, so

$$N(A,\chi,F_o(\mathfrak{V}))(\Gamma) \overset{\text{def}}{=} \sup_{u\in F_o(\mathfrak{V})} \ \inf_{[T]_s^A\in\Gamma} \ \inf_{\tilde{T}\in[T]_s^A} \left|\int_{\tilde{T}} d^c u\wedge\chi\right| \leq C\cdot\Lambda^{1/2}(A)(\Gamma).$$

Thus the proof of Theorem 7 is completed.

THEOREM 8. *For any open finite coverings $\mathcal{B}_1$ and $\mathcal{B}_2$ of a complex manifold M consisting of simply connected sets, we have*

$$N(A,\chi,F(\mathcal{B}_1))(\Gamma) = N(A,\chi,F(\mathcal{B}_2))(\Gamma), \quad \Gamma \in H^c_{2n-(k+\ell+1)}(M,A),$$

where A is an arbitrary open subset of M and χ is as in Definition 6.

Proof. Let $\mathcal{B}_1 = \{U_i;\ i=1,\dots,m\}$ and $\mathcal{B}_2 = \{V_j; j=1,\dots,r\}$. We construct a covering $\mathcal{C}$ of M in the following way:

$$\mathcal{C} = \{W_{ij};\ i=1,\dots,m,\ j=1,\dots,r\}$$

and $W_{ij} = U_i \cap V_j$. We shall prove that

$$N(A,\chi,F(\mathcal{B}_1))(\Gamma) = N(A,\chi,F(\mathcal{C}))(\Gamma),$$

$$N(A,\chi,F(\mathcal{B}_2))(\Gamma) = N(A,\chi,F(\mathcal{C}))(\Gamma), \Gamma \in H^c_{2n-(k+\ell+1)}(M,A)$$

for an arbitrary open subset A of M.

Let $A \in O(M)$. First, we obseve that the covering $\mathcal{C}$ is a subcovering of $\mathcal{B}_1$, so $F(\mathcal{B}_1) \subset F(\mathcal{C})$. Hence

$$N(A,\chi,F(\mathcal{B}_1))(\Gamma) \leq N(A,\chi,F(\mathcal{C}))(\Gamma), \ \Gamma \in H^c_{2n-(k+\ell+1)}(M.A). \tag{8}$$

To prove an inequality

$$N(A,\chi,F(\mathcal{C}))(\Gamma) \leq N(A,\chi,F(\mathcal{B}_1))(\Gamma), \quad \Gamma \in H^c_{2n-(k+\ell+1)}(M.A.)$$

we observe that for any $u \in F(\mathcal{C})$, there exists an element $\tilde{u} \in F(\mathcal{B}_1)$

such that $d^c u = d^c \tilde{u}$.

We construct this element in the following way. Let $u = \{u_{ij};\ i = 1,\dots,m,\ j = 1,\dots,r\}$ be an element of $F(\mathfrak{C})$. For a fixed $U_{i_*} \in \mathfrak{B}_1$, $i_* \in \{1,\dots,m\}$, the sets W_{i_*j}, $j = 1,\dots,r$, form a covering of U_{i_*}. The functions u_{i_*j}, $j = 1,\dots,r$, generate a form on U_{i_*}, which will be denoted by du_i. This form is closed since in each W_{i_*j} we have $du_i = du_{i_*j}$. The set U_{i_*} is simply connected, so there is a C^∞-function f_{i_*} on U_{i_*} such that $df_{i_*} = du_i$ and $d^c f_{i_*} = d^c u_i$. On each of sets W_{i_*j} we have $u_{i_*j}(x) - f_{i_*}(x) = c_{i_*j}$, where c_{i_*j} is a constant. Thus

$$\operatorname*{osc}_{W_{i_*j}} f_{i_*} \stackrel{\text{def}}{=} \sup_{x,y\in W_{i_*j}} |f_{i_*}(x) - f_{i_*}(y)| = \sup_{x,y\in W_{i_*j}} |[u_{i_*j}(x) - c_{i_*j}] -$$

$$-[u_{i_*j}(y) - c_{i_*j}]\,| = \sup_{x,y\in W_{i_*j}} |u_{i_*j}(x) - u_{i_*j}(y)| = \operatorname*{osc}_{W_{i_*j}} u_{i_*j} < +\infty$$

Hence

$$\operatorname*{osc}_{U_{i_*}} f_{i_*} < +\infty .$$

Now we set $\tilde{u}_{i*} = f_{i_*}$. It is a C^∞-function and the oscillation of $\tilde{u}_{i*}$ on U_{i*} is finite.

We repeat this construction for any $i = 1,\dots,m$. Finally we obtain an element $\tilde{u} = \{\tilde{u}_i;\ i = 1,\dots,m\}$ of $F(\mathfrak{B}_1)$ because on each set $U_{i_1} \cap U_{i_2} \neq \emptyset$ we have

$$d\tilde{u}_{i_1} = d(f_{i_1*}) = du_{i_1} = du_{i_2} = d(f_{i_2*}) = d\tilde{u}_{i_2}.$$

By the construction of u we get $d^c u = d^c \tilde{u}$, as desired.

Then for any $u \in F(\mathfrak{C})$ we have

$$\inf_{[T]_s^A \in \Gamma} [T]_s^A (d^c u \wedge \chi) = \inf_{[T]_s^A \in \Gamma} [T]_s^A (d^c \tilde{u} \wedge \chi),\ \Gamma \in H^c_{2n-(k+\ell+1)}(M,A).$$

and next

$$\sup_{u\in F(\mathfrak{C})} \inf_{[T]_s^A\in\Gamma} [T]_s^A(d^c u\wedge\chi) = \sup_{u\in F(\mathfrak{C})} \inf_{[T]_s^A\in\Gamma} [T]_s^A(d^c\tilde{u}\wedge\chi)$$

$$\le \sup_{\tilde{u}\in F(\mathfrak{B}_1)} \inf_{[T]_s^A\in\Gamma} [T]_s^A(d^c\tilde{u}\wedge\chi)$$

for any $\Gamma\in H^c_{2n-(k+\ell+1)}(M,A)$. Hence

$$N(A,\chi,F(\mathfrak{C}))(\Gamma) \le N(A,\chi,F(\mathfrak{B}_1))(\Gamma),\ \Gamma\in H^c_{2n-(k+\ell+1)}(M.A).$$

By (8) we obtain

$$N(A,\chi,F(\mathfrak{C}))(\Gamma) = N(A,\chi,F(\mathfrak{B}_1))(\Gamma),\ \Gamma\in H^c_{2n-(k+\ell+1)}(M.A).$$

By analogous considerations, we have

$$N(A,\chi,F(\mathfrak{C}))(\Gamma) = N(A,\chi,F(\mathfrak{B}_2)(\Gamma),\ \Gamma\in H^c_{2n-(k+\ell+1)}(M.A).$$

Thus the proof of Theorem 8 is completed.

COROLLARY 5. Let R be a compact Riemann surface without boundary. Then for any open finite covering $\mathfrak{B}$ of R consisting of simply connected sets, we have

$$N(A,\chi,F_o(\mathfrak{B}))(\Gamma) < +\infty\ ,\Gamma\in WH^c_1(R.A),$$

where A is an arbitrary open subset of R and χ is as in Definition 6.

P r o o f. By (6) and Theorems 7 and 8 it is obvious.

THEOREM 9. Let R be a compact Riemann surface without boundary and $\mathfrak{B}$ be an arbitrary open finite covering of R consisting of simply connected sets. Then $N(A,\chi,F_o(\mathfrak{B}))(\Gamma)$, $\Gamma\in WH^c_1(R,A)\cong WH^{sing}_1(R,A)$, is a semi-norm, i.e.,

(i) $0 \leq N(A,\chi,F_o(\mathcal{B}))(\Gamma) < +\infty$ for any $\Gamma \in WH_1^c(R,A)$;

(ii) $N(A,\chi,F_o(\mathcal{B}))(a\Gamma) = |a|\, N(A,\chi,F_o(\mathcal{B}))(\Gamma)$ for any $a \in \mathbb{R}$ and $\Gamma \in WH_1^c(R,A)$;

(iii) $N(A,\chi,F_o(\mathcal{B}))(\Gamma_1 + \Gamma_2) \leq N(A,\chi,F_o(\mathcal{B}))(\Gamma_1) + N(A,\chi,F_o(\mathcal{B}))(\Gamma_2)$ for any pair $\Gamma_1, \Gamma_2 \in WH_1^c(R,A)$.

for any open subset A of R and χ is as in Definition 6.

<u>Proof.</u> Let $A \in O(R)$, $\Gamma \in WH_1^c(R,A)$ and $\mathcal{B}$ be an open finite covering of R. The inequality $N(A,\chi,F_o(\mathcal{B}))(\Gamma) \geq 0$ is obvious. By Corollary 5 we have $N(A,\chi,F_o(\mathcal{B}))(\Gamma) < +\infty$. The conditions (ii) and (iii) can be prove in analogy to Theorem 2.

In the end we give another relation to Krasiński's semi-norms [4].

<u>Remark</u> 2. Suppose R is a compact Riemann surface without boundary. Let $\mathcal{B}$ be an open finite covering of R consisting of simply connected sets and $A \subset R$ be an arbitrary open subset of R. Let χ be a C^∞-function on R such that $\operatorname{supp} \chi \subset R \setminus A$. Denote by $F_o(\chi,\mathcal{B})$ the subfamily of $F_o(\mathcal{B})$ satisfying the additional condition:

(v) for any $i \in I$ we have $d[d^c v_i \wedge \chi] = 0$ on U_i, where $\{U_i\}_{i\in I} = \mathcal{B}$

Then, for any $\Gamma \in WH_1^c(R,A)$ we have

$$N(A,\chi,F_o(\chi,\mathcal{B}))(\Gamma) = N_{\chi,F_o(\chi,\mathcal{B})}(\gamma_T),$$

where $N_{\chi,F_o(\chi,\mathcal{B})}$ is a semi-norm in the sense of Krasiński over $H_1^c(R) \cong H_1^{sing}(R)$, and T is an arbitrary cycle on R such that $[T]_s^A \in \Gamma$.

<u>Proof.</u> Let $A \in O(R)$ and $\Gamma \in WH_1^c(R,A)$. By (v) we have

$$N(A,\chi,F_o(\chi,\mathcal{B}))(\Gamma) \overset{def}{=} \sup_{v\in F_o(\chi,\mathcal{B})} \inf_{[T]_s^A \in \Gamma} \inf_{T' \in [T]_s^A} |T'(d^c v \wedge \chi)|$$

$$= \sup\{[T]_s^A(d^c v \wedge \chi) : v \in F_o(\chi,\mathcal{B})\},$$

where $[T]_s^A$ is an arbitrary relative cycle of Γ. Now, owing to the fact that supp $(d^c u \wedge \chi) \subset R \setminus A$ for any $v \in F_0(\chi, \mathcal{B})$, we get

$$\sup\{[T]_s^A(d^c v \wedge \chi);\ v \in F_0(\chi, \mathcal{B})\} = \sup\{|T(d^c v \wedge \chi)| : v \in F_0(\chi, \mathcal{B})\},$$

where T is an arbitrary current on R from $[T]_s^A$. Thus we obtain

$$N(A, \chi, F_0(\chi, \mathcal{B}))(\Gamma) = \sup\{|T(d^c v \wedge \chi)| : v \in F_0(\chi, \mathcal{B})\},$$

where T is an arbitrary cycle on R such that $[T]_s^A \in \Gamma$. On the other hand, by (v), we have

$$\sup_{v \in F_0(\chi, \mathcal{B})} |T(d^c v \wedge \chi)| = \sup_{v \in F_0(\chi, \mathcal{B})} \inf_{T' \in \gamma_T} |T'(d^c v \wedge \chi)| = N_{\chi, F_0(\chi, \mathcal{B})}(\gamma_T).$$

Hence, the desired equality follows.

Finally, let us remark that if in Theorem 7 we put $A = \emptyset$ and $\chi = a = \text{const}$, then, by the considerations of Section 3 and by Remark 1, we get.

$$N_{a, F_0(a, \mathcal{B})}(\gamma) \le C \cdot \mathcal{L}^{1/2}(\gamma), \quad \gamma \in H_1^c(R),$$

where C is a constant independent of γ. This corresponds only to one part of the result of Krasiński [5] that the semi-norms $N_{a, F_0(a, \mathcal{B})}$ and $\mathcal{L}^{1/2}$ defined over $H_1^c(R)$ are equivalent. We belive that also the inequality

$$\Lambda^{1/2}(A)(\Gamma) \le C \cdot N(A, \chi, F(\mathcal{B}))(\Gamma)$$

is true for some $C > 0, \chi, \mathcal{B}$ and $F(\mathcal{B})$ but actually we have no proof of it.

R e f e r e n c e s

[1] ACCOLA, R.D.M.: Differentials and extremal length on Riemann, surfaces, Proc. Nat. Acad. Sci. USA 46(1960), pp. 540-543.

[2] AHLFORS, L.V. and L.SARIO: Riemann surfaces, Princeton Universtiy Press, Princeton 1960.

[3] CHERN, S.S., H.I.LEVINE and L.NIRENBERG: Intrinsic norms on complex manifolds, Global Analysis, Papers in honor of K.Kodaira, ed. by D.C. Spencer and S.Iynaga, Univ. of Tokyo Press and Princeton Univ. Press, Tokyo 1969, pp. 119-139.

[4] KRASIŃSKI, T.: Semi-norms on homology groups of complex manifolds, Proc. of the International Conference of Complex Analysis and Applications, Varna 1981 (in print).

[5] ——: On biholomorphic invariants related to homology groups, these Prodeedings, pp.

[6] KRÓLIKOWSKI, W.: Semi-norms and topologies on relative homology groups, to appear.

[7] RHAM, G. de: Variétés différentiables. Formes, courants, formes harmoniques, 2-ème éd., Hermann, Paris 1960.

[8] STOLL, W.: Value distribution of holomorphic maps into compact complex manifolds (Lecture Notes in Mathematics 135), Springer--Verlag, Berlin-Heidelberg-New York 1970.

Institute of Mathematics of the
Polish Academy of Sciences, Łódź Branch
Kilińskiego 86, PL-90-012 Łódź, Poland

CLOSE-TO-CONVEX FUNCTIONS WITH QUASICONFORMAL EXTENSION

Jan G. Krzyż (Lublin) and
Anil K. Soni (Bowling Green, OH)

Contents page

Summary

A sufficient condition for a close-to-convex function is obtained for the existence of a quasiconformal extension. The authors give also examples showing that this condition cannot be weakened essentially. Finally, an explicit construction of quasiconformal extension is given.

1. Introduction

Close-to-convex functions introduced independently and in a different way by Biernacki [3] and Kaplan [5]; also cf. [8], belong to the most important special conformal mappings. A function f regular in the unit disk $\mathbb{D}$ is said to be close-to-convex if there exists a $1:1$ conformal mapping g onto a convex domain G such that

(1) $f'(z)/g'(z) = p(z)$, $\operatorname{Re} p(z) > 0$ in $\mathbb{D}$.

This condition implies univalence of f in $\mathbb{D}$. If we put

(2) $h(w) = f \circ g^{-1}(w)$, $w \in G$

then obviously $h'(w) = p \circ g^{-1}(w) = q(w)$ is a function of positive real part in G. Thus a close-to-convex function f may be identified with a function h regular in a convex domain G whose derivative is a function of positive real part in G.

In the last twenty years, many papers have been published on functions regular and univalent in $\mathbb{D}$ which admit a quasiconformal extension to the whole plane. For a survey of this theory, see e.g. [2], or [6].

A natural question arises in connection with this, i.e., to characterize close-to-convex functions that admit a quasiconformal extension. In the next section we shall obtain a sufficient condition and give examples showing that this condition cannot be weakened essentially. The relevent results are due to both authors. In Section 3 (due to the first named author) an explicit construction of quasiconformal extension is given.

This research was done while the first named author held a visiting professorship at the Bowling Green State University and at the University of Delaware. The authors wish to thank Professors H. Al-Amiri and R.J.Libera for helpful discussions and criticism.

2. A sufficient condition for the existence of a quasiconformal extension

We shall now prove the following:

THEOREM 1. *A sufficient condition for a close-to-convex function f to have a quasiconformal extension to the whole plane is:*

1° *the associated convex univalent function g is bounded and*

2° *the set of values of the associated function p of positive real part is contained in a compact subset F of the open right half-plane.*

Proof. It is sufficient to show that f has a homeomorphic extension to the closed unit disk $\overline{\mathbb{D}}$ and $f(\partial\mathbb{D})$ is a quasicircle.

To this end, we consider the function $h = f \circ g^{-1}$ in a bounded convex domain G. By our assumptions $h'(w) = p \circ g^{-1}(w) = q(w)$ satisfies the following condition: there exist positive numbers m,M such that

$$(3) \qquad m \leq \operatorname{Re} q(w) \quad \text{and} \quad |q(w)| \leq M, \quad w \in G.$$

Since G is convex, we have

$$h(w_2)-h(w_1)=\int_{[w_1,w_2]} h'(w)\,dw=(w_2-w_1)\int_0^1 h'[w_1+t(w_2-w_1)]dt,$$

or

$$\frac{h(w_2)-h(w_1)}{w_2-w_1}=\int_0^1 h'[w_1+t(w_2-w_1)]dt. \tag{4}$$

From (3) and (4) it follows that

$$|h(w_2)-h(w_1)|/|w_2-w_1| \geq \operatorname{Re}\frac{h(w_2)-h(w_1)}{w_2-w_1} \geq m, \text{ i.e.}$$

$$|h(w_2)-h(w_1)| \geq m|w_2-w_1|.$$

Similarly

$$|h(w_2)-h(w_1)|/|w_2-w_1| \leq \int_0^1 |h'[w_1+t(w_2-w_1)]|dt \leq M,$$

so that finally

$$m|w_2-w_1| \leq |h(w_2)-h(w_1)| \leq M|w_2-w_1|. \tag{5}$$

From (5) it readily follows that h is univalent in G and has a homeomorphic extension on its closure $\overline{G}$ which also satisfies (5).

It is well known that the boundary, ∂G, of a bounded convex domain is a quasicircle, see e.g. [4].

According to Ahlfors [1], a Jordan curve J is a quasicircle if and only if

$$K(J)=\sup\frac{|w_1-w_2||w_3-w_4|+|w_1-w_4||w_2-w_3|}{|w_1-w_3||w_2-w_4|} \tag{6}$$

is finite, sup being taken over the set of all ordered quadruples w_1, w_2, w_3, w_4 of points on J. Now, since h is a homeomorphism on $\overline{G}$, $J_1 = h(\partial G)$ is also a Jordan curve and from (5) and (6) it follows immediately that

$$K(J_1) \leq (M^2/m^2)K(\partial G). \tag{7}$$

This proves that $K(J_1)$ is finite, i.e. J_1 is a quasicircle.

Consequently, h has a quasiconformal extension h^* to the whole plane. Now, a quasiconformal extension f^* of a close-to-convex function f satisfying 1° and 2° may be obtained as the composition $f^* = h^* \circ g^*$, where g^* is a quasiconformal extension of the bounded univalent function g. A construction of g^* was given in [4]. This ends the proof of Theorem 1.

It seems that the sufficient conditions 1° and 2° cannot be weakened essentially. In fact, the function $f(z) = -\log(1-z)$ is convex, unbounded and does not admit a quasiconformal extension since $f(\partial \mathbb{D})$ has a cusp with a zero angle at ∞. With $p(z) \equiv 1$ and $g = f$ we obtain a close-to-convex function f such that the associated functions p,q satisfy 2°, but not 1°. For the same f we can take $p(z) = (1-z)^{-1}$, $g(z) = z$ and now the associated functions satisfy 1°, but not 2°.

3. Construction of the quasiconformal extension

In view of the final remarks in the proof of Theorem 1, we need to construct a quasiconformal extension of the function h whose derivative has positive real part in a bounded convex domain G.

To this end, we shall first discuss some properties of bounded convex domains. The boundary, ∂G, of a bounded convex domain is a rectifiable Jordan curve which has a tangent vector a.e. The argument of the tangent vector is an increasing function of the arc length on ∂G whose total increment is 2π. Therefore, we may order the discontinuities of the tangent vector according to decreasing jumps of its argument and we obtain in this way the sequence $\{\tau_n\}$ of discontinuity points on ∂G. At the remaining points of ∂G both the tangent vector and the outward pointing normal exist. An outward pointing normal can also be defined at a discontinuity point τ_k as an outward pointing ray emanating from τ_k and contained inside the angle whose sides are outward pointing rays emanating from τ_k and orthogonal to one-sided tangents at τ_k.

We now introduce the function $\tau : \mathbb{C} \setminus G \to \partial G$ defined as follows: for a fixed $w \in \mathbb{C} \setminus G$, $\tau = \tau(w)$ is the unique point $\tau \in \partial G$, where $|\tau - w|$ attains its minimum as τ is ranging over $\overline{G}$. Obviously $\tau(w) \in \partial G$ and it is uniquely determined. In fact, if the minimum for some w is attained at τ_1 and τ_2, $\tau_1 \neq \tau_2$, then $\tau_o = \frac{1}{2}(\tau_1 + \tau_2) \in \overline{G}$ and $|w - \tau_o| < |w - \tau_1| = |w - \tau_2|$ which is a contradiction.

Moreover, $\tau(w)$ is continuous and even Lipschitzian. To prove this, take two points w_1, w_2 outside $\overline{G}$ and put $\tau_k = \tau(w_k)$, $k = 1,2$. Obviously, w_k lies on an outward pointing normal (in the sense given above) at τ_k. The outward pointing normals at τ_1, τ_2 lie outside the strip bounded by two parallel lined through τ_1, τ_2 which are perpendicular to the segment $[\tau_1, \tau_2]$. This implies that $|w_1 - w_2| \geq |\tau_1 - \tau_2|$.

We use the function $\tau(w)$ in the construction of a quasiconformal extension. We have namely

THEOREM 2. *Suppose G is a bounded convex domain and h is a function regular in G such that the values of h' are contained in a compact subset F of the open right half-plane.*

Let $w \in \mathbb{C} \setminus \overline{G}$ and let $\tau = \tau(w)$ be the unique point on ∂G where $|w-\tau|$ attains its minimum as τ ranges over $\overline{G}$. Then

$$(8) \qquad h^*(w) = \begin{cases} h(w), & w \in \overline{G} \\ h(\tau(w)) + w - \tau(w), & w \in \mathbb{C} \setminus \overline{G} \end{cases}$$

is a quasiconformal self-mapping of $\mathbb{C}$ whose complex dilatation $\mu(w)$ is bounded in absolute value by

$$(9) \qquad \mu_0 = \sup_{p \in F} \left| \frac{p-1}{p+1} \right| < 1.$$

P r o o f. As we know from Section 2, h is univalent in G and has a homeomorphic extension on $\overline{G}$ so the mapping (8) is well defined and continuous on $\mathbb{C}$. We shall approximate h^* by a sequence $\{H_n\}$ of quasiconformal mappings obtained as follows:

Let $\{G_n\}$ be a sequence of closed, convex polygons inscribed in ∂G which is uniformly convergent to $\overline{G}$, i.e. for any $\varepsilon > 0$ there exists n_0 such that all ∂G_n with $n > n_0$ are contained in the ε-neighborhood of ∂G. The mappings H_n are defined by the formula

$$(10) \qquad H_n(w) = \begin{cases} h(w), & w \in G_n \\ h[\tau_n(w)] + w - \tau_n(w), & w \in \mathbb{C} \setminus G_n, \end{cases}$$

where $\tau_n(w)$ is the unique point of G_n whose distance from w is a minimum.

It is easily verified that $h^*(w) = \lim H_n(w)$ for any fixed w.

We may assume that each opes side $\ell_n^{(k)}$ of G_n is contained inside G, whereas its end-points lie on ∂G. To each vertex $\tau_n^{(k)}$ of the polygonal line G_n there corresponds two outward pointing rays orthogonal to the adjacent sides with common end-point $\tau_n^{(k)}$. These rays subdivide $\mathbb{C} \setminus G_n$ into a finite number of non-overlapping domains of two kinds: angular domains $A_n^{(k)}$ whose boundaries consist of two rays emanating from $\tau_n^{(k)}$ and halfstrips $B_n^{(k)}$ whose boundaries consist of the side $\ell_n^{(k)}$ and two rays emanating from its end-points.

If w ranges over $A_n^{(k)}$, then $\tau_n(w) = \tau_n^{(k)} = \text{const}$ and consequently $H_n(w) = w + \text{const}$, i.e. H_n is $1:1$ conformal in each $A_n^{(k)}$.

Suppose now that w ranges over $B_n^{(k)}$. We may assume that the real axis Ou in the w-plane ($w=u+iv$) conincides with the side $\ell_n^{(k)}$ of $B_n^{(k)}$. As we know from section 2, $h(\ell_n^{(k)})$ is a simple, rectifiable (and even analytic) arc γ in the ζ-plane, where $\zeta = H_n(w)$. For $w = u + iv \in B_n^{(k)}$ we have $\tau_n(w) = u$ and from (10) we obtain

$$(11) \qquad \zeta = H_n(w) = h(u) + iv, \quad w = u + iv \in B_n^{(k)}.$$

We have $\operatorname{Re} h'(u) > 0$ for $u \in \ell_n^{(k)}$ which means that the local rotation of an infinitesimal segment on $\ell_n^{(k)}$ is less than $\frac{\pi}{2}$. Consequently, each line parallel to the imaginary axis in the ζ-plane intersects the arc $\gamma = h(\ell_n^{(k)})$ at at most one point. Note the each point of $\ell_n^{(k)}$ lies inside G, thus $h'(u)$ makes sense for $u \in \ell_n^{(k)}$. From (11) we see that the image curves of vertical rays in $B_n^{(k)}$ are rays emanating from the points of γ and parallel to the imaginary axis in the ζ-plane. Moreover, the image curves of horizontal segments in $B_n^{(k)}$ are arcs γ shifted by v in the direction of the imaginary axis. This implies that H_n is a homeomorphism on each half-strip $B_n^{(k)}$.

From (11) we can easily evaluate the complex dilatation μ_n of H_n in $B_n^{(k)}$. Using the notation $\partial H_n/\partial w = H_{n|w}$ etc., we have

(12) $$\mu_n(w) = \frac{H_{n|\bar{w}}}{H_{n|w}} = \frac{H_{n|u} + iH_{n|v}}{H_{n|u} - iH_{n|v}} = \frac{h'(u)-1}{h'(u)+1} .$$

From (12) we obtain

(13) $$|\mu_n(w)| \le \mu_0 = \sup_{p\in F} \left|\frac{p-1}{p+1}\right| < 1, \text{ a.e.}$$

since the values of $h' = p$ lie in a compact subset F of the open right half-plane.

The bound for $|\mu_n(w)|$ does not change under motion, hence the bound (13) is valid for all $B_n^{(k)}$.

If we assume $H_n(\infty) = \infty$, H_n becomes a local homeomorphism of the extended plane $\mathbb{C}$. Therefore, it is also a global homeomorphism whose complex dilatation is bounded a.e. by $\mu_0 < 1$ in absolute value. Thus, H_n is a quasiconformal self-mapping of $\mathbb{C}$.

It is well-known that the limit of a convergent sequence of K-quasiconformal mappings is either a constant, a mapping onto two points, or a K-quasiconformal mapping; cf. [6]. This implies that $h^* = \lim H_n$ is a quasiconformal self-mapping of $\mathbb{C}$ whose complex dilatation is also bounded a.e. by μ_0 in absolute value which ends the proof of Theorem 2.

Remark 1. It follows from our construction of H_n that in the presence of corner points τ on ∂G, h^* is conformal in any angular domain whose sides are outward pointing rays emanating from τ and orthogonal to the one-sided tangents at τ. This could be also deduced directly from (8).

Remark 2. As stated in section 2, a quasiconformal extension f^* of a close-to-convex function f may be obtained as a composition $f^* = h^* \circ g^*$. Since a construction of g^* has been given in [4], formula (8) permits us to construct f^* explicitly. Suppose that G is contained in a concentric annulus with radii r and R $(r < R)$. As shown in [4], the complex dilatation of g^* has a bound $[1-(r/R)^2]^{1/2}$. Hence the maximal dilatation of $f^* = h^* \circ g^*$ can be easily estimated in view of (9).

Remark 3. Taking $h^*(w) = h(\tau(w)) + C(w-\tau(w))$, with suitably chosen $C > 0$ instead of $C = 1$ as in (8), one could possibly improve the estimate (9) for some special functions h.

References

[1] AHLFORS, L. V.: Quasiconformal reflections, Acta Math. 109 (1963) 291-301.

[2] BECKER, J.: Conformal mappings with quasiconformal extensions. Aspects of contemporary complex analysis (Proc. NATO Adv. Study Inst., Univ. Durham, Durham 1979), Academic Press, London 1980, pp.37-77.

[3] BIERNACKI, M.: Sur la representation conforme des domaines linéairement accessibles, Prace Mat.-Fiz. 44(1936), 293-314.

[4] FAIT, M., J.KRZYŻ and J.ZYGMUNT,: Explicit quasiconformal extensions for some classes of univalent functions, Comment, Math. Helv. 51 (1976), 279-285.

[5] KAPLAN, W.: Close-to-convex schlicht functions, Michigan Math.J. 1 (1952), 169-185.

[6] LEHTO, O.: Conformal mappings and Teichmüller spaces, Technion Lecture Notes, Israel Institute of Technology, Haifa 1973.

[7] —— and K.VIRTANEN,: Quasiconformal mappings in the plane, Springer-Verlag, Berlin-Heidelberg-New York 1973.

[8] LEWANDOWSKI, Z.: Sur l'idéntité de certaines classes de fonctions univalentes I-II, Ann. Univ. Mariae Curie-Skłodowska 12 (1958), 131-146; 14 (1960), 19-46.

Institute of Mathematics
Marie Curie-Skłodowska University
Plac Marii Curie-Skłodowskiej 1
PL-20-031 Lublin, Poland
(J.G.Krzyż)

Department of Mathematics
and Statistics
Bowling Green State University
Bowling Green, OH 43403, U.S.A.
(A.K.Soni)

UN LIEN ENTRE L'ÉQUATION DE L'ÉLASTICITÉ ET L'ANALYSE COMPLEXE

Guy Laville (Paris)

Résumé. Il est possible de factoriser le bilaplacien Δ^2 à travers le système d'équations bien connu en analyse complexe: $\partial^2 u/\partial z_i \partial \bar{z}_j = \alpha_{ij}$ en utilisant non pas le calcul différentiel extérieur mais les algèbres de Clifford.

Abstract. The splitting of the bilaplacian Δ^2 is possible by combining twice the well known system of complex analysis $\partial^2 u/\partial z_i \partial \bar{z}_j = \alpha_{ij}$. Here we do not use exterior differential calculus but Clifford algebras.

1. Introduction. Il est connu que la recherche d'une fonction holomorphe de plusieurs variables complexes f telle que $f^{-1}(0)$ soit un ensemble analytique de codimension 1 donné, se ramène à l'étude du système d'équations

(1) $\partial^2 u/\partial z_i \partial \bar{z}_j = \alpha_{ij}$

où les α_{ij} sont des distributions données qui satisfont à la condition de compatibilité. Une méthode consiste à prendre la trace du système (1):

$$\Delta u = \sum_i \alpha_{ii}$$

et de trouver un potentiel de façon très judicieuse (méthode de P. Lelong, voir [6]). Une autre méthode consiste à utiliser le calcul différentiel extérieur et donc d'écrire (1) sous la forme

(2) $\partial \bar{\partial} u = \alpha$

avec

$$\alpha = \sum_{i,j} \alpha_{ij}\, dz_i \wedge d\bar{z}_j .$$

La condition de compatibilité s'écrivant ici: $d\alpha = 0$, pour ce faire on se ramène à la résolution de deux opérateurs le $\bar{\partial}$ et le d (voir [5]).

Cette méthode ne tient pas compte de deux choses: d'une part le problème est purement réel, d'autre part l'utilisation du calcul différentiel extérieur masque certaines propriétés algébriques du système (1). D'ailleurs, on peut se demander si on peut étudier directement la cohomologie du complexe $\xrightarrow{d} \xrightarrow{\partial\bar{\partial}} \xrightarrow{d}$ sans utiliser celle du $\bar{\partial}$.

On peut considérer que ce qui suit est une première tentative d'utiliser autre chose que le calcul différentiel extérieur en analyse complexe.

Notons que d'autres liens entre l'analyse complexe et le bilaplacien se trouvent dans [2] et [3].

2. Algèbre de Clifford et factorisation du Δ^2. Soit A, l'algèbre de Clifford associé à l'espace euclidien $\mathbb{R}^{n^2}$. Nous noterons

$$e_{pq}, \quad \text{pour} \quad 1 \leq p \leq n, \ 1 \leq q \leq n$$

les n^2 éléments de la base de A (base en tant qu'algèbre), en choisissant:

$$(3) \qquad \begin{aligned} &e_{pp}^2 = -1; \quad e_{pq}^2 = -2 \quad \text{pour} \quad p \neq q; \\ &e_{pq}\, e_{rs} + e_{rs}\, e_{pq} = 0 \quad \text{si} \quad (p,q) \neq (r,s), \end{aligned}$$

(Pour des raisons de calcul, nous ne prenons pas une base tout-à-fait orthonormée).

Dans A, nous avons l'involution: $\bar{e}_{pq} = -e_{pq}$.

Considérons à nouveau le système (1), il est équivalent à:

$$(4) \qquad \sum_{1 \leq p,q \leq n} e_{pq} \frac{\partial^2 u}{\partial z_p \partial \bar{z}_p} = \sum_{1 \leq p,q \leq n} e_{pq}\, \alpha_{pq}.$$

Ceci amène à introduire les opérateurs:

$$D_z = \sum_{1 \leq p,q \leq n} e_{pq} \frac{\partial^2}{\partial z_p \partial \bar{z}_q}, \quad \bar{D}_z = - \sum_{1 \leq p,q \leq n} e_{pq} \frac{\partial^2}{\partial \bar{z}_p \partial z_q}.$$

Remarquons que $\bar{D}_z$ est le conjugué de D_z à la fois au sens de l'involution de l'algèbre de Clifford et de la conjugaison complexe. D_z et $\bar{D}_z$ donnent en fait le même système d'équations, par exemple u est pluriharmonique si et seulement si $D_z u = 0$ (ou bien $\bar{D}_z u = 0$).

Soit:

$$\Delta^2 = \sum_{k=1}^{n} [(\partial^2/\partial x_k^2) + (\partial^2/\partial y_k^2)]^2$$

le bilaplacien, c'est-à-dire l'équation fondamentale de l'élasticité, de l'espace réel $\mathbb{R}^{2n}$ sous-jacent à $\mathbb{C}^n$. Du point de vue de l'analyse complexe, on a:

THÉORÈME. *Soit* u *une fonction à valeurs réelles, de classe* $\mathcal{C}^4$, Δ^2 *se factorise de la façon suivante*:

$$\Delta^2 u = 16\, D_z \overline{D}_z u.$$

Démonstration. On a:

$$D_z \overline{D}_z = \Big(\sum_k e_{kk} \frac{\partial^2}{\partial z_k \partial \bar{z}_k} + \sum_{p\neq q} e_{pq} \frac{\partial^2}{\partial z_p \partial \bar{z}_q}\Big)\Big(-\sum_h e_{hh} \frac{\partial^2}{\partial \bar{z}_h \partial z_h} - \sum_{r\neq s} e_{rs} \frac{\partial^2}{\partial \bar{z}_r \partial z_s}\Big)$$

$$= -\sum_{k,h} e_{kk} e_{hh} \frac{\partial^4}{\partial z_k \partial \bar{z}_k \partial \bar{z}_h \partial z_h} - \sum_{k,r\neq s} e_{kk} e_{rs} \frac{\partial^4}{\partial z_k \partial \bar{z}_k \partial \bar{z}_r \partial z_s}$$

$$\sum_{p\neq q,h} e_{pq} e_{hh} \frac{\partial^4}{\partial z_p \partial \bar{z}_q \partial \bar{z}_h \partial z_h} - \sum_{\substack{p\neq q\\ r\neq s}} e_{pq} e_{rs} \frac{\partial^4}{\partial z_p \partial \bar{z}_q \partial \bar{z}_r \partial z_s}.$$

D'après (3):

$$D_z \overline{D}_z = \sum_k \frac{\partial^4}{\partial z_k^2 \partial \bar{z}_k^2} - \frac{1}{2} \sum_{\substack{p\neq q\\ r\neq s}} e_{pq} e_{rs} \frac{\partial^4}{\partial z_p \partial \bar{z}_q \partial \bar{z}_r \partial z_s}$$

$$- \frac{1}{2} \sum_{\substack{r\neq s\\ q\neq p}} e_{rs} e_{pq} \frac{\partial^4}{\partial z_r \partial \bar{z}_s \partial \bar{z}_q \partial z_p} = \sum_k \frac{\partial^4}{\partial z_k^2 \partial \bar{z}_k^2} + 2 \sum_{p\neq q} \frac{\partial^4}{\partial z_p \partial \bar{z}_p \partial z_q \partial \bar{z}_q}$$

$$= [\sum_k (\partial^2/\partial z_k \partial \bar{z}_k)]^2 = \frac{1}{16} \Delta^2.$$

On a ainsi factorisé un opérateur elliptique à travers les opérateurs $\partial^2/\partial z_p \partial \bar{z}_q$. Nous pouvons aussi obtenir des opérateurs hyperboliques (ou plus exactement ultrahyperboliques), par exemple factoriser

$$(\partial^2/\partial z_1 \partial \bar{z}_1) - \sum_{k=2}^{n} (\partial^2/\partial z_k \partial \bar{z}_k),$$

en utilisant des algèbres de Clifford relatives à des formes quadratiques de signes alternés, par exemple, l'opérateur ci-dessus sera obtenu avec la forme quadratique de signes (+ - ... -).

Références

[1] BERNDTSSON, B.: Integral formulas and zeros of bounded holomorphic functions, Université de Göteborg, Suède, 1979.

[2] GARABEDIAN, P. R.: A Green's function in the theory of functions of several complex variables, Ann. of Math. 55, n° 1 (1952).

[3] GAVEAU, B. et P. MALLIAVIN: Courbure des surfaces de niveau des fonctions holomorphes bornées dans la boule, C.R.A.S. Paris 293 (1981), 135-138.

[4] HENKIN, G.: Solutions with estimates of the H. Lewy and Poincaré-Lelong equations, D.A.N. SSSR 225 (1975), 771-774.

[5] LAVILLE, G.: Résolution du $\partial\bar{\partial}$ avec croissance dans des ouverts pseudo-convexes étoilés, C.R.A.S. Paris 274 (1972), 554-556.

[6] LELONG, P.: Fonctions entières et fonctions plurisousharmoniques, J. Analyse Math. 12 (1964), 365-407.

[7] SKODA, H.: Valeurs au bord pour les solutions de l'opérateur d″ et caractérisation des zéros, Bull. Soc. Math. France 104 (1976), 225-299

Mathématiques, L. A. 213 du C. N. R. S.
Université Pierre et Marie Curie
4, Place Jussieu
F-75230 Paris Cedex 05, France

SUBSHEAVES IN BUNDLES ON $\mathbb{P}_n$ AND THE PENROSE TRANSFORM

Jürgen Leiterer (Berlin)

Contents

Summary

Let E be a holomorphic vector bundle on the n-dimensional complex projective space $\mathbb{P}_n$, $n \geq 2$, of generic splitting type $(a_1,\dots,a_r)$, $a_1 \geq \dots \geq a_r$. Let, for some $1 \leq i \leq r-1$, $a_i - a_{i+1} = 1$. Then the obstruction for the existence of a coherent subsheaf of type $(a_1,\dots,a_i)$ can be interpreted as a "part" $P_i(E)$ of the Penrose transform. We obtain that, for $n \geq 3$, the property "E does not contain a coherent subsheaf of generic splitting type $(a_1,\dots,a_i)$" is preserved under restriction to general hyperplanes. If $a_j - a_{j+1} = 1$ for all $j = 1,\dots,r-1$, then this property (fulfilled for $j = 1,\dots,r-1$) is equivalent to stability in the sense of Mumford and Takemoto.

Introduction

Let $\mathbb{P}_n$, $n \geq 2$, be the n-dimensional complex projective space, E a holomorphic vector bundle on $\mathbb{P}_n$ of generic splitting type $(a_1,\dots,a_r)$, $a_1 \geq \dots \geq a_r$. Let $1 \leq i \leq r-1$. If

$$a_i - a_{i+1} \geq 2,$$

then by the Grauert-Mülich-Spindler theorem [3,8,6] E contains a coherent subsheaf of generic splitting type $(a_1,\dots,a_i)$. In this paper we consider the case

$$a_i - a_{i+1} = 1.$$

It turns out that then the obstruction for the existence of a coherent subsheaf of generic splitting type $(a_1,\dots,a_i)$ can be interpreted as a "part" $P_i(E)$ of the Penrose transform $P(E)$ introduced in [4]. Locally $P_i(E)$ can be considered as a differential 1-form on open portions of the Grassmann manifold $\mathbb{G}_n$ of lines in $\mathbb{P}_n$, where, for $n \geq 3$ the restriction of vector bundles to the hyperplane $\mathbb{P}_{n-1} \subset \mathbb{P}_n$ corresponds to the restriction of differential forms to the respective Grassmannian $\mathbb{G}_{n-1} \subset \mathbb{G}_n$. By means of this correspondence one obtains that, for $n \geq 3$, the property "E does not contain a coherent subsheaf of generic splitting type $(a_1,\dots,a_i)$" is preserved under restriction to general hyperplanes. If

$$(0.1) \quad a_j - a_{j+1} = 1 \quad \text{for all } j = 1,\dots,r-1,$$

then this property (fulfilled for $j = 1,\dots,r-1$) is equivalent to stability in the sense of Mumford and Takemoto. Hence the restriction of a stable bundle of generic splitting type (0.1) to general hyperplanes is again stable. For $r = 2$ this is contained in the theorem of Bart [1] and Maruyama [5] (see also [6]). For $r = 3$ this is a result of Schneider [7,9].

A few comments on notation: No difference is made between a holomorphic vector bundle E over X and its sheaf of holomorphic sections. For $x \in X$, $E(x)$ is the fiber of the <u>bundle</u> E over x and E_x is the stalk of the <u>sheaf</u> E over x. If $f \in H^0(X,E)$ and $x \in X$, then $f(x)$ is the value of f in $E(x)$ and f_x is the value of f in E_x. If E,F are two holomorphic vector bundles over X, then we denote by $\mathrm{Hom}(E,F)$ the space of homomorphisms from E into F, that is, $\mathrm{Hom}(E,F) \cong H^0(X, F \otimes E^*)$.

1. A descente lemma

Here is given a modification of Lemma 2.1.2 in [6], Chapter 2. Let X,Y be complex manifolds, $\mu : X \to Y$ a surjective holomorphic submersion, Ω_Y^1 and Ω_X^1 the sheaves of holomorphic 1-forms on Y and X, $\Omega_\mu^1 := \Omega_X^1/\mu^*\Omega_Y^1$ the sheaf of relative holomorphic 1-forms with respect to μ. Define $d_\mu := \pi_\mu \circ d$, where d is the exterior differential operator and $\pi_\mu : \Omega^1_X \to \Omega^1_\mu$ is the quotient mapping. Let E be a holomorphic vector bundle over Y, F a holomorphic subbundle of the pull-back bundle μ^*E, $Q := \mu^*E/F$ its quotient. Since μ^*E can be described by transition functions which are constant along the fibres of μ, then d_μ makes sense also for holomorphic sections of μ^*E:

$$d_\mu : \mu^*E \to \mu^*E \otimes \Omega^1_\mu.$$

If f is a holomorphic section of μ^*E over the open set $U \subseteq X$ and φ is a holomorphic function on U, then we have the relative Leibniz rule

$$(1.1) \qquad d_\mu(\varphi f) = \varphi d_\mu f + f \otimes d_\mu \varphi \quad \text{on } U.$$

Let $\pi_Q : \mu^*E \otimes \Omega^1_\mu \to Q \otimes \Omega^1_\mu$ be the quotiet mapping. Setting $S_F := \pi_Q \circ d_\mu$ on F one obtains a map

$$S_F : F \to Q \otimes \Omega^1_\mu.$$

By the relative Leibniz rule (1.1), $S_F(\varphi f) = \varphi(S_F f)$. Hence

$$(1.2) \qquad S_F \in \mathrm{Hom}(F, Q \otimes \Omega^1_\mu).$$

1.1. LEMMA. *If the fibres of μ are connected, then the following conditions are equivaelnt:*

$$(1.3) \quad \begin{cases} \text{(i)} & F = \mu^*\tilde{F} \ \textit{for some holomorphic subbundle } \tilde{F} \textit{ of } E; \\ \text{(ii)} & d_\mu F \subseteq F \otimes \Omega^1_\mu; \\ \text{(iii)} & S_F = 0. \end{cases}$$

Proof. It is clear that (ii) $\Longleftrightarrow$ (iii) and (i) $\Longrightarrow$ (ii). We prove (ii) $\Longrightarrow$ (i). Since the fibres of μ are connected, we only have to show that for each point $x \in X$ there are a neighborhood V of X and a holomorphic frame $g_1, \dots, g_s$ of F over V such that

$$(1.4) \qquad d_\mu g_1 = \dots = d_\mu g_s = 0.$$

Fix $x_0 \in X$ and choose a holomorphic frame $f_1, \dots, f_r$ of E over a neighborhood U of $\mu(x_o)$. After a permutation we can assume that for some neighborhood $V = \mu^{-1}(U)$ of x_o

$$(1.5) \qquad E(x) = F(x) \oplus \mathrm{span}(\mu^* f_{s+1}(x), \dots, \mu^* f_r(x)), \quad x \in V.$$

Here "span" is an abbreviation for "$\mathbb{C}$-linear span of". Denote by $g_1(x), \dots, g_s(x)$ the projections of $\mu^* f_1(x), \dots, \mu^* f_s(x)$ to $F(x)$ along $\mathrm{span}(\mu^* f_{s+1}(x), \dots, \mu^* f_r(x))$. Then $g_1, \dots, g_s$ is a holomorphic frame of F over V and

$$g_j = \mu^* f_j + \sum_{k=s+1}^{r} \varphi_{kj} \mu^* f_k \quad \text{on} \quad V,$$

where φ_{kj} are holomorphic functions on V. Hence

$$d_\mu g_j = \sum_{k=s+1}^{r} f_k \otimes d_\mu \varphi_{kj} \quad \text{on} \quad V.$$

According to (1.3) and (1.5) this implies (1.4).

2. The obstructions $P_i(E)$

Let V be an $(n+1)$-dimensional complex vector space $(n \geq 2)$, $\mathbb{P} = \mathbb{P}(V)$ the projective space of 1-dimensional complex subspaces in V, $\mathbb{G} = \mathbb{G}(V)$ the Grassmann manifold of 2-dimensional complex subspaces in V, $\mathbb{F} = \mathbb{F}(V)$ the flag manifold of pairs (l, L) with $l \in \mathbb{P}$, $L \in \mathbb{G}$, $l \subseteq L$. The double fibration

$$(2.1) \qquad \begin{array}{ccc} \mathbb{F} & \xrightarrow{\ \nu\ } & \mathbb{G} \\ \downarrow{\scriptstyle \mu} & & \\ \mathbb{P} & & \end{array}$$

where $\mu(l, L) := l$ and $\nu(l, L) := L$, is called the standard diagram (cf., for example, [2,6]). The notations $\Omega^1_{\mathbb{P}}$, $\Omega^1_{\mathbb{F}}$, Ω^1_μ and d_μ are used as in Section 1 (with $X = \mathbb{F}$, $Y = \mathbb{P}$). Denote by $\mathcal{O}_{\mathbb{P}}(k)$ the k-th power of the hyperplane section bundle on $\mathbb{P}$ and put $\mathcal{O}_{\mathbb{F}}(k) := \mu^* \mathcal{O}_{\mathbb{P}}(k)$. If $Z \subseteq \mathbb{G}$, then we define

$$Z' := \nu^{-1}(Z) \quad \text{and} \quad Z'' := \mu(Z').$$

For a point $u \in \mathbb{G}$ we also write u' and u'' instead of $\{u\}'$ and $\{u\}''$. Observe that then u' as well as u'' are Riemannian spheres.

2.1. LEMMA (see [6], p.197, or [2], p.313). For each $u \in \mathbb{G}$, the restriction of $\mathcal{O}_{\mathbb{F}}(-1) \otimes \Omega^1_\mu$ to u' is the trivial bundle of rank $n-1$. Therefore, the 0-th direct image sheaf

$$(2.2)\qquad \Lambda_{\mathbb{G}} := \nu_*(\mathcal{O}_{\mathbb{F}}(-1) \otimes \Omega^1_\mu)$$

is locally free and of the rank $n-1$.

If $D \subseteq \mathbb{G}$ is open, then a holomorphic vector bundle on D'' is called D-uniform if there are integers $a_1 > \ldots > a_k$ and $r_1,\ldots,r_k \geq 1$ such that

$$E|u'' \cong \mathcal{O}_{u''}(a_i)^{\oplus r_1} \oplus \ldots \oplus \mathcal{O}_{u''}(a_k)^{\oplus r_k} \quad \text{for all } u \in D.$$

Then the collection $(a_1,r_1;\ldots;a_k,r_k)$ is called the splitting type of E.

2.2. LEMMA (see, for example, [6], p.60). Let $D \subseteq \mathbb{G}$ be an open set and E a holomorphic vector bundle over D''. Then E is D-uniform with splitting type $(a_1,r_1;\ldots;a_k r_k)$ if and only if there is a filtration

$$(2.3)\qquad 0 = F_0 \subset F_1 \subset \ldots \subset F_k = \mu^* E$$

by holomorphic subbundles F_i of μ^*E on D' such that

$$(2.4)\qquad F_i/F_{i-1} = \nu^* G_i \otimes \mathcal{O}_{\mathbb{F}}(a_i),$$

where G_i is a holomorphic vector bundle of rank r_i over D.

2.3. LEMMA. Let $D \subseteq \mathbb{G}$ be an open set, E a D-uniform holomorphic vector bundle on D'' of splitting type $(a_1,r_1;\ldots;a_k,r_k)$, $0 = F_0 \subset \ldots \subset F_k = \mu^* E$ the filtration from Lemma 2.2. Then, for $i = 1,\ldots,k-1$,

$$(2.5)\qquad d_\mu F_i \subseteq F_{i+1} \otimes \Omega^1_\mu$$

and, moreover,

$$(2.6)\qquad d_\mu F_i \subseteq F_i \otimes \Omega^1_\mu \quad \text{if } a_i - a_{i+1} \geq 2.$$

P r o o f. Fix $1 \leq i \leq k-1$. Let $p_j : F_k \otimes \Omega^1_\mu \to (F_k/F_j) \otimes \Omega^1_\mu$ be the quotient mapping. Then we have to prove that $p_{i+1} \circ d_\mu|F_i = 0$, and, moreover, $p_i \circ d_\mu|F_i = 0$ if $a_i - a_{i+1} \geq 2$. The relative Leibniz rule (1.1) shows that

$$p_j \circ d_\mu|F_i \in \operatorname{Hom}(F_i, (F_k/F_j) \otimes \Omega^1_\mu) \quad \text{for} \quad j = i, i+1.$$

Therefore it is sufficient to prove that for all $u \in D$

$$(2.7) \qquad H^0((F_k/F_{i+1}) \otimes \Omega^1_\mu \otimes F_i^*|u') = 0$$

and, moreover,

$$(2.8) \qquad H^0((F_k/F_i) \otimes \Omega^1_\mu \otimes F_i^*|u') = 0 \quad \text{if} \quad a_i - a_{i+1} \geq 2.$$

Since $a_1 > \ldots > a_k$, it follows from (2.3), (2.4) that for all $u \in D$

$$(2.9) \qquad F_1|u' \cong \mathcal{O}_{u'}(a_1)^{\oplus r_1} \oplus \ldots \oplus \mathcal{O}_{u'}(a_i)^{\oplus r_i},$$

$$(F_k/F_j)|u' \cong \mathcal{O}_{u'}(a_{j+1})^{\oplus r_{i+2}} \oplus \ldots \oplus \mathcal{O}_{u'}(a_k)^{\oplus r_k}.$$

Since $a_1 > \ldots > a_k$ and according to Lemma 2.1, this implies that $(F_k/F_{i+1}) \otimes \Omega^1_\mu \otimes F_i^*$ splits over each u', $u \in D$, into line bundles $\mathcal{O}_{u'}(b_m)$ with $b_m \leq -1$. The same is true for $(F_k/F_i) \otimes \Omega^1_\mu \otimes F_i^*$ if $a_i - a_{i+1} \geq 2$. This implies (2.7) and (2.8).

2.4. THEOREM (cf. [8] and [6], p.203). *Let $D \subseteq \mathbb{G}$ be an open set, E a D-uniform holomorphic vector bundle on D" of splitting type $(a_1, r_1; \ldots; a_k, r_k)$. If, for some $1 \leq i \leq k-1$, $a_i - a_{i+1} \geq 2$, then there exists a holomorphic subbundle of E which is D-uniform and of splitting type $(a_1, r_1; \ldots; a_i, r_i)$.*

P r o o f. Let $\{D_m\}$ be an open covering of D such that for all m the restriction $\mu|D_m$ has connected fibres. Then we obtain form Lemmas 1.1 and 2.3 holomorphic subbundles F_m of $E|D_m''$ such that $F_i|D_m' = \mu^* F_m|D_m'$, where $0 = F_0 \subset \ldots \subset F_k = \mu^* E$ is the filtration from Lemma 2.2. By (2.9), F_m is D_m-uniform and of splitting type $(a_1, r_1; \ldots; a_i, r_i)$. Since $a_1 > \ldots > a_k$ and $(a_1, r_1; \ldots; a_k, r_k)$ is the splitting type of E, for each $u \in D$, the subbundle in $E|u''$ which is isomorphic to $\mathcal{O}_{u''}(a_1)^{\oplus r_1} \oplus \ldots \oplus \mathcal{O}_{u''}(a_i)^{\oplus r_i}$ is uniquely determined. Hence $F_m = F_l$ on $D_m'' \cap D_l''$.

2.5. D e f i n i t i o n. Let $D \subseteq \mathbb{G}$ be an open set, E a D-uniform holomorphic vector bundle on D" of splitting type $(a_1, r_1; \ldots; a_k, r_k)$,

$0 = F_o \subset \ldots \subset F_k = \mu^* E$ the filtration from Lemma 2.2. Denote by $I_1(E)$ the set of all $1 \le i \le k-1$ such that

(2.10) $a_i - a_{i+1} = 1.$

According to Lemma 2.3, by the condition that the diagram

$$(2.11)\qquad \begin{array}{ccc} F_i & \xrightarrow{d_\mu} & F_{i+1} \otimes \Omega^1_\mu \\ \downarrow & & \downarrow \\ F_i/F_{i-1} & \xrightarrow{P'_i(E)} & (F_{i+1}/F_i) \otimes \Omega^1_\mu \end{array}$$

commutes, we can define maps $P'_i(E)$, $i \in I_1(E)$. It follows from the relative Leibniz rule (1.1) that $P'_i(E)$ is a homomorphism of vector bundles. Hence, by (2.4) and (2.9),

$$P'_i(E) \in H^o(D', \nu^* G_{i+1} \otimes \nu^* G_i^* \otimes \mathcal{O}_{\mathbb{F}}(-1) \otimes \Omega^1_\mu), \quad i \in I_1(E).$$

Let

$$P_i(E) \in H^o(D, G_{i+1} \otimes G_i^* \otimes \Lambda_{\mathbb{G}}), \quad i \in I_1(E),$$

be the induced sections (cf. (2.2)).

Remark. The section $P_i(E)$ are "parts" of the Penrose transform defined in [4] (for $n = 3$ and D contained in an affine portion of $\mathbb{G}$). Namely, if $[\{C_{pq}\}, \{\Theta_p\}]$ (terminology of [4], Section 4) represents the Penrose transform $P(E)$ of E, and if $C_{pq} = ((C_{pq})_{ml})^r_{m,l=1}$, $\Theta_p = ((\Theta_p)_{m,l})^r_{m,l=1}$, $r = r_1 + \ldots + r_k$, then the bundles G_i are represented by the cocycles $((C_{pq})_{ml})^{s_i}_{m,l=s_{i-1}+1}$, where $s_i = r_o + \ldots + r_i$, $r_o := 0$, and the sections $P_i(E)$ are represented by the forms $((\Theta_p)_{ml})^{m=s_{i-1}+1,\ldots,s_i}_{l=s_i+1,\ldots,s_{i+1}}$.

2.6. THEOREM. Let $D \subseteq \mathbb{G}$ be an open set, E a D-uniform holomorphic vector bundle on D'' of splitting type $(a_1, r_1; \ldots; a_k, r_k)$. If $1 \le i \le k-1$ such that $a_i - a_{i+1} = 1$*), then the following conditions are equivalent:

(i) $P_i(E) = 0$;

(ii) There is a holomorphic subbundle of E which is D-uniform and of splitting type $(a_1, r_1; \ldots; a_i, r_i)$

*) For $a_i - a_{i+1} \ge 2$ we have Theorem 2.4.

P r o o f. Since by Lemma 2.1 the restriction of $\nu^*G_{i+1} \otimes \nu^*G_i \otimes \mathcal{O}_{\mathbb{F}}(-1) \otimes \Omega^1_\mu$ to each u', $u \in D$, is trivial, we can see that $P_i(E) = 0$ iff $P_i'(E) = 0$. By (2.11), $P_i'(E) = 0$ iff $d_\mu F_i \subseteq F_i \otimes \Omega^1_\mu$. Now we conclude the proof by the same arguments as given in the proof of Theorem 2.4.

2.7. THEOREM. *Let E be a holomorphic vector bundle over $\mathbb{P}$, $S_E \subseteq \mathbb{G}$ the set of jump lines of E, and let $(a_1,r_1;\dots;a_k,r_k)$ be the splitting type of $E|\mathbb{G}\setminus S_E$. If $1 \leq i \leq k-1$ such that $a_i - a_{i+1} = 1$, then the following conditions are equivalent:*

(i) $P_i(E|(\mathbb{G}\setminus S_E)'') = 0$;

(ii) *There is a normal coherent subsheaf F in E whose restriction to $(\mathbb{G}\setminus S_E)''$ is a subbundle such that for all $u \in \mathbb{G}\setminus S_E$*

$$F|u'' = \mathcal{O}_{u''}(a_1)^{\oplus r_1} \oplus \dots \oplus \mathcal{O}_{u''}(a_i)^{\oplus r_i}. \tag{2.12}$$

P r o o f. By Theorem 2.6, condition (i) is equivalent to the existence of a holomorphic subbundle $\tilde{F}$ of $E|(\mathbb{G}\setminus S_E)''$ satisfying (2.12). That $\tilde{F}$ can be extended to a normal coherent subsheaf of E on the whole projective space $\mathbb{P}$ follows by the same arguments as given in [6], p. 203-204, in the proof of the Grauert-Mülich-Spindler theorem (which asserts that (ii) holds true if $a_i - a_{i+1} \geq 2$).

3. Restriction to hyperplanes

Throughout this section V is an $(n+1)$-dimensional complex vector space, $n \geq 3$. Denote by $\mathbb{P}^*(V)$ the space of all n-dimensional complex subspaces of V. If $W \in \mathbb{P}^*(V)$, then we have the canonical inclusion of the corresponding standard diagrams (2.1):

$$\begin{array}{ccccccc} \mathbb{F}(W) & \xrightarrow{\nu(W)} & \mathbb{G}(W) & & \mathbb{F}(V) & \xrightarrow{\nu(V)} & \mathbb{G}(V) \\ \downarrow{\scriptstyle \mu(W)} & & & \hookrightarrow & \downarrow{\scriptstyle \mu(V)} & & \\ \mathbb{P}(W) & & & & \mathbb{P}(V) & & \end{array} \tag{3.1}$$

and, for each open $D \subseteq \mathbb{G}(V)$, the restriction of differential forms $\mathbb{F}(V)$ to $\mathbb{F}(W)$ gives rises to a restriction map

$$\pi^V_W : H^0(D, \Lambda_{\mathbb{G}(V)}) \longrightarrow H^0(D \cap \mathbb{G}(W), \Lambda_{\mathbb{G}(W)}) \tag{3.2}$$

(see (2.2) for the definition of $\Lambda_{\mathbb{G}}$).

3.1. LEMMA. *Let* $U \subseteq \mathbb{P}^*(V)$ *be an open set,* D *a connected open subset of* $\mathbb{G}(V)$ *such that*

(3.3) $\quad \mathbb{G}(W) \cap D \neq \emptyset$ *for all* $W \in U$.

If $f \in H^0(D, \Lambda_{\mathbb{G}(V)})$ *such that*

(3.4) $\quad \pi_W^V f = 0$ *for all* $W \in U$,

then $f = 0$.

Proof. By definition of $\Lambda_{\mathbb{G}(V)}$, f can be identified with a section $f' \in H^0(\nu(V)^{-1}(D), \mathcal{O}_{\mathbb{F}(V)}(-1) \otimes \Omega^1_{\mu(V)})$. Put

$$\omega := \bigcup_{W \in U} \mathbb{G}(W) \cap D.$$

Since U is open and by (3.3), ω is a non-empty open subset of D. Since D is connected, it is therefore enough to prove that $f'(y) = 0$ for all $y \in \nu(V)^{-1}(\omega)$.

Fix $y_o = (l_o, L_o) \in \nu(V)^{-1}(\omega)$, where $l_o \in \mathbb{P}(V)$, $L_o \in \mathbb{G}(V)$, $l_o \subseteq L_o$. Choose $W_1 \in U$ such that $L_o \in \mathbb{G}(W_1)$. Since $\dim W_1 \geq 3$, $\dim L_o = 2$ and U is open, we can find $W_2 \in U$ such that $W_2 = W_1$ and also $L_o \in \mathbb{G}(W_2)$. Choose a linear basis $e_1, e_2, \ldots, e_{n+1}$ of V such that

$$l_o = \operatorname{span}(e_{n+1}),$$

(3.5) $$L_o = \operatorname{span}(e_n, e_{n+1}),$$

(3.6) $$W_1 = \operatorname{span}(e_1, e_3, \ldots, e_{n+1}),$$

(3.7) $$W_2 = \operatorname{span}(e_2, e_3, \ldots, e_{n+1}),$$

where "span" is an abbreviation for "$\mathbb{C}$-linear span of". Choose a differential from $\tilde{f}' \in H^0(Y, \mathcal{O}_{\mathbb{F}}(-1) \otimes \Omega^1)$ in a neighborhood Y of $y_o = (l_o, L_o)$ such that $\pi_{\mu(V)}(\tilde{f}') = f'$ on Y, where

$$\pi_{\mu(V)} : \mathcal{O}_{\mathbb{F}}(-1) \otimes \Omega^1 \longrightarrow \mathcal{O}_{\mathbb{F}}(-1) \otimes \Omega^1_{\mu(V)}$$

is the quotient mapping. In order to show that $f'(l_o, L_o) = 0$, we have to prove that the restricted differential form $\tilde{f}' | \mu(V)^{-1}(l_o) \cap Y$ vanishes at (l_o, L_o).

After shrinking Y if necessary, we have holomorphic coordinates $z_1,\ldots,z_{n-1}$ on $\mu(V)^{-1}(l_o)\cap Y$ such that, for each $y=(l_o,L)\in\mu(V)^{-1}(l_o)\cap Y$,

$$L = \operatorname{span}\Big(\sum_{j=1}^{n-1} z_j(y)e_j + e_n, e_{n+1}\Big).$$

Let

$$f'|\mu(V)^{-1}(l_o)\cap Y = \sum_{j=1}^{n-1} f_j dz_j,$$

where f_j are holomorphic functions on $\mu(V)^{-1}(l_o)\cap Y$. By (3.4), we have

$$(3.8)\qquad f'|\mu(W_1)^{-1}(l_o)\cap Y = 0$$

and

$$(3.9)\qquad f'|\mu(W_2)^{-1}(l_o)\cap Y = 0.$$

It follows from (3.6) and (3.8) that

$$f_1 = f_3 = \ldots = f_{n-1} = 0 \quad \text{if} \quad z_2 = 0,$$

and it follows from (3.6) and (3.9) that

$$f_2 = f_3 = \ldots = f_{n-1} = 0 \quad \text{if} \quad z_1 = 0.$$

Hence

$$f'|\mu(V)^{-1}(l_o)\cap Y = 0 \quad \text{if} \quad z_1 = z_2 = 0.$$

Since, by (3.5), $z_1(l_o,L_o) = z_2(l_o,L_o) = 0$, this completes the proof.

3.2. THEOREM. Let $D\subseteq \mathbb{G}(V)$ be a connected open set, E a D-uniform holomorphic vector bundle on D'' of splitting type $(a_1,r_1;\ldots;a_k,r_k)$, $1\le i\le k-1$. If there is an open set $U\subseteq \mathbb{P}^*(V)$ such that for all $W\in U$

$$(3.10)\qquad D\cap\mathbb{G}(W) \neq \emptyset$$

and $E|(D\cap\mathbb{G}(W))''$ contains a $D\cap\mathbb{G}(W)$-uniform holomorphic subbundle of splitting type $(a_1,r_1;\ldots;a_i,r_i)$, then E contains a D-uniform holomorphic subbundle of this splitting type.

P r o o f. By Theorem 2.4 we can assume that $a_i - a_{i+1} = 1$. Then by Theorem 2.6 we have to prove that the conditions

$$P_i(E) = 0$$

and

$$P_i(E|(D \cap \mathbb{G}(W))'') = 0 \quad \text{for all} \quad W \in U$$

are equivalent. According to the relation

$$P_i(E|(D \cap \mathbb{G}(W))'') = \pi_W^V P_i(E),$$

this follows from Lemma 3.1.

3.3. THEOREM. *Let* E *be a holomorphic vector bundle on* $\mathbb{P}(V)$, $S_E \subseteq \mathbb{G}(V)$ *the set of jump lines of* E,

$$E|u'' \cong \mathcal{O}_{u''}(a_1)^{\oplus r_1} \oplus \ldots \oplus \mathcal{O}_{u''}(a_k)^{\oplus r_k} \quad \text{for} \quad u \in \mathbb{G}(V) \setminus S_E,$$

where $a_1 > \ldots > a_k$. *Let* $1 \leq i \leq k-1$. *Suppose that there are open sets* $U \subseteq \mathbb{P}^*(V)$ and $D \subseteq \mathbb{G}(V) \setminus S_E$ such that for all $W \in U$

$$D \cap \mathbb{G}(W) \neq \emptyset$$

and there is a holomorphic subbundle F *of* $E|(D \cap \mathbb{G}(W))''$ *with*

$$F|u'' \cong \mathcal{O}_{u''}(a_1)^{\oplus r_1} \oplus \ldots \oplus \mathcal{O}_{u''}(a_i)^{\oplus r_i} \quad \text{for all} \quad u \in D \cap \mathbb{G}(W).$$

Then there exists a normal coherent subsheaf F *of* E *whose restriction to* $(\mathbb{G}(V) \setminus S_E)''$ *is a holomorphic subbundle with*

$$F|u'' \cong \mathcal{O}_{u''}(a_1)^{\oplus r_1} \oplus \ldots \oplus \mathcal{O}_{u''}(a_i)^{\oplus r_i} \quad \text{for all} \quad u \in \mathbb{G}(V) \setminus S_E. \tag{3.11}$$

P r o o f. After shrinking D we can assume that D is connected. Then by Theorem 3.2 there is a D-uniform holomorphic subbundle F_D of $E|D''$ with splitting type $(a_1, r_1; \ldots; a_i, r_i)$. Therefore, by Theorem 2.6, $P_i(E|D'') = 0$. Since $P_i(E|(\mathbb{G}(V) \setminus S_E)'')$ is a holomorphic section of a holomorphic vector bundle over $\mathbb{G}(V) \setminus S_E$ whose restriction to D is equal to $P_i(E|D'')$, this implies that $P_i(E|(\mathbb{G}(V) \setminus S_E)'') = 0$. Now the assertion follows from Theorem 2.7.

4. The splitting type $(k-1,k-2,\dots,0)$

We use the notion introduced in Section 2.

4.1. THEOREM. *Let* E *be a holomorphic vector bundle of rank* k *over* $\mathbb{P}(V)$, $S_E \subseteq \mathbb{G}(V)$ *the set of jump lines of* E, *and*

(4.1) $$E|u" \cong \mathcal{O}_{u"}(a_1) \oplus \dots \oplus \mathcal{O}_{u"}(a_k) \quad \textit{for} \quad u \in \mathbb{G}(V) \setminus S_E,$$

where

(4.2) $$a_j - a_{j+1} = 1 \quad \textit{for all} \quad j = 1,\dots,k-1.$$

Then the following two conditions are equivalent:

(i) E *is not stable (in the sense of Mumford and Takemoto).*

(ii) *For some* $1 \leq i \leq k-1$ *there exists a normal coherent subsheaf* F *of* E *whose restriction to* $(\mathbb{G}(V) \setminus S_E)"$ *is a holomorphic subbundle such that*

(4.3) $$F|u" \cong \mathcal{O}_{u"}(a_1) \oplus \dots \oplus \mathcal{O}_{u"}(a_i) \quad \text{for all} \quad u \in \mathbb{G}(V) \setminus S_E.$$

Proof. (ii) $\Longrightarrow$ (i) is trivial. Suppose E is not stable. Without loss of generality we can assume that

(4.4) $$a_1 = k-1, a_2 = k-2, \dots, a_k = 0.$$

Then we can find a connected open set $D \subseteq \mathbb{G}(V)$ and a D-uniform holomorphic subbundle $\tilde{F}$ of $E|D"$ such that

(4.5) $$F|u" \cong \mathcal{O}_{u"}(b_1) \oplus \dots \oplus \mathcal{O}_{u"}(b_s) \quad \text{for all} \quad u \in D,$$

where

(4.6) $$1 \leq s < k,$$

(4.7) $$b_1 \geq \dots \geq b_s,$$

(4.8) $$(b_1 + \dots + b_s)/s \geq (k-1)/2.$$

It follows from (4.4) and (4.1) that

(4.9) $b_1 \leq a_1 = k-1$.

By Theorem 2.4 we can assume that

(4.10) $b_j - b_{j+1} = 1$ for $j = 1, \ldots, s-1$.

(4.8)-(4.10) imply that

(4.11) $b_s \geq 1$.

Now let $0 = F_0 \subset F_1 \subset \ldots \subset F_k = \mu^* E|D'$ be the filtration form Lemma 2.2. Denote by $1 \leq i \leq k$ the integer for which

(4.12) $\mu^* \tilde{F}|D' \subseteq F_i$

but

(4.13) $\mu^* \tilde{F}|D' \not\subseteq F_{i-1}$.

Then by (4.7) and (4.11)

(4.14) $1 \leq i \leq k-1$.

<u>C l a i m</u>. $P_i(E|D") = 0$.

<u>P r o o f of the claim</u>. By definition of $P_i(E|D")$ we have to prove that

(4.15) $d_\mu f_i \subseteq F_i \otimes \Omega_\mu^1$.

Since rank $F_i = i$, rank $F_{i-1} = i-1$ and by (4.12), we can find an open set $\omega \subseteq D'$ and a holomorphic frame $f_1, \ldots, f_i$ of $F_i|\omega$ such that for all $x \in \omega$

(4.16) $f_1(x) \in \mu^* \tilde{F}(x)$

and

(4.17) $f_2(x), \ldots, f_i(x) \in F_{i-1}(x)$.

Since $d_\mu \mu^* \tilde{F} \subseteq \mu^* \tilde{F} \otimes \Omega^1_\mu$, it follows from (4.16) and (4.12) that

(4.18) $d_\mu f_1 \in H^0(\omega, F_i \otimes \Omega^1_\mu)$.

Further, from (4.17) and Lemma 2.3 one obtains

(4.19) $d_\mu f_2, \ldots, d_\mu f_i \in H^0(\omega, F_i \otimes \Omega^1_\mu)$.

Since $f_1, \ldots, f_i$ is a frame of $F_i|\omega$ and D' is connected, (4.18) and (4.19) imply (4.15). Hence the claim is proved.

Since $\mathbb{G}(V) \setminus S_E$ is connected, this implies that

$$P_i(E|(\mathbb{G}(V) \setminus S_E)") = 0.$$

Now condition (ii) follows from Theorem 2.7.

From Theorems 3.3 and 4.1 one obtains

4.2. COROLLARY. *Let E be as in Theorem 4.1. If E is stable (in the sense of Mumford and Takemoto), then the restriction of E to general hyperplanes is stable.*

Institut der Mathematik
Akademie der Wissenschaften der DDR
Mohrenstraße 39, DDR-1080 Berlin, DDR

References

[1] BARTH, W.: Some properties of stable rank-2 vector bundles on $\mathbb{P}_n$, Math. Ann. 226, 125-150 (1977).

[2] EASTWOOD, M.G., R.PENROSE and R.O.WELLS, Jr.: Cohomology and massless fields, Commun. Math. Phys. 78, 305-351 (1981).

[3] GRAUERT, H. and G.MÜLICH: Vektorbündel vom Rang 2 über dem n-dimensional komplex projektiven Raum, Manuscripta math. 16, 75-100 (1975).

[4] LEITERER, J.: The Penrose transform for bundles non-trivial on the general line, Preprint P-MATH-15/82, IMath der AdW der DDR, Berlin 1982, Math. Nachr., to appear.

[5] MARUYAMA, M.: Boundedness of semistable sheaves of small ranks, Preprint, Kyoto 1978.

[6] OKONEK, Ch., M.SCHNEIDER and H.SPINDLER: Vector bundles on complex projective spaces, Progress in Mathematics 3, Birkhäuser 1980.

[7] SCHNEIDER, M.: Chernklassen semi-stabiler Vektorraumbündel vom Rang 3 auf dem komplex-projektiven Raum, J. reine angew. Math. 315, 211-220 (1980).

[8] SPINDLER, H.: Der Satz von Grauert-Mülich Für beliebige semistabile holomorphe Vektorraumbündel über dem n-dimensionalen komplex-projektiven Raum, Math. Ann. 243, 131-141 (1979).

[9] SPINDLER, H.: Ein Satz über die Einschränkung holomorpher Vektorbündel auf $\mathbb{P}_n$ mit $c_1=0$ auf Hyperebenen, J. reine angew. Math. 327, 91-118 (1981).

APPLICATION OF OPTIMIZATION METHODS TO THE INVESTIGATION OF EXTREMAL PROBLEMS IN COMPLEX ANALYSIS

Leon Mikołajczyk (Łódź)

Contents

Summary

In the paper there have been presented some results being a consequence of the investigations of extremal problems in complex analysis by means of the methods of optimization theory. Suitable information about the methods applied can be found in papers [7], [9], [15] given in the references.

Introduction

In the last fifteen years there appeared many papers devoted to the investigations of extremal problems in complex analysis by means of the methods of optimal control theory. To those used most frequently belong: the mathematical programming method, the Pontryagin variational method, a general method of examining extremal problems in linear-topological spaces as well as a method of the extension of extremal problems. The examinations will be connected with complex functions of one complex variable. By an extremal problem we mean a problem for a conditional extremum under various side conditions.

Application of Optimization Methods in Complex Analysis

1. Information about the application of mathematical programming method

The mathematical programming method based on some properties of systems of Chemyshev functions and Stieltjes moments was successfully used to examine extremal problems in special classes of holomorphic functions ([10], [11], [22]). One may expect that the intensely developing theory of mathematical programming in the complex space will also find effective applications in the investigations of extremal problems in the spaces of holomorphic functions ([13], [1], [6]).

2. Notations and introductory notes

In the paper published in 1969 ([5]) A. Ja. Dubovitskii and A. A. Milyutin gave a general method of investigating extremal problems in finite-dimensional spaces and infinite-dimensional ones. The method developed by I. V. Girsanov ([7]) has found and still finds effective applications.

In the paper published in 1980 ([16]) we made an attempt of adopting this method to the investigations of extremal problems in certain special classes of complex functions.

Denote by C, K and X, respectively: a complex plane, the disc $\{z \in C: |z| < 1\}$ and an arbitrary compact Hausdorff space. We assume that the mapping $q: K \times X \to C$ possesses the following properties:

(i) for each $t \in X$, the mapping $z \to q(z,t)$ is analytic in K;
(ii) for each $z \in K$, the mapping $t \to q(z,t)$ is continuous on X;
(iii) for each r, $0 < r < 1$, there exists a number M_r such that $|q(z,t)| \le M_r$ for $|t| \le r$ and for $t \in X$.

Denote by P a set of probabilistic measures defined on Borel subsets of the space X. Let, for $\mu \in P$,

$$(1) \qquad f_\mu(z) = \int_X q(z,t)\,d\mu(t), \quad z \in K,$$

and let F denote the set $F = \{f_\mu: \mu \in P\}$.

It can be shown (cf. [4]) that:

(a) each function of the set F is analytic in the disc K;
(b) the mapping $\mu \to f_\mu$ is continuous (under the weak-$*$-topology induced from the space $C(X)^*$ and under the topology of almost uniform convergence on F);

(c) the set F is compact and it is a closed convex hull of the set of functions $\{z \to q(z,t);\ t \in X\}$;

(d) the only possible vertical points of the set F are the functions $z \to q(z,t)$, $t \in X$. If $t_o \in X$ and the condition

$$q(z,t_o) = \int_X q(z,t)d\mu(t)$$

holds only for $\mu = \delta_{t_o}$ where δ_{t_o} denotes a Dirac measure at the point t_o , then the function $z \to q(z,t_o)$ is a vertical point of the set F . In particular, if the mapping $\mu \to f_\mu$ is one-to-one, each function $z \to q(z,t)$, $t \in X$, is some vertical point of the set F .

As is known (cf. [21]) , many interesting classes of analytic functions possess the representation of form (1), where $X = [a,b]$, $a < b$, while $P = P(a,b)$ is a set of measures cumulated on the segment $[a,b]$ and such that

$$\int_a^b d\mu(t) = 1$$

In this case the mapping $\mu \to f_\mu$ is also, with the given q , one-to-one. Functions of the set F are now said to be representable by means of Stieltjes integral (cf. [8], p. 504), Moskva (1966)). In the sequel, we shall denote by F_q the set of such functions and, instead of (1), we shall write

$$f(z) = \int_a^b q(z,t)d\mu(t) \ , \quad z \in K . \tag{2}$$

As far as the mapping $q\colon K \times [a,b] \to C$ is concerned, various assumptions are made in case of need (cf. e.g. [19]), yet always such that the set F_q should be compact (in a definite topology), whereas the mapping $P(a,b) \ni \mu \to f \in F_q$ one-to-one.

Denote by $L_u(a,b)$, $a < b$, the set of functions $\varphi\colon [a,b] \to R$ which are measurable on the interval $[a,b]$ and satisfy the condition $\varphi(t) \in U$ for almost all $t \in [a,b]$, where U is a fixed subset of R . Denote by $\tilde{L}_u(a,b)$ the set of those functions $\varphi \in L_u(a,b)$ which satisfy the condition

$$\frac{1}{b-a} \int_a^b \varphi(t)dt = 1 . \tag{3}$$

Besides, denote by $F_q(M)$ a set of functions of the form

$$q(z) = \frac{1}{b-a} \int_a^b q(z,t)\varphi(t)dt \ , \tag{4}$$

where $M > 1$ is a fixed real number, the mapping $q: K \times [a,b] \to C$ possesses properties (i) - (iii), $\tilde{L}_u(a,b)$ and $U = [0,M]$. Functions belonging to many special classes of analytic functions possess the representation of form (4) under condition (3). If, for instance, $q(z,t) = (e^{it} + z)/(e^{it} - z)$, $a = -\pi$, $b = \pi$, then the set $F_q(M)$ is a subset of the set of Caratheodory functions (cf. [24], [25]). The above-mentioned representations of special classes of analytic functions constituted an inspiration for introducing the class $F_q(M)$. It appears that its properties are analogous to those of the class F_q.

Let A denote the complex linear space of all functions analytic in the disc K, with the topology of almost uniform convergence. Let Φ denote a functional defined on some subspace B of the space A and taking real values. Without loss of generality, let us assume that the set B consists of suitably normalized univalent functions. For $f \in A$, consider an $(n+1)$-dimensional complex vector $v(f)$ with components

$$f^{(o)}(z_o), f^{(1)}(z_o), \ldots, f^{(n)}(z_o)$$

where $f^{(o)}(z_o) \equiv f(z_o)$, z_o is a fixed point of the disc K, while n is a fixed positive integer.
Assume that the set B is compact and consider a set

$$V = \{v(f): f \in B\} \subset C^{n+1} .$$

This set is compact since the set B is compact.

Let us take into consideration a finite system of continuous functionals $\Phi_o, \ldots, \Phi_m$, $m \geq 1$, of the form

$$\Phi_j[f] = F_j(v(f)) = F_j(f(z_o), f^{(1)}(z_o), \ldots, f^{(n)}(z_o))$$

defined on the set B, where the functions F_j, $j = 0,1,\ldots,m$, are continuous and continuously differentiable in the real sense (cf [21], p. 184).

Problem 1. Determine the extremal values of the functional $\Phi_o[f]$ on the set B under the conditions $\Phi_j[f] \leq \lambda_j$, $j = 1,\ldots,m$, where λ_j, $j = 1,\ldots,m$, are fixed real numbers.
The function $f_o \in B$ is said to be a solution of Problem 1 in the set B if the functional $\Phi_o[f]$ takes its extremal value in the set $B_\lambda = \{f \in B: \Phi_j[f] \leq \lambda_j, j = 1,\ldots,m\}$ for $f = f_o$, with, that

$f_o \in B_\lambda$, $\lambda = \lambda(\lambda_1,\ldots,\lambda_m)$. With no loss of generality, we shall search for a minimal value of the functional $\Phi_o[f]$. We assume that Problem 1 has a solution.

3. Application of the Dubovitskii-Milyutin method

We shall formulate an extremal problem in the set $F_q(M)$. Let $B = F_q(M)$. The vector $v(g)$, where $g \in F_q(M)$, has now the components

$$c\int_a^b q(z_o,t)\varphi(t)\,dt \;,\ldots,\; c\int_a^b q^{(n)}(z_o,t)\varphi(t)\,dt$$

where $C = 1/(b-a)$.

It is readily seen that we may now formulate Problem 1 in the following way.

Problem 1′. Determine the extremal values (the minimal value) of the functional

$$\Phi_o[\varphi] = F_o\Big(c\int_a^b q(z_o,t)\varphi(t)\,dt,\ldots,c\int_a^b q^{(n)}(z_o,t)\varphi(t)\,dt\Big)$$

in the set of real functions φ , measurable on the interval $[a,b]$ and satisfying the conditions

$$\Phi_j[\varphi] = F_j\Big(c\int_a^b q(z_o,t)\varphi(t)\,dt,\ldots,c\int_a^b q^{(n)}(z_o,t)\varphi(t)\,dt\Big) \le \lambda_j$$

$$j = 1,\ldots,m \;,$$

$$\varphi(t) \in U \quad \text{for almost all} \quad t \in [a,b] \;,\; U = [0,M]$$

$$c\int_a^b \varphi(t)\,dt = 1 \;.$$

It is not difficult to see that if the function φ_o is a solution of Problem 1′ , then the function g_o defined by formula (4) is a solution of Problem 1. Conversely, if the function

$$g_o(z) = c\int_a^b q(z,t)\varphi_o(t)\,dt$$

is a solution of Problem 1, then the function φ_o is a solution of Problem 1′.

To examine Problem 1′, we have applied the Dubovitskii-Milyutin method in Girsanov's approach [7]. This approach allows one to carry on the considerations in the space $L_\infty(a,b)$ of real functions φ essentially bounded on the interval $[a,b]$, with norm

$$\|\varphi\| = \operatorname*{vrai\,sup}_{a\le t\le b} |\varphi(t)| .$$

We have obtained the following result.

THEOREM 1. *If the function*

$$g_o(z) = c \int_a^b q(z,t)\varphi_o(t)\,dt , \quad c = 1/(b-a) ,$$

is the argument of the extremum of the functional $\Phi_o[g]$ *in the set* $F_q(M)$ *under the conditions* $\Phi_j[g] \le \lambda_j$, $j = 1,\dots,m$, *then the function* φ_o *satisfies the condition*

$$a(t)\varphi_o(t) = \min_{\varphi\in U} a(t)\varphi$$

where $a(t)$ *is given by the formula*

$$a(t) = \sum_{j=0}^{m} l_j a_j(t) - l_{m+2} ;$$

$l_j \ge 0$, $-\infty < l_{m+2} < \infty$ *are given real numbers;*

$$a_j(t) = \frac{\partial \tilde F_j}{\partial x_1} (U_o^{(o)})\tilde U^{(o)} + \frac{\partial \tilde F_j}{\partial x_2} (\tilde V_o^{(o)})\tilde V^{(o)} + \dots +$$

$$+ \frac{\partial \tilde F_j}{\partial x_{2n+1}} (\tilde U_o^{(n)})\tilde U^{(n)} + \frac{\partial \tilde F_j}{\partial x_{2n+2}} (\tilde V_o^{(n)})\tilde V^{(n)} ,$$

$$\tilde F_j(x_1,x_2,\dots,x_{2n+2}) = F_j(x_1+ix_2,\dots,x_{2n+1}+ix_{2n+2}) ,$$

$$j = 0,1,\dots,m ;$$

$$\tilde U_o^{(k)} = c\int_a^b U^{(k)}(x_o,y_o,t)\varphi_o(t)\,dt$$

$$k = 0,1,\dots,n$$

$$\tilde V_o^{(k)} = c\int_a^b V^{(k)}(x_o,y_o,t)\varphi_o(t)\,dt$$

$$\tilde U^{(k)} = U^{(k)}(x_o,y_o,t)$$

$$k = 0,1,\dots,n$$

$$\tilde V^{(k)} = V^{(k)}(x_o,y_o,t) ;$$

$$U = \operatorname{Re} q , \quad V = \operatorname{Im} q , \quad x_o = \operatorname{Re} z_o , \quad y_o = \operatorname{Im} z_o .$$

4. Application of the extremum principle

In the sequel, we accept the notations and fact given on pages 3-5 of this paper.

Let m , $n_1,\dots,n_k$ be fixed positive integers. Consider a finite sequence of real functions $F_o, F_1, \ldots, F_m$ defined in the space R^{2n} where $n = n_1 + \ldots + n_k + k$. We assume that these functions are of class $C^{(1)}$.

Let $\zeta_1, \ldots, \zeta_k$ be fixed distinct points of the disc K. Denote by η a vector of the form

$$\eta(f) = [\mathrm{Re}\, f^{(o)}(\zeta_k), \mathrm{Im}\, f^{(o)}(\zeta_1), \ldots, \mathrm{Re}\, f^{(n_1)}(\zeta_1), \mathrm{Im}^{(n_1)}(\zeta_1),$$

(5)

$$\ldots, \mathrm{Re}\, f^{(o)}(\zeta_k), \mathrm{Im}\, f^{(o)}(\zeta_k), \ldots, \mathrm{Re}\, f^{(n_k)}(\zeta_k), \mathrm{Im}\, f^{(n_k)}(\zeta_k)]$$

where $f \in F$, $f \equiv f_\mu$, $f^{(o)} \equiv f$.

We introduce the notation

(6) $\mathrm{Re}\, f^{(i)}(\zeta_j) = u_j^i$, $\mathrm{Im}\, f^{(i)}(\zeta_j) = v_j^i$

for $i = 0,1,\ldots,n_j$ and $j = 1,2,\ldots,k$.

With notation (6) vector (5) has the form

$$\eta(f) = [u_1^o, v_1^o, \ldots, u_1^{n_1}, v_1^{n_1}, \ldots, u_k^o, v_k^o, \ldots, u_k^{n_k}, v_k^{n_k}] .$$

We construct a finite sequence of real functionals

$$F_j(f) = F_j(\eta(f)), \quad j = 0,1,\ldots,m,$$

defined on the family F.

In paper [16] we considered the following extremal problem.

Problem 1. Determine a minimum (maximum) of the functional $F_o(f)$ under the conditions $F_j(f) \le 0$, $j = 1,\ldots,m$, $f \in F$.

In order to examine Problem 1, the extremum principle and the theorem on extremal measures may be applied.

Let $\varphi_o, \varphi_1, \ldots, \varphi_m$ be a finite sequence of real functionals defined on some non-empty and convex subset U of a given space. Let us denote by L a function of the form

(7) $$L(u, \lambda_o, \lambda_1, \ldots, \lambda_m) = \sum_{i=1}^{m} \lambda_i \varphi_i(u)$$

where $u \in U$, while $\lambda_o, \lambda_1, \ldots, \lambda_m$ are given real numbers. Function (7) is called a Lagrange function, whereas the constants λ_i, $i = 0,1,\ldots,m$, are Lagrange multipliers.

THEOREM 2. (The extremum principle). If

I. $u_o \in U$ and $\varphi_i(u_o) \le 0$, $i = 1,2,\ldots,m$,

II. u_o <u>is the argument of the minimum of the functional</u> φ_o <u>on the set</u> U <u>under the constraints</u> $\varphi_i(u) \le 0$, $i = 1,2,\dots,m$,

III. <u>the vector function</u> $\varphi = [\varphi_o,\varphi_1,\dots,\varphi_m]$ <u>satisfies the following weak condition of convexity on the set</u> U . <u>For any</u> $u_1,u_2 \in U$ <u>and</u> $\alpha \in [0,1]$, <u>there exists</u> $u \in U$ <u>such that</u>

$$\varphi_i(u) \le \alpha\varphi_i(u_1) + (1-\alpha)\varphi_i(u_2) \quad \text{for} \quad i = 0,1,\dots,m ,$$

<u>then there exist Lagrange multipliers</u> $\lambda_i \ge 0$, $i = 0,1,\dots,m$, <u>not vanishing simultaneously and such that</u>

(a) $\quad L(u_o,\lambda_o,\lambda_1,\dots,\lambda_m) = \min_{u \in U} L(u,\lambda_o,\lambda_1,\dots,\lambda_m)$

(b) $\quad \lambda_i\varphi_i(u_o) = 0 \quad$ for $i = 1,\dots,m$.

The above theorem follows directly from Theorem 3 formulated in monograph [9] pp. 79-80.

Let $v = v(t)$ be a continuous real function defined on the space X .

Denote by E the following set $E = \{\tau \in X\colon v(\tau) = M\}$ where $M = \max_{t \in X} v(t)$.

<u>THEOREM 3.</u> <u>If</u>

$$\int_X v(t)\,d\mu_*(t) = \max_{\mu \in \mathcal{P}} \int_X v(t)\,d\mu(t) ,$$

<u>then the measure</u> μ_* <u>is cumulated on the set</u> E . (Proof-cf. [17] pp. 150).

Basing ourselves on Theorem 2 and 3, we prove a general necessary condition (given below) for the existence of the extremum in problem 1.

The Lagrange function for Problem 1 has the form

$$L(f,\lambda_o,\lambda_1,\dots,\lambda_m) = \sum_{j=0}^{m} \lambda_j F_j(f) .$$

Let f_*

$$f_*(z) = \int_X q(z,t)\,d\mu_*(t)$$

denote a solution of Problem 1.

Such a solution does exist provided that the set

$$W = \{f \in \mathcal{F}\colon F_j(f) \le 0 , \ j = 1,2,\dots,m\}$$

is not empty.

Let w be a function of the form

$$w(t) = \sum_{s=1}^{k} \sum_{i=0}^{n_s} (A_s^i \operatorname{Re} q^{(i)}(\zeta_s,t) + B_s^i \operatorname{Im} q^{(i)}(\zeta_s,t))$$

where $t \in X$,

$$A_s^i = \frac{\partial L(f,\lambda_o,\lambda_1,\dots,\lambda_m)}{\partial u_s^i}\Big|_{f=f_*}, \quad B_s^i = \frac{\partial L(f,\lambda_o,\lambda_1,\dots,\lambda_m)}{\partial v_s^i}\Big|_{f=f_*}$$

and

$$q^{(i)}(\zeta_s,t) = \frac{\partial^i}{\partial \zeta^i} q(\zeta,t)\Big|_{\zeta=\zeta_s}.$$

Denote by E_w a set of the form

$$E_w = \{t \in X: w(t) = \max_{\tau \in X} w(\tau)\}.$$

The set is not empty since the function w is continuous and the space X compact.

THEOREM 4. *If the function*

$$f_*(z) = \int_X q(z,t)d\mu_*(t)$$

is a solution of Problem 1 in which, for the function $F = [F_o,F_1,\dots,F_m]$, *the weak condition of convexity on the set* $\mathcal{F}$ *(condition III from Theorem 2) is satisfied, then the measure* μ_* *is cumulated on the set* E_w.

From theorem 4 follows

COROLLARY 1. *If* 1° $X \subset R$, 2° $w(t) \neq \text{const.}$, 3° *the mapping* $X \ni t \to q(z,t)$ *is analytic on the set* X, *then the extremal function* f_* *is of the form*

$$f_*(z) = \sum_{i=1}^{N} \alpha_i q(z,t_i)$$

where N *is some positive integer,* $\alpha_i \geq 0$, $i = 1,\dots,N$, $\sum_{j=1}^{N} \alpha_j = 1$, $t_i \in X$, $i = 1,\dots,N$.

5. Application of the maximum principle

We shall now give an example of the application of the Pontryagin maximum principle to the investigation of a certain extremal problem in the space of univalent functions in a half-plane.

Denote by H^1 the class of all functions which are holomorphic and univalent in the upper half-plane $P_z^+ = \{z: \text{Im } z > 0\}$ and which map this half plane on domains contained in the half-plane $P_w^+ = \{w: \text{Im } w > 0\}$. Moreover, those functions are normalized by the condition

$$\lim_{z \to \infty} [f(z) - z] = 0 \quad , \quad z \in P_z^+ .$$

It follows from the above definitions that the class in non-empty since the identity function certainly belongs to it.
It is proved that the class H^1 is a connected class in P_z^+ and is not a compact one (cf. [20]).

Let H_L^1 stand for the set of all functions $f \in H^1$ for which the complement $f(P_z^+)$ of P_w^+ is bounded. In virtue of Schwarz'z symmetry principle, each function $f \in H_L^1$ has an analytic continuation in $P_z^- = \{z: \text{Im } z < 0\}$ according to the formula $f(z) = \overline{f(\bar{z})}$, $z \in P_z^-$. Moreover, the function $f \in H_L^1$ is, by continuity, uniquely continuable onto the whole real axis $\partial P_z^+ = \{z: \text{Im } z = 0\}$, with the exception of some bounded part of ∂P_z^+. The function f thus continued is holomorphic in some ring

$$K_r = \{z: r < |z| < \infty , r > 0\} .$$

Expanding the function f in K_r in a Laurent series, we obtain

$$(8) \qquad f(z) = z + \sum_{k=1}^{\infty} c_k z^{-k}$$

where all the coefficients c_k are real numbers.

Let $\tilde{H}_L^1$ denote the set of all functions $f \in H_L^1$, each of which transforms the half-plane P_z^+ onto a domain $G_o = f(P_z^+)$ of the complex plane C_w, obtained from the half-plane P_w^+ by removing a finite number of pairwise disjoint Jordan arcs. It can be shown that the class $\tilde{H}_L^1$ is non-empty and dense in H_L^1 in the sense of the topology of uniform convergence inside P_z^+ (cf. [18]).

Making use of the ideas included in papers [14], [12] and [6], one can prove the following

<u>THEOREM 5.</u> *Let the differential equation*

$$(9) \qquad \frac{\partial f(z,t)}{\partial t} + \frac{1}{u(t)-z} \frac{\partial f(z,t)}{\partial z} = 0$$

be given, where $(z,t) \in P_z^+ \times [0,t_o]$, *while* $u(t)$ *is a measurable function on the interval* $[0,t_o]$. *If* $\varphi(z)$ *is a holomorphic function in the half-plane* P_z^+ *, whereas* $f(z,t)$ *is a solution of equation* (9) *with the initial condition*

(10) $f(z,0) = \varphi(z)$,

then there exists exactly one holomorphic and univalent function

$$g: P_z^+ \times [0,t_o] \to P_z^+$$

such that

(11) $f(z,t) = \varphi(g(z,t))$ for $t \in [0,t_o]$.

Let $\varphi \in \tilde{H}_L^1$, and let it possess the expansion in a Laurent series given by formula (8). It turns out that, for this function, there exist: a number $t_o > 0$ and a real function $u = u(t)$ piecewise continuous on the interval $[0,t_o]$, with no points of discontinuity of the second kind, continuous on the right at the point 0 and on the left at the point t_o , such that $\varphi(z) = f(z,0)$ where $f(z,t) \in \tilde{H}_L^1$ with any $t \in [0,t_o]$ is the integral of equation (9) with condition (10). Since, with each $t \in [0,t_o]$, the function $f(z,t)$ belongs to $\tilde{H}_L^1$, therefore this function, in the neighbourhood of infinity, has the expansion

(12) $$f(z,t) = z + \sum_{k=1}^{\infty} x_k(t) z^{-k}$$

with real coefficients $x_k = x_k(t)$ $(k = 1,2,\ldots)$. Substituting (12) in equation (9), we obtain, for the coefficients $x_k(t)$, $k = 1,2,\ldots$, the following system of equations:

$$\dot{x}_1 = 1$$

$$\dot{x}_2 = u$$

$$\dot{x}_3 = u^2 - x_1$$

$$\cdots\cdots\cdots$$

$$\dot{x}_k = u^{k-1} - \sum_{m=1}^{k-2} m\, x_m\, u^{k-m-2}$$

$$\cdots\cdots\cdots\cdots\cdots\cdots$$

for almost all $t \in [0,t_o]$.

If c_k $(k = 1,2,\ldots)$ are coefficients in the expansion of the function φ in a Laurent series, then from condition (10) we have

(13) $x_k(0) = c_k$, $k = 1,2,\ldots$.

It can be shown (cf. [3]) that $c_1 \leq 0$, with that $c_1 = 0$ if and only if $\varphi(z) \equiv z$. Besides, $x_k = x(t_o) = 0$, $k = 1,2,\ldots$. From this and from condition (13) it follows that $t_o = -c_1$.

From among all functions of class H_L^1 we choose those for which the first two coefficients c_1 and c_2 of the expansion in a Laurent series are known. Let us consider the following extremal problem:

P r o b l e m 1. Determine the extremal values of the coefficient c_3 in expansion (8), with the first coefficients c_1 and c_2 fixed.

Let a point move in the space of variables (x_1, x_2, x_3) according to the law

$$\begin{aligned} &\dot{x}_1 = 1 , \quad \dot{x}_2 = u , \quad \dot{x}_3 = u^2 - x_1 , \quad t \in [0, t_o] , \\ &x_i(0) = c_i , \quad x_i(t_o) = 0 , \quad i = 1,2,3. \end{aligned} \tag{14}$$

We accept that the function $u = u(t)$ is a control function. Note that

$$\int_0^{t_o} (u^2(t) - x_1(t))dt = \int_0^{t_o} \dot{x}_3(t)dt =$$

$$= x_3(t_o) - x_3(0) = -x_3(0) = -c_3 ,$$

and thus

$$x_3(0) = c_3 = -\int_0^{t_o} (u^2(t) - x_1(t))dt .$$

Let us consider

P r o b l e m 2. Determine a control u that carries a point from the position (c_1, c_2, x_3) at the instant $t = 0$ to the position $(0,0,0)$ at the instant t_o in such a way that the functional $J_1 = x_3(0)$ should take its maximal value or, which is equivalent, that the functional

$$J(u) = \int_0^{t_o} (u^2(t) - x_1(t))dt \tag{15}$$

should attain its minimal value.

Applying the transformation $\tilde{x}_1 = x_1 - t$, $\tilde{x}_2 = x_2$, from (14) and (15) we obtain

$$\dot{\tilde{x}}_1 = 0 , \quad \dot{\tilde{x}}_2 = u , \quad \tilde{x}_1(t_o) = -t_o , \quad \tilde{x}_2(t_o) = 0 ,$$

$$\tilde{x}_1(0) = c_i , \quad i = 1,2,$$

$$(16) \quad J(u) = \int_0^{t_0} (u^2(t) - \tilde{x}_1(t) - t)\,dt .$$

In this situation, problem 2 can be formulated in the following way:

Problem 3. Determine a control u that carries a point from the position c_1, c_2 at the instant $t = 0$ to the position $(-t_0, 0)$ at the instant t_0 in such a way that functional (16) should attain its minimum.

Of course, the optimal control found for problem 3 is optimal for problem 2.

Making use of the theorems given in monograph [15] on pages 226-234 (including the maximum principle), we state that problem 3 does possess a solution. Besides, if the function $f \in \tilde{H}^1_L$ has expansion (8) with c_1 and c_2 fixed, then the precise estimate

$$c_3 \le c_2^2/c_1 - c_1^2/2$$

takes place. This estimate was obtained, in some other way, by V.V. Sobolev and T.N. Sellyakhova in 1974 [23].

The result obtained constitutes a confirmation of the efficacy of simultaneous application of the methods of complex analysis and optimal control to investigating extremal problems in the spaces of holomorphic and univalent functions in a half-plane.

References

[1] ABRAMS R.A.: Nonlinear programming in complex space, Doctoral thesis in Applied Mathematics, Northwestern Univ., Evanston III, 1969.

[2] ABRAMS R.A. BEN-ISRAEL A.: Nonlinear programming in complex space: necessary conditions, SIAM J. Control, Vol. 9, No 4 (1977) 606-620.

[3] ALEKSANDROV J.A.: Parametriseskie prodolzenia w teorii odnolistnych funksii, Moskva (1976).

[4] BRICKMAN L., MACGREGOR T.H., WILKEN D.R.: Convex hulls of some classical families of univalent functions. Trans. of the Amer. Marh. Soc. 156 (1971), 91-107.

[5] DUBOVICKII A.Ja., MILJUTIN A.A.: Extremum problems in the presence of restrictions, Z. Vychisl. Mat. i Mat. Fiz. 5 (1965) 395-453 (in Russian; English transl. U.S.S.R. Comput. Math. and Math. Phys. 5 (1965), 1-80).

[6] FRIEDLAND S., SCHIFFER M.: On coefficient regions of univalent functions, J. d'Analyse Mathématique, Vol. 31 (1977), 125-168.

[7] GIRSANOV J.V.: Lectures on mathematical theory of extremum problems. Lecture Notes in Economics and Mathematical Systems 67 (1972).

[8] GOLUZIN G.M.: Geometricheskaya teoria funksii kompleksnovo-peremennovo

[9] IOFFE A.D., TIHOMIROV V.M.: Theory of extremal problems, Studies in Math. and Ibs. Appl. 6 (1979).

[10] KASANIUK S.A.: Matematicheskoye programirovanie i zadachi uslovnego ekstremuma v spetsialnych klassakh analiticheskikh funksii, Sibir. Matem. Zuhrn. 10 (1969), 1224-1233.

[11] KASANIUK S.A., FINOGENOVA W.G.: O oblastiakh znachenii veshchestvennykh funktsii s fiksirovannymi koeffisientami i o oblastiakh znacheni system ikh koeffisientov, Isv. Vys. Uceb. Zav. No 4 (71) (1968), 40-47.

[12] KUFAREV P.P., SOBOLEV V.V., SPORYSHEVA L.V.: Ob odnom metode issledovania ekstremalnykh zadachi dla funksii odnolistnych v poluploskosti, Tr. Tomsk. Gos. Univ. Vol. 200 (1968), 142-164.

[13] LEVINSON N.: Linear programming in complex space, J. Math. Anal. Appl. 14 (1966), 44-62.

[14] LÖWNER K.: Untersuchungen über schlichte konforme Abbildungen des Einheitskreises I, Math. Ann. 89, (1923), 103-121.

[15] LEE E.B., MARKUS L.: Osnovy teorii optymalnovo upravlenia, Moskva (1972).

[16] MIKOŁAJCZYK L., WALCZAK S.: On application of the Dubovitskii--Milyutin method to investigating certain extremal problems, Demonstratio Mathematica, Vol. XIII, No 2, (1980), 509-529.

[17] MIKOŁAJCZYK L., WALCZAK S.: Application of the extremum principle to investigating certain extremal problems, Trans. of the American Math. Soc., Vol. 259, No 1 (1980), 147-155.

[18] MOSKVIN W.G., SELLJATHOVA T.N., SOBOLEV W.W.: Extremalnye svoistva nekotorych klasov funksii s fixirovanymi pervymi coeffisientamii, otobrazajustikh konformo poluploskost v sebia, Sibir. Mat. Żur., Vol. 21, No 2, (1980), 139-154.

[19] PFALTZGRAFF J.A., PINCHUK B.: A variational method for classes of meromorphic functions, Journal d'Analyse Math. XXIV, (1971).

[20] PIOTROWSKA J.: Extremal problems in some classes of univalent functions in a half-plane, Annales Polonici Math. 34 (1977), 201-220.

[21] POMMERENKE Ch.: Univalent functions, Gottingen (1975).

[22] PLATYNOWICZ B.: O pewnych zadaniach ekstremalnych w specjalnych klasach funkcji holomorficznych, Rozprawa doktorska, Uniwersytet Łódzki (1981).

[23] SELLIAKHOVA T.N., SOBOLEV V.V.: O vzajemnom roste koeffisientov funksii odnolistnykh v poluploskostii, Dokl. A.N. S.S.S.R. Vol. 218, No 4, (1974), 768-770.

[24] WALCZAK S.: Method of examining conditional extrema in some families of complex functions, Bull. Acad. Polon. Sci. Ser. Math. Astronom Phys. 11 (1976), 961-968.

[25] WALCZAK S.: On some problem of the variation calculus, Bull. Sci. Lett. Łódź, Vol. 24, 11 (1974), 1-10.

Institute of Mathematics
University of Łódź
Banacha 22, PL-90-258 Łódź , Poland

POTENTIAL THEORY IN NEVANLINNA THEORY AND ANALYTIC GEOMETRY

Robert E. Molzon (Lexington, KY)

1. INTRODUCTION

Let $F : M \to N$ be a holomorphic mapping between complex manifolds. In classical function theory M is a domain in $\mathbb{C}$ or a Riemann surface and N is $\mathbb{C}$ or $\mathbb{P}\mathbb{C}$. The potential theory of the Laplacian plays a decisive role in this situation. If M is a bounded domain in $\mathbb{C}^n$ and $N = \mathbb{C}$ then many different potentials have been used to study such mappings. Associated with the domain are various kernels and singular integrals which depend on the domain itself. A basic technique is to obtain properties of the function by studying the singular integral associated with the domain.

In these notes I would like to discuss another aspect of potential theory in several complex variables which, as far as applications are concerned, has emerged rather recently. This potential theory takes place on the range manifold N and in case $N = \mathbb{P}\mathbb{C}$ reduces to the elliptic capacity of Tsuji. If $N = \mathbb{P}^n\mathbb{C}$ then there have been several different capacities which have found interesting applications. It is not yet clear what the connection between these various capacities are. In Section 2 some of these theories are recalled and in Section 3 the capacity of Example A is used to study the value distribution and boundary behavior of a meromorphic mapping $F : \Omega \to \mathbb{P}^m\mathbb{C}$. In case Ω is a bounded domain in $\mathbb{C}^n$ with smooth boundary and Ω admits a logarithmic pseudoconvex exhaustion then a result given in Section 3 says that a meromorphic function $f : \Omega \to \mathbb{P}\mathbb{C}$ which omits a set of positive elliptic capacity is in the Nevanlinna class. In particular if f is holomorphic then f has admissible boundary values almost everywhere on $\partial\Omega$.

2. REVIEW OF POTENTIAL THEORY

I give in this section a brief review of various potential theories which have played a recent role in value distribution theory and analytic geometry. One may think of these potential theories as being defined on the range of a holomorphic mapping. Basically the potentials considered arise in one of the following ways: singular integrals, extremal functions associated with the Monge-Ampère operator, and extremal functions associated with an algebra of plurisubharmonic functions. The examples together with an idea of applications follow.

EXAMPLE A. $\log(\|Z\|\,\|W\|/|\langle Z,W\rangle|)$: Suppose $E \subset \mathbb{P}^n\mathbb{C}$ is compact and $\mu \in P(E)$ is a probability measure supported on E. If Z and W denote points in $\mathbb{P}^n\mathbb{C}$, define a potential by:

$$V_\mu(Z) = \int_E \log(\|Z\|\,\|W\|/|\langle Z,W\rangle|)d_\mu(W) \quad . \tag{1}$$

Let

$$V(E) = \inf_{\mu\in P(E)} \sup_{Z\in \mathbb{P}^n\mathbb{C}} V_\mu(Z) \; . \tag{2}$$

If V(E) is finite put $C_1(E) = 1/V(E)$, otherwise put $C_1(E)=0$. C_1 has the properties of an inner capacity for sets in $\mathbb{P}^n\mathbb{C}$; some of the properties are given below.

(i) If $E \subset \mathbb{P}^n\mathbb{C}$ is a compact nonlocally pluripolar set then $C_1(E)>0$.

(ii) If σ is a nondegenerate smooth curve in $\mathbb{P}^n\mathbb{C}$ then the image of σ in $\mathbb{P}^n\mathbb{C}$ has positive capacity. By a nondegenerate smooth curve we mean a smooth mapping $\sigma : [0,1] \to \mathbb{P}^n\mathbb{C}$ such that there exists a lifting $\tilde{\sigma} : [0,1] \to \mathbb{C}^{n+1}$ which is smooth and such that $\{\tilde{\sigma}^{(k)}(t)\}_{k\geq 1}$ spans $\mathbb{C}^{n+1}$ for every $t \in [0,1]$. $\tilde{\sigma}^{(k)}(t)$ denotes the kth derivative of $\tilde{\sigma}$ at t.

(iii) As a special case of the above, any real analytic arc not contained in a hyperplane in $\mathbb{P}^n\mathbb{C}$ has positive capacity.

The C_1 capacity has an application to analytic geometry which I will briefly discuss. Suppose $X \subset \mathbb{C}^n$ is a closed analytic subvariety of $\mathbb{C}^{n+1}$ of pure dimension k. The geometric problem I want to consider is to estimate the growth of X in terms of the growth of the intersection of X with hyperplanes in $\mathbb{C}^{n+1}$. Specifically the growth of X may be measured by its order which is measured by the growth of the volume of X in a ball of radius r. Now hyperplanes through the origin in $\mathbb{C}^{n+1}$ may be parameterized by points W in $\mathbb{P}^n\mathbb{C}$. Write H^W for the hyperplane corresponding to W. The following result relating the order of X and the order of $X \cap H^W$ holds.

THEOREM 1 [10]. *Suppose $E \subset \mathbb{P}^n\mathbb{C}$ with $C_1(E) > 0$ and $X \subset \mathbb{C}^{n+1}$ is an analytic variety of pure dimension. Then if μ denotes the equilibrium measure on E,*

$$\text{order}\,(X) = \text{order}\,(X \cap H^W)$$

for μ *almost all* $W \in E$.

Another result related to the above theorem is the following.

THEOREM 2 [10]. *Suppose* $E \subset \mathbb{P}^n\mathbb{C}$ *has* $C_1(E) > 0$ *and* $X \subset \mathbb{C}^{n+1}$ *is a one dimensional analytic variety. If* $X \cap H^W$ *is finite for all* $W \in E$ *then* X *is algebraic.*

In particular only very small sets of hyperplanes are required to determine the growth of an analytic variety in $\mathbb{C}^n$ through intersection.

EXAMPLE B. $\log(\|Z\|^d / |p(Z)|)$: Suppose p is a homogeneous polynomial of degree d on $\mathbb{P}^n\mathbb{C}$, hence a polynomial in n+1 variables. Denote by $S_{n+1,d}$ the set of all such polynomials. $S_{n+1,d}$ may be thought of as dual to $\otimes_s^d \mathbb{C}^{n+1}$, the d-fold symmetric tensor product of $\mathbb{C}^{n+1}$. Let $\mathbb{P}S_{n+1,d}$ denote the associated projective space of such polynomials. The Euclidean norm on $\mathbb{C}^{n+1}$ induces a norm on $S_{n+1,d}$. Let N_d denote the unit sphere with respect to this norm. If $K \subset N_d$ is a compact circled set, define a potential on $\mathbb{P}^n\mathbb{C}$ by

$$U_{d,\mu}(Z) = \int_{p\in K} \log(\|Z\|^d / |p(Z)|)\,d\mu(p) \tag{3}$$

where $\mu \in P(K)$ is a probability measure supported on K. Associating K with a compact subset of $\mathbb{P}S_{n+1,d}$ one may equally as well write the potential as

$$U_{d,\nu}(Z) = \int_{P\in E} \log \frac{\|Z\|^d \|P\|}{|P(Z)|}\,d\nu(P) \tag{4}$$

where $\nu \in P(E)$ and we think of P as being given in homogeneous coordinates. If $\phi : \mathbb{P}^n\mathbb{C} \to \mathbb{P}S^*_{n+1,d}$ denotes the Veronese map then the potential may be written as

$$U_{d,\nu}(Z) = \int_{P\in E} \log \frac{\|\phi(Z)\|\,\|P\|}{|\langle\phi(Z),P\rangle|}\,d\nu(P)\ . \tag{5}$$

If d=1 we are back to the singular integral of Example A. Define now a capacity by:

$$U_d(E) = \inf_{\nu\in P(E)} \sup_{Z\in\mathbb{P}^n\mathbb{C}} U_{d,\nu}(Z) \tag{6}$$

$$C_d(E) = 1 / U_d(E) \tag{7}$$

with the convention that $C_d(E)=0$ if $U_d(E) = +\infty$. Now using the form (5) above for the potential, one can obtain properties for $C_d(E)$ directly from the properties given for $C_1(E)$ in Example A.

Applications to analytic geometry are similar to those indicated in Example A except that we are now able to study intersection with an algebraic variety of degree d. We have for example the following result.

THEOREM 3 [12]. *Suppose* $E \subset \mathbb{P}S_{n+1,d}$ *is a set of polynomials of positive* C_d *capacity so* $C_d(E) > 0$. *If* $P \in \mathbb{P}S_{n+1,d}$, *let* Y^P *denote the affine algebraic variety in* $\mathbb{C}^{n+1}$ *associated with* P. *Suppose* $X \subset \mathbb{C}^{n+1}$ *is an analytic variety of pure dimension. If* $X \cap Y^P$ *is algebraic for all* $P \in E$ *then* X *is algebraic.*

EXAMPLE C. $(dd^c u)^n$: Let $\Omega \subset \mathbb{C}^n$ be a bounded strictly pseudoconvex domain with smooth boundary and let PSH(Ω) denote the plurisubharmonic functions defined on Ω. If $E \subset \Omega$ is compact, associate with E an extremal function defined by ([3]):

$$u_E(z) = \sup \{\phi(z) : \phi \in \mathrm{PSH}(\Omega),\ \phi \le 0 \text{ on } E, \text{ and } \phi < 1 \text{ on } \Omega\}$$

$$\hat{u}_E(z) = \limsup_{\zeta \to z} u_E(\zeta) .$$

Define a capacity function on compact subsets of Ω by:

$$C(E) = \int_\Omega (dd^c \hat{u}_E)^n .$$

Since $\hat{u}_E$ is a bounded plurisubharmonic function, this definition makes sense. The Monge-Ampère operator acting on $\hat{u}_E$ has the following properties:

(i) Support $(dd^c \hat{u}_E)^n \subset E$.

(ii) If E is locally pluripolar then $(dd^c \hat{u}_E)^n \equiv 0$.

(iii) If E is not locally pluripolar then $(dd^c \hat{u}_E)^n$ is a positive measure on E so that

$$\int_\Omega (dd^c \hat{u}_E)^n = \int_E (dd^c \hat{u}_E)^n > 0 .$$

The extremal function $\hat{u}_E$ may thus be used to give a characterization of pluripolar sets in $\mathbb{C}^n$.

I now turn to a global version of the above extremal function and capacity. Let

$$L = \{\psi \in \mathrm{PSH}(C^n) : \psi(z) \le C + \log(1+|z|) \text{ for some constant } C\} .$$

If $E \subset \mathbb{C}^n$ is compact then we associate with E the extremal function defined by Siciak [18]:

$$\Psi_E(z) = \sup \{\psi(z) : \psi \in L \text{ and } \psi \leq 0 \text{ on } E\}$$

$$\hat{\Psi}_E(z) = \limsup_{\zeta \to z} \Psi_E(\zeta) .$$

If $E \subset \mathbb{C}^n$ is compact and locally pluripolar then $\Psi_E \equiv +\infty$. Now we wish to apply the Monge-Ampère operator to $\hat{\Psi}_E$. In case E is locally pluripolar so $\hat{\Psi}_E \equiv +\infty$ use the convention $(dd^c\hat{\Psi}_E)^n \equiv 0$. If E is non-locally pluripolar so $\hat{\Psi}_E$ is locally bounded, $(dd^c\hat{\Psi}_E)^n$ makes sense. For E compact define

$$C(E) = \int_{\mathbb{C}^n} (dd^c\hat{\Psi}_E)^n .$$

We record the following properties of this capacity.

(i) Support $(dd^c\hat{\Psi}_E)^n \subset E$.

(ii) If E is non-locally pluripolar then $(dd^c\hat{\Psi}_E)^n$ is positive so that

$$\int_{\mathbb{C}^n} (dd^c\hat{\Psi}_E)^n > 0 .$$

We now mention two applications of this capacity to geometry and function theory. The first application is a result of Gruman [5] which measures the growth of the intersection of analytic varieties in $\mathbb{C}^n$ with linear subspaces of $\mathbb{C}^n$ and the second application is an average Blaschke condition for holomorphic mappings due to the author [11].

If $X \subset \mathbb{C}^n$ is a closed analytic subvariety of $\mathbb{C}^n$ let $\theta_\ell(X)$ denote the current of integration of $X \cap \ell$ with $\ell \in G_q(\mathbb{C}^n)$ an ℓ-plane in $\mathbb{C}^n$. $G_q(\mathbb{C}^n)$ denotes the Grassmannian of all ℓ-planes through 0 in $\mathbb{C}^n$. Let $\beta = dd^c|z|^2$ and

$$\sigma_X(\ell,r) = (\theta_\ell(X)\big|_{B^n(r)} , \beta^{p+q-n})$$

where $q \geq n-p$ and X is assumed to have pure dimension p. Let

$$\sigma_X(r) = \int_{X \cap B^n(r)} \beta^p .$$

We then have the following result which relates the growth of X to the growth of $X \cap \ell$ for all ℓ except perhaps a locally pluripolar set.

THEOREM 4 [5]. *Let* $X \subset \mathbb{C}^n$ *be an analytic variety of pure dimension* p *and* $q \geq n-p$. *Let* $\varepsilon > 0$ *and* $\beta > 1$ *be constants. Then the set*

$$E(\varepsilon,\beta) = \left\{ \ell \in G_q(\mathbb{C}^n) : \limsup_{r\to\infty} \frac{\sigma_X(\ell,r)\cdot r^{2(n-q)}}{(\log r)^{\beta}\cdot \sigma_X((1+\varepsilon)r)} \neq 0 \right\}$$

is locally pluripolar in $G_q(\mathbb{C}^n)$.

REMARK. What the above theorem roughly says is that the growth of $X \cap \ell$ is larger than the growth of X for at most a set of ℓ which is locally pluripolar in $G_q(\mathbb{C}^n)$.

PROOF. Although the proof is too involved to be given here we remark that the main idea is to use the extremal function $\hat{\Psi}_E$ to produce a measure $d\mu = (dd^c\hat{\Psi}_E)^n$ associated to a nonlocally pluripolar set in $G_q(\mathbb{C}^n)$. An estimate for the average of $\sigma_X(\ell,r)$ over this non-locally pluripolar set with respect to μ is then obtained which leads to the result via an argument by contradiction. □

As a second application we mention the following result on average Blaschke conditions for holomorphic mappings. Let $F : B^n \to \mathbb{C}^n$ be a nondegenerate holomorphic mapping of the unit ball to $\mathbb{C}^n$; hence, $F^{-1}(a)$ is discrete for each $a \in \mathbb{C}^n$. Let $n(r,a) = \text{card}(F^{-1}(a) \cap B^n(r))$, $B^n(r)$ denoting a ball of radius r. The following average Blaschke condition then holds.

THEOREM 5 [11]. *Let* $F : B^n \to \mathbb{C}^n$ *be a bounded nondegenerate holomorphic mapping. If* $E \subset \mathbb{C}^n$ *is compact and nonlocally pluripolar, there exists on* E *a probability measure* μ, *depending only on* E, *such that*

$$\int_{r_o}^{r} \frac{dt_1}{t_1} \cdots \int_{r_o}^{r} \frac{dt_n}{t_n} \int_E n(t_n,a)\,d\mu(a) \leq M < +\infty ,$$

for some constant M.

REMARK. The probability measure μ in the above result is constructed from $(dd^c\hat{\Psi}_E)^n$ where E is a nonlocally pluripolar set.

EXAMPLE D. $\tau(E)$, Tchebycheff constant: As a final example we recall the Tchebycheff constant of H. Alexander which is associated to compact sets of $\mathbb{P}^n\mathbb{C}$. A compact set E in $\mathbb{P}^n\mathbb{C}$ may be associated with a compact circled set K in $S = \{z \in \mathbb{C}^{n+1} : |z| = 1\}$. If p is a homogeneous polynomial of degree d on $\mathbb{C}^{n+1}$, we say that p is normalized if

$$\int_S \log |p| d\sigma = d \int_S \log |z_1| d\sigma$$

where $d\sigma$ denotes the usual volume element on the sphere. Let P_d denote the normalized polynomials of degree d. If K is compact and circled in S then

$$m_d(K) = \inf_{p \in P_d} \sup_{z \in K} |p(z)|$$

$$\tau(K) = \lim_{d \to \infty} [m_d(K)]^{\frac{1}{d}} .$$

One property of this Tchebycheff constant is that a compact set E in $\mathbb{P}^n\mathbb{C}$ is locally pluripolar if and only if $\tau(E)=0$. We note that τ makes sense as a function on sets in $\mathbb{P}^n\mathbb{C}$ by the identification with circled sets in S. The characterization of locally pluripolar sets in $\mathbb{P}^n\mathbb{C}$ by the Tchebycheff constant leads to the following question. Is it possible to directly relate the Tchebycheff constant of a compact circled set in S to the capacity of the set defined by the Monge-Ampère operator? In other words, do there exist universal constants c_1 and c_2 such that an equality of the form

$$c_1 \, \tau(K) \le C(K) \le c_2 \, \tau(K)$$

holds for compact circled sets $K \subset S$?

3. BOUNDARY VALUES OF MEROMORPHIC MAPPINGS

This section deals with an application of the capacity of Example A in Section 2 to function theory on bounded domains in $\mathbb{C}^n$. In order to first present the ideas in a simple setting we consider the case of meromorphic functions. Suppose $\Omega \subset \mathbb{C}^n$ is a strictly pseudoconvex bounded domain with smooth boundary. Let τ denote a finite exhaustion function on Ω. Denote

$$\alpha = dd^c \log \tau$$

$$\beta = dd^c \tau$$

$$\sigma = d^c \log \tau \wedge \alpha^{n-1} .$$

We assume τ is proper, smooth up to the boundary with $\partial\Omega = \{z : \tau(z) = 1\}$, α and β positive on Ω and Hausdorff 2n-1 measure absolutely continuous with respect to σ on $\{z : \tau(z) = r\}$. Let

$$\Omega(r) = \{z : \tau(z) = r\}$$

$$\Omega[r_0, r) = \{z : r_0 \le \tau(z) < r\}$$

etc.

We say that such an exhaustion is logarithmic pseudoconvex following Stoll. If $\alpha^n \equiv 0$ we say that the exhaustion τ is parabolic.

Suppose ν is a divisor on Ω. Put

$$n_\nu(r) = \int_{\Omega[0,r)} \nu\alpha^{n-1} + n_\nu(0) = r^{2-2n} \int_{\Omega[r]} \nu\beta^{n-1}$$

$$N_\nu(r_0, r) = \int_{r_0}^{r} \frac{n_\nu(t)}{t}\, dt .$$

Let f be a meromorphic function on Ω. If $U \subset \Omega$ is an open set then $\tilde{f}$ is a reduced representation for f on U if $\pi \circ \tilde{f} = f$ on $U - \tilde{f}^{-1}(0) \cup S$ where $S \subset \Omega$ is a thin analytic set such that $f : \Omega - S \to \mathbb{P}\mathbb{C}$ is a holomorphic map and $\tilde{f} : U \to \mathbb{C}^2$. Here $\pi : \mathbb{C}^2 - \{0\} \to \mathbb{P}\mathbb{C}$ denotes the natural projection. This concept of a reduced representation of a meromorphic function will be used to discuss the relationships between various forms of the order function of a meromorphic function. If $F : \Omega \to \mathbb{P}^m\mathbb{C}$ is a meromorphic mapping the concept of a reduced representation may be extended to cover this situation.

Now let f be a meromorphic function and let ν denote the total divisor, zero devisor plus polar divisor, of f. Jensen's formula then tells us that

$$N_\nu(r_0, r) = \int_{\partial\Omega(r)} \log |f|\sigma - \int_{\partial\Omega(r_0)} \log |f|\sigma - \int_{\Omega(r_0, r]} \log |f|\alpha^n .$$

Now by rewriting the terms in Jensen's formula one obtains the First Main Theorem (FMT) of Nevanlinna. Let ω denote the Kähler form of the Fubini-Study metric on $\mathbb{P}\mathbb{C}$. Define

$$A(r) = \frac{1}{r^{2(n-1)}} \int_{\Omega[r]} f^*\omega \wedge \beta^{n-1}$$

$$T(r) = \int_{r_0} \frac{A(t)}{t}\, dt .$$

$T(r)$ is called the order function of a meromorphic function. Now when one rewrites the terms in Jensen's formula, the boundary term which occurs is

$$m(r,a) = \int_{\partial\Omega(r)} f^* \log \frac{1}{\|Z:a\|} \sigma$$

where $\|Z:a\|$ denotes the projective distance from Z to a. We also obtain the error term

$$S(r,a) = \int_{\Omega(r_o,r]} f^* \log \frac{1}{\|Z:a\|} \alpha^n .$$

If $a \in \mathbb{P}^n\mathbb{C}$ let ν_a denote the zero divisor of f-a. Let

$$n(r,a) = r^{2-2n} \int_{\Omega[r]} \nu_a \beta^{n-1}$$

$$N(r,a) = \int_{r_o}^{r} \frac{n(t,a)}{t} dt .$$

Applying Jensen's formula and using the above definitions gives the FMT:

$$N(r,a) + m(r,a) - m(r_o,a) - S(r,a) = T(r) .$$

We will say that the meromorphic function f is in the Nevanlinna class provided $T(r) \le O(1)$. Suppose $f : \Omega \to \mathbb{C}$ is holomorphic. One may think of f as being a holomorphic mapping into $\mathbb{P}\mathbb{C}$ with reduced representation $\tilde{f} = (1,f)$. Rewriting the order function one finds

$$T(r) = \int_{\partial\Omega(r)} \log \|\tilde{f}\| \sigma - \int_{\partial\Omega(r_o)} \log \|\tilde{f}\| \sigma - \int_{\Omega(r_o,r]} \log \|\tilde{f}\| \alpha^n .$$

Now from this representation of the order function one sees that $T(r) \le O(1)$ provided

$$\limsup_{r\to 1} \int_{\partial\Omega(r)} \log \sqrt{1+|f|^2}\, \sigma \le O(1) .$$

The definition of the Nevanlinna class, $T(r) \le O(1)$, therefore agrees with the usual definition for holomorphic functions on bounded domains.

We return to the case of a meromorphic function $f : \Omega \to \mathbb{P}\mathbb{C}$ and find a sufficient condition which says that f must be in the Nevanlinna class. We apply the potential defined in Example A which for $\mathbb{P}\mathbb{C}$ reduces to the standard elliptic capacity. Suppose $E \subset \mathbb{P}\mathbb{C}$ is a set of positive elliptic capacity. Then there exists an equilibrium measure μ supported on E such that

$$u_E(Z) = \int_E \log \frac{1}{||Z:a||} d\mu(a) \leq O(1) .$$

We recall that the projective distance between Z and a in $\mathbb{P}\mathbb{C}$ is given by

$$||Z:a|| = \frac{||Z||\,||a||}{|\langle Z,a^{\perp}\rangle|}$$

where on the right-hand side we think of Z and a as being representatives for the points in $\mathbb{P}\mathbb{C}$ given in homogeneous coordinates $[Z_0:Z_1]$ and $[a_0:a_1]$. The point $a^{\perp}$ is represented by $[-a_1:a_0]$. We make this distinction in the case of $\mathbb{P}\mathbb{C}$ because we want to think of a as a point, not a hyperplane, in the FMT. Keeping the above remarks in mind, we have the folllowing result.

THEOREM 6. *Suppose* Ω *is a bounded domain with smooth boundary and logarithmic pseudoconvex exhaustion. Let* $f : \Omega \to \mathbb{P}\mathbb{C}$ *be a meromorphic function. If* f *omits a set* $E \subset \mathbb{P}\mathbb{C}$ *of positive elliptic capacity then* f *is in the Nevanlinna class, that is* $T(r) \leq O(1)$.

PROOF. Let μ be an equilibrium measure supported on E. We know

$$T(r) \leq m(r,a) + N(r,a) .$$

Integrate the above inequality with respect to μ to get

$$T(r) \leq \int_E m(r,a) d\mu(a)$$

where we have used the fact that μ is a probability measure. Rewriting the right-hand side gives

$$T(r) \leq \int_{\partial\Omega(r)} f^* \left(\int_E \log \frac{1}{||Z:a||} d\mu(a) \right) \sigma$$

$$\leq \int_{\partial\Omega(r)} f^* u_E(Z)\sigma \quad \int_{\partial\Omega(r)} \text{const. } \sigma \leq O(1) . \quad \square$$

Now suppose Ω is a strictly pseudoconvex domain with smooth boundary and $H^2(\Omega,\mathbb{Z})=0$. Skoda [15] and Henkin [7] have shown that a meromorphic function in the Nevanlinna class may be written as the quotient of two holomorphic functions on Ω which are in the Nevanlinna class. Furthermore Stein [20] has shown that a holomorphic function in the Nevanlinna class on a bounded domain in $\mathbb{C}^n$ with smooth boundary has admissible limits almost everywhere on $\partial\Omega$. Putting these two results together leads to the following corollary of the above result.

COROLLARY 7. *Suppose $\Omega \subset \mathbb{C}^n$ is a bounded strictly pseudoconvex domain with smooth boundary and with parabolic exhaustion. Suppose further that $H^2(\Omega,\mathbb{Z})=0$. If $f : \Omega \to \mathbb{P}\mathbb{C}$ is a meromorphic function which omits a set $E \subset \mathbb{P}\mathbb{C}$ of positive elliptic capacity then f has admissible limits almost everywhere on $\partial\Omega$.*

PROOF. By the above remarks and Theorem 6 we may write $f = g/h$ where g and h are holomorphic functions in the Nevanlinna class on Ω. Our problem is to show that the set of boundary points at which a holomorphic function in the Nevanlinna class has a given limit, say α, is of measure zero. Let $A_\alpha(\zeta)$ denote an admissible approach region at $\zeta \in \partial\Omega$ where $\alpha > 0$. Let

$$A_\alpha(\zeta,r) = A_\alpha(\zeta) \cap \{z \in \Omega : \tau(z) > r\} .$$

Now since we are assuming that the exhaustion function τ is parabolic we have $\alpha^n \equiv 0$. The error term $S(r,a)$ is therefore identically zero. This gives the inequality

$$T(r) \geq m(r,a) - m(r_o,a)$$

or

$$T(r) \geq \int_{\partial\Omega(r)} \log \frac{1}{\|f(z):a\|} - C_o$$

where C_o is a constant independent of r. Now suppose f(z) has limit α as $z \to \zeta$ in admissible approach regions $A_\alpha(\zeta)$ for $\alpha > 0$ and all ζ in a set K of positive measure on $\partial\Omega$. Then there exists a subset $K_o \subset K$ of positive measure on which the convergence is uniform. Hence given $\varepsilon > 0$ there exists $R < 1$ such that $\|f(z):a\| < \varepsilon$ provided $z \in A_\alpha(\zeta,r)$ if $\zeta \in K_o$ and $r > R$. This gives the inequality

$$T(r) \geq \log (1/\varepsilon) \int_{\partial\Omega(r)\cap E_o(\alpha,R)} \sigma - C_o$$

where $E_o(\alpha,R) = \cup_{\zeta\in K_o} A_\alpha(\zeta,R)$, for $r > R$. Since ε may be chosen arbitrarily small and

$$\int_{\partial\Omega(r)\cap E_o(\alpha,R)} \sigma \geq \text{const.}$$

independent of $R < 1$, it follows that $T(r)$ must be unbounded, a contradiction. We therefore conclude that if f is holomorphic on Ω and is in the Nevanlinna class, then for at most a set $\zeta \in \partial\Omega$ of measure zero, does f have admissible limit a at ζ. In particular f can have admissible limit 0 or ∞ at most on a set of measure zero which implies the conclusion of the corollary. □

We now discuss analogues of the above results for meromorphic mappings

$$F : \Omega \to \mathbb{P}^m\mathbb{C} .$$

If Ω admits a logarithmic pseudoconvex exhaustion function then there is a FMT for F as follows. Let ν_A denote the divisor of the meromorphic function $\langle F(z),A\rangle$ where $A \in \mathbb{P}^m\mathbb{C}$. The divisor is nonzero whenever the image of F intersects the hyperplane defined by A. Let

$$n(r,A) = r^{2-2n} \int_{\Omega[r]} \nu_A \beta^{n-1}$$

$$N(r,A) = \int_{r_o}^{r} \frac{n(t,A)}{t}\, dt .$$

With

$$||A:B|| = \frac{|\langle A,B\rangle|}{||A||\,||B||}$$

put

$$m(r,A) = \int_{\partial\Omega(r)} F^* \left(\log \frac{1}{||Z:A||}\right) \sigma$$

$$S(r,A) = \int_{\Omega(r_o,r]} F^* \left(\log \frac{1}{||Z:A||}\right) \alpha^n .$$

If ω is the Kähler form of the Fubini-Study metric then

$$T(r) = \int_{r_o}^{r} \left[\int_{\Omega[t]} F^*\omega \wedge \beta^{n-1} + \nu(0) \right] \frac{dt}{t}$$

where $\nu(0)$ is the indeterminant multiplicity of F at 0. The FMT then states:

$$N(r,A) + m(r,A) - m(r_o,A) - S(r,A) = T(r) .$$

We will say that the meromorphic mapping F is in the Nevanlinna class provided $T(r) \le O(1)$. We then have the following result.

THEOREM 8. Suppose $\Omega \subset \mathbb{C}^n$ is a bounded domain with smooth boundary and logarithmic exhaustion. If $F : \Omega \to \mathbb{P}^m\mathbb{C}$ is a meromorphic mapping which fails to intersect a set $E \subset \mathbb{P}^m\mathbb{C}$ of hyperplanes of positive capacity, so $C_1(E) > 0$, the F is in the Nevanlinna class.

PROOF. The proof follows the idea of the proof of Theorem 6; one obtains an estimate for T(r) by integrating the inequality

$$T(r) \le N(r,A) + m(r,A)$$

with respect to an equilibrium measure for the set E parameterizing the omitted hyperplanes. Since $C_1(E) > 0$ there exists an equilibrium measure μ such that

$$u_E(Z) = \int_E \log \frac{1}{\|Z:A\|} \, d\mu(A) \le O(1) .$$

The proof then proceeds in exactly the same way as in Theorem 5. □

The objective of the above discussion was to provide an introduction to some of the techniques of potential theory which lead to results in function theory and geometry. There seem to be many more possibilities for applications of the capacities which were considered in Section 2 and we hope that effort and interest in this subject will increase.

Acknowledgement: The author would like to express his thanks to the Sonderforschungsbereich 40 at Bonn for supporting his research.

REFERENCES

[1] H. Alexander: Projective capacity, Conference on Several Complex Variables, Ann. Math. Studies 100, Princeton University Press, Princeton 1981.

[2] ____________: A note on projective capacity (manuscript).

[3] E. Bedford and B.A. Taylor: Potential theoretic properties of plurisubharmonic functions, Acta Math., to appear.

[4] J.P. Demailly: Formulas de Jensen en plusieurs variables et applications arithmétiques (manuscript).

[5] L. Gruman: Ensembles exceptionnel pour les applications holomorphic dans $\mathbb{C}^n$, Seminaire Lelong-Skoda 1981, to appear.

[6] W. Hayman: Meromorphic Functions, Oxford University Press, Oxford 1968.

[7] G.M. Henkin: Solutions with estimates of the H. Lewy and Poincaré-Lelong equations. Construction of functions in the Nevanlinna class with prescribed zeros in strictly pseudoconvex domains, Dokl. Akad. Nauk. SSSR 225, 771-774 (1975).

[8] J.R. King: The currents defined by analytic varieties, Acta Math. 127, 185-220 (1971).

[9] F. Leja: Sur les suites des polynômes bornés presque partout sur la frontiere l'un domaine, Math. Ann. 108, 517-524 (1933).

[10] R. Molzon, B. Shiffman, and N. Sibony: Average growth estimates for hyperplane sections of entire analytic sets, Math. Ann. 257, 43-59 (1981).

[11] R. Molzon: Blaschke conditions for holomorphic mappings (manuscript).

[12] ________: Potential theory on $\mathbb{P}^n\mathbb{C}$: Applications to characterization of pluripolar sets and growth of analytic varieties (manuscript).

[13] R. Nevanlinna: Analytic Functions, Springer-Verlag, Berlin, Heidelberg, New York 1970.

[14] W. Rudin: Function Theory in the Unit Ball in $\mathbb{C}^n$, Springer-Verlag, Berlin, Heidelberg, New York 1980.

[15] H. Skoda: Valeurs au bord pour les solutions de l'operateur d'', et caracterisation des zéros des fonctions de la classe de Nevanlinna, Bull. Soc. Math. France 104, 225-299 (1976).

[16] ________: Solution à croissance du second problème de Cousin dans $\mathbb{C}^n$, Ann. de L'Institut Fourier 21, 11-23 (1971).

[17] J. Siciak: Extremal plurisubharmonic functions in $\mathbb{C}^n$, Ann. Polonici Math. 39, 175-211 (1981).

[18] ________: Extremal plurisubharmonic functions and capacities in $\mathbb{C}^n$ (manuscript).

[19] J. Siciak: On some extremal functions and their applications in the theory of analytic functions of several complex variables, Trans. Amer. Math. Soc. 105, 322-357, (1962).

[20] E. Stein: Boundary Behavior of Holomorphic Functions of Several Complex Variables, Princeton University Press, Princeton 1972.

[21] W. Stoll: Value Distribution on Parabolic Spaces, Springer-Verlag, Berlin, Heidelberg, New York 1977.

[22] M. Tsuji: Potential Theory in Modern Function Theory, Maruzen Co., Tokyo 1958.

Department of Mathematics
University of Kentucky
Lexington, KY 40506, U.S.A.

APPLICATIONS OF THE EXISTENCE OF WELL GROWING HOLOMORPHIC FUNCTIONS

Peter Pflug (Osnabrück)

Contents

Introduction

The author gives a construction of the hulls of holomorphy and completeness results for invariant distances on pseudoconvex domains by developing the method based upon the problem of existence of well growing holomorphic functions. His main aim is to simplify the arguments confirming that the found maximal extension domain is indeed the hull of holomorphy.

1. A construction of the hulls of holomorphy

We begin with recalling that any domain in $\mathbb{C}^1$ is a domain of holomorphy; otherwise in the higher-dimensional situation. Hence one is led to define for a domain $G \subset \mathbb{C}^n$ the largest domain G' containing G such that any holomorphic function on G can be analytically continued into G' as the hull of holomorphy $H(G)$. It is know that $H(G)$ need not be a schlicht domain.

To determine the shape of the hull of holomorphy for special domains has become very important in theoretical physics, for example, in relativistic quantum field theory [23]. So far, theoretical physicistis and mathematicians have attacked this problem in two steps, namely

a) find a maximal extension domain for a subclass of holomorphic functions by real methods,

b) confirm this maximal extension domain is the hull of holomorphy.

The procedure in Step b) depends on the concrete situation and is, in general, very complicated. Our aim is to simplify the arguments. Before going into details we start with a simple example to demonstrate what is meant by the term "real methods".

E x a m p l e. Assume $B \subset \mathbb{R}^n$ to be a domain, then it is well know that $H(\mathbb{R}^n + iB) = \mathbb{R}^n + i\ \mathrm{conv}(B)$. This has been proved by Bochner. Here we give another proof to explain the above remarks.

Assume f to be a holomorphic function on $\mathbb{R}^n + iB$ such that f satisfies the following growth condition: for $\forall$ compact set $K \subset B\ \exists m_K \in \mathbf{N},\ M_k > 0$ such that for $z \in \mathbb{R}^n + iK$ it holds that

$$|f(z)| \leq M_k (1 + |x|)^{m_k}.$$

It is easy to extend these functions to $\mathbb{R}^n + i\ \mathrm{conv}(B)$ using real methods; it is a simple procedure to find a distribution $T \in (D')$ such that for any $y \in B$ $e^{-\xi y} T_\xi \in (S')$ and $f(x+iy) = F_\xi [e^{-\xi\, y} T_\xi]_x$ on $\mathbb{R}^n + iB$. By an easy calculation one can obtain that for $y \in \mathrm{conv}(B)$ the distribution $e^{-\xi\, y} T_\xi$ belongs to (S'). Hence $F[e^{-\xi y} T_\xi]_x$ gives the holomorphic continuation of the function f.

At this point we leave the proof and summarize what has been done so far and what is still left to do.

S i t u a t i o n. Assume that a given domain G is contained in a domain of holomorphy G' (in our example: $G = \mathbb{R}^n + iB$ and $G' = \mathbb{R}^n + i\times \mathrm{conv}(B)$) and that for a family F of holomorphic functions on G it is known that any $f \in F$ can be analytically continued into G'. The remaining question is: does the envelope of holomorphy $H(G)$ of G equals G'?

R e m a r k. Of course it is not enough to take $F = H^\infty_G$ as the rough example of Sibony shows.

Looking at all results found by the physicists by real methods one is led to the following:

D e f i n i t i o n. For a domain G and a positive number N a holomorphic function $f : G \longrightarrow \mathbb{C}$ is called <u>of polynomial growth of order</u> N iff one has the following estimation:

$$\sup_{z\in G} |f(z)|\Delta_G^N(z) < \infty,$$

where

$$\Delta_G(z) = (1+|z|^2)^{-1/2} \min(1, \operatorname{dist}(z,\partial G)).$$

Using this notion of polynomial growth one can obtain, applying a famous theorem of Skoda, the following

THEOREM [14]. *If z^o is a boundary point of a pseudoconvex domain $G\subset\mathbb{C}^n$ and if $z^\nu \in G$, $z^\nu \longrightarrow z^o$, then, for $\varepsilon > 0$ there exists a holomorphic function f on G of polynomial growth of order $n+\varepsilon$ such that the sequence $\{f(z_\nu)\}$ is unbounded.*

Remark. Similar results have been also obtained by Cnop [3], Sibony [19] Ferrier [6], and Jarnicki [9] but with weaker growth conditions.

As an immediate consequence one can receive the following

COROLLARY [14]. *Suppose that any function f, holomorphic in a domain $G\subset\mathbb{C}^n$, of polynomial growth of order $n+\varepsilon$ ($\varepsilon > 0$ fixed) can be analytically continued into a domain of holomorphy $G'\supset G$ and, in addition, assume $H(G)$ to be schlicht, then $H(G) = G'$.*

In order to use this corollary one has to decide whether the hull of holomorphy of a given domain is schlicht. Unfortunately, only a few criterions are known to answer this question. Therefore it is important to ask if it is possible to cancel the assumption on the schlichtness of $H(G)$. And, indeed, the following result can be verified.

THEOREM [16,18]. *Suppose that for two domains $G\subset G'$ the following assumptions are fulfilled:*

a) G' is a domain of holomorphy,

b) any holomorphic function f on G possesses a holomorphic continuation into G', provided that f is of polynomial growth of order $4n$.

Then $H(G)$ is schlicht and $H(G) = G'$.

Remark. Weaker results have been obtained by Jarnicki [9].

Proof. Assume $(H(G),\Pi)$ to be the hull of holomorphy as a Riemann domain over $\mathbb{C}^n$ and define

$$\Delta_{H(G)}(x) = \min(\delta_{H(G)}(x), 1+|\Pi(x)|^2{}^{-1/2})$$

where $\delta_{H(G)}$ is the usual boundary distance on H(G). Then, taking the family

$$F:=\{F \in H_{H(G)} : \sup_{x\in H(G)} |f(x)| \Delta^{4n}_{H(G)}(x) < \infty\}$$

it is easy to see that whenever $f \in F$, then $f|_G$ is of polynomial growth of order 4n. Applying $\bar{\partial}$-methods it can also be shown that F separates the points of H(G).

From those items of information it is clear that H(G) is schlicht and hence, by the above corollary, H(G) equals G'. Hence the theorem is proved.

Now we have reached the point when we should come back to the second step in the proof of Bochner's theorem. The only point which can be easily verified is that any function holomorphic on $\mathbb{R}^n + iB$ which is of polynomial growth of order 4n belongs to the functions described in the first step. Using the above theorem Bochner's result is obtained.

To conclude this section we should repeat how it is possible to calculate the hulls of holomorphy:

(a) real methods (Fourier-Laplace transform) to extend functions of good growth,

b) applying the above theorem.

We should also mention that if the reader is interested in more difficult examples, the book of Vladimirov [22] can show him many of them.

After having discussed how to construct the hulls of holomorphy when applying functions of polynomial growth, we turn now to disscuss L^2-holomorphic functions as well as bounded holomorphic functions.

2. Completeness-results for invariant distances on pseudoconvex domains

We will start with some general definitions.

Let $D \subset \mathbb{C}^n$ be a domain and $d : D\times D \to \mathbb{R}_{\geq 0}$ a distance inducing the topology of D. The following notions are known:

D e f i n i t i o n, a) D is called d-complete iff any Cauchy-sequence w.r.t. d converges to a point in D.

b) D is called finitely compact w.r.t. d iff any ball w.r.t. d is a relatively compact subset of D.

Remark: 1) It's a simple fact that b) implies a).

2. If d is assumed to be an inner distance then by a theorem of Hopf-Rinow a) and b) are equivalent.

What is the connection of this notion "complete" and complex analysis ? For example, it can be made clear by the following theorem of Grauert [7]:

THEOREM: Any pseudoconvex domain is complete w.r.t. a distance which can be induced by a Kähler metric.

Remark. We should mention that the converse of Grauert's theorem is also true if the domain is assumed to be a fat domain [11], but we do not deal with this problem.

So far, the metric was not very explicit. On the other hand any domain is connected with different notions of so called invariant distances. Thus the question arises: what about the completeness w.r.t. those explicit distances ?

We start with recalling the definitions of the distances we shall deal with.

The Carathéodory distance of $z,w \in D$ is defined by the formula

$$C_D(z,w) = \sup \{\tfrac{1}{2}\log \frac{1+|f(w)|}{1-|f(w)|} : f : D \longrightarrow E,\ f(z) = 0\},$$

where E is the unit disc. The question of the completeness w.r.t. C_D can be regarded as the question of measuring how many bounded holomorphic functions exist on D.

Remark. Due to an example by Vigué [21] there is a Cara-finitely compact pseudoconvex domain on which C_D is not an inner distance. Simpler examples have been given before by T. Barth. Hence, one has to distinguish the two notions given in the introduction.

We turn now to the notion of the Bergman distance. Denote by $K_D(z,w) = K(z,w)$ the Bergman kernel of D and then define the Kähler metric

$$ds^2(z,X) = \sum_{\nu,\mu=1}^{n} \frac{\partial^2 \log K(z,z)}{\partial z_\nu \partial \bar{z}_\mu} X_\nu \bar{X}_\mu$$

which is called the Bergman metric. The Bergman distance d_B of $z,w \in D$ is defined as usually:

$$d_B(z,w) = \inf \int_0^1 ds(\gamma(t), \dot{\gamma}(t))dt,$$

where the infimum is taken over all piecewise C^1-curves joining z to w in D. By a result of Hahn [8] one has the following inequalities (the comparison result): $C_D(z,w) \le C_D^i(z,w) \le d_B(z,w)$.

For the sake of completeness we should mention another explicit invariant distance - the so called Kobayashi distance $d_k(z,w)$ of w, $z \in D$, for which it is known that $C_D(z,w) \le d_k(z,w)$. Hence we give the following remark: the completeness w.r.t. C_D implies the completeness w.r.t. d_B and d_K.

Now the following question arises: which pseudoconvex domains are complete w.r.t. C_D or d_B or d_K? To discuss it let us start with a few remarks on C_D.

LEMMA. D is finitely-compact w.r.t. C_D iff D is strictly H^∞-convex, which is defined as follows: for fixed $z^* \in D$ and any sequence $z^\nu \in D$, $z^\nu \longrightarrow z^o \in \partial D$, there exists a holomorphic function $f : D \longrightarrow E$ with

$$E(z^*) = 0 \quad \text{and} \quad \sup |f(z^\nu)| = 1.$$

This lemma is due to Løw [17].

Hence one obtains what follows: if D is finitely compact w.r.t. C, then D is H^∞-convex and a H^∞-domain of holomorphy. Both properties are true for pseudoconvex domains with a C^∞-boundary (Catlin, Sibony). Therefore it is reasonable to ask whether any pseudoconvex domain with C^∞-boundary is finitely-compact w.r.t. C?

In the case of strictly pseudoconvex domains and pseudoconvex domains with real analytic boundary in $\mathbb{C}^2$ the answer is affirmative because any boundary point is a peak point for the algebra $C(\overline{D}) \cap H_D$ (cf. Bedford and Fornaess [1]). Yet looking for peak points we see that our formulation is too strong as the following result can show.

THEOREM [17]. Let D be a bounded Reinhardt domain in $\mathbb{C}^n$ with $0 \in D$. Then the following conditions are equivalent:

1) D is finitely compact w.r.t. C,
2) D is complete w.r.t. C,
3) D is pseudoconvex.

Remark.

Barth has informed us that Vigué [21] has obtained the same result but only in dimension 2.

Proof. 3 → 1. Without loss of generality D is assumed to lie in the unit polycylinder. Now, when assuming D not to be finitely compact there exists a sequence $z^\nu \in D$, $z^\nu \longrightarrow z^o \in \partial D$, and number $0 < M < 1$ such that for any holomorphic function $f : D \longrightarrow E$, $f(0) = 0$, $|f(z^\nu)| \leq M$ holds for all $\nu \in \mathbf{N}$.

For the simplicity's sake suppose that $z_1^o, \ldots, z_n^o \neq 0$. Using the pseudoconvexity of D one can find a linear functional L on $\mathbb{R}^n$ such that for all $x = (\log|z_1|, \ldots, \log|z_n|)$ $(z \in D)$ the inequality

$$L(x) = \sum_{i=1}^{n} \xi_i x_i < L(x^o)$$

with $x^o = (\log|z_1^o|, \ldots, \log|z_n^o|) =: \log|z^o|$ is true.

Hence $\xi_i \geq 0$ and therefore $C := L(x^o) \leq 0$. Assume $\xi_{i_1} > 0, \ldots, \xi_{i_k} > 0$ and the other $\xi_\nu = 0$, then using the Dirichlet pigeon hole principle one can get that for any $N \in \mathbb{N}$ $\exists \beta_{1,N}, \ldots, \beta_{k,N} \in \mathbb{Z}$ and $1 \leq k_N \leq N^k$ such that

$$|\beta_{\nu,N}/k_N - \xi_{i,\nu}| \leq 1/Nk_N$$

Then define

$$f_N(z) = e^{-Ck_N} z_{i_1}^{\beta_{1,N}} \ldots z_{i_k}^{\beta_{k,N}}$$

to get a holomorphic function:

$$g_N(z) := f_N(z)/\|f_N\|$$

with the estimation along the sequence z^ν:

$$|g_N(z^\nu)| \geq \exp[-N^k(C-L(\log|\overset{\nu}{z}|)) - kM^*/N - 2k|C|/N],$$

contradicting the assumption $|g_N(z^\nu)| \leq M < 1$.

As an immediate consequence one can conclude:

COROLLARY. *Any pseudoconvex bounded Reinhardt-domain is complete w.r.t. the Bergman distance.*

Remarks: a) This has been also obtained by Skwarczyński [20], and recently, by Nakajima [11]).
b) From this result it can be seen that, in general, H^∞ is much bigger than $C^0(\bar{D}) \cap H_D$.

To finish the discussion on the Carathéodory-distance it should be mentioned that the Sibony example is not complete w.r.t. the Carathéodory distance but it is locally finitely compact (compare Eastwood [5]).

The last problem lies in deciding what pseudoconvex domains are d_B-complete. It turns out that there is a fairly general result. Yet first a few words on the boundary behaviour of the Bergman kernel are neccessary.

Assume D to be a bounded domain in $\mathbb{C}^n$ then the Bergman kernel can be described as follows.

$$K_D(z,z) = \sup\{|f(z)|^2/\|f\|^2_{L^2(D)} : f \in L^2_h(D),\ f \neq 0\}.$$

The boundary behaviour of K_D can be given by the following

THEOREM [16] . *Assume D as above and, in addition, to be pseudoconvex with C^2-boundary. Then, for any $0 < \lambda < 1$, there exists $\delta = \delta(\lambda) > 0$ such that for all $z \in D$ with $\operatorname{dist}(z,D) < \delta$ one gets*

$$K_D(z,z) \geq 1/\operatorname{dist}(z,\partial D)^{2\lambda}.$$

Proof. Suppose the statement is false. Then there exists $\lambda_o \in (0,1)$ and a sequence $z^\nu \in D$, $z^\nu \longrightarrow z^o \in \partial D$ such that

$$K_D(z^\nu, Z^\nu) < 1/\operatorname{dist}(z^\nu,\partial D)^{2\lambda o}.$$

If $\nu >> 1$, take points $w^\nu \notin \bar{D}$ with

$$\operatorname{dist}(z^\nu,\partial D)=\operatorname{dist}(w^\nu,\partial D)=1/2|z^\nu-w^\nu|$$

and using Skoda's theorem construct functions $h_i^\nu \in L_h^2(D)$ such that

a) $1=\sum_{i=1}^{n} h_i^\nu(z)(z_i-w_i^\nu)$ on D

and

b) $||h_i^\nu||_{L^2(D)} \leq C \operatorname{dist}(w^\nu,D)^{-\frac{\delta-1}{2}}$

where $\delta>1$ and $1-\frac{\delta-1}{2}>\lambda_o$. Then by the above description of the kernel we conclude the proof by

$$1 \leq \sum_i |h_i^\nu(z^\nu)|\ |w_i^\nu-z_i^\nu|$$

$$\leq \sum \sqrt{K_D(z^\nu,z^\nu)}\ ||h_i^\nu||_{L^2(D)} |z_i^\nu-w_i^\nu|$$

$$\leq \frac{1}{\operatorname{dist}(z^\nu,\partial D)^{\lambda_o}}\ C \operatorname{dist}(w^\nu,\partial D)^{-\frac{\delta-1}{2}} \operatorname{dist}(z^\nu,\partial D) \xrightarrow[\nu\to\infty]{} 0,$$

which yields the needed contradiction.

R e m a r k s: 1) Using a more general boundary regularity including C^1-boundary one can obtain for any boundary sequence $z^\nu \in D$, $z^\nu \longrightarrow z^o \in \partial D$, a function $f \in L_h^2(D)$, $||f||_2=1$ and $|f(z^\nu)| \to \infty$, from which $K(z^\nu,z^\nu) \to \infty$ [15].

2) Assuming D to be strongly pseudoconvex the following boundary behaviour of the Bergman kernel is known:

$$K_D(z,z) \approx 1/\operatorname{dist}(z,\partial D)^{n+1}.$$

Hence the exponent can be regarded as a special case of the following:

$$\dim \mathbb{C}^n + 1 - \dim[\text{zero space of Levi-form}] - \varepsilon.$$

The following theorem of Ohsawa [13] shows that this interpretation is correct.

THEOREM. D *is assumed to be a pseudoconvex domain in* $\mathbb{C}^n$ *with a*

C^2-*boundary*. *The zero space of the Levi form at* $z^o \in \partial D$ *is supposed to be* k. *Then one gets the estimate*

$$\inf_{z\in D} K_D(z,z)\,|z-z^o|^{n+1-k-\varepsilon} \geq C_\varepsilon > 0 \quad \text{for any} \quad \varepsilon > 0.$$

R e m a r k. The proof of this result is much more complicated than the above one; it involves the $\bar\partial$-problem on complete Kähler manifolds.

Now we return to our original question. Using a criterion of Kobayashi (the existence of this criterion has been given by E.Ligocka) one can obtain the following general result [16 ,12].

THEOREM. *Any pseudoconvex domain with* C^1-*boundary is complete w.r.t.* d_B.

P r o o f. Assume D not to be complete. Then there exists a sequence $z^\nu \in D$ which is Cauchy w.r.t. d_B. By an elementary argument, different from Kobayashi's proof, one can construct a subsequence denoted again by z^ν and reals Θ_ν such that

$$K(\ ,z^\nu)\exp(i\Theta_\nu)[K(z^\nu,z^\nu)]^{-1/2}$$

is a Cauchy-sequence in $L^2_h(D)$. That is why it converges to $f \in L^2_h(D)$ with $\|f\|_{L^2(D)} = 1$, from which it follows that

$|f(z^\nu)|/[K_D(z_\nu,z_\nu)]^{1/2} \longrightarrow 1$. On the other hand there exists a neighborhood U of z^o and a sequence $g_\mu \in H^\infty(D\cap U)$ with

$$\|g_\mu - f\|_{L^2(U\cap D)} \longrightarrow 0 \ \text{in}\ L^2(D\cap U).$$

Using the extremal property of the kernel and a localization result for the kernel one obtains:

$$\|f-g_\mu\|_{L^2(U\cap D)} \geq \frac{|f(z^\nu)|}{\sqrt{K_{D\cap U}(z^\nu,z^\nu)}} - \frac{|g_\mu(z^\nu)|}{\sqrt{K_{D\cap U}(z^\nu z^\nu)}}$$

$$\geq \frac{\alpha|f(z^\nu)|}{\sqrt{K_D(z^\nu,z^\nu)}} - \frac{|g_\mu(z^\nu)|}{\sqrt{K_{D\cap U}(z^\nu z^\nu)}} \longrightarrow \alpha > 0$$

as μ and ν tend to ∞, which contradicts the L^2-approximation.

Using ideas of Ohsawa's proof [12] for the above C^1 case it is possible to prove the following localization result:

THEOREM: Assume D to be a bounded pseudoconvex domain in $\mathbb{C}^n$, $z^o \in \partial D$ and $U = U(z^o)$ such that there is no Cauchy-sequence $z^\nu \in U \cap D$, $z^\nu \longrightarrow z^o$, w.r.t. the invariant Skwarczyński's distance [20] $\lambda_{U\cap D}$ on $U \cap D$. Then no Cauchy-sequence $w^\nu \in D$ w.r.t. d_B^D with $w^\nu \longrightarrow z^o$ can exist.

P r o o f. Assume that there exists a Cauchy-sequence $z^\nu \in D$ w.r.t. d_B^D with $z^\nu \longrightarrow z^o$. As above, for a subsequence denoted again by z^ν and reals Θ_ν one obtains a Cauchy-sequence

$$K(\ ,z^\nu)\exp(i\Theta_\nu)[K(z^\nu,z^\nu)]^{-1/2} \quad \text{in} \quad L^2(D).$$

Hence

$$1 - \frac{|K(z^\nu,z^\mu)|}{\sqrt{K(z^\nu z^\nu)}\ \sqrt{K(z^\mu,z^\mu)}} \longrightarrow 0 \quad \text{as } \mu \text{ and } \nu \text{ tend to } \infty,$$

which implies using an appropriate complete orthonormal basis for any $\varepsilon > 0$, that there exists a number $N(\varepsilon)$ such that for any $\nu,\mu \geq N(\varepsilon)$ there is no $f \in L_h^2(D)$ with $||f||_{L^2(D)} = 1$, $f(z^\nu) = 0$ and

$$|f(z^\mu)| \geq \varepsilon |K(z^\nu,z^\mu)|/[K(z^\nu,z^\nu)]^{1/2}.$$

By assumption it follows that

$$(*)_{\nu_\mu} = 1 - \frac{|K_{D\cap U}(z^\nu,z^\mu)|}{\sqrt{K_{D\cap U}(z^\nu,z^\nu)}\ \sqrt{K_{D\cap U}(z^\mu,z^\mu)}} \not\longrightarrow 0 \quad \text{as } \mu \text{ and } \nu \text{ tend to } \infty.$$

Hence, there is a number $\delta > 0$ such that for any $N \in \mathbf{N}$ there are $\nu,\mu \geq N$ with $(*)_{\nu_\mu} \geq \delta$. Using

$$F_{\nu_\mu} = \frac{K_{D\cap U}(\ ,z^\mu)}{\sqrt{K_{D\cap U}(z^\mu,z^\mu)}} - \frac{K_{D\cap U}(z^\nu,z^\mu)}{\sqrt{K_{D\cap U}(z^\nu,z^\nu)}\ \sqrt{K_{D\cap U}(z^\mu,z^\mu)}}\ \frac{K_{D\cap U}(\ ,z^\nu)}{\sqrt{K_{D\cap U}(z^\nu,z^\nu)}}$$

and denoting $\dfrac{K_{D\cap U}(z^\nu,z^\mu)}{\sqrt{K_{D\cap U}(z^\nu,z^\nu)}\ \sqrt{K_{D\cap U}(z^\mu,z^\mu)}}$ by a_{ν_μ} one can see that

$$F_{\nu_\mu}(z^\nu) = 0$$

$$|F_{\nu_\mu}(z^\mu)| \geq (1-|a_{\nu_\mu}|)K_{D\cap U}(z^\mu,z^\mu)/[K_{D\cap U}(z^\mu,z^\mu)]^{1/2}.$$

Applying Hörmander's L^2-techniques one can construct functions $f_{\nu_\mu} \in L^2_h(D)$ with

$$f_{\nu_\mu}(z^\nu) = 0, \quad ||f_{\nu_\mu}||_{L^2(D)} = 1$$

$$|f_{\nu_\mu}(z^\mu)| \geq (\delta/c)^{1/2}|K(z^\nu,z^\mu)\ /[K(z^\nu,z^\nu)]^{1/2},$$

inducing the expected contradiction. Hence the theorem is proved.

Immediately, this result yields the following

COROLLARY. The Sibony example [19] is d_B-complete.

P r o o f. The only thing which has to be mentioned is that Sibony's example is locally Caratheodory complete. This implies that Skwarczyński-Cauchy sequences converging to a boundary point do not exist locally (compare Burbea).

We conclude by the following application:

THEOREM (Ligocka [10]). If D is a pseudoconvex domain with C^1-boundary, $D \subset \mathbb{C}^n_z \times \mathbb{C}^m_w$, then the Bergman kernel cannot be locally written as the product $K_D((z,w),(z,w)) = K_1(z)\ K_2(w)$ of two real analytic functions.

Fachbereich Mathematik der Universität Osnabrück
Abt. Vechta (Driverstraße 22)
Postfach 1379, D-2848 Vechta, BRD

R e f e r e n c e s

[1] BEDFORD, E. and J.E. FORNAESS: A construction of peak points on weakly pseudoconvex domains, Ann. of Math. 107 (1978), 555-568.

[2] BURBEA, J.: On metric and distortion theorems, Princeton University Press: Recent Developments in Several Complex Variables (1981), 65-92.

[3] CNOP, I.: Spectral study of holomorphic functions with bounded growth, Ann. Inst. Fourier 22 (1972), 293-310.

[4] DIEDERICH, K. and P. PFLUG: Über Gebiete mit vollständiger Kählermetrik, Math. Ann. 257. (1981), 191-198.

[5] EASTWOOD, A.: À propos des variétés hyperboliques completes, C.R. Acad. Paris 280 (1975), 1071-1075.

[6] FERRIER, J.P.: Spectral theory and complex analysis, North-Holland 1973.

[7] GRAUERT, H.: Charakterisierung der Holomorphiegebiete durch die vollständige Kählersche Metrik, Math. Ann. 131 (1965), 38-75.

[8] HAHN, K.T.: On the completeness of the Bergmann metric and its subordinate metrics II, Pac. Journ. Math. 68, (1977), 437-446.

[9] JARNICKI, M.: Holomorphic functions with bounded growth on Riemann domains over $\mathbb{C}^n$, Bull. Acad. Polon. Sci. Sér. Sci. Math. Astronom. Phys. 27 (1979), 675-680.

[10] LIGOCKA, E.: Domains of existence of real analytic functions and the inverse of the Bremermann theorem, Bull Acad. Polon. Sci. Sér. Math. Astronom. Phys. 26. (1978), 495-499.

[11] NAKAJIMA, K.: On the completeness of bounded Reinhardt domains with respect to Bergman metric. Preprint 1982.

[12] OHSAWA, T.: A remark on the completeness of the Bergmann metric, Proc. Japan Acad. 57 (1981) 238-240.

[13] ——: Boundary behaviour of the Bergmann kernel function on pseudoconvex domains, preprint 1981.

[14] PFLUG P.: Über polynomiale Funktionen auf Holomorphiegebieten, Math. Zeitschrift 139 (1974), 133-139.

[15] ——: Quadratintegrable holomorphe Funktionen und die Serre- - Vermutung, Math. Ann. 216 (1975), 285-288,

[16] ——: Various applications of the existence of well growing holomorphic functions, Functional Analysis, Holomorphy and Approximation Theory, North Holland. Math. Studies 71 (1982), 391-412.

[17] ——: About the Caratheodory Completeness of all Reinhardt domain, Advances in Functional Analysis, Holomorphy and Approximation Theory 1981, to appear in Lecture Notes in Pure and Applied Math.

[18] ——: Eine Bemerkung über die Konstruktion von Holomorphiehüllen, Zeszyty Nauk Uniw. Jag. 23 (1982), 21-22.

[19] SIBONY, N.: Prolongement des fonctions holomorphes bornées et metrique de Carathéodory, Universite Paris XI, no 73.

[20] SKWARCZYŃSKI, M.: Biholomorphic invariants related to the Bergman functions, Dissertationes Mathematicae 173 (1980).

[21] VIGUÉ, J.P.: La distance de Carathéodory n'est pas intérieure, to appear in Resultate der Mathematik.

[22] VLADIMIROV, V.S.: Methods of the theory of functions of many complex variables, M.I.T. Press 1966.

[23] ——: Analytic functions of several complex variables and quantum field theory, Proc. Steklov Institute of Mathematics 135 (1978), 69-81.

SUR LES DÉRIVATIONS DES ANNEAUX DES SÉRIES CONVERGENTES

Arkadiusz Płoski (Kielce)

Résumé. Nous donnons une démonstration de l'inegalité de S. Łojasiewicz pour les fonctions analytiques ayant un zéro isolé.

1. Soit $C_n = K\{x\}$ l'anneau des séries convergentes des variables $(x_1,\dots,x_n) = x$ à coefficients dans un corps valué K de caractéristique zéro. On note $m(C_n)$ l'idéal maximal de C_n, $\mathrm{Der}(C_n)$ le C_n-module des dérivations de l'anneau C_n dans C_n. Alors $\mathrm{Der}(C_n)$ est engendré par les dérivations partielles $\partial/\partial x_i$. Pour tout idéal I de C_n nous poserons $\mathrm{Der}(I) = \{D \in \mathrm{Der}(C_n) : D(I) \subset I\}$.

1.1. THÉORÈME. *Soit* I *un idéal premier de* C_n. *On a:*

(i) *si* $f \in m(C_n)$ *et* $Df \equiv 0(I)$ *pour toute dérivation* $D \in \mathrm{Der}(I)$ *alors* $f \equiv 0(I)$;

(ii) *si* $f, g \in m(C_n)$ *et* $DfD'g - D'fDg = 0(I)$ *pour toutes dérivations* $D, D' \in \mathrm{Der}(I)$ *alors il existe une série* $P(u,v) \neq 0$ *dans* $K\{u,v\}$ *telle que* $P(f,g) \equiv 0(I)$.

La démonstration du théorème repose sur la proposition suivante:

1.2. PROPOSITION. Soit I un idéal premier de C_n, $d = \dim(C_n/I) > 0$ *et soit* $h = (h_1,\dots,h_d)$ *un système de paramètres* $\mathrm{mod}\, I$ (*c'est-à-dire: l'idéal* $(h_1,\dots,h_d)C_n + I$ *est un idéal de définition dans l'anneau* C_n). *Alors il existe des dérivations* $D_1,\dots,D_d \in \mathrm{Der}(I)$ *telles que* $D_i h_j \equiv 0(I)$ *pour* $i \neq j$ *et* $D_i h_i \not\equiv 0(I)$ *pour* $i = 1,\dots,d$.

Démonstration de la proposition Posons $J = \{f(t,x) \in K\{t,x\} : f(h(x),x) \equiv 0(I)\}$. Alors J est un idéal premier de $K\{t,x\}$, les variables $t = (t_1,\dots,t_d)$ forment un système de paramètres $\mathrm{mod}\, I$, $I \subset J$ et $t_i \equiv h_i(x)(J)$ pour $i = 1,\dots,d$. Il suffit de démontrer (1.2) pour l'idéal J et le système de paramètres $t = (t_1,\dots,t_d)$. En effet supposons qu'il existe des dérivations

$$\overline{D}_i = \sum_k a_{ik}(t,x)\frac{\partial}{\partial t_k} + \sum_l b_{il}(t,x)\frac{\partial}{\partial x_l}$$

telles que $\bar{D}_i \in \mathrm{Der}(J)$, $\bar{D}_i t_j \equiv 0\,(J)$ pour $i \neq j$ et $\bar{D}_i t_i \not\equiv 0\,(J)$. Posons

$$D_i = \sum_l b_{il}(h(x), x) \frac{\partial}{\partial x_l}$$

pour $i = 1,\ldots,d$, un simple calcul montre que les dérivations D_i vérifient les conditions de (1.2). Pour construire les dérivations $\bar{D}_i$ rappelons la description d'un idéal premier de $K\{t,x\}$.

1.3. LEMME (cf. [2], chapitre IV, théorème 5.1). *Soit J un idéal premier de* $K\{t,x\}$ *tel que les variables* $t = (t_1,\ldots,t_d)$ *forment un système de paramètres* mod J. *Alors il existe une suite de polynômes*

$$P_i = P_i(t, x_1,\ldots,x_i) = c_{i0}(t)x_i^{m_i} + \sum_{j=1}^{m_i} c_{ij}(t, x_1,\ldots,x_{i-1})x_i^{m_i - j}$$

telle que

a) $P_i \equiv 0\,(\mathrm{mod}\,J)$, $m_i > 0$, $c_{i0}(t) \not\equiv 0\,(\mathrm{mod}\,J)$, $\dfrac{\partial P_i}{\partial x_i} \not\equiv 0\,(\mathrm{mod}\,J)$ *pour* $i = 1,\ldots,n$;

b) *si* $c(t) = \prod_i c_{i0}(t)$ *alors pour toute série* $f(t,x) = 0\,(J)$ *il existe un entier* $N > 0$ *tel que* $c(t)^N f(t,x) = 0\,(\mathrm{mod}(P_1,\ldots,P_n)K\{t,x\})$.

Dans [2] on démontre le lemme 1.3 dans le cas formel en utilisant le fait que l'application canonique $K\{t\} \longrightarrow K\{t,x\}/J$ est injective et finie. Le cas analytique se traite de manière analogue.

Posons maintenant

$$\bar{D}_i g = \frac{\partial(P_1,\ldots,P_n, g)}{\partial(t_i, x_1,\ldots,x_n)} \quad \text{pour } i = 1,\ldots,d.$$

Les conditions a) entraînent les relations

$$\bar{D}_i t_i = (-1)^{n+1} \frac{\partial P_1}{\partial x_1} \cdots \frac{\partial P_n}{\partial x_n} \quad \text{pour } i = 1,\ldots,d \text{ et } \bar{D}_i t_j = 0$$

dans $K\{t,x\}$ pour $i \neq j$. Il reste à prouver que $\bar{D}_i \in \mathrm{Der}(J)$ pour $i = 1,\ldots,d$. Dans ce but supposons que $g \equiv 0\,(\mathrm{mod}\,J)$ alors d'après la condition b) du lemme il existe des séries $Q_1,\ldots,Q_n \in K\{t,x\}$ telles que $c(t)^N g \equiv Q_1 P_1 + \ldots + Q_n P_n$ ce qui, après dérivation par rapport à t_k, x_l conduit à

$$c(t)^N \frac{\partial g}{\partial t_k} \equiv Q_1 \frac{\partial P_1}{\partial t_k} + \ldots + Q_n \frac{\partial P_n}{\partial t_k} (\mathrm{mod}\,J),$$

$$c(t)^N \frac{\partial g}{\partial x_l} \equiv Q_1 \frac{\partial P_1}{\partial x_l} + \ldots + Q_n \frac{\partial P_n}{\partial x_l} (\mathrm{mod}\,J).$$

Ce dernier système de relations implique que $c(t)^N \overline{D}_i g \equiv 0 \pmod J$, alors $\overline{D}_i g \equiv 0 \pmod J$ ce qui démontre la proposition.

D é m o n s t r a t i o n du théorème. Soit $h = (h_1, \ldots, h_d)$ un système de paramètres mod I. Pour toute série $f \in m(C_n)$ notons $P_f = P_f(t: v) \in K\{t\}[v]$ le polynôme minimal de $f \bmod I$ (cf. [1], chapitre IV, pp. 195-6). Alors P_f est un polynôme distingué tel que:

(a) $P_f(h,f) \equiv 0\,(I)$,

(b) pour toute série $Q(t,v) \in K\{t,v\}$ telle que $Q(h,f) \equiv 0\,(I)$ le polynôme P_f divise Q dans $K\{t,v\}$. Soit $D_1, \ldots, D_d \in \operatorname{Der} I$ un système de dérivations tel que dans (1.2). Nous monterons d'abord que pour toute série $f \in m(C_n)$ et pour tout $i = 1, \ldots, d$ la condition $D_i f \equiv 0 \pmod I$ entraîne $(\partial/\partial t_i) P_f = 0$ dans $K\{t,v\}$. En effet, d'après (a) on a $D_i(P_f(h,f)) \equiv 0 \pmod I$, alors

$$\frac{\partial P_f}{\partial t_i}(h,f)\, D_i h_i + \frac{\partial P_f}{\partial v}(h,f)\, D_i f \equiv 0 \pmod I.$$

Il résulte de là et de l'hypothèse $D_i f \equiv 0 \pmod I$ que $(\partial/\partial t_i) P_f(h,f) \equiv 0 \pmod I$ ce qui implique d'après b) que P_f divise $(\partial/\partial t_i) P_f$; alors $(\partial/\partial t_i) P_f = 0$ dans $K\{t,v\}$ car $\deg_v[(\partial/\partial t_i) P_f] < \deg_v(P_f)$.

Soit maintenant $f \in m(C_n)$ une série telle que $Df \equiv 0 \pmod I$ pour $D \in \operatorname{Der} I$, en particulier $D_i f \equiv 0 \pmod I$ pour $i = 1, \ldots, d$; alors $(\partial/\partial t_i) P_f = 0$ dans $K\{t,v\}$ pour tout $i \in \{1, \ldots, d\}$, d'où $P_f(t,v) = v$ ce qui démontre (i). Pour vérifier (ii) considérons des séries f, g $m(C_n)$ telles que $Df\,D'g - D'f\,Dg \equiv 0 \pmod I$ pour $D, D' \in \operatorname{Der} I$. On peut supposer $g \not\equiv 0 \pmod I$, alors il existe un système de paramètres $h_1, \ldots, h_d$ avec $h_1 = g$. Les rélations $D_i f\, D_j g - D_j f\, D_i g \equiv 0 \pmod I$ impliquent $D_i f \equiv 0 \pmod I$ pour $i = 2, \ldots, d$, d'où $(\partial/\partial t_i) P_f = 0$ pour $i > 1$, alors $P_f = P(t_1, v)$. Ceci prouve (ii).

2. Nous donnerons maintenant des conséquences du théorème 1.1.

2.1. COROLLAIRE (cf. [4], chapitre II, cor. 8.3 à la page 50). *Pour toute série* $f \in m(C_n)$ *on a*:

$$f \in \operatorname{rad}\left(\frac{\partial f}{\partial x_1}, \ldots, \frac{\partial f}{\partial x_n}\right) C_n.$$

D é m o n s t r a t i o n. D'après (i) pour tout idéal premier I: $(\partial/\partial x_1) f, \ldots, (\partial/\partial x_n) f \in I \Longrightarrow f \in I$. Il suffit de rappeler le fait que le nilradical d'un idéal est une intersection d'idéaux premiers.

2.2. COROLLAIRE. *Si* $f, g \in m(C_n)$ *alors il existe une série* $P(u,v) \neq 0$ *dans* $K\{u,v\}$ *telle que*

$$P(f,g) \in \left(\frac{\partial(f,g)}{\partial(x_1,x_2)}, \ldots \frac{\partial(f,g)}{\partial(x_{n-1},x_n)}\right)C_n.$$

Démonstration. On a:

$$\mathrm{rad}\left(\frac{\partial(f,g)}{\partial(x_i,x_j)}\right)C_n = I_1 \cap \ldots \cap I_s,$$

où $I_1,\ldots,I_s$ sont des idéaux premiers. Si $D, D' \in \mathrm{Der}\, C_n$, alors

$$D f D' g - D' f D g \equiv 0, \ \mathrm{mod}\left(\frac{\partial(f,g)}{\partial(x_i,x_j)}\right)C_n \subset I_l \quad \text{pour } l = 1,\ldots,s.$$

D'après (ii) il existe une série $P_l(u,v) \neq 0$ telle que $P_l(f,g) \equiv O(I_l)$, donc $\prod_l P_l(f,g) \in \bigcap_l I_l$. Par définition du nilradical il existe un entier $p > 0$ telle que la série $(\prod_l P_l(f,g))^p$ appartienne à l'idéal engendré par les jacobiens $\partial(f,g)/\partial(x_i,x_j)$. Il suffit alors de poser $P(u,v) = (\prod_l P_l(u,v))^p$.

2.3. *Exemple*. Nous déduisons du corollaire 2.2 un cas particulier de l'inegalité de Łojasiewicz (cf. e.g. [3]): *Soit* $f \in \mathbb{R}\{x_1,\ldots, x_n\}$ *une série sans terme constant convergente dans un voisinage* U *de zéro de* $\mathbb{R}^n$ *telle que* $f(c) > 0$ *pour* $c \in U \setminus \{0\}$. *Alors il existe des constantes* $A > 0$, $q > 0$ *telles que* $f(c) \geq A|c|^q$ *pour* $c \in U$ *assez petit*.

Démonstration. Posons $g(x) = x_1^2 + \ldots + x_n^2$ et soit $P(u,v)$ une série telle que

$$(*) \qquad P(f(x), g(x)) = \sum_{i<j} q_{ij}(x)\left(x_i \frac{\partial f}{\partial x_j} - x_j \frac{\partial f}{\partial x_i}\right).$$

En diminuant U si nécessaire on peut supposer que les séries $q_{ij}(x)$ sont convergentes dans U et que la série $P(u,v)$ converge dans un voisinage de l'ensemble $\{(f(c), g(c)) : c \in U\}$. Soit $m_f(r) = \min\{f(c) : |c| = r\}$; alors $f(c) \geq m_f(|c|)$ pour $c \in U$ assez petit. Pour tout $r > 0$ tel que $\{c \in \mathbb{R}^n : |c| = r\} \subset U$ on a $m_f(r) = f(c(r))$ où $|c(r)| = r$ et

$$c_i(r) \frac{\partial f}{\partial x_j}(c(r)) - c_j(r) \frac{\partial f}{\partial x_i}(c(r)) = 0 \quad \text{pour } i,j = 1,\ldots,n;$$

alors $P(m_f(r), r^2) = 0$ d'après (*). Le théorème de Puiseux affirme l'existence de constantes $A > 0$ et $q > 0$ telles que $m_f(r) > A r^q$ pour r assez petit, ce qui achève la démonstration de l'inegalité.

Ouvrages cités

[1] ABHYANKAR, S. S.: Local analytic geometry, Academic Press, New York - London 1964.

[2] LEFSCHETZ, S.: Algebraic geometry, Princeton Univ. Press, Princeton, NJ, 1953.

[3] ŁOJASIEWICZ, S.: Semi-analytic sets, Global analysis and its applications III, International Atomic Energy Agency, Vienna 1974, pp. 25-29.

[4] TOUGERON, J. C.: Idéaux de fonctions différentiables, Springer-Verlag, Berlin - Heidelberg - New York 1972.

Staszica 4 m. 3
PL-25-008 Kielce
Poland

A DISTORTION THEOREM FOR A CLASS OF POLYNOMIAL MAPPINGS

Przemysław Skibiński (Łódź)

Contents | page

Summary

In the present paper some polynomial mappings of two variables are considered. There has been obtained a geometric relationship between zeros of the Jacobian of the mapping under consideration and values of the mapping at these zeros.

Introduction

Charzyński and Kozłowski [1] have considered the polynomial mappings of the form

$$(*)\qquad \zeta = P(w) = w + C_2 w^2 + \dots + C_{m+1} w^{m+1}$$

and they have obtained geometric relationship between the zeros

$$(**)\qquad w_k\ ,\ k = 1,\dots$$

of the derivative $P'(w)$, i.e. the singular points of the mapping $(*)$, and the corresponding values

$$\zeta_k = P(w_k)\ ,\quad k = 1,\dots,$$

being the singular points of the inverse mapping of $(*)$. They have proved, in particular, that the following inequality holds:

$$\inf_k |w_k| \geq \frac{1}{4} \inf_k |\zeta_k| \ .$$

A Distortion Theorem for a Class of Polynomial Mappings

The relation belongs to the range of the so called distortion theorems, and states that the singular points (**) of the polynomial (*) may not approach arbitrarily close to zero, with the simultaneous distant from zero position of the singular points of the inverse algebraic mapping of (*).

In the present paper a result is obtained which is an analogue of the above fact for some mappings of two variables.

1. Notation

$\mathbb{C}$, $\mathbb{C}^2$ will denote, respectively, the topological field of complex numbers z and the vector space over $\mathbb{C}$ of pairs $w = (x,y)$, $x \in \mathbb{C}$, $y \in \mathbb{C}$. We shall look upon these as being provided with their natural topology with norm $|z|$ and $|w| = \text{Max}\,(|x| , |y|)$ for $\mathbb{C}$ and $\mathbb{C}^2$, respectively. We assume the analogous convention for functions and mappings with arguments and values belonging to the above described field and vector space. We shall denote by single letters or - in justificable situations - with the additionally indicated indeterminate argument functions and mappings. For the given holomorphic mapping F from $\mathbb{C}^2$ into $\mathbb{C}^2$ the corresponding Jacobian will be denoted by J_F , and the set of zeros of the Jacobian J_F - by Ω_F . At last, we shall treat the empty set, in terms of distance, as lying at the distance $+\infty$ from the origin.

2. Some polynomial mappings

Here we shall consider the mappings of the form

$$(1) \qquad F = (P,Q) = \sum_{i=1}^{m} (c_iP_i, d_iQ_i) ,$$

where

$$(2) \qquad P_i , Q_i , \quad i = 1,\dots,m$$

are given, fixed for the sequel, homogenous polynomials of variables x and y , $x \in \mathbb{C}$, $y \in \mathbb{C}$, of degree i , and such that the Jacobians

$$(2') \qquad P_{ix}Q_{jy} - P_{iy}Q_{jx} , \quad i,j = 1,\dots,m ,$$

do not vanish identically, whereas

$$(3) \qquad c_2 ,\dots, c_m , \quad d_2 ,\dots, d_m$$

are arbitrary variable complex coefficients, and

$$(3') \qquad c_1 = 1 , \quad d_1 = 1$$

3. The main result

We proceed to estimate the distance of the set Ω_F from the origin for the mapping (1).

THEOREM (on distortion) For the above described mappings of the type (1) there exists a positive constant α depending on the base (2) and independent of the coefficients (3) such that for every mapping (1) the following inequality holds:

$$(4)\qquad \inf_{w \in \Omega_F} |w| \geq \alpha \inf_{\zeta \in F(\Omega_F)} |\zeta| .$$

P r o o f. At first we shall show that we may confine ourselves to the special mappings $\dot{F}$ from among those considered here for which there is

$$\inf_{\dot{\zeta} \in \dot{F}(\Omega_{\dot{F}})} |\dot{\zeta}| = 1 .$$

In fact, suppose that for the mappings in question there exists the above described constant α and let F be any mapping of the form (1). Then putting

$$(5)\qquad \dot{F}(\dot{w}) = \frac{1}{\varkappa} F(\varkappa \dot{w}) \quad , \text{ where } \varkappa = \inf_{\zeta \in F(\Omega_F)} |\zeta| ,$$

it is easy to check that the set Ω_F is the image of the set $\Omega_{\dot{F}}$ under the homothety with the coefficient $\varkappa$; at the same time the set $F(\Omega_F)$ is the image of the set $\dot{F}(\Omega_{\dot{F}})$ under the same homothety. From this it follows easily that

$$(6)\qquad \inf_{\dot{\zeta} \in \dot{F}(\Omega_{\dot{F}})} |\dot{\zeta}| = 1 ,$$

so $\dot{F}$ is the above mentioned special mapping. Therefore according to the hypothesis we have

$$(7)\qquad \inf_{\dot{w} \in \Omega_{\dot{F}}} |\dot{w}| \geq \alpha \cdot 1$$

Hence, in virtue of the above mentioned relations between Ω_F and $\Omega_{\dot{F}}$, as well as between $F(\Omega_F)$ and $\dot{F}(\Omega_{\dot{F}})$, we have

$$\inf_{w \in \Omega_F} |w| = \varkappa \inf_{\dot{w} \in \Omega_{\dot{F}}} |\dot{w}| \quad \text{and} \quad \varkappa \inf_{\dot{\zeta} \in \dot{F}(\Omega_{\dot{F}})} |\dot{\zeta}| = \inf_{\zeta \in F(\Omega_F)} |\zeta| ,$$

which, according to (6) and (7), gives already the relation (5) for arbitrarily chosen, thus for every mapping F of the type (1).

Now, let us suppose that for the special mappings the above mentioned constant does not exist. This means that for every positive integer n there exist: a special mapping $\dot{F}_n$ and a number $\dot{\rho}_n$ such that

(8) $$\dot{\rho}_n = \operatorname*{Inf}_{\dot{w} \in \Omega_{\dot{F}_n}} |\dot{w}| < \frac{1}{n} .$$

At the same time, since $\dot{F}_n$ is special, we have

(9) $$|\dot{F}_n(\dot{w})| \geq 1 \quad \text{for} \quad \dot{w} \in \Omega_{\dot{F}_n} .$$

Let us put now

(10) $$R_n(t) = \frac{1}{\dot{\rho}_n} = \dot{F}_n(\dot{\rho}_n t) , \quad t \in \mathbb{C}^2$$

Taking into account (10) it is easy to verify that Ω_{R_n} is the image of $\Omega_{\dot{F}_n}$ under the homothety with the coefficient $1/\dot{\rho}_n$. Hence together with (9) and (10) it follows that

(11) $$|t| \geq 1 \quad \text{and} \quad |\dot{\rho}_n R_n(t)| \geq 1 \quad \text{for} \quad t \in \Omega_{R_n} .$$

At last the inequalities (8) and (11) give the convergence

(12) $$\lim_{n \to \infty} |R_n(t)| = +\infty \quad \text{for} \quad t \in \Omega_{R_n} ,$$

uniform with respect to t.

Since the set $\Omega_{\dot{F}_n}$ is closed, there exist points w_n^* there such that $|w_n^*| = \dot{\rho}_n$; thus

(13) $$t_n^* = w_n^*/\dot{\rho}_n \in \Omega_{R_n} \quad \text{and} \quad |t_n^*| = 1 .$$

Let us notice now that the mappings (10) are of the type (1), so they can be represented in the form

$$R_n = \sum_{i=1}^{m} (\dot{c}_{in} P_i , \dot{d}_{in} Q_n) , \quad \text{where} \quad c_{1n} = 1 = d_{1n} .$$

Let us put

(14) $$\dot{C}_n = \operatorname{Max} (|\dot{c}_{1n}|, \ldots, |\dot{c}_{mn}|), \quad \dot{D}_n = \operatorname{Max} (|\dot{d}_{1n}|, \ldots, |\dot{d}_{mn}|) .$$

Then, by virtue of (12) and (13), we have

(15) $$\lim_{n \to} |\dot{C}_n| = +\infty \quad \text{or} \quad \lim_{n \to} |\dot{D}_n| = +\infty .$$

Consider now the normalized mappings

(16) $$S_n = \sum_{i=1}^{m} ((\dot{c}_{in}/\dot{C}_n)P_i \, , \, (\dot{d}_{in}/\dot{D}_n)Q_i) \, , \quad n = 1,\ldots .$$

For any n the coefficients of the coordinates in the terms of the sum (16) are bounded by 1 . At the same time at least one of the coefficients for both of the coordinates is equal to 1 with respect to the absolute value. Therefore choosing, if needed, the subsequence we can assume that the sequence (16) is convergent to a certain mapping:

$$S = (\sum_{i=k}^{m} \gamma_i P_i \, , \, \sum_{j=l}^{m} \delta_j Q_j) \, , \quad 1 \le k \le m \, , \quad 1 \le l \le m$$
$$\gamma_k \neq 0 \, , \quad \delta_l \neq 0 .$$

We observe that $k > 1$ or $l > 1$, which is the consequence of (3') and (15). From this it follows immediately that the point $(0,0)$ is a zero of the Jacobian J_S .

Let us take now an arbitrary "direction" $\sigma = (\alpha,\beta)$ such that

(17) $$\gamma_k \delta_l P_{kx}(\sigma) Q_{ly}(\sigma) - \gamma_k \delta_l P_{ky}(\sigma) Q_{lx}(\sigma) \neq 0 .$$

Such a "direction" exists according to the above observation and (2'). Let us consider Jacobian J_S on this "direction", i.e. the polynomial $J_S(\sigma\zeta)$ of one variable $\zeta \in \mathbb{C}$. It is easy to check that the expansion of the polynomial in powers of ζ begins with the power ζ^{k+l-2} with a non-vanishing coefficient, which is equal to the right-hand side of (17). Thus the point 0 is an isolated zero of $J_S(\sigma\zeta)$. Therefore, according to the convergence of the sequence of J_{S_n} to J_S and, consequently, the convergence of the sequence $J_{S_n}(\sigma\zeta)$ to $J_S(\sigma\zeta)$ we see (by the Hurwitz's theorem) that for n sufficiently large the polynomial $J_{S_n}(\sigma\zeta)$ has at some points ζ_n zeros arbitrarily close to the point 0 . This means that the Jacobian J_{S_n} has at the points $\sigma\zeta_n$ zeros arbitrarily close to the point $(0,0)$. Since the above mentioned zeros of the Jacobian J_{S_n} coincide with those of the Jacobian J_{R_n} , we finally conclude that for n sufficiently large the Jacobian J_{R_n} has the zeros $\sigma\zeta_n$ arbitrarily close to $(0,0)$.

On the other hand the relation (11) asserts that the zeros of the Jacobian J_{R_n} are lying beyond the unit ball.

Thus we have a contradiction which ends the proof.

Finally let us notice that it would be interesting to examine the existence and the properties of the infinite sequences of mappings analogous to (2), for which there exists a universal positive constant such that for every mapping of the type (1), obtained by cutting off the above mentioned sequence, the relation (4) holds.

It is a pleasure to thank Professor Z. Charzyński for having outlined the idea of this note as well as for his many helpful remarks during preparation of the paper.

References

[1] CHARZYŃSKI,Z. and KOZŁOWSKI,A.: Geometry of polynomials II, Bull. Soc. Sci. Lettres Łódź 28, 6 (1978), 1-10.

[2] NARASIMHAN,R.: Several complex variables, University of Chicago Press, Chicago and London 1971.

Institute of Mathematics
University of Łódź
ul. Banacha 22
PL-90-238 Łódź, Poland

SOME CLASSES OF REGULAR FUNCTIONS DEFINED BY CONVOLUTION

Jan Stankiewicz and Zofia Stankiewicz (Rzeszów)

Summary. S. Ruscheweyh, using Hadamard convolution, introduced new definitions for some classes of regular functions. The definitions are used here to prove several properties of these classes.

Ruscheweyh [4, 5] gave some new criteria that a regular normalized function f belongs to some known class of functions. Using the concept of convolution he presented the new definitions for the class S of univalent functions and for the class St of univalent starlike functions.

Let N denote the class of all regular normalized functions

(1) $$f(z) = z + a_2 z^2 + \dots, \quad z \in U = \{z : |z| < 1\}.$$

The convolution of Hadamard product $f * g$ of two functions

$$f(z) = a_0 + a_1 z + a_2 z^2 + \dots, \quad g(z) = b_0 + b_1 z + b_2 z^2 + \dots$$

is defined as follows:

(2) $$(f * g)(z) = a_0 b_0 + a_1 b_1 z + a_2 b_2 z^2 + \dots .$$

We need the following notation:

$S = \{f \in N :\ f$ is univalent in $U\}$,

$St = \{f \in N :\ f$ is univalent and starlike in $U\}$,

$Sc = \{f \in N :\ f$ is univalent and convex in $U\}$,

$St_\alpha = \{f \in N :\ \mathrm{Re}(zf'(z)/f(z)) > \alpha,$ for $z \in U\}$, $0 \le \alpha < 1$,

$Sc_\alpha = \{f \in N :\ \mathrm{Re}(1 + zf''(z)/f'(z)) > \alpha,$ for $z \in U\}$, $0 \le \alpha < 1$,

$S_\beta = \{f \in N :\ \mathrm{Re}(e^{i\beta} zf'(z)/f(z)) > 0,$ for $z \in U\}$, $-\pi/2 < \beta < \pi/2$,

$S(\alpha) = \{f \in N :\ |\arg(zf'(z)/f(z))| < \alpha\pi/2,$ for $z \in U\}$, $0 < \alpha \le 1$.

These classes are very often the investigated classes in geometrical theory of univalent functions. Using the properties of convolution

S. Ruscheweyh gave the definitions of the classes S and St in a different way. For a given class of functions he determined a certain one- or two-parametrical family of functions so that the convolution of every function f from the first class with every function h of the second class is different from zero for all $z \in U_0 = \{z : 0 < |z| < 1\}$.

Let

$$S' = \{h(z) = \frac{z}{(1-xz)(1-yz)} : |x| \leq 1, \quad |y| \leq 1\},$$

$$St' = \{h(z) = (\frac{z}{(1-z)^2} + it\frac{z}{1-z})/(1+it) : t \in \mathbb{R}\}.$$

THEOREM A (Ruscheweyh [4]). For $f \in N$ we have

$$f \in S \iff \forall_{z \in U_0} \ \forall_{h \in S'} \ (f * h)(z) \neq 0,$$

$$f \in St \iff \forall_{z \in U_0} \ \forall_{h \in St'} \ (f * h)(z) \neq 0.$$

Ruscheweyh introduced and investigated also a new class M which can be defined as follows:

$$M = \{f \in N : \forall_{z \in U_0} \ \forall_{h \in St} \ (f * h)(z) \neq 0\}.$$

It is known that M contains all close-to-convex functions; every function $f \in M$ is univalent and its coefficients satisfy the Bieberbach conjecture.

In [6] the authors gave also definitions of this kind for the classes St_α, Sc_α and $\check{S}_\beta$. These results can be expressed by the corresponding classes St_α', Sc_α' and S_β' as follows:

$$St_\alpha' = \{h(z) = (z + \frac{x+2\alpha-1}{1-\alpha} z^2)/(1-z)^2 : |x| = 1\},$$

$$Sc_\alpha' = \{h(z) = (z + \frac{x+\alpha}{1-\alpha} z^2)/(1-z)^3 : |x| = 1\},$$

$$\check{S}_\beta' = \{h(z) = [z + (x - e^{-2i})z^2/(1+e^{-2i})]/(1-z)^2 : |x| = 1\}.$$

These results can be generalized on many other classes of regular functions. We give some definitions and present a general idea how to proceed in other cases. We start with some special cases.

THEOREM 1. *Let* $S(\alpha)$ *be the class of* α-*angularly starlike functions. For* $f \in N$ *we have*

$$(3) \qquad f \in S(\alpha) \Leftrightarrow \underset{z \in U_0}{\forall} \ \underset{h \in S(\alpha)'}{\forall} \ (f * h)(z) \neq 0,$$

where

$$S(\alpha)' = \{h(z) = [\frac{z}{(1-z)^2} - (\frac{1+x}{1-x})^{\alpha} \frac{z}{1-z}]/[1 - (\frac{1+x}{1-x})^{\alpha}] : |x| = 1, \ x \neq 1\}.$$

Proof. To prove this theorem we need the following simple properties of the convolution:

(i) $f(z) * (z/(1-z)) = f(z)$,

(ii) $f(z) * (z/(1-z)^2) = zf'(z)$,

(iii) $f * (h_1 + h_2) = f * h_1 + f * h_2$,

(iv) $f * (ch) = c(f * h)$, c - constant.

Thus, for $f \in N$ and $h \in S(\alpha)'$ we have

$$(4) \qquad (f * h)(z) = [f(z) * \frac{z}{(1-z)^2} - (\frac{1+x}{1-x})^{\alpha} f(z) * \frac{z}{1-z}]/[1 - (\frac{1+x}{1-x})^{\alpha}]$$

$$= [\frac{zf'(z)}{f(z)} - (\frac{1+x}{1-x})^{\alpha}] f(z)/(1 - (\frac{1+x}{1-x})^{\alpha}).$$

Therefore

$$(5) \qquad (f * h)(z) \neq 0 \Leftrightarrow zf'(z)/f(z) \neq (\frac{1+x}{1-x})^{\alpha}.$$

The function $[(1+z)/(1-z)]^{\alpha}$ is univalent and maps conformally the unit disc U onto the angular domain $B = \{w : |\arg w| < \alpha\pi/2\}$ and that is why its boundary values $[(1+x)/(1-x)]^{\alpha}$ cover the boundary of B. Thus, the function $zf'(z)/f(z)$ does not take any boundary value of the domain B and therefore all its values lie in B (because $zf'(z)/f(z)$ is equal to 1 for $z = 0$ and to 1 within B). Thus, we have

$$(6) \qquad |\arg(zf'(z)/f(z))| < \alpha\pi/2$$

and therefore $f \in S(\alpha)$.

Now conversely, if $f \in S(\alpha)$ then by (6) the function $zf'(z)/f(z)$ does not take any value $[(1+x)/(1-x)]^{\alpha}$ and by (5) we have $(f * h)(z) \neq 0$ for every $z \in U_0$ and $h \in S(\alpha)'$.

In the papers by Janowski [2], Stankiewicz and Waniurski [7], and some others the following classes were investigated:

$$P(A,B) = \{p(z) = 1 + p_1 z + p_2 z^2 + \dots : p(z) \prec \frac{1+Az}{1-Bz}\},$$

where $|A| \leq 1$, $|B| \leq 1$, $A+B \neq 0$ and $\prec$ means the subordination of the functions in the unit disc. By the classes $P(A,B)$ we can define some subclasses of univalent functions:

$$S(A,B) = \{f \in N : zf'(z)/f(z) \in P(A,B)\} = \{f \in N : zf'(z)/f(z) \prec \frac{1+Az}{1-Bz}\}.$$

This two-parametrical family of functions is a general family and by some specification of parameters A and B we can obtain many well known subclasses of starlike functions. In particular, if $A = B = 1$ then we get the class $St(S(1,1) = St)$.

We can also obtain the definition for the class $S(A,B)$ by the use of the convolution.

THEOREM 2. *For* $f \in N$ *we have*

$$f \in S(A,B) \iff \forall_{z \in U_0} \ \forall_{h \in S(A,B)'} \ (f * h)(z) \neq 0,$$

where

$$S(A,B)' = \{h(z) = (z - \frac{A+x}{A+B} z^2)/(1-z)^2 : |x| = 1\}$$

$$= \{h(z) = (z/(1-z)^2 - \frac{1+Ax}{1-Bx} z/(1-z))/(1 - \frac{1+Ax}{1-Bx}) : |x| = 1\}.$$

Proof. If $h \in S(A,B)'$ and $f \in N$, then

$$(7) \qquad 0 \neq (f * h)(z) = (\frac{zf'(z)}{f(z)} - \frac{1+Ax}{1-Bx}) f(z)/(1 - \frac{1+Ax}{1-Bx}) \iff \frac{zf'(z)}{f(z)} \neq \frac{1+Ax}{1-Bx}.$$

It means that $zf'(z)/f(z)$ does not take any boundary value of the domain $H(U)$, where $H(z) = (1+Az)/(1-Bz)$. Since the function $zf'(z)/f(z)$ for $z = 0$ takes the value $1 = H(0)$ and H is univalent, then $zf'(z)/f(z) \prec H(z)$ and therefore $f \in S(A,B)$.

Conversely, if $f \in S(A,B)$ then $zf'(z)/f(z) \prec H(z)$ and therefore $zf'(z)/f(z)$ does not take any value $H(x) = (1+Ax)/(1-Bx)$, $|x| = 1$. By (7) we have $(f * h)(z) \neq 0$ for $z \in U_0$ and every $h \in S(A,B)'$. The proof is complete.

The class $S(A,B)$ can be generalized. We replace the function $H(z) = (1+Az)/(1-Bz)$ by an arbitrary univalent (convex) function $G(z)$.

We usually suppose that $G(0) = 1$ and $G(z) \neq 0$ for all $z \in U$. Such classes of functions were investigated by Goodman [1].

DEFINITION. Let G be some fixed function which is univalent (and convex) in the unit disc U and such that $G(0) = 1$, $0 \notin G(U)$. We define

$$S(G) = \{f \in N : zf'(z)/f(z) \prec G(z)\}.$$

When $G(z) = (1+z)/(1-z)$ then $S(G) = St = S(1,1)$; when $G(z) = (1+Az)/(1-Bz)$ then $S(G) = S(A,B)$.

THEOREM 3. *For* $f \in N$ *we have*

$$f \in S(G) \iff \underset{z \in U_0}{\forall} \ \underset{h \in S(G)'}{\forall} \ (f * h)(z) \neq 0,$$

where

$$S(G)' = \{h(z) = \left(\frac{z}{(1-z)^2} - G(x)\frac{z}{1-z}\right)/(1 - G(x) \ : |x| = 1 ,$$

and $G(x)$ are the boundary values of the function $G(z)\}$.

Proof. For $f \in N$ and $h \in S(G)'$ we have

(8) $0 \neq (f * h)(z) = (zf'(z)/f(z) - G(x))\, f(z)/(1 - G(x)) \iff zf'(z)/f(z)$

$\neq G(x)$ for $z \in U$ and for every boundary value $G(x)$.

It means that $zf'(z)/f(z)$ takes values only in $G(U)$. Thus, $zf'(z)/f(z) \prec G(z)$ and therefore $f \in S(G)$.

Conversely, if $f \in S(G)$ then $zf'(z)/f(z) \prec G(z)$ and $zf'(z)/f(z)$ does not take any boundary value $G(x)$. That is why for every $h \in S(G)'$ and $z \in U_0$ we get by (8) that $(f * h)(z) \neq 0$. The proof is completed.

The latter theorem is the general one and it shows how to obtain the analogous definitions for almost all subclasses of starlike functions and the related classes of functions.

Now, we give some analogous definitions for the subclasses of convex functions.

Let us put

$$Sc(G) = \{f \in N : 1 + zf''(z)/f(z) \prec G(z)\},$$

and in a special case

$$Sc(A,B) = \{f \in N : 1 + zf''(z)/f(z) \prec (1+Az)/(1-Bz)\}.$$

THEOREM 4. For $f \in N$ we have

$$f \in Sc(G) \iff \forall_{z \in U_0} \forall_{h \in Sc(G)'} (f * h)(z) \neq 0,$$

where

$$Sc(G) = \{h(z) = \left|\frac{z+z^2}{(1-z)^2} - G(x)\frac{z}{(1-z)^2}\right| / (1 - G(x)) : |x| = 1,$$

and $G(x)$ are the boundary values of $G(z)\}$.

Proof. Firstly, we notice that

(9) $f \in Sc(G) \iff zf'(z) \in S(G)$,

(10) $h \in Sc(G)' \iff zh'(z) \in S(G)'$,

(11) $(zf'(z)) * h(z) = f(z) * (zh'(z)) = z[(f * h)(z)]'$.

Now, by Theorem 3, and by (9), (10) and (11) we have

$$f \in Sc(G) \iff zf'(z) \in S(G), \quad \forall_{z \in U_0} \forall_{h \in S(G)'} zf'(z) * h(z) \neq 0$$

$$\iff \forall_{z \in U_0} \forall_{h \in S(G)'} f(z) * (zh'(z)) \neq 0 \iff \forall_{z \in U_0} \forall_{h \in Sc(G)'} (f * h)(z) \neq 0.$$

The theorem is proved.

Now, we recall the following classes:

$R_\alpha = \{f \in N : \operatorname{Re} f'(z) > \alpha, \ z \in U\}, \quad 0 \leq \alpha < 1,$
$Q = \{f \in N : \operatorname{Re}(f(z)/z) > \alpha, \ z \in U\}, \quad 0 \leq \alpha < 1,$
$\tilde{S} = \{f \in N : f(z)$ is locally univalent in $U\}$.

THEOREM 5. Let Q be one of the classes R_α, Q_α or $\tilde{S}$. For $f \in N$ we have

$$f \in Q \iff \forall_{z \in U_0} \forall_{h \in Q'} (f * h)(z) \neq 0,$$

where

$$R_\alpha' = \{h(z) = z/(1-z)^2 - (\alpha + it)z/(1 - \alpha - it) : t \in \mathbb{R}\},$$
$$Q_\alpha' = \{h(z) = z/(1-z) - (\alpha + it)z/(1 - \alpha - it) : t \in \mathbb{R}\},$$
$$\tilde{S} = \{h(z) = z/(1-z)^2\}.$$

Proof. 1° $Q = R_\alpha'$. For $f \in N$, $h \in R_\alpha'$ we have

$$0 \neq (f * h)(z) = z(f'(z) - \alpha - it)/(1 - \alpha - it)$$

and therefore $f'(z) \neq \alpha + it$. Since $f'(0) = 1$, then we get $\operatorname{Re} f'(z) > \alpha$ for $z \in U$.

2° $Q = Q_{\alpha}'$. Now, for $f \in N$ and $h \in Q_{\alpha}$ we get

$$0 \neq (f * h)(z) = (f(z) - (\alpha + it)z)/(1 - \alpha - it),$$

and therefore $f(z)/z \neq \alpha + it$. Thus, again $\mathrm{Re}(f(z)/z) > \alpha$.

3° $Q = \tilde{S}$. For every $f \in N$ we have $f(z) * (z/(1-z)^2) = zf'(z)$. That is why $f'(z) \neq 0$ in U and f are locally univalent in U.

The Theorem 5 can be generalized. We replace the classes R_{α} and Q_{α} by $R(G)$ and $Q(G)$ respectively, where

$$R(G) = \{f \in N : f'(z) \prec G(z)\},$$
$$Q(G) = \{f \in N : f(z)/z \prec G(z)\}.$$

The function G is the same as in the definition of the class $S(G)$. The corresponding classes $R(G)'$, $Q(G)'$ are defined as follows:

$$R(G)' = \{h(z) = (z/(1-z)^2 - G(x)z)/(1 - G(x)) : |x| = 1\},$$
$$Q(G)' = \{h(z) = (z/(1-z) - G(x)z)/(1 - G(x)) : x = 1\}.$$

We can introduce some special metrics $\rho_l(f,g)$ to the class N of all regular normalized functions in a following way: let $f(z) = z + a_2z^2 + \dots$ and $g(z) = z + b_2z^2 + \dots$ belong to N, and let l be a fixed positive integer (or a positive number). We define

$$\rho_l(f,g) := 2^l|a_2 - b_2| + 3^l|a_3 - b_3| + \dots = \sum_{k=2}^{\infty} k^l|a_k - b_k|.$$

If the last series is not convergent, then we put $\rho_l(f,g) = +\infty$.

Using this metric we define the neighbourhood of the function $f(z)$

$$\mathcal{N}_{l,\delta}(f) := \{g \in N : \rho_l(f,g) < \delta\}.$$

For a given function $f \in N$ we define

$$f_{n,\varepsilon}(z) = \begin{cases} \dfrac{f(z) + \varepsilon z}{1 + \varepsilon} & \text{for } n = 1, \\ f(z) + \varepsilon z^n & \text{for } n = 2,3,\dots \,. \end{cases}$$

THEOREM 6. *If for some positive integer n and for every complex number ε, $|\varepsilon| < \delta$, the functions $f_{n,\varepsilon}(z)$ belong to St_{α}, then the neighbourhood*

$$\mathcal{N}_{1,\delta(1-\alpha)}(f) \subset St_{\alpha}.$$

For $n = 1$, $\alpha = 0$ we obtain the result of Ruscheweyh [5]. For n, α we get the result of Rahman and Stankiewicz [3].

Proof. The condition $f_{n,\varepsilon} \in St_\alpha$ can be written in a form

$$(f_{n,\varepsilon} * h)(z) \neq 0 \quad \text{for} \quad z \in U_0, \quad h \in St'_\alpha.$$

or equivalently

$$(f * h)(z) \neq \varepsilon z^n * h(z) = \varepsilon h_n z^n,$$

where $h(z) = z + h_2 z^2 + \ldots$. It means that $(f * h)(z)/z^n$ does not take any value in the disc $\{w : |w| < \delta \inf_{h \in St'_\alpha} |h_n|\} = \{w : |w| < \delta(1-\alpha)\}$. Thus

$$|z^n/(f * h)(z)| \leq \frac{1}{\delta(1-\alpha)},$$

and by Schwarz Lemma

$$|z^n/(f * h)(z)| \leq \frac{|z|^{n-1}}{\delta(1-\alpha)} \quad \text{or equivalently} \quad \left|\frac{(f * h)(z)}{z}\right| \leq \delta(1-\alpha).$$

Now, let $g \in \mathcal{N}_{1,\delta(1-\alpha)}(f)$. Then

$$|(g * h)(z)/z| = |(f * h)(z)/z - ((f-g) * h)(z)/z|$$

$$\geq |(f * h)(z)/z| - |((f-g) * h)(z)/z|$$

$$\geq \delta(1-\alpha) - \sum_{k=2}^{\infty} |a_k - b_k||h_k||z|^{k-1} \geq \delta(1-\alpha) - |z| \sum_{k=2}^{\infty} k((1-\alpha)|a_k - b_k|$$

$$\geq \delta(1-\alpha) - |z|(1-\alpha)\rho_1(f,g) \geq \delta(1-\alpha)(1-|z|) > 0 \quad \text{for} \quad z \in U, \; h \in St'_\alpha.$$

Thus $(g * h)(z) \neq 0$ for $z \in U_0$ and $h \in St'_\alpha$, and therefore $g \in St_\alpha$. The theorem is proved.

THEOREM 7. If for some positive integer n and for every ε, $|\varepsilon| < \gamma$ the functions $f_{n,\varepsilon}(z) \in Sc_\alpha$, then $\mathcal{N}_{2,n\gamma(1-\alpha)}(f) \subset Sc_\alpha$.

Proof. By the relation (9), $f_{n,\varepsilon} \in Sc_\alpha$ implies that

$$z f'_{n,\varepsilon}(z) = zf'(z) + \varepsilon n z^n \in St_\alpha.$$

Using the Theorem 6 with $\delta = n\gamma$ we obtain

$$\mathcal{N}_{1,n\gamma(1-\alpha)}(zf') \subset St_\alpha.$$

Since $\rho_1(zf', zg') = \rho_2(f,g)$, then

$$\mathcal{N}_{2,n\gamma(1-\alpha)} \subset Sc_\alpha,$$

as desired.

References

[1] GOODMAN, A.W.: Analytic functions that take values in a convex region, Proc. Amer. Math. Soc. 14 (1963), 60-64.

[2] JANOWSKI, W.: Some extremal problems for certain families of analytic functions, Ann. Polon. Math. 28 (1973), 297-326.

[3] RAHMAN, Q.I. and J. STANKIEWICZ: On the Hadamard products of schlicht functions, to appear.

[4] RUSCHEWEYH, S.: Linear operators between classes of prestarlike functions, Comment. Math. Helvetici 52 (1977), 497-507.

[5] ——: Neighbourhoods of univalent functions, Proc. Amer. Math. Soc., to appear.

[6] SILVERMAN, H., E.M. SILVIA, and D. TELLAGE: Convolution conditions for convexity, starlikeness and spiral-likeness, Math. Z. 162 (1978), 125-130.

[7] STANKIEWICZ, J. and J. WANIURSKI: Some classes of functions subordinate to linear transformation and their applications, Ann. Univ. Mariae Curie-Skłodowska 28 (1974), 85-94.

Institute of Mathematics and Physics
Rzeszów Technical High School
Poznańska 2, PL-35-084 Rzeszów, Poland

A DIFFERENTIAL GEOMETRIC QUANTUM FIELD THEORY ON A MANIFOLD I

Osamu Suzuki (Tokyo)

Contents

Introduction

In this paper a concept of a differential geometric quantum field is introduced and some well known quantizations are discussed in a unified manner: Boson fields can be obtained by using a canonical symplectic structure on a cotangent bundle of a manifold, and Fermion fields can be obtained by using exterior differential forms, and a quantization of a Hamilton-Jacobi equation can be obtained by using a metrical connection of a Riemannian manifold. Our quantum fields are given by using connections, so it may be hoped that some differential geometric properties of a manifold are obtained by quantum fields. In fact, it can be proved that existence of Boson fields characterises flatness of the cotangent bundle of a manifold.

In physics quantizations are obtained by replacing operators for classical physical variables in a classical system. For example, a Schrödinger equation is obtained by replacing certain operators for canonical variables of a classical Hamiltonian in a well known manner. A second quantization of an electromagnetic field is obtained by replacing Boson operators for Fourier coefficients of a solution of Maxwell equations. In this way we obtain an algebra of operators in the second quantization. It may be a remarkable fact that an algebra obtained above must be one of the following two algebras, a canonical commutation algebra or an anti-commutation algebra.

Recently, Kostant [3] constructed "geometric quantization theory" by constructing a Lie homomorphism from a Poisson algebra of classical variables to an algebra of operators on some Hilbert space and applied his theory to unitary representation theory.

In this paper we shall give another description of quantization process which is motivated by the second quantization method in physical field theory. We define creation and annihilation operators by using connections and make a field theory by following a formalism of physical field theory. Hence it may be said that our field theory is rather *-algebraic approach to quantization problems. Therefore our theory does not treat in general Lie homomorphisms as in the Kostant theory. Moreover, our quantizations depend deeply on curvature properties, so we can get some differential geometric properties by the field operators. The second part of the paper will contain a complex manifold approach to the topic.

Now we shall describe the contents of this paper in a more detail. In Section 1 we define a differential geometric quantum field in the following manner. If an affine connection is given on a manifold, then we can define a covariant differentiation ∇. We make a formally adjoint operator ∇^* with respect to some inner products of covariant tensor fields. Then we can define a Hamiltonian operator by

$$\mathcal{H}_p = \nabla_p^* \nabla_p + \nabla_{p-1} \nabla_{p-1}^*,$$

where the suffix p denotes a degree of tensor fields on which ∇ operators. The eigen tensor fields of this operator are called "pure states" of the field, if they exist. Following the notions in physical field theory, we call ∇_p (resp. ∇_p^*) creation (resp. annihilation) operates on p-particle states. Time evolutions of ∇_p and ∇_p^* are defined by using Heisenberg euqations as in physical field theory. We shall call $\{\nabla_p, \nabla_p^*, \mathcal{H}_p\}$ a differential geometric quantum field induced by a connection, or simply, a quantization of a connection. We call this a quantization of a connection because of the following reason: If we make a classical limit in some sense, we obtain an equation of a geodesic which is defined by a given connection in some cases. At the end of Section 1, we treat harmonic oscillator and show our basis idea in this simple case. In Section 2, by using a connection which is defined by a canonical symplectic form on a cotangent of a manifold, we shall construct a Boson field and characterize flatness of a cotangent bundle by using this field. This construction of a field is very similar to Kostant's quantization. At the end of this section we can see that its classical limit gives an equation of harmonic oscillator. In Section 3, we shall construct a Fermion field on a compact Riemannian manifold. This field can be obtained by using exterior differential forms. In this case a Hamiltonian is nothing but a Laplace--Beltrami operator of differential forms. In the case of a compact kähler manifold we obtain another Fermion field by using $\bar{\partial}$-operator and its formally adjoint operator. We remark that pure states of this field are nothing but solution of wave equations of Dirac type of spin $\frac{1}{2}$ in the case of a two-dimansional complex torus. In a final section, we shall give a quantization of a Riemannian connection. By using this covariant differentiation and its formally adjoint operator with respect to a canonical inner product, we obtain a field. In this case a Hamiltonian is nothing but a Lichnerowicz Laplacian. If we apply this method to a compact semi-simple Lie group, we obtain a Casimir operator as a Hamiltonian. This fact leads us to a construction of unitary representations of a compact semi-simple Lie group by using our differential geometric quantum field theory. If this Hamiltonian has a quasi-classical limit, then the Schrödinger equation of this Hamiltonian gives a Hamilton-Jacobi equation of a motion without a potential.

By above examples, we may say that our quantizations by using connections include many important examples and natural construction of quantum fields. Thus we may hope that our idea will be investigated furthermore and this quantization method let us solve differential geometric problems on manifolds.

The author would like to express his hearty thanks to Professors A.Aomoto, H.Araki, A.Inoue, M.Itoh, S.Sakai, H.Takai, A.Takeshita and J.Yamashita for many various discussions and suggestions,and to Professor J.Ławrynowicz for his interest in the author's work and for sending him Professor J.Czyż's paper on geometric quantization.

1. Definition of a differential geometric quantum field

In this section we give a concept of quantum fields in general form and next we give a differential geometric quantum field by using a connection. Finally we demonstrate our basic idea by treating a simple example of a harmonic oscillator.

Let H_p $(p = 1,2,\ldots,N)$ $(N \leq +\infty)$ be a sequence of Hilbert spaces over $\mathbb{C}$ whose positive definite hermitian inner product is denoted by $(\ ,\ )_p$. In order to include many examples we assume that another continuous sesquilinear form $((\ ,\ ))_p$ is given on H_p. This may be identical with the original inner product on H_p as a special case. Let $\nabla_p : H_p \to H_{p+1}$ be a densely defined closed operator. Let $\nabla_p^* : H_{p+1} \to H_p$ be its formally adjoint operator with respect to $((\ ,\ ))_p$, which is defined by

$$(1.1) \quad ((\nabla_p f, g))_{p+1} = ((f, \nabla_p^* g))_p$$

for well defined f and g. In the following we do not explain domains of operators. Thus one reads that, formulas hold for well defined elements of domains. We define a Hamiltonian $\mathcal{H}_p$ of degree p by

$$\mathcal{H}_p = \nabla_p^* \nabla_p + \nabla_{p-1} \nabla_{p-1}^* .$$

Then $\mathcal{H}_p$ is an operator $\mathcal{H}_p : H_p \to H_p$. With these preliminaries we define a quantum field as follows:

<u>Definition (1.2)</u>. A quartet $\{H_p, \nabla_p, \nabla_p^*, \mathcal{H}_p\}$ is called a <u>quantum field</u>. ∇_p is called a <u>creation operator</u> and ∇_p^* an <u>annihilation operator</u>. $\mathcal{H}_p$ is called a <u>Hamiltonian operator</u>. We define a <u>commutation relation</u> of degree p by

$$\mathcal{R}_p = \nabla_p^* \nabla_p - \nabla_{p-1} \nabla_{p-1}^* .$$

If $\mathcal{H}_p$ has an eigen vector in H_p, then it is called a <u>pure state</u> of degree p of the field.

<u>Remark</u>. Terminology used above comes from the second quantization theory of physical field theory. We must remark that although terminology is the same, our definitions are more or less different from the ones in physics. Thus it is better to regard our definitions to be pure mathematical ones.

As in physics, we formulate the following definition:

<u>Definition (1.3)</u>. (1) A quantum field is called a <u>Boson field</u> if and only if

$$\mathcal{R}_p = c_p \cdot 1_p$$

holds for every p with some constant c_p which may depend on p, where 1_p is the identity operator in H_p.

(2) A quantum field is called a <u>Fermion field</u> if and only if

$$\nabla_{p+1}\nabla_p = 0 \quad \text{and} \quad \nabla^*_{p-1}\nabla^*_p = 0$$

hold for every p.

Remark. If we take ∇_p identically equal zero, then we get a Fermion field. In the following we omit this trivial Fermion field.

In Section 2 and 3 we will consider existence problems of Boson and Fermion fields.

Here we shall define time evolutions of ∇_p and ∇^*_p by using Heisenberg equations in the physical field theory [1]:

$$(1.4) \qquad \frac{d\nabla_p}{dt} = \frac{c}{\hbar}[\nabla_p \mathcal{H}_p - \mathcal{H}_{p+1}\nabla_p], \quad \frac{d\nabla^*_p}{dt} = \frac{c}{\hbar}[\nabla^*_p \mathcal{H}_p - \mathcal{H}_{p-1}\nabla^*_p]$$

with some constant c, where $\hbar$ is called Planck's constant devided by 2π. We define a Schödinger equation for $\mathcal{H}_p$ by

$$\frac{\hbar}{i}\frac{\partial \overline{\Psi}_p}{\partial t} = \mathcal{H}_p \overline{\Psi}_p$$

and a solution $\overline{\Psi}_p$ is called a wave equation of the field. In the case where $\mathcal{H}_p$ is a self-adjoint operator, we can define a time evolution operator for a Schrödinger equation by

$$T^p_t = \exp(\frac{i}{\hbar}\mathcal{H}_p t) : H_p \longrightarrow H_p.$$

In this case the solution of (1.4) is given by

$$\nabla_p(t) = T^{p+1}_t \circ \nabla_p \circ T^{p-1}_t$$

with $c = i$ (see [1]).

Now we shall construct a differential geometric quantum field on a manifold. Let M be a manifold of C^∞-class and let $\pi : E \longrightarrow M$ be a vector bundle over M. We fix notation as follows: $T^*(M)$ denotes the cotangent bundle of M and $T^{*p}(M)$ denotes p-times tensor product of $T^*(M)$, $\Gamma(M, E \otimes T^{*p}(M))$ denotes sections of $E \otimes T^{*p}(M)$ of C^∞--class. $D_p(E)$ denotes sections of $E \otimes T^{*p}(M)$ with compact supports. In the case where E is a trivial bundle, we write D_p for $D_p(E)$. Now we assume that a connection ∇ is given on $E \otimes T^*$:

$$\nabla : \Gamma(M, E) \longrightarrow \Gamma(M, E \otimes T^*).$$

If metrics are given on E and $T^*(M)$, then we have a sequence of Hilbert spaces

$$H_p = \mathcal{L}^2(M, E \otimes T^{*p}(M)).$$

By using the connection ∇, we obtain

$$\nabla_p : \Gamma(M, E \otimes T^{*p}(M)) \to \Gamma(M, E \otimes T^{*p+1}(M)).$$

Then we obtain a densely defined closed operator by extending it to an operator in a sense of distribution:

$$\nabla_p : H_p \to H_{p+1}.$$

Suppose that a sesquilinear form $((\ ,\ ))_p$ is given on H_p. Then we obtain

$$\nabla_p^* : H_{p+1} \to H_p$$

as in (1.1). From this we obtain a quantum field $\{H_p, \nabla_p, \nabla_p^*, \mathcal{H}_p\}$. Thus we have the following definition:

Definition (1.5). The quantum field obtained above is called a quantization of a connection ∇ (with respect to $((\ ,\ ))_p$).

We call this field a quantization of a connection, because a classical limit of this operator gives an equation of geodesics in some cases (see §2 and §4). We do not know in general if there exists a quantization of a connection ∇ whose classical limit gives an equation of geodesics defined by ∇.

In the case of a differential geometric quantum field, we can define a selection rule of the field as follows: We may treat covariant differentiations as operators between the same degrees: For a vector field X,

$$\hat{\nabla}_{X_p} : \Gamma(M, E \otimes T^{*p}(M)) \to \Gamma(M, E \otimes T^{*p}(M)).$$

Then we can define the formally adjoint operator with respect to $((\ ,\ ))_p$, which is denoted by $\hat{\nabla}^*_{X_p}$. We formulate the following definition:

Definition (1.6). In the quantum field $\{H_p, \nabla_p, \nabla_p^*, \mathcal{H}_p\}$, the pair of $\hat{\nabla}_p$ and $\hat{\nabla}_p^*$ is called a selection rule of the field.

Hence creation and annihilation operators have a deep relationship with the selection rule. It may be said that Konstant's quantization is nothing but a selection rule of a field which is defined by a symplectic structure. If we deal with a unitary representation theory of Lie groups, we must consider a selection rule. In fact, in the case of a compact semi-simple Lie group, the differential representation of a unitary representation can be described in terms of a selection rule of a certain field (see §4).

We finish this section with giving a simple example and demonstrating our basic idea of the differential geometric quantum field. We choose $\mathbb{R}$ as a base manifold M and a trivial line bundle as E, whose fibre is denoted by $\xi\,(\xi \in \mathbb{R})$. As a metric of E we take $\exp(-\varphi)$ with a function φ on $\mathbb{R}$ of C^{∞}-class. Choosing the canonical flat metric on $\mathbb{R}$, we obtain the metrical connection

$$\nabla f = \left(\frac{df}{dx} - \frac{d\varphi}{dx} f\right) dx,$$

where $f \in \Gamma(\mathbb{R})$. We choose $\mathcal{L}^2(\mathbb{R}, T^{*p}(\mathbb{R}))$ as H_p. We take $f \in D_p$ and $g \in D_{p+1}$ and make the formal adjoint operator ∇_p^* by

$$\int_{\mathbb{R}} \nabla_p f \cdot g \, d\mu = \int_{\mathbb{R}} f \cdot \nabla_p^* g \, d\mu,$$

where $d\mu$ is a canonical Lebesgue measure on $\mathbb{R}$. Then we obtain

$$\nabla_p^* g = -\left(\frac{dg'}{dx} + \frac{d\varphi}{dx} g'\right) \underbrace{dx \otimes dx \otimes \ldots \otimes dx}_{p-1 \text{ times}} \quad \text{for} \quad g = g' \underbrace{dx \otimes dx \otimes \ldots \otimes dx}_{p\text{-times}}.$$

If we choose $\varphi = x^2/2$, then

$$\nabla_p f = \left(\frac{df'}{dx} - x f'\right) dx \otimes dx \otimes \ldots \otimes dx \quad \text{for} \quad f = f' \, dx \otimes dx \otimes \ldots \otimes dx,$$

$$\nabla_p^* g = -\left(\frac{dg'}{dx} + x g'\right) dx \otimes dx \otimes \ldots \otimes dx \quad \text{for} \quad g = g' \, dx \otimes dx \otimes \ldots \otimes dx.$$

Referring to notations in physics, we use ∇_p and ∇_p^* for $-\sqrt{1/2}\,\nabla_p$ and $-\sqrt{1/2}\,\nabla_p^*$, respectively. Then we get the commutation relation:

$$R_p f = f \quad (p \geqq 1)$$

and Hamiltonians:

$$\mathcal{H}_0 = \frac{1}{2}(- \frac{d^2}{dx^2} + x^2 + 1), \quad \mathcal{H}_p = - \frac{d^2}{dx^2} + x^2 \quad (p \geq 1).$$

The pure states of $\mathcal{H}_p$ $(p \geq 1)$ can be expressed as

$$\phi_n^p = \exp(- x^2/2) H_n(x) \; dx \otimes dx \otimes \ldots \otimes dx,$$

where H_n is a Hermite polynomial of degree n, and its eigen-value is $\lambda_n = n + 1/2$. The selection rule of this field is given for $p \geq 1$ by

$$\hat{\nabla}_p \phi_n^p = \sqrt{n+1} \; \phi_{n+1}^p \quad \text{or} \quad \hat{\nabla}_p^* \phi_n^p = \sqrt{n} \phi_{n-1}^p.$$

This is nothing but a Heisenberg matrix representation of quantization of a harmonic oscillator.

2. Existence of Boson fields and flatness of a cotangent bundle of a manifold

In this section we shall prove that existence of Boson fields characterizes flatness of a cotangent bundle of a manifold. Firstly, we shall prove the following

THEOREM I. *If a cotangent bundle of a manifold admits a flat and torsion free connection, then there exists a Boson field on the cotangent bundle.*

In order to prove the converse of Theorem I, we introduce a canonical symplectic quantum field (see Def. (2.2)). Then we can prove the following

THEOREM II. *A canonical symplectic quantum field is a Boson field if and only if a cotangent bundle admits a flat and torsion free connection.*

Remark. Theorem II is suggested by Prof. H.Takai. The author would like to express his hearty thanks to him for his suggestion.

In the remaining part of this section we shall give a concept of canonical symplectic quantum fields and give proofs of Theorems I and II.

Let M be a manifold of C^∞-class and $T^*(M)$ be a cotangent bundle of M with a natural projection $\pi : T^*(M) \rightarrow M$. In the following we write $M^* = T^*(M)$. We choose a system of local coordinates $q^1, q^2, \ldots, q^n$

and denote canonical fibre coordinates on $\pi^{-1}(U)$ by $p_1, p_2, \ldots, p_n$, where $n = \dim M$. Then

$$\theta = \sum_{i=1}^{n} p_i dq^i$$

is a globally defined one-form on M^* and $\Omega = d\theta$ is a closed two-form on M^*. It can be written as

$$\Omega = \sum_{i=1}^{n} dp_i \wedge dq^i.$$

In the following we denote $p_1, p_2, \ldots, p_n, q^1, q^2, \ldots, q^n$ by u^i $(1 \leqq i \leqq 2n)$. We make the following notations:

$$u^{\bar{i}} = p_i \quad \text{if} \quad u^i = q^i, \quad u^{\bar{i}} = q^i \quad \text{if} \quad u^i = p_i.$$

Then if we express Ω as

$$\Omega = \Sigma g_{ij}\, du^i \otimes du^j, \tag{2.1}$$

we can see that

$$g_{i\bar{j}} = \delta_{ij} \quad \text{and} \quad g_{ij} = -g_{ji} \tag{2.2}$$

hold. We denote the inverse matrix of (g_{ij}) by

$$(g^{ij}). \tag{2.3}$$

We set

$$\mathcal{T}^*(M) = \bigoplus_{p=0}^{\infty} \Gamma(M, T^{*p}(M)), \quad \mathcal{T}^*_S(M) = \bigoplus_{p=0}^{\infty} \Gamma(M, T_S^{*p}(M)),$$

where $T_S^{*p}(M)$ denotes a sheaf of germs of symmetric tensor fields of degree p.

With these preliminaries we shall give a canonical symplectic quantum field on M^*. We follow field construction methods in §1. We choose for E a trivial line bundle $\tau: F \longrightarrow M$. Then $F = \mathbb{C} \times M^*$ holds and ξ denotes its fibre coordinate. We choose a C^∞-function φ_λ on each local coordinates U_λ which depends only on q^i_λ $(i = 1, 2, \ldots, n)$. Thus if we denote a new fibre coordinate ξ_λ on U_λ by

$$\xi_\lambda = \exp \varphi_\lambda \cdot \xi,$$

we get the system of patching functions $\{f_{\lambda\mu}\}$ on $U_\lambda \cap U_\mu$ by

$$\xi_\lambda = f_{\lambda\mu}\xi_\mu, \quad f_{\lambda\mu} = \exp \varphi_\lambda \cdot \exp(-\varphi_\mu).$$

Here we assume that an affine connection ∇^o is given on M^*. Then we can define a connection on $F \otimes T^*(M^*)$

$$\nabla : \Gamma(M^*, F) \longrightarrow \Gamma(M^*, F \otimes T^*(M^*))$$

by

$$\nabla = \hbar\, \nabla^o + \sqrt{-1}\, \theta.$$

Then for $f \in \Gamma(M^*, F \otimes T^{*p}(M^*))$, given by

$$f = \Sigma f_{\lambda\, i_1 i_2 \ldots i_p}\, du_\lambda^{i_1} \otimes du_\lambda^{i_2} \ldots \otimes du_\lambda^{i_p},$$

we can write

$$\nabla f = \Sigma \Delta_k\, f_{\lambda i_1 i_2 \ldots i_p}\, du_\lambda^k \otimes du_\lambda^{i_1} \ldots \otimes du_\lambda^{i_p},$$

where

$$\nabla_k f_{\lambda i_1 i_2 \ldots i_p} = \begin{cases} \hbar\, \nabla^o_{q_\lambda^k} + (\sqrt{-1}\, p_k^\lambda - \dfrac{\partial \varphi_\lambda}{\partial q_\lambda^k}) f_{\lambda i_1 \ldots i_p} & (u_\lambda^k = q_\lambda^k), \\ \hbar\, \nabla^o_{p_k^\lambda}, & (u_\lambda^k = p_k^\lambda). \end{cases}$$

Thus it is easily seen that the following relation holds for $f \in \Gamma(U_\lambda, F)$ with respect to local coordinates u_λ^k:

$$(2.4) \qquad (\nabla_{p_i^\lambda} \nabla_{q_\lambda^j} - \nabla_{q_\lambda^j} \nabla_{p_i^\lambda}) f = \hbar^2 (\nabla^o_{p_i^\lambda} \nabla^o_{q_\lambda^j} - \nabla^o_{q_\lambda^j} \nabla^o_{p_i^\lambda}) f + \sqrt{-1}\, \hbar \delta_{ij} f$$

and for other u_λ^i and u_λ^j:

$$(2.5) \qquad (\nabla_{u_\lambda^i} \nabla_{u_\lambda^j} - \nabla_{u_\lambda^j} \nabla_{u_\lambda^i}) f = \hbar^2 (\nabla^o_{u_\lambda^i} \nabla^o_{u_\lambda^j} - \nabla^o_{u_\lambda^j} \nabla^o_{u_\lambda^i}) f.$$

If we choose an arbitrary positive definite metric on M^*, then we have a norm of sections $\Gamma(M^*, F \otimes T^{*p}(M^*))$. In what follows we restrict our considerations to symmetic tensor fields. We write a section $f \in \Gamma(M^*, F \otimes T_S^{*p}(M^*))$ as follows:

$$f = \Sigma f_{\lambda i_1 i_2 \ldots i_p} S(du_\lambda^{i_1} \otimes du_\lambda^{i_2} \otimes \ldots \otimes du_\lambda^{i_p}),$$

where S means the symmetrization of tensors. Then for a symmetric tensor, we can define a norm and so we obtain a sequence of Hilbert spaces:

$$H_p = \mathcal{L}^2(M^*, F \otimes T_S^{*p}(M^*)) \quad (p = 0, 1, 2, \ldots).$$

For $f \in \Gamma(M^*, F \otimes T^{*p}(M^*))$, we set

$$f_\lambda^{j_1 j_2 \ldots j_p} = \Sigma g^{j_1 i_1} g^{j_2 i_2} \ldots g^{j_p i_p} f_{\lambda i_1 i_2 \ldots i_p},$$

(g^{ij} has been defined in (2.3)). Now we define a sesquilinear form on H_p by

$$((f,g))_p = \Sigma \int_{M^*} \exp(-2\varphi_\lambda) f^{i_1 i_2 \ldots i_p} \overline{g_{i_1 i_2 \ldots i_p}} \, dV,$$

($dV = \Lambda^n \Omega$ and Ω has been defined in (2.1)). With respect to this inner product, we define a formally adjoint operator ∇_p^*

$$\nabla_p^* : D(M^*, F \otimes T_S^{*p+1}(M^*)) \longrightarrow D(M^*, F \otimes T_S^{*p}(M^*))$$

by $((\nabla_p f, g))_{p+1} = ((f, \nabla_p^* g))_p$. Hence, by using a construction method of fields in §1, we obtain a quantum field.

<u>D e f i n i t i o n (2.5)</u>. The quantum field $\{H_p, \nabla_p, \nabla_p^*, \mathcal{H}_p\}$ obtained above is called a <u>canonical</u> <u>symplectic</u> <u>quantum</u> field on M^*.

<u>R e m a r k</u>. Since $((\ ,\ ))_p$ is not hermitian, a Hamiltionian $\mathcal{H}_p$ has not a pure state in general. ∇_p^* does not depend on choices of Hilbert spaces H_p.

In the remaining part of this section we shall give proofs of Theorems I and II. We begin with the following proposition:

<u>PROPOSITION (2.7)</u>. <u>For</u> $g = \Sigma g_{\lambda i_0 i_1 \ldots i_p} S(du_\lambda^{i_0} \otimes du_\lambda^{i_1} \otimes \ldots \otimes du_\lambda^{i_p})$, we have

$$\nabla_p^* g = - \sum_{k=0}^{p} g^{i_k \alpha} \nabla_\alpha g_{i_0 i_1 \dots i_p} S(du_\lambda^{i_0} \otimes du_\lambda^{i_1} \otimes \dots \otimes \overset{\vee}{du_\lambda^{i_k}} \otimes \dots \otimes du_\lambda^{i_p}),$$

<u>where</u> $\vee$ <u>means taking out</u> $du_\lambda^{i_k}$.

This propositon can be easily proved in a well known manner, so it may be omitted.

Now we calculate the following commutation relation (cf. (1.2)):

$$\mathcal{R}_p f = (\nabla_p^* \nabla_p - \nabla_{p-1} \nabla_{p-1}^*) f.$$

In order to prove Theorems I and II, it is sufficient to prove the following proposition:

<u>PROPOSITION (2.8)</u>. <u>In a canonical symplectic field,</u> $\mathcal{R}_p f = c_p f$ <u>holds if and only if the connection</u> ∇^o <u>on</u> M^* <u>is flat and torsion-free</u>.

<u>Proof</u>. We choose $g \in D(M^*, T_S^{*p}(M^*))$,

$$g = \Sigma g_{i_1 i_2 \dots i_p} S(du^{i_1} \otimes du^{i_2} \dots \otimes du^{i_p}).$$

Then the following holds:

$$\mathcal{R}_p g = - \sum_{j_0 \beta} g^{j_0 \beta} \nabla_\beta \nabla_{j_0} g_{j_1 j_2 \dots j_p} S(du^{j_1} \otimes \dots \otimes du^{j_p})$$
$$- \sum_{j_0 \beta} \sum_{k=1}^{p} g^{j_k \beta} \nabla_\beta \nabla_{j_0} g_{j_1 j_2 \dots j_p} S(du^{j_0} \otimes \dots \otimes \overset{\vee}{du^{j_k}} \otimes \dots \otimes du^{j_p})$$
$$+ \sum_{j_0 \beta} \sum_{k=1}^{p} g^{j_k \beta} \nabla_{j_0} \nabla_\beta g_{j_1 j_2 \dots j_p} S(du^{j_0} \otimes \dots \otimes \overset{\vee}{du^{j_k}} \otimes \dots \otimes du^{j_p}).$$

Here we write

$$\mathcal{R}_p g = I + II,$$

where

$$I = - \Sigma g^{j_0 \beta} \nabla_\beta \nabla_{j_0} g_{j_1 j_2 \dots j_p} S(du^{j_1} \otimes du^{j_2} \otimes \dots \otimes du^{j_p})$$

and

$$II = \Sigma g^{j_k \beta} (\nabla_{j_0} \nabla_\beta - \nabla_\beta \nabla_{j_0}) g_{j_1 j_2 \dots j_p} S(du^{j_0} \otimes \dots \otimes \overset{\vee}{du^{j_k}} \otimes \dots \otimes du^{j_p}).$$

Firstly, we calculate I. By using (2.2), we can see that $g^{j_o\beta} \neq 0$ if and only if $\beta = \bar{j}_o$. Therefore we have

$$I = -\Sigma g^{j_o\bar{j}_o} \nabla_{\bar{j}_o} \nabla_{j_o} g_{j_1 j_2 \cdots j_p} S(du^{j_1} \otimes du^{j_2} \otimes \ldots \otimes du^{j_p}).$$

Referring to $g^{\alpha\beta} = - g^{\beta\alpha}$, we obtain

$$I = - \frac{1}{2} \Sigma g^{j_o\bar{j}_o} (\nabla_{j_o} \nabla_{\bar{j}_o} - \nabla_{\bar{j}_o} \nabla_{j_o}) g_{j_1 j_2 \cdots j_p} S(du^{j_1} \otimes du^{j_2} \otimes \ldots \otimes du^{j_p})$$

Similarly, by using $g^{\alpha\beta} \neq 0$ when $\beta = \bar{\alpha}$, we get

$$II = \Sigma g^{j_k\bar{j}_k} (\nabla_{j_o} \nabla_{\bar{j}_k} - \nabla_{\bar{j}_k} \nabla_{j_o}) g_{j_1 j_2 \cdots j_p} S(du^{i_o} \otimes \ldots \check{d}u^{i_k} \ldots \otimes du^{i_p})$$

By using (2.4) and (2.5), we obtain

$$(2.9) \quad I = \frac{1}{2} \Sigma g^{j_o\bar{j}_o} \hbar^2 (\nabla^o_{j_o} \nabla^o_{\bar{j}_o} - \nabla^o_{\bar{j}_o} \nabla^o_{j_o}) g_{j_1 j_2 \cdots j_p} S(du^{j_1} \otimes \ldots \otimes du^{j_p})$$

$$- \frac{1}{2} \Sigma g^{j_o\bar{j}_o} \hbar\sqrt{-1}\, g_{j_1 j_2 \cdots j_p} S(du^{j_1} \otimes \ldots \otimes du^{j_p})$$

$$(2.10) \quad II = \Sigma g^{j_k\bar{j}_k} \hbar^2 (\nabla^o_{j_o} \nabla^o_{\bar{j}_k} - \nabla^o_{\bar{j}_k} \nabla^o_{j_o}) g_{j_1 j_2 \cdots j_p} S(du^{j_o} \otimes \ldots \check{du}^{j_k} \ldots \otimes du^{j_p})$$

$$+ \Sigma g^{j_k\bar{j}_k} \hbar\sqrt{-1}\, g_{j_1 j_2 \cdots j_p} S(du^{j_1} \otimes du^{j_2} \otimes \ldots \otimes du^{j_p}).$$

Now we assume that the given field is a Boson field. Then for $g \in D(M^*, T_S^{*p}(M^*))$ we have

$$(2.11) \quad \mathcal{R}_p g = c_p g$$

We notice that $\mathcal{R}_p$ is depending on a choice of $\hbar$. Thus we write $\mathcal{R}_p(\hbar)$. Referring to (2.9) and (2.10), we can see that for $\hbar$ and $\hbar'$ the following holds:

$$\frac{\mathcal{R}_p(\hbar)}{\hbar^2} - \frac{\mathcal{R}_p(\hbar')}{\hbar'^2} = c_p(\hbar,\hbar') \cdot 1_p,$$

where $c_p(\hbar,\hbar')$ is a constant which depende on $\hbar,\hbar'$ and p. Hence, if (2.11) holds for some $\hbar$, then it also holds for any $\hbar$ $(\hbar \neq 0)$. Hence, chosing suitable $\hbar$, we may assume that

$$(\overset{o}{\nabla}_j\overset{o}{\nabla}_k - \overset{o}{\nabla}_k\overset{o}{\nabla}_j) g_{j_1 j_2 \dots j_p} = 0$$

holds for every j and k. This implies that the curvature and torsion of ∇^o vanish and proves one part of Proposition (2.8). Thus it can be proved easily that if the curvature and torsion of ∇^o vanish, then we obtain a Boson field by (2.9) and (2.10). Hence, the remaining part of Proposition (2.8) is proved.

Hereby we complete our proof of Theorem I and II.

Now we make time evolutions of operators ∇_p and ∇_p^*. We choose $c = i\hbar/c_p$ in (1.4) (c_p has been obtained in (2.11)). By using the assumption of Boson field, we can see that

$$\frac{d\nabla_p}{dt} = -i\nabla_p, \quad \frac{d\nabla_p^*}{dt} = i\nabla_p^*.$$

Integrating these equations, we obtain

$$\nabla_p(t) = e^{-it}\nabla_p, \quad \nabla_p^*(t) = e^{it}\nabla_p^*.$$

Hence by using the definition of ∇, we can see that

$$\nabla_p(t) = e^{-it}(\hbar\nabla_p^o + i\theta).$$

If we choose $\varphi_\lambda = -\frac{1}{2}\Sigma q_\lambda^{i2}$ and set

$$\nabla_p(t) = e^{-it}\hbar\nabla_p^o + \sum_{i=1}^{n} (Q_i(t) + \sqrt{-1}\, P_i(t))dq^i,$$

we can easily see that

$$\frac{dQ_i}{dt} = P_i \quad \text{and} \quad \frac{dP_i}{dt} = -Q_i.$$

This is nothing but an equation of harmonic oscillator. Hence we may say that our canonical quantum field is just a quantization of harmonic oscillator.

Remark. As is expected, by using this field, we may consider a quantization of a motion with a more general Hamiltonian. This will be discussed elsewhere.

3. Fermion fields on a manifold

In this section we prove existence of a Fermion field on a Riemannian manifold by using differentiations and their formally adjoint operators. Namely, we have

THEOREM III. There exists a Fermion field on a Riemannian manifold.

This theorem tells us that although the existence of the Boson field characterises a manifold strictly, the existence of a Fermion field does not characterise a base manifold.

Let M be a Riemannian manifold of C^∞-class of dimension n. Let

$$A = \bigoplus_{p=0}^{n} A^p, \quad A^p = \wedge^p T^*(M)$$

be a sheaf of germs of exterior differential forms. By the original metric, we can define a hermitian inner product on section of A by

$$(f,g)_p = \int_M f \wedge \bar{*}\, g \, dV \quad \text{for} \quad f,g \in \Gamma(M,A^p),$$

where $*$ is the usual star operator and dV is the volume form defined by the metric. From this we obtain the Hilbert spaces

$$H_p = \mathcal{L}^2(M,A^p).$$

Let $d : \Gamma(M,A^p) \longrightarrow \Gamma(M,A^{p+1})$ be the exterior differention in the distribution sense and $\delta : \Delta(M,A^{p+1}) \longrightarrow \Gamma(M,A^p)$ be its formally adjoint operator with respect to the above inner product. Now we follow the method of defining quantum fields in Section 1 and we obtain a field by using $H_p, d, \delta, \mathcal{H}_p$. In this case $\mathcal{H}_p = d\delta + \delta d$ is nothing but the Laplace-Beltrami operator. Here we make the following definition:

Definition (3.1). The quantum field obtained in the above manner is called a quantum field defined by exterior forms.

Referring to the fact that $d^2 = 0$ and $\delta^2 = 0$, we see easily that this is a Fermion field. So we have proved Theorem III.

Here we shall obtain another expression of the above field which is similar to the physical expression. In the following we assume that a manifold is compact. Since $\mathcal{H}_p$ is a self-adjoint and strongly elliptic operator, H_p can be expressed as a direct sum of eigenspaces $H_p(\varepsilon)$:

$$H_p(\varepsilon) = \{\phi \in H_p : \mathcal{H}_p\phi = \varepsilon\phi\}.$$

Moreover, ε may take only countably many non-negative values and each $H_p(\varepsilon)$ is of finite dimension. By using

$$d\mathcal{H}_p = \mathcal{H}_p d \quad \text{and} \quad \delta\mathcal{H}_p = \mathcal{H}_p\delta,$$

we see that

$$d : H_p(\varepsilon) \longrightarrow H_{p+1}(\varepsilon), \quad \delta : H_p(\varepsilon) \longrightarrow H_{p-1}(\varepsilon).$$

Hence restricting our field to $H_p^+ = \bigoplus_{\varepsilon>0} H_p(\varepsilon)$, we obtain also a Fermion field. We define

$$a_p(\varepsilon) = \begin{cases} 1/\sqrt{\varepsilon}\cdot d & \text{on } H_p(\varepsilon), \\ 0 & \text{otherwise;} \end{cases} \qquad a_p^*(\varepsilon) = \begin{cases} 1/\sqrt{\varepsilon}\cdot\delta & \text{on } H_p(\varepsilon), \\ 0 & \text{otherwise.} \end{cases}$$

Then we obtain

$$a_{p+1}(\varepsilon')a_p(\varepsilon) = 0, \quad a_{p-1}^*(\varepsilon')a_p^*(\varepsilon) = 0,$$

$$a_{p+1}^*(\varepsilon')a_p(\varepsilon) + a_{p-1}(\varepsilon')a_p^*(\varepsilon) = \delta_{\varepsilon\varepsilon'},$$

which is nothing but a usual physical expression of a Fermion field.

Here we make comments on some relationships between our Fermion field and the equation of Dirac of a particle with spin $\frac{1}{2}$. In the case where M is a Kähler manifold, we can obtain another Fermion field by considering

$$\bar{\partial} : \mathcal{L}^2(M,\wedge^{pq}T^*(M)) \longrightarrow \mathcal{L}^2(M,\wedge^{pq+1}T^*(M))$$

and its formally adjoint operator

$$\theta : \mathcal{L}^2(M,\wedge^{pq+1}T^*(M)) \longrightarrow \mathcal{L}^2(M,\wedge^{pq}T^*(M)),$$

where $\wedge^{pq}T^*(M) = \wedge^p T^*(M)\wedge^q\overline{T^*}(M)$ and $T^*(M)$ is the holomorphic cotangent bundle of M. Now we take as M a complex torus $M = \mathbb{C}^2/\Gamma$, where Γ is a group which is a group which is generated by

$$\begin{cases} z_1' = z_1 + L, \\ z_2' = z_2; \end{cases} \qquad \begin{cases} z_1' = z_1, \\ z_2' = z_2 + L, \end{cases}$$

where z_1 and z_2 are the coordinates in $\mathbb{C}^2$. Choosing the flat metric on M, we make a Fermion field. Then pure states can be obtained by

$$(\bar{\partial}\theta + \theta\bar{\partial})\phi = \varepsilon\phi.$$

For a positive ε, we define η by

$$\sqrt{2\varepsilon}\,\eta = (\bar{\partial} + \theta)\phi.$$

Then

$$(\bar{\partial} + \theta)\eta = \sqrt{2\varepsilon}\,\phi.$$

Hence we see that these equations are nothing but the Weyl representation of wave functions of the Diract equation of spin $\frac{1}{2}$ with positive mass $\sqrt{2\varepsilon}$. Thus we arrive at

PROPOSITION (3.2) *On a 2-dimensional complex torus we obtain a Fermion field whose pure states satisfy the Dirac equation of spin $\frac{1}{2}$ with positive masses.*

4. Quantum field theory on a Riemannian manifold and unitary representations of a compact semi-simple Lie group

In this section we shall treat a quantum field theory on a Riemannian manifold and state well known results of unitary representations from our quantum field theoretic view point.

Let (M,g) be a Riemannian manifold with a metric g. Let ∇ be a metrical connection of g. With respect to usual Riemannian inner products of tensor fields, we can obtain a formally adjoint operator ∇^*. Hence following the field construction method in Section 1, we obtain a quantization of a Riemannian connection. We can show the following

THEOREM IV. *Hamiltonians of the above described field are Lichnerowicz laplacians of* M. *If the following Schödinger equation of the Hamiltonian of degree* 0

$$\frac{\hbar}{i}\frac{\partial\bar{\psi}}{\partial t} = \mathcal{H}_o\bar{\psi}$$

<u>admits a quasi-classical limit, then we obtain the Hamilton-Jacobi equation of the type</u>

$$\frac{\partial S}{\partial t} = \Sigma g^{ij}\frac{\partial S}{\partial q^i}\frac{\partial S}{\partial q^j}$$

<u>as a classical limit</u>.

From this equation we can easily obtain an equation of a geodesic. Hence we may say that our quantization is regarded as a quantization of a classical geodesic equation.

Now we apply this quantization on a compact semi-simple Lie group G. On G we have a positive definite metric which is induced by the Killing form. So we can obtain a quantum field in the above manner. Then we obtain the following

THEOREM V. <u>A quantum field which is obtained by the Killing form on a compact semi-simple Lie group has Casimir operators as Hamiltonians and any differential representation of a unitary representation can be obtained by a selection rule of this field</u>.

We may say that this theorem is a generalization of the following well known result: In the case $SO(3)$, every differential representation of a unitary representation can be obtained by a selection rule of a quantum field which is defined by a total angular momentum.

Now we give proofs of Theorems IV and V.

Let M be a Riemannian manifold with a metric g:

$$ds^2 = \Sigma g_{ij}dq^i \otimes dq^j.$$

Let ∇_p be a metrical connection of g:

$$\nabla_p : \Gamma(M, T^{*p}(M)) \longrightarrow \Gamma(M, T^{*p+1}(M)).$$

For a section $f \in \Gamma(M, T^{*p}(M))$, expressed as

$$f = \Sigma f_{i_1 i_2 \ldots i_p} dq^{i_1} \otimes dq^{i_2} \otimes \ldots \otimes dq^{i_p},$$

we set

$$f^{j_1 j_2 \ldots j_p} = \Sigma g^{j_1 i_1} g^{j_2 i_2} \ldots g^{j_p i_p} f_{i_1 i_2 \ldots i_p}.$$

Then for $f, g \in \Gamma(M, T^{*p}(M))$, we have a positive definite symmetric inner product

$$(f,g) = \int_M (f.g)\,dV,$$

where

$$(f,g) = f_{i_1 i_2 \ldots i_p} \overline{g^{i_1 i_2 \ldots i_p}}$$

and

$$dV = \wedge^n \Omega, \quad \Omega = \Sigma g_{ij}\, dq^i \wedge dq^j.$$

Then for support compact section f and g, we can define a formally adjoint operator of ∇_p by

$$(\nabla_p f, g) = (f, \nabla_p^* g).$$

It is easily seen that for $g \in \Gamma(M, T^{*p}(M))$,

$$g = \Sigma g_{j_o j_1 \ldots j_p}\, dq^{j_o} \otimes dq^{j_1} \otimes \ldots \otimes dq^{j_p},$$

$$\nabla_p^* g = -\Sigma g^{\alpha j_o} \nabla_\alpha g_{j_o j_1 \ldots j_p}\, dq^{j_1} \otimes dq^{j_2} \ldots \otimes dq^{j_p}.$$

Here we set

$$H_p = \mathcal{L}^2(M, T^{*p}).$$

Then we have a quantization of ∇, $\{H_p, \nabla_p, \nabla_p^*, \mathcal{H}_p\}$. In this field the Hamiltonian

$$\mathcal{H}_p = \nabla_p^* \nabla_p + \nabla_{p-1} \nabla_{p-1}^*$$

is called a Lichnerowicz laplacian.

Now we consider a quasi-classical limit in the case $p = 0$. In this case $\mathcal{H}_o$ is nothing but a rough Laplacian

$$\mathcal{H}_o = -\Sigma g^{ij} \nabla_i \nabla_j.$$

Hence the Schrödinger equation of $\mathcal{H}_o$ can be written as

$$\frac{\hbar}{i}\frac{\partial\bar{\psi}}{\partial t} = -\Sigma g^{ij}\nabla_i\nabla_j\bar{\psi}.$$

Now we assume that a quasi-classical limit solution of the following type (cf. [1], § 31) is given:

$$\bar{\psi} = \exp(iS/h)A,$$

where S and A are functions of positions and time. Choosing $\hbar \to 0$, we see that

$$(4.1) \qquad \frac{\partial S}{\partial t} = \Sigma g^{ij}\frac{\partial S}{\partial q^i}\frac{\partial S}{\partial q^j}.$$

This is nothing but the Hamilton-Jacobi equation of a motion with the classical Hamiltonian

$$\mathcal{H}_c = \Sigma g^{ij}p_i p_j.$$

A solution of (4.1) is generated by motions which are defined by the following Hamilton equations:

$$\frac{dp_i}{dt} = \frac{\partial\mathcal{H}_c}{\partial q^i}, \quad \frac{dq^i}{dt} = -\frac{\partial\mathcal{H}_c}{\partial p^i},$$

where q^i and p^i are canonical variables. These are equations of a geodesic with respect to the metric g. Hence we have proved Theorem IV.

Now we assume that the base manifold is a compact semi-simple Lie group G. Then, using the Killing form, we define a metric of G and we obtain a quantum field as in the above manner. By using a well known result [2] we see that the Hamiltonian $\mathcal{H}_p$ is the Casimir operator. We consider pure states of $\mathcal{H}_o$:

$$\mathcal{H}_o\phi = \varepsilon\phi.$$

Eigenvalues make a discrete set $\{\varepsilon_n\}$ and pure states for a eigenvalue ε_n make a finite dimensional vector space H_n so that

$$\mathcal{L}^2(G) = \bigoplus_n H_n$$

holds. In this case the selection rule $\hat{\nabla}_{X^0} : \Gamma(G,C^\infty) \to \Gamma(G,C^\infty)$ is a derivation with respect to an invariant vector field X. Thus we have exactly a differential representation of a regular representation. Since every H_n is invariant under the selection rule, so we arrive at the conclusion of Theorem V.

R e m a r k. The statement holds also in the case of a compact Lie group by applying the above discussion to a metric which is invariant under the adjoint representation (cf. [2]).

R e f e r e n c e s

[1] DIRAC, P.A.M.: The principles of quantum mechanics, 4th-ed., Oxford Univ. Press, Oxford 1958.

[2] KOISO, N.: Rigidity and stability of Einstein metrics - The case of compact symmetric spaces, Osaka J. Math. 17 (1980), 51-73.

[3] KOSTANT, B.: Quantization and unitary representations (Lecture Notes in Math. 170), Springer Verlag, Berlin-Heidelberg-New York 1970, pp. 87-208.

Department of Mathematics
College of Humanities and Sciences,
Nihon University, Tokyo 156, Japan

ON THE FIRST TWO EVEN-ODD LINEAR FUNCTIONALS OF BOUNDED REAL UNIVALENT FUNCTIONS

Olli Tammi (Helsinki)

Contents

page

1. Introduction

Let us denote by $S(b)$ the class of bounded univalent functions f defined in the unit disc $U: |z| < 1$,

$$S(b) = \{f \mid f(z) = b(z + a_2 z^2 + \ldots), \quad |f(z)| < 1, \ 0 < b < 1\}.$$

$S_R(b)$ is the subclass of this with all the coefficients real.

We want to maximize the linear combinations $a_2 + \mu a_3$ and $a_4 + \mu a_3$ in the class $S_R(b)$. For such even-odd indexed functionals K. Zyskowska has proved the following interesting result: For any $\mu > 0$ there exists an interval $0 < b < b_\mu$ where the left radial-slit-mapping ($f(U)$ consists of the unit disc minus a radial slit along the negative real axis) maximizes the functional [7]. This generalizes the result of Z. Jakubowski and others concerning the case $\mu = 0$ [1]. In what follows we shall maximize the first two even-odd functionals for a large set of the μ-values. The exact Zyskowska-boundary, $\sup_{S_R(b)} b_\mu$, will also be determined.

Our estimations are based on two inequalities: The Power inequality and the inequality of Jokinen. The former condition is a well-known Grunsky-type condition for the $S(b)$-functions [3]. The latter holds especially for the $S_R(b)$-functions reading

$$a_4 - 2a_2a_3 + a_2^3 - b^2a_2 + 2\lambda(a_3 - a_2^2 + 1 - b^2) \leqq \frac{2}{3}(1 + \lambda)^3, \quad -1 \leqq \lambda \leqq 0.$$

This condition is proved by Jokinen in [2] by aid of Löwner-functions. The proof must be omitted here.

2. $a_2 + \mu a_3$

It is well known that by properly estimating the combination $\mathrm{Re}\,(a_3 - a_2^2 + 2\lambda a_2)$ in $S(b)$ the coefficient body (a_2, a_3) can be completely determined [4]. Especially in $S_R(b)$ this coefficient body can be presented in explicit form and graphs in the a_2a_3-plane can be drawn (cf. [4] pp. 149, 160-161).

Because a_3 can be sharply estimated in a_2 we obtain for $a_2 + \mu a_3$ a sharp estimation in a_2. By maximizing this in the interval $|a_2| \leqq 2(1 - b)$ we get

$$\max_{S_R(b)} (a_2 + \mu a_3).$$

The exact calculations and the result, for $a_3 + \lambda a_2$, are presented in [4] and [5]. It is convenient to express the result in terms of the extremal domains. They are characterized by the symbol $\alpha{:}\beta$, where α is the amount of starting points and β the amount of end points of the slits (cf. e.g. Figure 13, p. 55, in [4]). The result is symmetric in μ and hence we may take $\mu \geqq 0$ for which the following extremal domains hold:

$$0 \leqq \mu \leqq \frac{1}{4b}; \quad \text{left radial-slit-mapping},$$

$$\frac{1}{4b} \leqq \mu \leqq \frac{1}{4b(1 + \ln b)}; \quad 1{:}2,$$

$$\frac{1}{4b(1 + \ln b)} \leqq \mu < +\infty; \quad 2{:}2.$$

Thus, the Zyskowska boundary in the present case is

$$\sup_{S_R(b)} b_\mu = \begin{cases} 1, & 0 \leqq \mu \leqq \frac{1}{4}, \\ \frac{1}{4\mu}, & \frac{1}{4} \leqq \mu < +\infty. \end{cases}$$

3. The estimation of $a_4 + \mu a_3$

The Power inequality estimating a_4 reads

$$a_4 \leqq \frac{2}{3}(1 - b^3) - \frac{b}{2}a_2^2 - \frac{13}{12}a_2^3 + 2a_2a_3 - 2\lambda(a_3 - \frac{3}{4}a_2^2 + ba_2) + \lambda^2[2(1 - b) - a_2], \ \lambda \in R.$$

Choose here λ so that the right side is minimized:

$$\lambda = \frac{a_3 - \frac{3}{4}a_2^2 + ba_2}{2(1 - b) - a_2}.$$

This yields the optimized form of the Power inequality, giving for a_4 a sharp estimate and thus determining the coefficient body (a_2,a_3,a_4) in the part I $\subset (a_2,a_3)$ (cf. Figure 38, p. 149, in [4]). Add μa_3 on both sides of the optimized inequality and rewrite the right side in the form which includes a_3 in a perfect square term:

$$a_4 + \mu a_3 \leqq \frac{2}{3}(1 - b^3) - \frac{b}{2}a_2^2 - \frac{13}{12}a_2^3 - (\frac{\mu}{2} + a_2)\left[-\frac{a_2^2}{2} + (4b - 2)a_2 - \frac{\mu}{2}(2(1 - b) - a_2)\right]$$

$$- \frac{1}{2(1 - b) - a_2}\left[a_3 + ba_2 - \frac{3}{4}a_2^2 - (\frac{\mu}{2} + a_2)(2(1 - b) - a_2)\right]^2.$$

The omission of the last non-positive term leads to an estimate in a_2:

$$a_4 + \ a_3 \leqq \frac{2}{3}(1 - b^3) + \frac{1}{2}(1 - b\ )\mu^2 + (2 - 3b - \frac{\mu}{4})a_2 + (2 - \frac{9}{2}b - \frac{\mu}{4})a_2^2 - \frac{7}{12}a_2^3 = F_1.$$

This upper bound is sharp on the parabola

$$1^\circ: \quad a_3 = -\frac{a_2^2}{4} + (2 - 3b - \frac{\mu}{2})a_2 + \mu(1 - b)$$

which runs through the corner point $(2(1 - b), 3 - 8b + 5b^2)$ of the coefficient body (a_2,a_3).

We can proceed exactly in the same manner in using the Jokinen inequality. For a_4 we obtain first an optimized condition. This extends the range of the coefficient body (a_2,a_3,a_4) to concern all the algebraic functions and defined in the region I $\cup$ II $\subset (a_2,a_3)$ (again, cf. Figure 38, p. 149, in [4]). When adding μa_3 on both sides and then maximizing in a_3 we find a sharp estimate in a_2. Through comparisons which are to be omitted here we obtain the following upper bound in a_2:

$$a_4 + \mu a_3 \leqq \begin{cases} F_1 = F_2 - \frac{9}{4}(a_2 + \frac{2}{3}b + \frac{\mu}{3})^3, & -\frac{2}{3}b - \frac{\mu}{3} \leqq a_2 \leqq 2(1 - b), \\ F_2 = F_3 + \frac{2}{3}(a_2 + 1 + \frac{\mu}{2})^3, & -1 - \frac{\mu}{2} \leqq a_2 \leqq -\frac{2}{3}b - \frac{\mu}{3}, \\ F_3 = a_2^3 + (3b^2 - 2)a_2 + \mu(a_2^2 - 1 + b^2), & -2(1 - b) \leqq a_2 \leqq -1 - \frac{\mu}{2}. \end{cases}$$

As was mentioned above, F_1 is a sharp upper bound on 1°. Similarly, F_2 is sharp on the parabola

$$2^\circ: \quad a_3 = 2a_2^2 + (2 + \mu)a_2 + \left(b + \frac{\mu}{2}\right)^2$$

and F_3 is sharp on the lower boundary arc of the coefficient body:

$$3^\circ: \quad a_3 = a_2^2 - 1 + b^2.$$

Observe that the above validity intervals depend on μ.

4. The existence of the extremal functions

If the arc $1^\circ \cup 2^\circ \cup 3^\circ \subset \mathrm{I} \cup \mathrm{II} \subset (a_2, a_3)$ then the maximum found is sharp. When checking this we may proceed as follows.

The equality functions of the above inequalities can be expressed in Löwner's $\kappa(u) = e^{-i\vartheta(u)}$, $b \leqq u \leqq 1$. In $S_R(b)$ already $\cos\vartheta(u)$ determines the function f. In the case of the Power inequality the equality function is described in Figures 30-31, pp. 128-129, of [4]. Similarly, the inequality of Jokinen has the equality function presented by Figure 37, p. 146, of [4]. It appears that in the optimized maximum case $\lambda = a_2 + \frac{\mu}{2}$. Taking this into account we arrive at the following necessary and sufficient existence conditions for the parameters σ, σ_1 and σ_2 determining $\cos\vartheta(u)$:

$$\begin{cases} 8\sigma + (6a_2 - 2 + 3\mu)\sigma^{-1/2} - (9a_2 + 6b + 3\mu) = 0, \\ \frac{1}{3} - \frac{4}{3}\sigma^{3/2} \leqq a_2 + \frac{\mu}{2} \leqq \frac{1}{3} + \frac{8}{3}\sigma^{3/2}, \\ b \leqq \sigma \leqq 1; \end{cases}$$

$$\begin{cases} \sigma_2 = \left(\dfrac{1 - 3a_2 - 1.5\mu}{4}\right)^{2/3}, \\ \sigma_1 = \sigma_2 + \dfrac{3a_2 + 2b + \mu}{4}, \\ b \leqq \sigma_1 \leqq \sigma_2 \leqq 1. \end{cases}$$

If the maximizing value a_2 is such that these conditions hold we are sure that $a_4 + \mu a_3$ is maximized for the corresponding algebraic function. On the other hand, because all algebraic extremal functions are covered by the above inequalities, the unsharp cases are necessarily elliptic. For them, we do not have sharp inequalities available.

5. The regions of the extremal domains

The expressions of the maxima of the F_ν-functions depend on μ and the same holds for the validity intervals as well as the existence of the maximizing $\cos \vartheta(u)$. The checking of all these simultaneous conditions is best performed by aid of a computer. The result yields altogether five adjoining regions for five algebraic types of the extremal domains. The boundary curves of these regions are algebraic. The result will be described in detail in [6].

Let us mention finally, that also in the present case the Zyskowska boundary can be completely determined. This consists of the following arcs:

$$b = \frac{6 + \mu}{11} - \frac{\sqrt{100 + 4\mu + 15\mu^2}}{22} \quad \left(\mu^2 - 4(1 - 2b)\mu + 4(1 - b)(11b - 1) = 0\right)$$

for

$$2 - 2\sqrt{2} \leqq \mu \leqq -30 + 4\sqrt{66}$$

and

$$\mu = \frac{1}{2b} - 2 + \frac{15}{4}b$$

for

$$-30 + 4\sqrt{66} \leqq \mu < +\infty.$$

For $\mu \geqq -30 + 4\sqrt{66}$ the result follows already from the Power inequality. This inequality might be strong enough to determine parts of the Zyskowska boundary for some higher indexes, too. Similarly, in $S(b)$ analogous information can possibly be obtained from the Power inequality for $\text{Re}\,(a_4 + \mu a_3)$.

References

[1] JAKUBOWSKI, Z.J., ZIELIŃSKA, A., ZYSKOWSKA, K.: Sharp estimation of even coefficients of bounded symmetric univalent functions. - Ann. Polon. Math., to appear.

[2] JOKINEN, O.: On the use of Löwner identities for bounded univalent functions. - Ann. Acad. Sci. Fenn. Ser. A I Math., Dissertationes n:o 41, 1982, 52 p.

[3] TAMMI, O.: Extremum problems for bounded univalent functions. - Lecture Notes in Mathematics 646, Springer-Verlag, Berlin-Heidelberg-New York, 1978, 313 p.

[4] TAMMI, O.: Extremum problems for bounded univalent functions II. - Ibid. 913, 1982, 168 p.

[5] TAMMI, O.: On maximizing certain fourth-order functionals of bounded univalent functions. - To appear in Ann. Univ. Marie Curie-Skłodowska.

[6] TAMMI, O.: On maximizing $a_4 + \mu a_3$ for real bounded univalent functions. - To appear in Ann. Acad. Sci. Fenn.

[7] ZYSKOWSKA, K.: On general estimation of coefficients of bounded symmetric univalent functions. - To appear.

Department of Mathematics
University of Helsinki
Hallituskatu 15
SF-00100 Helsinki 10
Finland

GENERALIZED-ANALYTIC COVERINGS IN THE MAXIMAL IDEAL SPACE

Toma V. Tonev (Sofia)

Summary. We give some conditions under which on certain subsets of the maximal ideal space of a uniform algebra there exist structures of generalized-analytic coverings in the sense of R. Gunning and H. Rossi.

Let $S=\{p\}$ be an additive subsemigroup of $R_+ = \mathrm{Rat}\,[0,\infty)$, containing 0. Denote by Γ_d the group generated by $S\cup(-S)$ and provided with discrete topology, and by G - the dual group of Γ_d. G is a compact connected group and $\hat{G}\cong\Gamma_d$. The big plane is the cone $\mathbb{C}_G=[0,\infty)\times G/\{0\}\times G$ over G with the peak $*=\{0\}\times G/\{0\}\times G$ and a big disc with radius c if it is the set $\Delta_G(c)=\{(\lambda,g)\in\mathbb{C}_G \mid \lambda<c\}$. We call generalized polynomials the linear combinations over $\mathbb{C}$ of functions $\chi^p(\lambda,g) = \lambda^p\chi^p(g)$, $p\in S$, where $\chi^p\in\hat{G}$ are the characters $\chi^p(g)=g(p)$ for all $g\in G$. Because of $S\subset R_+$, we have $\chi^p(\lambda,g)=(\chi^1(\lambda,g))^p$, so that all the functions χ^p have arbitrary rational powers. Given an open set $U\subset\mathbb{C}_G$, we denote by $A_G(U)$ the algebra of all generalized-analytic functions, i.e. of all complex-valued functions on U, which are approximable locally by gen.-polynomials on compact subsets of U. For a compact set $K\subset\mathbb{C}_G$ we denote by $A_G(K)$ the algebra of all continuous functions on K, which are gen.-analytic on Int K. The algebra $A_G(K)$ of generalized analytic functions for the case $K=\Delta_G(1)$ was introduced by Arens and Singer in 1956 in [1], where they considered $A_G(\Delta_G(1))$ as a function algebra on G in a slightly more general situation. Only recently gen.-analytic functions have been studied on arbitrary sets in $\mathbb{C}_G$ [6, 7, 11, 12]. Given $S=Z_+$ we obtain $G\cong S^1$, $\mathbb{C}_G\cong\mathbb{C}^1$ and $\chi^n(\lambda,0)=(\lambda^n,n\theta)$, i.e. $\chi^n(z)=z^n$ and $A_G(U)=\mathrm{Hol}(U)$, $U\subset\mathbb{C}^1$.

Let A be a uniform algebra on the compact set X and $\mathrm{sp}\,A$ its maximal ideal space. Let $\{f^p\}_{p\in S}$ be a multilicative semigroup of elements from A, where S is as shown above. We call spectral mapping of S the mapping $\Phi_S : \mathrm{sp}\,A \to \mathbb{C}_G : \Phi_S(x) = (|f_1(x)|, g_x)$, where $g_x \in G$ is defined as follows: $g_x(p) = f^p(x)/|f^p(x)|$; $g_x(-p) = \overline{g_x(p)}$. An immediate corollary to this definition is the property

$$\widehat{f^p}(x) = \chi^p(\Phi_S(x)),$$

where $x \in \mathrm{sp}\,A$ and $\widehat{f^p}$ stands for the Gelfand transform of f^p. In the classical case when $S \cong \mathbb{Z}_+$, $\Phi_S(\varphi) = \hat{f}_1(\varphi)$; the last property is simply $\widehat{f^n}(x) = (\widehat{f^1}(x))^n$. We call <u>spectrum</u> $\sigma(S)$ of a semigroup $S \subset A$ the image of $\Phi : \sigma(S) = \Phi_S(\mathrm{sp}\,A)$. In [12] and [13] several aspects of the spectrum $\sigma(S)$ are discussed.

In [4] Bishop proved the following theorem (see also [14]): Let A be a uniform algebra on a space X and M its maximal ideal space. Let $f \in A$ and W be components of $\mathbb{C} \setminus f(X)$. Assume that there exists a set of positive plane measure $G \subset W$, such that $f^{-1}(\lambda) = \{p \in M | f(p) = \lambda\}$ is a finite set for each $\lambda \in G$. Then each point p in $f^{-1}(W) = \{p \in M | f(p) \in W\}$ has a neighbourhood in M which is a finite union of analytic discs through p. In fact, the set $f^{-1}(W)$ can be provided with structure of an analytic covering in the sense of Gunning and Rossi (cf. [8]), on which the functions $h \circ \hat{f}^{-1}$ are analytic for each $h \in A$. This result was generalized afterwards in several directions. Thus Basener [2] replaced the finiteness condition by countableness one. In [3], in order to find a structure of n-dimensional analytic covering in the maximal ideal space he replaced the single element f by an n-vector of elements of A, say $(f_1,\dots,f_n)$, and the Gelfand transform of f — by the mapping $F(x) = (f_1(x),\dots,f_n(x)) : \mathrm{sp}\,A \to \mathbb{C}^n$.

We consider some cases when the set $f^{-1}(\lambda)$ is neither finite nor countable. The roles of single elements of A are taken here by multiplicative semigroups in A, the Gelfand transform — by the spectral mapping of the semigroup, and, roughly speaking, the "analyticity" — by "gen.-analyticity".

In [12] the following theorem is proved: If A is linearly generated by a semigroup $S = \{f^p\}$, then the spectral mapping Φ is one-to-one, the spectrum $\sigma(S)$ is gen.-polynomially convex and $A \cong P_G(\sigma(S))$. As a corollary, the functions $h \circ \Phi^{-1}$, $h \in A$, are gen.-analytic on $\mathrm{Int}\,\sigma(S)$. We extend this result in the following way:

<u>THEOREM 1</u>. <u>Let</u> W <u>be a component of</u> $\mathrm{Int}\,\sigma(S) \setminus \Phi(X)$ <u>and</u> Φ <u>is one-to-one on</u> $\Phi^{-1}(W)$. <u>Then</u> $h \circ \Phi^{-1}$ <u>are gen.-analytic functions on</u> W <u>for any</u> $h \in A$, <u>i.e.</u> $A|_{\Phi^{-1}(W)} \subseteq A_G(W)$.

P r o o f. Without loss of generality one can consider that $S \cong R_+$. The topology of $\mathbb{C}_G$, i.e. the factor-topology induced in $\mathbb{C}_G$ from $[0,\infty)\times G$, is equivalent to the topology generated by the convergence of χ^p-values of the points of $\mathbb{C}_G$, $p\in S$. The base neighbourhoods relative to this topology are the connected sets $Q((\lambda_o,g_o),\varepsilon,p) = \{(\lambda,g)\in\mathbb{C}_G \mid |\chi^p(\lambda,g)-\chi^p(\lambda_o,g_o)|<\varepsilon\}$. We shall need the following generalization of the theorem of Rudin (cf. [14]).

LEMMA 1. Let $\mathcal{L}$ be an algebra of continuous functions on a base neighbourhood $Q((\lambda_o,g_o),\varepsilon,p)$, such that:

1° all the functions χ^p belong to $\mathcal{L}$.

2° $\mathcal{L}$ satisfies a maximum modulus principle relative to bQ:

$$|F(\lambda,g)| \le \max_{bQ}|F| \quad \text{for all} \quad (\lambda,g)\in\bar{Q},\quad F\in\mathcal{L}.$$

Then $\mathcal{L}\subset A_G(\bar{Q})$.

Now, let $Q\subset\bar{Q}\subset W$ be a base neighbourhood in W centred at $(\lambda_1,g_1)\in W$, and let $f\in A$. The function $F=f\circ\Phi^{-1}$ is continuous on Q, where $f=F(\Phi)$. We shall show that $F\in A_G(Q)$. Let $\mathcal{L}$ be the algebra of restrictions of $F=f\circ\Phi^{-1}$, $f\in A$ on $\bar{Q}$. Φ^{-1} is an open subset of $sp\,A$ with topological boundary contained in $\Phi^{-1}(bQ)$. According to the local maximum modulus principle [5, 14] if $h\in A$, from $\partial A\cap\Phi^{-1}(\bar{Q})=\emptyset$, we get

$$|h(\Phi^{-1}(x))| \le \max_{(\partial A\cap\Phi^{-1}(Q))\cup b\Phi^{-1}(Q)} |h| \le \max_{\Phi^{-1}(bQ)} |h|.$$

Consequently, $|H(x)|\le\max_{bQ}|H|$ for any function H from $\mathcal{L}$. Because for any $\varphi\in\Phi^{-1}(\bar{Q})$ we have $\hat{f}^p(\varphi)=\chi^p(\Phi(\varphi))$, so $\chi^p(\lambda,g)=\hat{f}^p(\Phi^{-1}(\lambda,g))$, where $(\lambda,g)=\Phi(\varphi)$. Hence $\chi^p\in\mathcal{L}$ for any $p\in S$. According to Lemma 1, the functions from $\mathcal{L}$ as well as the functions $F=f(\Phi^{-1})$, $f\in A$, are gen.-analytic. Thus F is gen.-analytic on W, Q.E.D.

Now we introduce, analogically to the concept of classical analytic covering (in the sense of Gunning and Rossi [8]), the concept of a generalized-analytic covering.

Let D be a domain in $\mathbb{C}_G$ and let Λ be a subset in D. We call Λ a negligible set if Λ is nowhere dense and if for any subdomain $D'\subset D$ and for every gen.-analytic function f on $D'\setminus\Lambda$, locally bounded in D', the function f admits a unique gen.-analytic extension to the whole domain D'.

Generalized-Analytic Coverings in the Maximal Ideal Space

Definition 1. We call a generalized-analytic covering any triple (X,π,U), for which

1° X is a locally compact Hausdorff space;

2° U is a domain in $\mathbb{C}_G$;

3° π is a proper continuous mapping of X onto U, for which the set $\pi^{-1}(\lambda,g)$ is discrete for any $(\lambda,g)\in U$;

4° there exist a negligible set $\Lambda\subset U$ and an integer λ, such that π is a λ-sheeted covering mapping of $X\setminus\pi^{-1}(\Lambda)$ onto $U\setminus\Lambda$;

5° $X\setminus\pi^{-1}(\Lambda)$ is dense in X.

Sometimes X is called a covering onto U and Λ — its critical space.

Definition 2. A continuous complex function f defined on an open subset V of a gen.-analytic covering X is called gen.-holomorphic if for any open subset $V'\subset X\setminus\pi^{-1}(\Lambda)$ on which π is homeomorphic, the function $(f|_{V'})\circ\pi^{-1}$ is gen.-analytic on $\pi(V')$.

THEOREM 2. Let $X\supset\Phi^{-1}(b\Phi(\operatorname{sp}A))\cup\partial A$, W be a connected component of $\mathbb{C}_G\setminus\Phi(X)$ and let there exist a $k<\infty$, such that the number $\#\Phi^{-1}(\lambda,g)$ of elements of the set $=\Phi^{-1}(\lambda,g)$ is $\leq k$ for any $(\lambda,g)\in W$. Then the set $\Phi^{-1}(W)$ has the structure of a k_1-sheeted $(k_1\leq k)$ gen.-analytic covering over W and f is a gen.-holomorphic function on this covering for any $f\in A$.

For the proof we shall need several auxiliary results.

LEMMA 2. Let $(\lambda_o,g_o)\in W$, where $\Phi:\operatorname{sp}A\to\mathbb{C}_G$ is a spectral mapping and W is as in Theorem 2. If Φ takes the value (λ_o,g_o) on $\operatorname{sp}A$, then $W\subset\Phi(\operatorname{sp}A)$.

Proof. The set $W_1=\{(\lambda,g)\in W\mid \nexists\, x\in\operatorname{sp}A:\ \Phi(x)=(\lambda,g)\}=W\setminus\Phi(\operatorname{sp}A)$ is open in W. If $y\in\overline{W}_1\cap\Phi(\operatorname{sp}A)$, then $y\in b\Phi(\operatorname{sp}A)\subset\Phi(X)$, i.e. $bW_1\subset\Phi(X)$, i.e. $\overline{W}_1=W_1$, and hence W_1 is a closed subset of W. Because $W\setminus W_1=W\cap\Phi(\operatorname{sp}A)\neq\emptyset$, so the set $W\setminus W_1$ coincides with $W\cap\Phi(\operatorname{sp}A)$, and consequently $W\subset\Phi(\operatorname{sp}A)$.

LEMMA 3. Let B be a uniform algebra on X, Φ be a spectral mapping $\Phi:\operatorname{sp}B\to\mathbb{C}_G$, and W be a component of $\mathbb{C}_G\setminus\Phi(X)$. Let $Q\subset\overline{Q}\subset W$ be a base neighbourhood of $(\lambda_o,g_o)\in W$ and let J be a connected component of $\Phi^{-1}(\overline{Q})$, for which $\Phi(J)\cap Q\neq\emptyset$. Then $\Phi(J)\subset Q$.

The proof is close to the classical one (cf. [14], Lemma 11.3) and we shall omit it.

Now, denote by W_1 the set $\{(\lambda,g)\in W\mid \#\Phi^{-1}(\lambda,g)=1\}$. For a fixed point $(\lambda_o, g_o) \in W_k$, let $p_1, \ldots, p_k$ be the elements of $\Phi^{-1}(\lambda_o, g_o)$. By J_ν we shall denote the component of $\Phi^{-1}(\bar{Q})$ containing p_ν, $\nu = 1, \ldots, k$, $(\lambda_o, g_o) \in Q \subset \bar{Q} \subset W$.

LEMMA 4. *Let* Φ *be a spectral mapping* $\Phi : \mathrm{sp}\, A \to \mathbb{C}_G$.

1) *If* $(\lambda_o, g_o) \in W_k$ *and* Q *is a base neighbourhood of* (λ_o, g_o) *contained in* W, *then* $\Phi^{-1}(Q) \subset \bigcup_{\nu=1}^{k} J_\nu$;

2 *If* $Q \subset W_k$, *then*

i) Φ *maps* $J_\nu \cap \Phi^{-1}(Q)$ *injectively onto* Q *for any* ν;

ii) *for any element* f *of* A *there exists a gen.-analytic function* F *on* Q, *for which* $f = F_\nu(\Phi)$ *on* $J_\nu \cap \Phi^{-1}(Q)$, $1 \le \nu \le k$.

Proof. 1) Suppose that there exists an $x \in \Phi^{-1}(Q)$, for which $x \notin \bigcup_\nu J_\nu$, and let K be the connected component of $\Phi^{-1}(\bar{Q})$ containing x. K has no common points with J for any ν. According to Lemma 3, (λ_o, g_o) belongs to $\Phi(K)$ and hence $\Phi^{-1}(\lambda_o, g_o)$ contains more than k points, which contradict $\# \Phi^{-1}(\lambda_o, g_o) \le k$.

2i) Let $(\lambda_1, g_1) \in Q$. For any ν, $(\lambda_o, g_o) = \Phi(p_\nu) \in \Phi(J_\nu)$ and consequently for any $(\lambda_1, g_1) \in Q$ there exists a point in every J_ν, which is mapped by Φ onto (λ_1, g_1). Because $(\lambda_1, g_1) \in W_k$, so there exists only one point q_ν in any J_ν with $\Phi(q_\nu) = (\lambda_1, g_1)$, Q.E.D.

2ii) Fix ν and let $\Phi_\nu = \Phi_{J_\nu}$. Then Φ_ν is also a spectral mapping (for the restricted semigroup on J_ν) and maps injectively $J \cap \Phi^{-1}(Q)$ onto Q. We can consider $A(J_\nu)$ as a uniform algebra on $X_1 = \Phi_\nu^{-1}(bQ)$. Now $\Phi_\nu(\Phi_\nu^{-1}(bQ)) = bQ$ and according to the local maximum modulus principle $\partial A(J_\nu) \subset bJ_\nu = \Phi_\nu^{-1}(bQ)$. We can see that Φ_ν maps injectively $J_\nu \cap \Phi^{-1}(Q)$ onto the connected component Q of $\mathbb{C}_G \setminus \Phi_\nu(bJ) = \mathbb{C}_G \setminus bQ$. Thus, Theorem 1 applied to algebra $A(J_\nu)$ completes the proof of 2ii).

Proof of Theorem 3. Without loss of generality we may assume that $W_k \ne \emptyset$. Let $(\lambda, g) \in W_k$ and $p_1(\lambda, g), \ldots, p_k(\lambda, g)$ be the points of $\Phi^{-1}(\lambda, g)$. Fix a (λ_1, g_1) from W_k and choose a $g \in A$, taking different values on the set $\{p_1(\lambda_1, g_1), \ldots, p_k(\lambda_1, g_1)\} = \Phi^{-1}(\lambda_1, g_1)$. The function

$$H(\lambda, g) = \prod_{i<j} (g(p_i(\lambda, g)) - g(p_j(\lambda, g)))^2$$

is continuous on W_k, $\mathrm{supp}\, H \subset W_k$ and $H(\lambda_1, g_1) \ne 0$. According to Lemma 4, if $(\lambda, g) \in Q \subset \bar{Q} \subset \mathrm{supp}\, H \subset W_k$, then $\Phi^{-1}(Q) = \bigcup_{\nu=1}^{k} \mathrm{Int}\, J_\nu$, $\mathrm{Int}\, J_i \cap \mathrm{Int}\, J_k = \emptyset$. We may assume that $p_\nu(\lambda, g) \in \mathrm{Int}\, J_\nu$ for any $(\lambda, g) \in Q$. According to Lemma 4, the function g can be expressed on $\mathrm{Int}\, J_\nu$ as $g = G_\nu(\Phi_\nu)$ for some gen.-analytic function G_ν on Q. Consequently, $g(p) =$

$g(\Phi_\nu^{-1}\Phi_\nu(p)) = g(p_\nu(\Phi_\nu(p))) = (g \circ p_\nu \circ \Phi_\nu)(p) = G_\nu \circ \Phi_\nu(p)$, and hence $g \circ p_\nu = G_\nu$. We conclude that $g \circ p_\nu$ are gen.-analytic functions on Q and consequently they are gen.-analytic on $\operatorname{supp} H$. If (λ_0, g_0) is a boundary point of W_k in W, then $\Phi^{-1}(\lambda_0, g_0) < k$. It means that if $(\lambda,g) \to (\lambda_0, g_0)$, $(\lambda,g) \in W_k$, then $H(\lambda,g) \longrightarrow 0$. We define H as 0 on $W \setminus W_k$. Now H is a function, continuous on W and gen.-analytic on $\{(\lambda,g) | H(\lambda,g) \neq 0\}$ and, according to a generalized version of Radó theorem, H can be extended to W as a gen.-analytic function, i.e. H itself is gen.-analytic on W. Hence, the set $\Lambda = W \setminus W_k$ is contained in the zero-set of a gen.-analytic non-zero function, which is negligible according to a result of Grigorjan [7]. Now, let us consider the triple $(\Phi^{-1}(W), \Phi, W)$, where $\Phi^{-1}(W)$ is a locally compact set, W is a domain in $\mathbb{C}_G$, and Φ is a proper continuous mapping from $\Phi^{-1}(W)$ onto W. The set $\Phi^{-1}(\lambda,g)$ is finite (and hence discrete) for any $(\lambda,g) \in W$, the set Λ is negligible and outside $\Phi^{-1}(\Lambda)$ (i.e. in $\Phi^{-1}(W_k)$), and Φ is a k-sheeted covering mapping from $\Phi^{-1}(W) \setminus \Phi^{-1}(\Lambda)$ onto $W \setminus \Lambda \subset W_k$ — according to Lemma 4. The set $\Phi^{-1}(W) \setminus \Phi^{-1}(W \setminus W_k)$ is dense in $\Phi^{-1}(W)$ for Φ is an open mapping on the set $\Phi^{-1}(W)$. Consequently, $\Phi^{-1}(W)$ is a gen.-analytic covering over W with $\Lambda = W \setminus W_k$ as a critical set. Consider now the restrictions $f|_{\Phi^{-1}(W)}$ of elements of A. As above (for the function H), one can see that for any neighbourhood $Q \subset W_k$, the function $f|_{\operatorname{Int} J_\nu}$ can be expressed as $F_\nu \circ \Phi$ for some gen.-analytic function F_ν on $Q = \Phi_\nu(\operatorname{Int} J_\nu)$. It shows that f is a gen.-holomorphic function on the covering $\Phi^{-1}(W)$. The theorem is proved.

Interesting questions arise in connection with Theorem 2, for which not much is known. For instance: When the condition $\# \Phi^{-1}(\lambda,g) \leq k$ is fulfilled on W?

References

[1] ARENS, R. and I. SINGER, Generalized analytic functions, Trans. Amer. Math. Soc. 81 (1956), 379-393.

[2] BASENER, R., A condition for analytic structure, Proc. Amer. Math. Soc. 36 (1972).

[3] ——, A generalized Šilov boundary and analytic structure, Proc. Amer. Math. Soc. 47 (1975).

[4] BISHOP, E., Holomorphic completions, analytic continuations and the interpolation of seminorm, Ann. Math. 78 (1963), 468-500.

[5] GAMELIN, T., Uniform algebras, Prentice-Hall, Englewood Cliffs, N.J. 1969.

[6] GRIGORJAN, S., On algebras generated by analytic functions in the sense of Arens-Singer, Doklady Acad. Nauk Armjan. SSR 68 (1979), 146-148.

[7] ——, On singularities of generalized-analytic functions, Doklady Acad. Nauk Armjan. SSR 71 (1980), 65-68.

[8] GUNNING, R. and H. ROSSI, Analytic functions of several complex variables, Prentice-Hall, Englewood Cliffs, N.J. 1965.

[9] HOFFMAN, K. and I. SINGER, Maximal algebras of continuous functions, Acta Math. 103 (1960), 217-241.

[10] KATO, T., Perturbation theory for linear operators, Springer-Verlag, New York-Heidelberg-Berlin 1966.

[11] TONEV, T., Generalized-analytic functions — recent results, Compt. Rend. de l'Acad. Bulg. des Sci. 34 (1981), 1061-1064.

[12] ——, Some properties of generalized-analytic functions, Ann. de l'Univ. de Sofia, Fac. de Math. et Méch., to appear.

[13] ——, Commutative Banach algebras and analytic functions of countable-many variables, to appear.

[14] WERMER, J., Banach algebras and several complex variables, Springer-Verlag, New York-Heidelberg-Berlin 1976.

Institute of Mathematics of the
Bulgarian Academy of Sciences
BG-1090 Sofia, P.O.Box 373, Bulgaria

ON THE DEFICIENCIES OF MEROMORPHIC FUNCTIONS OF SMOOTH GROWTH

Sakari Toppila (Helsinki)

Contents

Summary. For meromorphic functions of smooth growth, Theorems 1 and 2, formulated in Section 1, are proved.

1. Introduction and statement of results

We shall use the usual notation of the Nevanlinna theory. It is proved in [2] that if f is a transcendental meromorphic function in the plane such that

$$(1.1)\quad \lim_{r\to\infty} \frac{T(2r,f)}{T(r,f)} = 1$$

then

$$(1.2)\quad \delta(\infty,f') \geq 2\,\delta(\infty,f) - 1,$$

and

$$(1.3)\quad \Delta(\infty,f') \geq \frac{\Delta(\infty,f)}{2-\Delta(\infty,f)}$$

with equality in (1.3) if f has only simple poles.

Let $f^{(k)}$ be the k-th derivative of f. We shall prove the following extension for (1.2) and (1.3).

THEOREM 1. *Let f be a transcendental meromorphic function satis-*

fying (1.1) *and* k *a positive integer. Then*

$$(1.4)\quad \delta(\infty,f^{(k)}) \geq \frac{2\delta(\infty,f)-1}{k-(k-1)\delta(\infty,f)}$$

and

$$(1.5)\quad \Delta(\infty,f^{(k)}) \geq \frac{\Delta(\infty,f)}{k+1-k\Delta(\infty,f)}$$

with equality in (1.5) *if* f *has only simple poles;* $\Delta(0,f^{(k)}) = 0$ *if* $\Delta(\infty,f) \geq 1/2$, *and if* $0 \leq \delta(\infty,f) \leq 1/2$ *then*

$$(1.6)\quad \Delta(0,f^{(k)}) \leq \frac{1-2\delta(\infty,f)}{1-\delta(\infty,f)} \limsup_{r\to\infty} \frac{N(r,f)}{N(r,f^{(k)})}.$$

If f is a meromorphic function which satisfies

$$(1.7)\quad T(r,f) = O((\log r)^2) \quad (r \to \infty),$$

then it also satisfies the condition (1.1). The following example shows that (1.4) and (1.6) are sharp.

THEOREM 2. Let $0 < d < 1$ *and let* $\varphi(r)$ *be a positive and increasing function of* r *such that* $\varphi(r) \to \infty$ *as* $r \to \infty$. *There exists a transcendental meromorphic function* f *satisfying*

$$(1.8)\quad T(r,f) = O(\varphi(r)\log r) \quad (r \to \infty),$$

so that $\delta(\infty,f) = d$ *and for any positive integer* k,

$$(1.9)\quad \delta(\infty,f^{(k)}) = \frac{2d-1}{k-(k-1)d}$$

if $1/2 \leq d < 1$, *and*

$$(1.10)\quad \Delta(0,f^{(k)}) = \frac{1-2d}{1-d} \limsup_{r\to\infty} \frac{N(r,f)}{N(r,f^{(k)})}$$

if $0 < d \leq 1/2$.

2. Some lemmas

Following Anderson [1], we can say that a countable union of discs in the plane is a *slim set* if the sum of the radii of those discs intersecting the annulus $r < |z| < 2r$ is $o(r)$ as $r \to \infty$.

The following lemma is proved in [2].

LEMMA A. *Let* f *be a transcendental meromorphic function satisfying* (1.1). *Then* f' *satisfies* (1.1),

$$n(r,a,f) = o(T(r,f)) \quad (r \to \infty) \tag{2.1}$$

and

$$N(2r,a,f) = N(r,a,f) + o(T(r,f)) \quad (r \to \infty) \tag{2.2}$$

whenever a *is a complex number*,

$$\log|f(z)| = N(|z|,0,f) - N(|z|,\infty,f) + o(T(|z|,f)) \tag{2.3}$$

as $z \to \infty$ *outside a slim set*,

$$m(r,f) \leq \sup\{m(t,f') : 1 \leq t \leq r\} + o(T(r,f)) \quad (r \to \infty) \tag{2.4}$$

and

$$1/2 + o(1) \leq \frac{T(r,f')}{T(r,f)} \leq 2 + o(1) \quad (r \to \infty). \tag{2.5}$$

We prove the following

LEMMA 1. *Let* f *be a transcendental meromorphic function satisfying* (1.1) *and* k *a positive integer. Then*

$$m(r,f) \leq N(r,f) + N(r,0,f^{(k)}) - N(r,f^{(k)}) + o(T(r,f)) \quad (r \to \infty). \tag{2.6}$$

Proof. Let E be the union of the exceptional slim sets for f and $f^{(k)}$ as in Lemma A. Let F be such a set of values of r, for which the circle $|z| = r$ lies outside E. Since for any large r, there exists $t \in F$ such that $r \leq t \leq 2r$, it follows from Lemma A that, in order to prove (2.6), it is sufficient to show that (2.6) holds as $r \to \infty$ through F.

We choose d, $0 < d < 1/2$, and a complex value a such that $|a| = 1$, $f^{(k)}(a) \neq 0$, and the disc $|z-a| \leq 2d$ does not contain any pole of f.

Let $r > 10$, $r \in F$, and let z_0 be a point lying on $|z| = r$. We set

$$P(z) = f(z_0) + \sum_{n=1}^{k-1} \frac{f^{(n)}(z_0)}{n!}(z-z_0)^n.$$

Using Taylor's formula

$$f(z) = P(z) + \frac{1}{(k-1)!} \int_{z_0}^{z} f^{(k)}(w)(z-w)^{k-1}dw$$

and integrating along the circle $|z| = r$, we deduce that

(2.7) $\quad |f(z) - P(z)| \le (\pi r)^k M(r, f^{(k)})$,

for all z lying on $|z| = r$. Let b_s be the poles of f. Applying the Poisson-Jensen formula, we get from (2.7) the estimate

$$\log|f(z) - P(z)| \le (2\pi)^{-1} \int_0^{2\pi} \frac{(r^2 - x^2)\log|f(re^{i\varphi}) - P(re^{i\varphi})|}{r^2 + x^2 - 2rx\cos(\varphi - \alpha)} d\varphi$$

$$+ \sum_{|b_s|<r} \log\left|\frac{r^2 - \bar{b}_s z}{r(z - b_s)}\right| \le (1 + o(1))\log M(r, f^{(k)}) + O(\log r)$$

$$+ n(3,\infty,f)\log\frac{2r}{d} + \sum_{3<|b_s|<r} \log\frac{2r}{|b_s| - |z|} \le (1 + o(1))\log M(r, f^{(k)})$$

$$+ N(r,f) + O(n(r,\infty,f)) + O(\log r),$$

for any $z = xe^{i\alpha}$ lying on $|z - a| \le d$.

Together with Lemma A it implies that

(2.8) $\quad \log|f(z) - P(z)| \le N(r,0,f^{(k)}) - N(r,f^{(k)}) + N(r,f) + o(T(r,f))$

for all z lying on $|z - a| \le d$.

We write

$$M_1 = \max\{|f(z) - P(z)| : |z - a| \le d\}.$$

Since P is a polynomial and the degree of P is at most $k - 1$, we can write

$$P(z) = A_0 + A_1(z - a) + \ldots + A_k(z - a)^k,$$

where $A_k = 0$. For $s = 0, 1, \ldots, k$ we get

$$\left|A_s - \frac{f^{(s)}(a)}{s!}\right| = \left|(2\pi i)^{-1} \int_{|z-a|=d} \frac{P(z) - f(z)}{(z-a)^{s+1}} dz\right| \le M_1/d^s \le M_1/d^k,$$

which implies that

$$|f(z_0)| = |P(z_0)| \leq \sum_{s=0}^{k-1} \left(\frac{M_1}{d^k} + \left| \frac{f^{(s)}(a)}{s!} \right| \right) |z_0 - a|^s \leq M_1 d^{-k} (2r)^k + O(r^k)$$

and

$$M_1 \geq \frac{d^k |f^{(k)}(a)|}{k!}.$$

These estimates imply that

$$\log^+ |f(z_0)| \leq \log^+ M_1 + O(\log r) \leq \log M_1 + O(\log r),$$

and we deduce from (2.8) that

$$\log^+ |f(z_0)| \leq N(r,0,f^{(k)}) - N(r,f^{(k)}) + N(r,f) + o(T(r,f)),$$

for all z_0 lying on $|z| = r$. Continuingly, this implies that

$$m(r,f) \leq N(r,0,f^{(k)}) - N(r,f^{(k)}) + N(r,f) + o(T(r,f)),$$

as $r \to \infty$ through F. Thus, Lemma 1 is proved.

3. Proof of Theorem 1

Let f and k be as in Theorem 1. Then

$$(3.1) \quad N(r,f) \leq N(r,f^{(k)}) \leq (k+1) N(r,f)$$

for $r \geq 1$. If $0 \leq \delta(\infty,f) \leq 1/2$ then (1.4) holds. Let us suppose that $\delta(\infty,f) > 1/2$. Then from the formula (3.1), Lemma 1, and Lemma A we deduce

$$\frac{m(r,f^{(k)})}{T(r,f^{(k)})} = \frac{m(r,f^{(k)})}{N(r,f^{(k)}) + m(r,f^{(k)})}$$

$$\geq \frac{N(r,0,f^{(k)}) - N(r,f^{(k)}) + o(T(r,f))}{N(r,f^{(k)}) + (N(r,0,f^{(k)}) - N(r,f^{(k)}))}$$

$$\geq \frac{m(r,f) - N(r,f)}{(k+1)N(r,f) + m(r,f) - N(r,f)} + o(1)$$

$$\geq \frac{m(r,f) - N(r,f)}{T(r,f) + (k-1)N(r,f)} + o(1)$$

$$\geq \frac{\delta(\infty,f) - (1 - \delta(\infty,f))}{1 + (k-1)(1 - \delta(\infty,f))} + o(1)$$

$$= \frac{2\delta(\infty,f)-1}{k-(k-1)\delta(\infty,f)} + o(1) \quad (r \to \infty),$$

which proves (1.4) and together with Lemma A shows that $\Delta(0,f^{(k)}) = 0$.

Let us suppose that $0 \leq \delta(\infty,f) \leq 1/2$. We deduce from Lemma A and Lemma 1 that

$$\begin{aligned} m(r,0,f^{(k)}) &\leq \max(0, N(r,f^{(k)}) - N(r,0,f^{(k)})) + o(T(r,f)) \\ &\leq \max(0, N(r,f) - m(r,f)) + o(T(r,f)) \\ &\leq \left(1 - \frac{\delta(\infty,f)}{1-\delta(\infty,f)}\right) N(r,f) + o(T(r,f^{(k)})) \quad (r \to \infty), \end{aligned}$$

which implies that

$$\frac{m(r,0,f^{(k)})}{T(r,f^{(k)})} \leq \frac{m(r,0,f^{(k)})}{N(r,f^{(k)})} + o(1) \leq \frac{1-2\delta(\infty,f)}{1-\delta(\infty,f)} \frac{N(r,f)}{N(r,f^{(k)})} + o(1)$$

as $r \to \infty$. This proves (1.6).

If $\Delta(\infty,f) = 0$ then (1.5) holds. Let us suppose that $\Delta(\infty,f) > 0$. We choose a sequence r_p, $1 < r_1 < r_2 < \ldots$, such that $r_p \to \infty$ as $p \to \infty$ and

(3.2) $m(r_p,f) = (\Delta(\infty,f) + o(1))T(r_p,f) \quad (p \to \infty)$.

For any p, we choose t_p, $1 \leq t_p \leq r_p$, such that

$$m(t_p,f^{(k)}) = \sup\{m(t,f^{(k)}) : 1 \leq t \leq r_p\}.$$

Applying (2.4) subsequently, we deduce that

$$m(t_p,f^{(k)}) \geq m(r_p,f) + o(T(r_p,f)) \quad (p \to \infty),$$

which together with (3.2) implies that $t_p \to \infty$ as $p \to \infty$, and together with (3.1) and (3.2) shows that

$$\begin{aligned} \frac{m(t_p,f^{(k)})}{T(t_p,f^{(k)})} &\geq \frac{m(t_p,f^{(k)})}{m(t_p,f^{(k)}) + N(r_p,f^{(k)})} + o(1) \\ &\geq \frac{m(r_p,f)}{m(r_p,f) + (k+1)N(r_p,f)} + o(1) \end{aligned}$$

$$\geq \frac{\Delta(\infty,f)}{\Delta(\infty,f)+(k+1)(1-\Delta(\infty,f))}+o(1)$$

$$=\frac{\Delta(\infty,f)}{k+1-k\Delta(\infty,f)}+o(1) \quad (p\to\infty),$$

which proves (1.5).

Suppose now that f has only simple poles. Then

$$N(r,f^{(k)})=(k+1)N(r,f)$$

for $r>0$, and from the theorem on the logarithmic derivative we get

$$m(r,f^{(k)})\leq m(r,f)+o(T(r,f)) \quad (r\to\infty).$$

All these estimates taken together imply that

$$\frac{m(r,f^{(k)})}{T(r,f^{(k)})}=\frac{m(r,f^{(k)})}{(k+1)N(r,f)+m(r,f^{(k)})}\leq\frac{m(r,f)}{m(r,f)+(k+1)N(r,f)}+o(1)$$

$$\leq\frac{\Delta(\infty,f)}{\Delta(\infty,f)+(k+1)(1-\Delta(\infty,f))}+o(1)$$

$$=\frac{\Delta(\infty,f)}{k+1-k\Delta(\infty,f)}+o(1) \quad (r\to\infty),$$

which shows that the equality is valid in (1.5) if f has only simple poles. Thus, Theorem 1 is proved.

4. Proof of Theorem 2

Let d and $\varphi(r)$ be as in Theorem 2. We choose $r_0=100$, and, for $p\geq 1$, r_p, s_p and t_p such that

(4.1) $s_p=p!$,

(4.2) $\log r_p > s_{p+1}r_{p-1}$,

(4.3) $\varphi(r_p)>s_p$,

and

(4.4) $\log t_p = d\log r_{p+1}$.

We set

$$g(z) = \sum_{p=1}^{\infty} (z/r_p)^{s_p}, \quad h(z) = \sum_{p=1}^{\infty} (z/t_p)^{s_p},$$

and choose a finite complex value b such that the function $f(z) = g(z)/(b + h(z))$ has only simple poles.

It follows from (4.1), (4.2) and (4.4) that the functions g and h satisfy

$$\log M(r) \leq (1 + o(1)) s_p \log r \quad (p \to \infty)$$

for $r_p \leq r \leq r_{p+1}$, which implies that they satisfy (1.7), and shows together with (4.3) that f satisfies (1.8).

For $p \geq 2$, we choose r'_p and t'_p such that

$$\text{(4.5)} \quad s_{p-1} \log(r'_p/r_{p-1}) = s_p \log(r'_p/r_p)$$

and

$$\text{(4.6)} \quad s_{p-1} \log(t'_p/t_{p-1}) = s_p \log(t'_p/t_p).$$

Applying Lemma A, we deduce that

$$\text{(4.7)} \quad \log|h(z)| = (1 + o(1)) \log M(|z|, h) = (1 + o(1)) N(|z|, \infty, f)$$
$$= (1 + o(1)) s_p \log(|z|/t_p)$$

for $t'_p \leq |z| \leq t'_{p+1}$ when z lies outside a slim set, and similarly, that

$$\log|g(z)| = (1 + o(1)) \log M(|z|, g) = (1 + o(1)) s_p \log(|z|/r_p)$$

when $r'_p \leq |z| \leq r'_{p+1}$ and z lies outside a slim set.

Let $r'_p \leq r \leq t'_p$. It follows from (4.8), (4.7), (4.5), (4.4) and (4.2) that

$$\text{(4.9)} \quad \frac{m(r,f)}{T(r,f)} = \frac{s_p \log(r/r_p) - s_{p-1} \log(r/t_{p-1})}{s_p \log(r/r_p)} + o(1)$$

$$\geq 1 - \frac{s_{p-1} \log(r'_p/t_{p-1})}{s_p \log(r'_p/r_p)} + o(1) = \frac{m(r'_p, f)}{T(r'_p, f)} + o(1)$$

$$= 1 - \frac{s_{p-1} \log(r'_p/t_{p-1})}{s_{p-1} \log(r'_p/r_{p-1})} + o(1)$$

$$= 1 - \frac{\log(r_p'/r_p) + \log r_p - \log t_{p-1}}{\log(r_p'/r_p) + \log r_p - \log r_{p-1}} + o(1) = d + o(1) \quad (p \to \infty).$$

For $t_p' \le r \le r_{p+1}'$, we get from (4.8), (4.7) and (4.9):

$$\frac{m(r,f)}{T(r,f)} = \frac{s_p \log(r/r_p) - s_p \log(r/t_p)}{s_p \log(r/r_p)} + o(1) = 1 - \frac{\log(r/t_p)}{\log(r/r_p)} + o(1)$$

$$\ge 1 - \frac{\log(r_{p+1}'/t_p)}{\log(r_{p+1}'/r_p)} + o(1) = \frac{m(r_{p+1}',f)}{T(r_{p+1}',f)} + o(1) = d + o(1) \quad (p \to \infty).$$

These estimates imply that $\delta(\infty, f) = d$.

Let k be a positive integer. Let $||z| - r_{p+1}| \le 1$. It follows from (4.1), (4.2), (4.7) and (4.4) that

$$\log|f(z) - (t_p/r_p)^{s_p}| = \log|\sum_{n \ne p} (z/r_n)^{s_n} - (t_p/r_p)^{s_p}(b + \sum_{n \ne p} (z/t_n)^{s_n})|$$

$$- \log|b + h(z)|$$

$$\le (2 + o(1)) s_{p-1} \log r_{p+1} + (1 + o(1)) s_p \log t_p$$

$$- (1 + o(1)) s_p \log(r_{p+1}/t_p)$$

$$= 2 s_p \log t_p - (1 + o(1)) s_p \log r_{p+1}$$

$$= (2d - 1 + o(1)) s_p \log r_{p+1} \quad (p \to \infty).$$

For any z lying on the circle $|z| = r_{p+1}$, we get

$$(4.10) \quad \log|f^{(k)}(z)| = \log\left|\frac{k!}{2\pi i} \int_{|w-z|=1} \frac{f(w) - (t_p/r_p)^{s_p}}{(w-z)^{k+1}} dw\right|$$

$$\le (2d - 1 + o(1)) s_p \log r_{p+1}$$

as $p \to \infty$.

Since f has only simple poles, it follows from (4.7) and (4.4) that

$$(4.11) \quad N(r_{p+1}, f^{(k)}) = (k+1) N(r_{p+1}, f) = (k + 1 + o(1)) s_p \log(r_{p+1}/t_p)$$

$$= (k+1+o(1))(1-d)s_p \log r_{p+1}$$

as $p \to \infty$.

If $d \geq 1/2$, we deduce from (4.10) and (4.11) that

$$\delta(\infty, f^{(k)}) \leq \frac{2d-1}{(k+1)(1-d)+2d-1} = \frac{2d-1}{k-(k-1)d},$$

which together with Theorem 1 proves (1.9). If $d \leq 1/2$ then (4.10) and (4.11) imply that

$$\Delta(0, f^{(k)}) \geq \frac{1-2d}{(1-d)(k+1)} = \frac{1-2d}{1-d} \limsup_{r \to \infty} \frac{N(r,f)}{N(r,f^{(k)})},$$

which together with Theorem 1 proves (1.10). Thus Theorem 2 is proved.

References

[1] ANDERSON, J.M.: Asymptotic values of meromorphic functions of smooth growth, Glasgow Math. J. 20 (1979), 155-162.

[2] TOPPILA, S.: On the deficiencies of a meromorphic function and of its derivative, J. London Math. Soc. (2), 25 (1982), 273-287.

University of Helsinki
Department of Mathematics
SF-00100 Helsinki 10, Finland

CAUCHY PROBLEMS WITH MONOGENIC INITIAL VALUES

Wolfgang Tutschke (Halle an der Saale)

Summary. Using an abstract version of the Cauchy-Kovalevska theorem, the initial value problem $(\partial/\partial t)w=\mathcal{L}w$, $w(x,0)=w_0(x)$ is solved, where the initial function $w_0=w_0(x)$ is monogenic in x and $\mathcal{L}$ transforms the set of all monogenic functions into itself; t being a real variable. The constructed solution is monogenic for every t.

In view of the classical Cauchy-Kovalevska theorem the initial value problem

$$(\partial/\partial t)w = \mathcal{L}w, \quad w(z,0) = w_0(z)$$

is solvable, where the initial function $w_0 = w_0(z)$ is holomorphic in z and $\mathcal{L}$ transforms the set of all holomorphic functions into itself. The function $w = w(z,t)$ we are looking for, depends on z and a (real) variable t, interpreted as time. Using an abstract version of the Cauchy-Kovalevska theorem (see Ovsjannikov [6] and Trèves [8]), the given holomorphic initial values can be replaced by generalized analytic ones. This generalization allows us to solve initial value problems for differential operators $\mathcal{L}$ that are more general than those transforming holomorphic functions into themselves. If $\mathcal{L}$ transforms a set of generalized analytic functions into itself, the Cauchy problem with Hölder continuous initial values is solved in [11]. Generalized analytic initial values belonging to $\mathcal{L}_p$ are considered in [5]. Furthermore, the generalized analytic functions in the sense of I. N. Vekua (see [16] and also [9, 10]) can be replaced by pseudoholomorphic functions in the sense of Bers [1]. Initial value problems with pseudoholomorphic initial functions are solved in [14].

On the other hand the concept of holomorphy and its generalizations can also be defined in higher-dimensional real spaces. Thus the corresponding initial value problems can be studied for initial functions with properties similar to those of holomorphic functions (the Cauchy-Kovalevska theorem in the case of holomorphic functions of several variables in the usual sense does not differ from the complex one-dimensional case). Potential vectors in $\mathbb{R}^3$ as initial vectors are considered in [12], while the case of generalized potential vectors is studied in [14]. Besides, the Clifford analysis (see, for instance, [2-4]) leads to an important generalization of the concept of holomorphy to higher dimensional real Euclidean spaces: the so-called (left and right, resp.) monogenic functions generalize directly the classical concept of holomorphy.

Again, using the abstract version of the Cauchy-Kovalevska theorem, in the present paper initial value problems are solved for monogenic initial functions. The constructed solution is monogenic for every t.

Another approach to a monogenic generalization of the classical Cauchy-Kovalevska theorem is given by Sommen [7]: he constructs a monogenic extension into optimal domains, if the initial values are the restriction to $\mathbb{R}^n$ of a holomorphic function in $\mathbb{C}^n$.

Let $\mathcal{A}$ be the Clifford algebra generated by $e_0, e_1, \ldots, e_n$, where $e_0 = 1$, $e_\alpha^2 = -1$, $e_\alpha e_\beta + e_\beta e_\alpha = 0$, and $e_\alpha(e_\beta e_\gamma) = (e_\alpha e_\beta)e_\gamma$ for $\alpha, \beta, \gamma = 1, \ldots, n$. Thus the elements of $\mathcal{A}$ are linear combinations $a = \Sigma_A a_A e_A$, where A is an arbitrary combination $\{\alpha_0, \alpha_1, \ldots, \alpha_h\}$ of a subset of $\{0, 1, \ldots, n\}$, $e_A = e_{\alpha_0} \ldots e_{\alpha_h}$, $e_\emptyset = e_0$, and a_A are real or complex numbers.

Moreover, let G be a domain in $\mathbb{R}^{n+1}$ and $x = (x_0, x_1, \ldots, x_n) \in G$. Define the differential operators

$$D = \frac{\partial}{\partial x_0} + e_1 \frac{\partial}{\partial x_1} + \ldots + e_n \frac{\partial}{\partial x_n}, \quad \overline{D} = \frac{\partial}{\partial x_0} - e_1 \frac{\partial}{\partial x_1} - \ldots - e_n \frac{\partial}{\partial x_n}.$$

If u_A are real- (or complex-) valued functions defined in G, then $u = \Sigma_A u_A e_A$ is an $\mathcal{A}$-valued function in G. It is called <u>left monogenic</u> if $Du = 0$. Since

$$(1) \qquad D\overline{D} = \overline{D}D = \Delta,$$

every component u_A of a monogenic function is harmonic.

Now, suppose that G is a bounded domain. The space $\mathcal{C}(\overline{G})$ of $\mathcal{A}$-valued functions u, continuous in $\overline{G}$, can be equipped with the sup-norm

$$\|u\|_{\mathcal{C}(\overline{G})} = \max_A \sup_G |u_A|.$$

Next, let K be any compact subset of G situated at least at the distance $\delta > 0$ from the boundary of G. Applying the Poisson integral formula to a ball of radius δ, centred at a point of K, one gets

$$(2) \qquad \|(\partial/\partial x_i)u\|_{\mathscr{C}(K)} \leq (\mathrm{const}/\delta)\|u\|_{\mathscr{C}(\overline{G})}.$$

Together with the given domain G we consider a family of its subdomains G_s, $0 < s \leq 1$, such that $G_1 = G$ and, moreover,

$$\mathrm{dist}(G_s, \partial G_{s'}) \geq \mathrm{const}/(s' - s) \quad \text{for} \quad s' > s.$$

The set of all $\mathcal{A}$-valued functions, monogenic in G_s and continuous on $\overline{G}_s$ is denoted by $\mathcal{W}_s$. Further, let $\mathcal{L}$ be a linear operator of the first order with bounded $\mathcal{A}$-valued coefficients. Then $\mathcal{L}u$ is defined in G_s for each $u \in \mathcal{W}_s$. By (2) we arrive at the estimate

$$(3) \qquad \|\mathcal{L}u\|_{\mathcal{W}_s} \leq [\mathrm{const}/(s' - s)]\|u\|_{\mathcal{W}_{s'}} \quad \text{for} \quad u \in \mathcal{W}_{s'} \text{ and } s < s'.$$

Observe that, in general, $\mathcal{L}u$ is not monogenic.

In the following, we consider only such operators $\mathcal{L}$ which transform monogenic functions into monogenic functions. By (1) the operator $\mathcal{L} = \overline{D}$ has this property. Another example is given by

$$\mathcal{L} = \partial/\partial x_k,$$

where k is a fixed number.

In this paper we do not formulate sufficient conditions under which an operator $\mathcal{L}$ transforms monogenic functions into themselves. We only mention some further examples.

Namely, in the case $n = 2$ and

$$(4) \qquad u = u_0 - u_1 e_1 - u_2 e_2,$$

monogenic functions are identical with potential vectors $u = (u_1, u_2, u_3)$, defined by

$$\mathrm{div}\, u = 0, \qquad \mathrm{rot}\, u = 0.$$

In the paper [12] we have determined vector-valued operators $\mathcal{L}$ transforming the set of all monogenic functions of the type (4) into itself. (A general result concerning operators in the plane, which transform generalized analytic functions into themselves, is obtained in [13]. An operator transforming (F,G)-pseudoholomorphic functions into themselves is derived in [14].)

Assume now that $\mathcal{L}$ is any first order differential operator with bounded coefficients, that transforms monogenic functions into themselves. By (3) $\mathcal{L}$ is a bounded operator mapping $\mathcal{W}_{s'}$ into $\mathcal{W}_s$, and its norm is not larger than $\text{const}/(s'-s)$. On the other hand the Banach spaces $\mathcal{W}_s$ form such a scale, that the abstract version of the Cauchy-Kovalevska theorem is applicable. Thus the following theorem holds:

THEOREM. Let G be a bounded domain in $\mathbb{R}^{n+1}$, $x \in G$, and let $u_0 = u_0(x)$ be monogenic in G. Moreover, let $\mathcal{L}$ be a first order differential operator transforming monogenic functions into themselves. Then there exists an $\mathcal{A}$-valued solution $u = u(x,t)$ of the differential equation

$$(\partial/\partial t)u = \mathcal{L}u,$$

satisfying the initial conditiom

$$u(x,0) = u_0(x)$$

and being monogenic for every t. The solution can be constructed with the help of the method of successive approximations. In G_s, $0 < s < 1$, the solution exists for $0 \leq t \leq T_s$, where T_s depends on G_s.

Finally, we remark that the solution $u = u(x,t)$ looked for is proved to be also a solution of the integro-differential equation

$$u(x,t) = u_0(x) + \int_{\tau=0}^{t} \mathcal{L}u(x,\tau)\,d\tau.$$

The successive approximations are defined with the help of the right-hand side of this integro-differential equation.

In the case $n = 1$ monogenic functions are classical holomorphic functions in $z = x_0 + i x_1$, where $x = (x_0, x_1)$ and e_1 is denoted by i. Thus the theorem formulated above contains as a special case the linear case of the classical Cauchy-Kovalevska theorem.

Bibliography

[1] BERS, L.: Theory of pseudo-analytic functions, New York University, 1953.

[2] DELANGHE, R. and F. BRACKX: Hypercomplex function theory and Hilbert modules with reproducing kernel, Proc. London Math. Soc. 37 (1978), 545-576.

[3] GOLDSCHMIDT, B.: Verallgemeinerte analytische Vektoren im $\mathbb{R}^n$, Universität Halle, B-Dissertationsschrift, Halle (Saale) 1980.

[4] ——: Properties of generalized analytic vectors in $\mathbb{R}^n$, Math.

Nachr. 103 (1981), 245-254.

[5] МАНДЖАВИДЗЕ, Т.Ф. и В. ТУЧКЕ: Некоторые оценки норм производных голоморфных функций в подобластях и одно применение к задачам с начальными значениями, ZAA, in print.

[6] ОВСЯННИКОВ, Л.В.: Задача Коши в школе банаховых пространств аналитических функций. Труды симпозиума по механике сплошной среды и родственным проблемам анализа II, Тбилиси 1971, pp. 219-229.

[7] SOMMEN, F.: Some connections between Clifford analysis and complex analysis. Complex variables: Theory and application, to appear.

[8] TRÈVES, F.: Basic linear partial differential equations, New York - San Francisco - London 1975.

[9] TUTSCHKE, W.: Partielle komplexe Differentialgleichungen in einer und in mehreren komplexen Variablen, Berlin 1977.

[10] ——: Vorlesungen über partielle Differentialgleichungen, Leipzig 1978.

[11] ——: Задача с начальными значениями для обобщенных аналитических функций, зависящих от времени, ДАН ССР 262, No.5(1982), 1081-1085.

[12] ——: Потенциальные векторы, зависящие от времени. Труды симпозиума по случаю 75-летия со дня рождения академика И.Н. Векуа, Тбилиси 1982, in print.

[13] ——: Ассоциированные операторы комплексного анализа, Сообщ. АН Груз. ССР, 107 (1982), 481-484.

[14] ——: Решение задачи Коши в классах функций, являющихся пространственными обобщениями обобщенных аналитических функций, in preparation.

[15] —— and C. WITHALM: The Cauchy-Kovalevska theorem for pseudo-holomorphic functions in the sense of L. Bers. Complex variables: Theory and applications, in print.

[16] VEKUA, I. N.: Verallgemeinerte analytische Funktionen [trans. from Russian], Berlin 1963.

Sektion-Mathematik der Martin-
Luther-Universität Halle-Wittenberg
Universitätsplatz 8/9
DDR-4020 Halle an der Saale, DDR

NONLINEAR QUASICONFORMAL GLUE THEOREMS

Wen Guo-chun (Peking)

Contents

Summary

Let $\Gamma_0, \Gamma_1, \ldots, \Gamma_N, L_1, \ldots, L_M$ be the boundary contours of an $(N+M+1)$-connected plane domain D, where $\Gamma_1, \ldots, \Gamma_N, L_1, \ldots, L_M$ are situated inside Γ_0. In D there are some mutually exclusive contours $\gamma_1, \ldots, \gamma_n, l_1, \ldots, l_m$. We assume that

$$(0.1)\quad \Gamma = \bigcup_{j=0}^{N} \Gamma_j,\quad L = \bigcup_{j=1}^{M} L_j,\quad l = \bigcup_{j=1}^{m} l_j,\quad \gamma = \bigcup_{j=1}^{n} \gamma_j \in C_\mu \quad (0<\mu<1)$$

and denote

$$(0.2)\quad D_\gamma^- = \bigcup_{j=1}^{n} D_{\gamma_j},\quad D_l^- = \bigcup_{j=1}^{m} D_{l_j},\quad D^- = (D_\gamma^- \cup D_l^-) \cap D,\quad D^+ = D \setminus \overline{D^-},$$

where D_{γ_j} and D_{l_j} are the domains surrounded by γ_j and l_j, respectively. We deal with the nonlinear uniformly elliptic complex equation of the first order $w_{\bar{z}} = F(z, w, w_z)$, $F = Q(z, w, w_z)w_z$, $z \in D$. We suppose that it satisfies the condition C: 1) $Q(z,w,s)$ is continuous in $w \in \mathbb{E}$ (the whole plane) for almost every point $z \in D$ and $s \in \mathbb{E}$, and is measurable in $z \in D$ for all continuous functions $w(z)$ and all measurable functions $s(z)$ in $D^+ \setminus \{z_0\}$, $z_0 \in D^+$; 2) the equation satisfies the uniformly elliptic condition. We prove then that the equation has a homeomorphic solution $w(z)$ which maps quasiconformally D^+ and D^- onto G^+ and G^-, respectively, with $w(z_0) = \infty$, $z_0 \in D^+$, and satisfies the gluing conditions

$$(0.3)\quad \begin{aligned} &w^+[\alpha(t)] = w^-(t),\ t\in\gamma;\quad w^+[\alpha(t)] = \overline{w^-(t)},\ t\in l;\\ &w^+[\alpha(t)] = w^+(t),\ t\in L;\quad w^+[\alpha(t)] = \overline{w^+(t)} + i\,h(t),\ t\in\Gamma, \end{aligned}$$

where $\alpha(t)$ maps each of γ_j, l_j, L_j, and Γ_j topologically onto itself; they give positive shifts on $\gamma\cup\Gamma$ and reverse shifts on $l\cup L$, etc.

1. Introduction

In this paper, we consider the quasiconformal glue theorems for the nonlinear uniformly elliptic complex equation of the first order

$$(1.1)\quad w_{\bar z} = F(z, w, w_z),\quad F = Q(z, w, w_z)w_z$$

in a multiply connected domain. Without loss of generality, we suppose that D is an $(N+M+1)$-connected domain with the boundary contours $\Gamma_0, \Gamma_1, \dots, \Gamma_N, L_1, \dots, L_M$, where $\Gamma_1, \dots, \Gamma_N, L_1, \dots, L_M$ are situated inside Γ_0. In D there are some mutually exclusive contours $\gamma_1, \dots, \gamma_n, l_1, \dots, l_m$. We assume the relations (0.1) and accept the notation (0.2), where D_{γ_j} and D_{l_j} are the domains surrounded by γ_j and l_j, respectively. We suppose that Eq. (1.1) satisfies the condition C:

1) $Q(z,w,s)$ is continuous in $w\in\mathbb{E}$ (the whole plane) for almost every point $z\in D$ and $s\in\mathbb{E}$, and is measurable in $z\in D$ for all continuous functions $w(z)$ and all measurable functions $s(z)$ in $D^+\setminus\{z_0\}$, where $z_0\in D^+$;

2) Eq. (1.1) satisfies the uniformly elliptic condition, i.e.

$$(1.2)\quad |F(z,w,s_1) - F(z,w,s_2)| \le q|s_1 - s_2|,\quad 0\le q<1,$$

for almost every point $z\in D$ and $w, s_1, s_2\in\mathbb{E}$.

The general quasiconformal glue problems for the nonlinear complex equation (1.1) in $D^\pm$ require assuming the continuity of a solution $w(z)$ in $\overline{D^\pm}\setminus\{z_0\}$ $(w(z_0)=\infty)$, which also has to satisfy the glue conditions (0.3), where $\alpha(t)$ maps each of γ_j, l_j, L_j, and Γ_j topologically onto itself. They give positive shifts on $\gamma\cup\Gamma$ and reverse shifts on $l\cup L$, where $\alpha[\alpha(t)] = t$ for $t\subset L\cup\Gamma$, $\alpha(t)$ has the fixed points $a_j\in\Gamma_j$, $j = 0, 1, \dots, N$;

$$(1.3)\quad C^1_\mu[\alpha(t), \partial D^\pm] \le d < +\infty,\quad |\alpha'(t)| \ge d^{-1} > 0,$$

and $h(t) = 0$, $t\in\Gamma_0$; $h(t) = h_j$, $t\in\Gamma_j$, $j = 1,\dots,N$. Here h_j $(j = 1,\dots, N)$ are all unknown constants to be determined appropriately. The above problem will be called the problem Q.

By using the theorem of conformal gluing, the method of elimination, a priori estimates of the homeomorphic solutions for Eq. (1.1), and the Leray-Schauder theorem, we are going to obtain the following main theorem:

THEOREM 1.1. Let Eq. (1.1) satisfy the condition C. Then there exists a homeomorphic solution $w(z) = w_1(z)$, $z \in D^+$, $w(z) = w_2(z)$, $z \in D^-$, of the problem Q for Eq. (1.1), which maps quasiconformally D^+, $D^- \cap D_{\gamma_j}$, and $D^- \cap D_{1_j}$ onto the domains G^+, G_{γ_j} $(j = 1,\dots,n)$, and G_{1_j} $(j = 1,\dots,m)$, respectively, so that $w(z_0) = \infty$.

Besides, we can also prove the quasiconformal glue theorems with each condition of (0.3) for the nonlinear complex equation (1.1).

2. The general conformal glue theorem for univalent meromorphic functions

In this section we discuss the problem Q for Eq. (1.1) with $F(z, w, w_z) = 0$, i.e. for univalent meromorphic functions.

THEOREM 2.1. The problem Q for univalent meromorphic functions has a homeomorphic solution.

Proof. In analogy to Refs. [1]-[5], we apply the conformal glue function $\zeta(z)$ with the first three glue conditions in (0.3) and the method of elimination. Thus the glue conditions (0.3) of the problem Q for the univalent meromorphic function $\Phi(z)$ can be transformed into the following boundary conditions of the boundary value problem R for meromorphic functions:

$$(2.1)\quad \Psi^+[\beta(\zeta)] = G(\zeta)\overline{\Psi^+(\zeta)} + i\,H(\zeta),\quad \zeta \in \zeta(\Gamma),$$

in which $\beta(\zeta)$ has similar properties to $\alpha(t)$, $G(\zeta) = \overline{X(\zeta)}/X[\beta(\zeta)]$,

$$H(\zeta)X[\beta(\zeta)] = 0,\ \zeta \in \zeta(\Gamma_0);\quad H(\zeta)X[\beta(\zeta)] = h_j,\ \zeta \in \zeta(\Gamma_j),\ j = 1,\dots,N,$$

and where $X(\zeta)$ is the typical solution for the method of elimination, $X(\zeta) \neq 0$ for $\zeta \in \zeta(\Gamma)$, and $G(\zeta) \in C_\nu[\zeta(\Gamma)]$, $0 < \nu < 1$. Then we can find an analytic function $\phi(\zeta)$ which is a solution of the boundary value problem S with the boundary condition

$$(2.2)\quad \phi^+[\beta(\zeta)] = G(\zeta)\overline{\phi^+(\zeta)} + g(\zeta) + i\,H(\zeta),\quad \zeta \in \zeta(\Gamma),$$

where $g(\zeta) = G(\zeta)/\bar{\zeta} - 1/\beta(\zeta)$. Hence the meromorphic function $\Psi(\zeta) = \phi(\zeta) + 1/\zeta$ is a solution of the above problem R, and so

$$(2.3)\quad w = \Phi(z) = X[\zeta(z)]\,\Psi[\zeta(z)]$$

is a sectionally meromorphic function which has a pole of the first order at $z = z_0$ and satisfies the glue conditions (0.3). Moreover, it is clear that the function

$$(2.4)\quad \tilde{\Phi}(z) = \begin{cases} X[\zeta(z)]\,\Psi[\zeta(z)], & z \in (D^+ \cup D^-) \setminus D_1^-, \\ \overline{X[\zeta(z)]\,\Psi[\zeta(z)]}, & z \in D_1^- \cap D^-, \end{cases}$$

maps the boundary L and Γ onto some Jordan arcs $\tilde{L}$ and some parallel slits $\tilde{\Gamma}$, respectively, as well as γ, l onto some closed Jordan curves $\tilde{\gamma}$, $\tilde{l}$ in the w-plane. By using the method like that in Refs. [2] and [1], we may verify the univalence of $\Phi(z)$ in D^-. In order to prove that Φ is univalent in D^+, we choose arbitrarily $w_1 \in \tilde{L} \cup \tilde{\Gamma} \cup \tilde{\Phi}(D^-)$ and $w_1 \neq \infty$. It follows from the argument principle that

$$(2.5)\quad \Delta_{\partial D^+} \arg[\Phi(t) - w_1] = 0.$$

Hence, there exists a point $z_1 \in D$ such that $\Phi(z_1) = w_1$. In addition, it is not difficult to see that the image of $\Phi(z)$ in D^+ cannot contain any point on $\tilde{L} \cup \tilde{\Gamma} \cup \tilde{\Phi}(D^-)$. This completes the proof.

3. The quasiconformal glue theorem for Eq. $w_{\bar{z}} = F(z,w,w_z)$, $F = Q(z,w,w_z)w_z$

In the following, firstly we give a priori estimates for solutions of the problem Q for Eq. (1.1). Afterwards, on the basis of the above result and the Leray - Schauder fixed-point theorem, we derive the solvability of the problem Q for Eq. (1.1).

THEOREM 3.1. *Suppose that the nonlinear complex equation* (1.1) *satisfies the condition* C. *Then the homeomorphic solution of the problem* Q *for* Eq. (1.1) *with the conditions*

$$(3.1)\quad w(z_0) = \infty,\ w(z_1) = 0,\ z_1\ (\neq z_0) \in D^+,\ |w(a_0)| = 1$$

satisfies the estimates

$$(3.2)\quad C_\beta[W(z), \overline{D^\pm}] \le M_1,\quad \| |W_{\bar{z}}| + |W_z| \|_{L_{p_0}(\overline{D^\pm})} \le M_2,$$

in which $W(z) = w(z)\zeta(z)$, $\zeta(z)$ *is a homeomorphic solution of the corresponding Beltrami equation*, $\zeta(z_0) = z_0$, $\zeta(a_0) = a_0$, $\beta = 1 - 2/p_0$, *where* p_0 (>2) *is an appropriate real constant, and* $M_j = M_j(q, p, D^\pm, \alpha(t))$, $j = 1, 2$.

Proof. Let us substitute the stated solution of the problem Q for $w(z)$ in Eq. (1.1) and the glue conditions (0.3). Thus the constants h_j $(j=1,\dots,N)$ are determined. By using the method as in Refs. [4] and [5], we obtain

$$(3.3)\quad |h_j| \le M_3 = M_3(q, p, D^{\pm}, \alpha(t)),\ j=1,\dots,N,$$

and can derive from it that the solution $w(z)$ satisfies the estimate

$$(3.4)\quad C[W(z), D^{\pm}] \le M_4 = M_4(q, p, D^{\pm}, \alpha(t)),$$

where $W(z) = w(z)\zeta(z)$ and $\zeta(z)$ is stated as above. Afterwards, we can prove that the estimate (3.1) holds.

<u>Proof of Theorem 1.1</u>. Firstly, we assume for Eq. (1.1) that the function $F(z, w, w_z)$ vanishes in a neighbourhood $|z| < \varepsilon_n$, where $\varepsilon_n \to 0$ as $n \to \infty$. We write such a complex equation in the form

$$(3.5)\quad w_{\bar z} = F_n(z, w, w_z),\ F_n = Q_n(z, w, w_z)w_z.$$

Now, let $w_n(z)$ be a solution of the problem Q for Eq. (3.5). Then $w_n(z)$ can be expressed as follows:

$$(3.6)\quad w_n(z) = \Phi_n(z) + \Psi_n(z),\ \Psi_n(z) = T\omega_n,$$

where $\omega_n(z) \in L_p(\overline{D^{\pm}})$, $p>2$, and $\Phi_n(z)$ is a meromorphic function which has a pole of the first order at $z=z_0$. According to Theorem 3.1, it is easy to see that $\Phi_n(z)$, $\Psi_n(z)$, $\omega_n(z)$, and $w_n(z)$ satisfy the estimates

$$(3.7)\quad C_\beta[\Phi_n(z), \overline{D^{\pm}_*}] \le M_5,\ C_\beta[\Psi_n(z), \overline{D^{\pm}_*}] \le M_6;$$

$$(3.8)\quad \|\omega_n\|_{L_p(\overline{D^{\pm}_*})} = \|w_{\bar z}\|_{L_p(\overline{D^{\pm}_*})} < M_7,\ \|w_z\|_{L_p(\overline{D^{\pm}_*})} \le M_8,$$

where $D^{\pm}_* = D^{\pm} \setminus \{|z| \le \varepsilon_n\}$, $M_j = M_j(q, p, D^{\pm}, \alpha(t), \varepsilon_n)$, $j=5,\dots,8$.

Secondly, by using the Leray - Schauder theorem, we can prove that the problem Q for Eq. (3.5) has a solution $w_n(z)$ which satisfies the conditions (3.1) and has the form (3.6). Similarly to the proof of Theorem 3.1, we learn that the above solution is univalent in $D^{\pm}$ and so it is a homeomorphic solution of the problem Q for Eq. (3.5).

Finally, we observe that the solution $w_n(z)$ in question satisfies the estimate (3.2) and that $w_n(z)$ can be written in the form

$$(3.9)\quad w_n(z) = \Phi_n[\zeta_n(z)],\ \zeta_n(z_0) = z_0,\ \zeta_n(a_0) = a_0.$$

Thus we apply the principle of compactness for a sequence of solutions

$\{w_n(z)\}$ of the complex equations (3.5) $(n = 1, 2, \dots)$ and select a subsequence of $\{w_n(z)\}$, which uniformly converges to the solution $w_o(z)$ of Eq. (1.1) in any compact subset of $D^{\pm} \setminus \{0\}$; $w_o(z)$ is a homeomorphic solution of the problem Q for Eq. (1.1), as desired.

References

[1] HUANG Hai-yang: On compound problems of analytic functions with shift, to appear.

[2] LITVENČUK, G. S.: Boundary value problems and integral equations with shift [in Russian], Izdat. "Nauka" Fiz.-Mat. Lit., Moscow 1977.

[3] LU Chien-ke: On compound boundary problems, Scientia Sinica 14 (1965), 1545-1555.

[4] WEN Guo-chun: The singular case of Riemann - Hilbert boundary value problems, Acta Sci. Natur. Univ. Pekin. 1981, No. 4, 1-14.

[5] ——: On linear and nonlinear compound boundary value problems with shift, ibid. 1982, No. 2, 1-12.

Institute of Mathematics
Peking University, Beijing
People's Republic of China

PROBLEMS IN THE THEORY OF FUNCTIONS OF ONE COMPLEX VARIABLE

collected by Olli Tammi (Helsinki) and prepared by Julian Ławrynowicz (Łódź)

Here, a few open problems in the theory of functions of one complex variable are collected. The problems were posed by the authors at the problem session on August 23. They seem to be either original or less well known to the experts, but are included here since they might merit attention.

SUBORDINATION

Jan Stankiewicz
Institute of Mathematics and Physics, Rzeszów Technical High School
Poznańska 2, PL-35-084 Rzeszów, Poland

PROBLEM 1. For φ and ϕ convex, $f \prec \varphi$ and $g \prec \phi$, is it true that $f * g \prec \varphi * \phi$? (The answer is positive if we suppose that f or g is convex.)

PROBLEM 2. It is known that if $f(z) + \varepsilon z^n \in St$ for every ε, $|\varepsilon| < \delta$, then $[f(z) + \varepsilon z]/(1 + \varepsilon) \in St$. If $[f(z) + \varepsilon z]/(1 + \varepsilon) \in St$ for every ε, $|\varepsilon| < \delta$, then $f(z) + \varepsilon z^n \in St$ for $|\varepsilon| < \delta/n$. Is it true that if $f(z) + \varepsilon z^n \in St$ for $|\varepsilon| < \delta/n$, then $[f(z) + \varepsilon z]/(1 + \varepsilon) \in St$ for $|\varepsilon| < \delta/n$?

PROBLEM 3. It is known that if $Bf = \int_0^z \zeta^{-1} f(\zeta)d\zeta$, where $f \prec F$ and F is convex, then $Bf \prec BF$ and $(B + \alpha B^4)f \prec (B + \alpha B^4)F$ for some α. It would be interesting to find some α such that if $f \prec F$ and F is convex, then $(B + \alpha B^k)f \prec (B + \alpha B^k)F$ for $k = 2, 3$.

THE CLASS S(b)

Jan Szynal
Institute of Mathematics, Maria Curie-Skłodowska University
Plac Marii Curie-Skłodowskiej 1, PL-20-031 Lublin, Poland

Let $S(b)$, $0 < b \leq 1$, be the class of holomorphic and univalent functions in the unit disk $|z| < 1$ which have the form $f(z) = b(z + a_2 z^2 + \dots)$ and satisfy the condition $|f(z)| < 1$ for $|z| < 1$.

For arbitrary real (or complex) numbers μ and λ find

$$(*) \qquad \max\{\mathrm{Re}[(a_4 - 2a_2a_3 + a_2^3) + \mu(a_3 - a_2^2) + \lambda a_2] : f \in S(b)\}.$$

The exact value of (*) should lead to the sharp bound for $|a_4|$ in $S(b)$.

THE CLASS $S_R(b)$

Olli Tammi

Department of Mathematics, University of Helsinki
Hallituskatu 15, SF-00100 Helsinki 10, Finland

Please, try to maximize $a_4 - 2a_2a_3 + a_2^3 - b^2a_2 + 2\lambda(a_3 - a_2^2 + 1 - b^2)$, $\lambda \in \mathbb{R}$, in $S_R(b)$ by aid of the variational method, in order to confirm the inequality of O. Jokinen [Ann. Acad. Sci. Fenn. Ser. A I Math., Dissertationes 41 (1982), 52 pp.].

PROBLEMS IN THE THEORY OF QUASICONFORMAL MAPPINGS

collected by Makoto Ohtsuka (Tokyo) and prepared by Julian Ławrynowicz (Łódź)

Here, a few open problems in the theory of quasiconformal mappings are collected. The problems were posed by the authors at the problem session on August 25. They seem to be either original or less well known to the experts, but are included here since they might merit attention.

CONFORMAL MAPPINGS WITH QUASICONFORMAL EXTENSION

Jan G. Krzyż
Institute of Mathematics, Maria Curie-Skłodowska University
Plac Marii Curie-Skłodowskiej 1, PL-20-031 Lublin, Poland

Let f be a function close-to-convex in the unit disk $\mathbb{D}$, such that a convex univalent function g and a function p of positive real part in $\mathbb{D}$ are associated with p by the relation: $f' = p\,g'$. The following sufficient condition for f to have a qc (= quasiconformal) extension to $\mathbb{C}$ was given by J. G. Krzyż and A. K. Soni [these Proceedings, pp. 320-327]:

(i) $g(\mathbb{D})$ is a bounded convex domain and

(ii) the closure of $p(\mathbb{D})$ is a compact subset of the open right half-plane.

PROBLEM 1. Prove (or disprove) the following necessary condition for f to have a qc extension to $\mathbb{C}$: there exists a decomposition $f' = p\,g'$ such that (i) and (ii) hold.

PROBLEM 2. Give a characterization of Bazilevič functions admitting a qc extension to $\mathbb{C}$. [For the definition of Bazilevič functions which are more general than close-to-convex functions see p. 130 of these Proceedings].

PROBLEM 3. Find the construction of the qc extension for Bazilevič functions admitting qc extension.

LIMITING PROPERTY OF EXTREMAL LENGTH

Makoto Ohtsuka

Department of Mathematics, Faculty of Science, Gakushuin University
1-5-1 Mejiro, Toshimu-ku, J-171 Tokyo, Japan

Let $p>1$. Let G be an open set, and K_0, K_1 be mutually disjoint compact sets in $\mathbb{R}^n$. We call a Borel measurable function $\rho \geq 0$ <u>admissible</u> if $\int_\gamma \rho\, ds \geq 1$ for every rectifiable curve γ connecting K_0 and K_1 in G. We define

$$1/\lambda_p(K_0, K_1; G) = \inf\{\int \rho^p dx;\ \rho \text{ admissible}\}.$$

Let $\{K_0^{(k)}\}$ and $\{K_1^{(k)}\}$ be sequences of compact sets decreasing to K_0 and K_1, respectively, and define $\lambda_p(K_0^{(k)}, K_1^{(k)}; G)$ as above.

<u>PROBLEM</u>. Is it always true that $\lim_{k\to\infty} \lambda_p(K_0^{(k)}, K_1^{(k)}; G) = \lambda_p(K_0, K_1; G)$?

CAPACITY AND GEOMETRY OF RIEMANNIAN MANIFOLDS

Antoni Pierzchalski

Institute of Mathematics of the Polish Academy of Sciences
Łódź Branch, Kilińskiego 86, PL-90-012 Łódź, Poland

It is well known that the geometry of a Riemannian manifold M of dimension n can be determined by the volume of small spherical balls: under some additional assumptions, if the volume of an arbitrary small ball on M agrees with the Euclidean volume of a ball of the same radius in $\mathbb{R}^n$, then M is flat. Can the geometry of M be determined by a capacity of rings? More precisely, let $R(a,b)$ be a small spherical ring with radii a and b on M and let C_a and C_b be its boundary components. Define the capacity of $\mathbb{R}$ as follows:

$$\operatorname{cap} \mathbb{R}(a,b) = \inf_u \int_M |\operatorname{grad} u|^2,$$

where the infimum is taken over all C^1-functions u on M such that $u|C_a = 0$ and $u|C_b = 1$. Denote by $c(a,b)$ the capacity of $R(a,b)$ in $\mathbb{R}^n$. What can we say about the geometry of M if $\operatorname{cap} R(a,b) = c(a,b)$ for all (small) rings $R(a,b)$ on M? Has such a manifold to be flat?

PROBLEMS IN THE THEORY OF FUNCTIONS OF SEVERAL COMPLEX VARIABLES AND IN INFINITE-DIMENSIONAL COMPLEX ANALYSIS

collected and prepared by Christer O. Kiselman (Uppsala)

At a problem session on August 24 during the 8th Conference on Analytic Functions at Błażejewko, fourteen speakers presented open problems in the theory of analytic functions of several complex variables and in infinite-dimensional complex analysis. These problems (as well as a few others submitted during the conference) are described below. In many cases, the problems are original, i.e. discovered and posed by the authors. Other problems are more or less well known to experts in the field, but are included here since they seemed to merit attention.

The texts were prepared at Błażejewko without access to a library, and this fact explains that references are sometimes lacking. However, anybody who wants to work on a particular problem should be able to track the references and is in any event encouraged to contact the authors for further information.

INTERPOLATION

Bo Berndtsson
CTH, Dept. of Mathematics, S-412 96 Göteborg, Sweden

Let f be an entire function in $\mathbb{C}^n$ of exponential type, i.e. $|f(z)| \leq C \exp(A|z|)$ for some constants C, A. Denote by V the set $\{z \in \mathbb{C}^n; f(z) = 0\}$ and assume $\partial f \neq 0$ on V. When does the following property hold:

(*) Any holomorphic function of exponential type on V can be extended to an entire function of exponential type.

When $n = 1$ the following estimate is necessary and sufficient:

(**) $\exists\ \varepsilon, C \quad |\partial f(z)| \geq \varepsilon \exp(-C|z|)$ on V.

When $n > 1$ there are known several sufficient conditions, but very little is known about necessary ones. (**) is sufficient but not necessary when $n > 1$. Sufficient conditions are given in works of Berenstein-Taylor, Demailly, Jarnicki, Nishimura, Yoshioka and myself.

EMBEDDING RIEMANN SURFACES IN $\mathbb{C}^2$

P.M. Gauthier, Université de Montréal, Mathématiques, Montréal H3C 3J7, Canada

Since any open Riemann surface is Stein, it can be properly embedded in $\mathbb{C}^3$. Nishino showed that the unit disc can be properly embedded in $\mathbb{C}^2$. Laufer showed that an annulus can be properly embedded in $\mathbb{C}^2$, and Alexander showed the same is true of the punctured disc. R. Narasimhan discussed Nishino's proof in his lecture at the international conference held in Paris in 1975 in honor of H. Cartan. Narasimhan claimed that Nishino's proof works for finite Riemann surfaces. Kito has also considered this question.

PROBLEM. Can every open Riemann surface be properly embedded into $\mathbb{C}^2$?

UNIFORM APPROXIMATION ON UNBOUNDED SETS

P.M. Gauthier, address as above

Carleman's theorem (1927) states that for every pair of continuous functions f and ε on the real line $\mathbb{R}$, with ε positive, there is an entire function g on $\mathbb{C}$ with

$$|f(x) - g(x)| < \varepsilon(x), \quad x \in \mathbb{R}.$$

PROBLEM. To what extent can this result be extended and generalized in $\mathbb{C}^n$?

REFERENCES

1) Alexander, Math. Scand. 45 (1979).
2) Aupetit, in Complex Approximation, ed. B. Aupetit.
3) Droste, Math. Ann. 257 (1981).
4) Kasten and Schneider, Math. Z. 178 (1981).
5) Nunemacher, Math. Ann. 224 (1976).
6) Nunemacher, in Proc. Symp. Pure Math. 30.
7) Sakai, Math. Ann. 253 (1980), no. 2.
8) Scheinberg, J. Analyse Math. 29 (1976).
9) Stout, in Complex Analysis, ed. B. Aupetit.

QUESTIONS CONCERNING THE CARATHÉODORY AND KOBAYASHI PSEUDODISTANCES

Valentin Z. Hristov

Institute of Mathematics, Bulgarian Academy of Sciences
P.O. Box 373, BG-1090 Sofia, Bulgaria

Let X be a complex (analytic) space and let c_X and k_X be the Carathéodory and the Kobayashi pseudodistances on X respectively.

QUESTION 1. Are there any examples of spaces X for which c_X is locally nondegenerate (in the sense that every point $p \in X$ has a neighbourhood U_p on which $c_{X|U_p}$ is an actual distance, i.e. $p \neq q \in U_p \Rightarrow c_X(p,q) > 0$) but in the whole X there are different points at zero distance ?

This question is related to the new definition of Carathéodory-hyperbolicity given by Kobayashi in his survey in Bull. Amer. Math. Soc. 82 (1976)(p. 367). More generally we can ask the following

QUESTION 2. Are there any new examples of spaces X on which c_X or k_X is a typical pseudodistance - not trivial and not an actual distance ?

Now, using any pseudodistance d on X one can define an equivalence relation R by

$$p \overset{R}{\sim} q \quad \text{iff} \quad d(p,q) = 0 .$$

Denote by $R(p)$ the equivalence class of any point $p \in X$. If d is c_X or k_X, R will be denoted by R^c or R^k respectively.

It is known that $R^c(p)$ are analytic subsets of X defined by the family of all bounded holomorphic functions (C.R. Acad. Bulg. Sci. 30 (1977) p. 643), but the $R^k(p)$ are in general not analytic if X is compact.

QUESTION 3. If X is non-compact, are the sets $R^k(p)$ analytic subsets of X ?

A well-known theorem of Cartan gives necessary and sufficient conditions for the existence of a complex structure on X/R if R is a proper equivalence relation (by definition the natural projection $\pi: X \to X/R$ is a proper mapping). We proved that R^k is proper if X is a compact space (to appear in Proc. Intern. Conf. on Complex Analysis and Appl. - Varna 1981) and Cartan's theorem is applicable. There are generalizations of Cartan's result due to Wiegmann and Furushima concerning semi-proper equivalence relations only (by definition for every compact subset K of X/R there is a compact subset L of X with $\pi(L) = K$) . In order to use these generalizations for non-compact spaces X we ask the following

QUESTION 4. If X is non-compact, is the relation R^k semi-proper ?

More generally, we state

QUESTION 5. What kind of topological properties do the relations R^c and R^k have and under which conditions ?

For example, one can consider here not only the property of R^c or R^k to be proper or semi-proper but also to be open (π is an open mapping), or closed (analogously), etc. Let us note that if R is open, it is also semi-proper.

Finally, there is another question related to the coincidence of the original topology on X and the topology induced by the Carathéodory distance c_X in C.R. Acad. Bulg. Sci. 33 (1980), p. 1602.

THE EXISTENCE OF DECOMPOSITION OPERATORS IN SPACES OF HOLOMORPHIC FUNCTIONS IN STRICTLY PSEUDOCONVEX DOMAINS

Piotr Jakóbczak, Institute of Mathematics, Jagiellonian University
Reymonta 4, PL-30-059 Kraków, Poland

1. Let D be a strictly pseudoconvex domain in $\mathbb{C}^n$ (with smooth boundary), and suppose that there exist: a neighbourhood $\tilde{D}$ of $\overline{D}$, an integer $m \geq n$, a mapping $\psi: \tilde{D} \to \mathbb{C}^m$ which maps $\tilde{D}$ biholomorphically onto a closed complex submanifold $\psi(\tilde{D})$ of $\mathbb{C}^m$, and a strictly convex domain $C \subset \mathbb{C}^m$ (with smooth boundary) such that $\psi(D) \subset C$, $\psi(\tilde{D} \setminus \overline{D}) \subset \mathbb{C}^m \setminus \overline{C}$, and $\psi(\tilde{D})$ intersects ∂C transversally. (Given a strictly pseudoconvex domain D, such m, $\tilde{D}$, ψ and C always exist, in virtue of the Fornaess embedding theorem). Denote by $A_d(D \times D)$ the space of all functions which are holomorphic in $D \times D$ and continuous in the set $(\overline{D} \times \overline{D}) \setminus \{(z,z);\ z \in \partial D\}$, and let $(A_d)_0\ (D \times D)$ be the space of all functions in $A_d(D \times D)$ which vanish on the diagonal of $D \times D$.

PROBLEM. Show that there exists an operator

$$R: (A_d)_0(\psi(D) \times \psi(D)) \longrightarrow (A_d)_0(C \times C),$$

such that for every $f \in (A_d)_0(\psi(D) \times \psi(D))$, $Rf|\psi(D) \times \psi(D) \equiv f$.

(Such a result would allow us to prove the following theorem: If D is a strictly pseudoconvex domain in $\mathbb{C}^n$ (with smooth boundary), then for every $f \in (A_d)_0(D \times D)$ there exist functions $f_1(z,s), \ldots, f_n(z,s) \in A_d(D \times D)$, such that

$$f(z,s) = \sum_{i=1}^{n} (z_i - s_i) f_i(z,s).$$

This is true if we assume that D is a strictly convex domain in $\mathbb{C}^n$.)

2. Let D be a strictly pseudoconvex domain in $\mathbb{C}^n$ (with smooth boundary). Denote by $A(D \times D)$ the space of all functions which are holomorphic in $D \times D$ and continuous in $\overline{D} \times \overline{D}$, and let $A_0(D \times D)$ be the space of all functions in $A(D \times D)$ which vanish on the diagonal of $D \times D$. Let $g_1(z,s), \ldots, g_N(z,s) \in A_0(D \times D)$ be such that $\{(z,z);\ z \in \overline{D}\} = \{(z,s) \in \overline{D} \times \overline{D};\ g_1(z,s) = \ldots = g_N(z,s) = 0\}$, and for each $z \in D$ the germs of the functions $g_1, \ldots, g_N$ at (z,z) generate the ideal of germs at (z,z) of holomorphic functions, which vanish on the diagonal of $D \times D$.

PROBLEM. Prove that for every $f \in A_0(D \times D)$ there exist functions $f_1, \ldots, f_N$ which are holomorphic in $D \times D$ and continuous in the set $(\overline{D} \times \overline{D}) \setminus \{(z,z);\ z \in \partial D\}$ such that

$$f(z,s) = \Sigma_{i=1}^{N} g_i(z,s) f_i(z,s) .$$

(A result of this kind was proved in the case when D is the unit disk in $\mathbb{C}$, with an additional assumption on the functions $g_1, \ldots, g_N$. A related result was proved in N. Øvrelid, Generators of the maximal ideals of $A(D)$, Pacific J. Math. 39 (1971), 219-223).

AUTOMORPHISMS WITH EXPONENTIAL GROWTH

Marek Jarnicki, Institute of Mathematics
Jagiellonian University, Reymonta 4, PL-30-059 Kraków, Poland

Let $F: \mathbb{C}^n \to \mathbb{C}^n$ be a biholomorphic mapping (onto $\mathbb{C}^n$) such that:

(i) for some $\mu, A, B > 0$: $\|F(x)\| \leq A \exp(B\|x\|^{\mu})$, $x \in \mathbb{C}^n$,

(ii) $\det(d_x F) \equiv 1$, $x \in \mathbb{C}^n$.

Do there exist constants $A', B' > 0$ such that:

$$\|F^{-1}(x)\| \leq A' \exp(B'\|x\|^{\mu}), \quad x \in \mathbb{C}^n \ ?$$

THE MODULUS OF CONTINUITY OF ANALYTIC FUNCTIONS IN A DOMAIN AND ON ITS ŠILOV BOUNDARY

Burglind Jöricke, Institut der Mathematik
Akademie der Wissenschaften der DDR, Mohrenstraße 39, DDR-1080 Berlin, DDR

A well-known theorem, which was proved by Tamrazov and independently by three American authors, Rubel, Shields and Taylor, says the following: If $G \subset \mathbb{C}$ is a domain in the complex plane which is not very bad and f

is a function analytic in G and continuous in $\overline{G}$ such that the modulus of continuity of the restriction of f to the boundary ∂G is not greater than a given nonnegative nondecreasing semiadditive function ω, then the modulus of continuity of the function in the whole domain is not greater than $C\omega$ for some constant C not depending on f. In the case of several complex variables we pose the following problem.

PROBLEM. Suppose that $G \subset \mathbb{C}^n$ is not very bad (for example has piecewise smooth boundary). For which subsets $S \subset \partial G$ is the following true: If f is analytic in G and continuous in $\overline{G}$ and the restriction $f|S$ has a modulus of continuity not greater than ω then the function f in the whole domain has a modulus of continuity not greater than $C\cdot\omega$?

For nice domains we can take $S = \partial G$ (Ščehorskiĭ, preprint). One could expect that it is possible to take $S = \Delta(G)$, where $\Delta(G)$ is the Šilov boundary of the algebra $A(G)$ of functions analytic in G and continuous in $\overline{G}$, but this is not true in general (see B. Jöricke, "Comparison of the modulus of continuity of an analytic function in a domain in $\mathbb{C}^n$ and on the Šilov boundary", to appear). For strongly pseudoconvex domains it is, of course, true, and so it is for regular Weil domains (ibid.). For arbitrary domains we have the following

CONJECTURE. Let $G \subset \mathbb{C}^n$ have a smooth boundary and let $U \subset \partial G$ be a relatively open subset of the boundary which contains the Šilov boundary $\Delta(G)$. Then every function $f \in A(G)$ with $f|U$ having a modulus of continuity not greater than ω has a modulus of continuity not greater than $C\omega$.

Jerzy Kalina's contribution appears on page 16.

ESCEPTAROJ EN NEFINIE DIMENSIA KOMPLEKSA ANALITIKO

Christer O. Kiselman
Upsala universitato, matematika instituto
Thunbergsvägen 3, S-752 38 Uppsala, Sweden

La polusaj aroj en $\mathbb{C}^n$, t.e. la aroj kie plursubharmona funkcio kiu ne identas al $-\infty$ prenas tiun ĉi valoron, aperas kiel esceptaroj ĉe multaj problemoj en la finie dimensia kompleksa analitiko. En multaj kazoj estas ankaŭ konate, ke ili konsistigas la ĝustan klason, t.e. aro apereblas kiel esceptaro ĉe iu eco se kaj nur se ĝi estas polusa. En nefiniaj dimensioj, aliaj klasoj da esceptaroj, kiel Lebesgue-aj neniomaj aroj kaj aroj kun Newton-a kapacito nulo, ne havas facile difineblajn analogojn, dum polusaj aroj estas same facile difineblaj kiel en finiaj dimensioj. Kontraste al malgrasaj aroj, la polusaj aroj respektas la kompleksan strukturon de la spaco. Pro tiuj ĉi kaŭzoj la polusaj aroj ŝajnas almenaŭ tiom taŭgaj kiom la malgrasaj aroj kiel klaso da esceptaroj en la nefinie dimensia kompleksa analitiko, kaj multaj rezultoj apogantaj tiun ĉi pretendon estas konataj.

Ni diru ke holomorfa funkcio f en malferma aro Ω en $\mathbb{C}^n$ estas pluigebla, simbole $f \in P$, se ekzistas punkto z en Ω tia ke la Taylor-a serio de f en z havas konverĝan radiuson kiu superas la distancon de z al $\mathbb{C}^n \setminus \Omega$. Se Ω estas holomorfregiono, la pluigeblaj funkcioj devus esti esceptaj. Ekzemple estas konate (M. Downarowicz) ke se Ω estas balanca kaj havas iun kroman econ, tiam P estas malgrasa, eĉ rara, por iu ege forta topologio en $\mathcal{O}(\Omega)$. Estas ankaŭ konate (P. Lelong) ke por barita aro M en $\mathcal{O}(\Omega)$, $P \cap M$ estas polusa en $\mathcal{O}(\Omega)$. Sed la metodoj ŝajnas ne sufiĉi por pruvi ke P estas polusa en $\mathcal{O}(\Omega)$ kun ĝia kutima Freŝea topologio.

Por preni alian ekzemplon ni konsideru la kreskon de entjeraj funkcioj. Estas konate ke la entjeraj funkcioj kun ordo malpli ol pozitiva nombro ρ formas polusan aron en la Banach-a spaco E_ρ de ĉiuj entjeraj funkcioj plenumantaj

$$\|f\|_\rho = \sup_z |f(z)|e^{-|z|^\rho} < +\infty \ .$$

Sed ĉi tie, ankoraŭfoje, la metodoj ŝajne ne sufiĉas por pruvi ke la aro de ĉiuj funkcioj kun ordo malpli ol ρ estas polusa en la Freŝea spaco de ĉiuj funkcioj kun ordo ne pli ol ρ, topologiizita per la normoj $\|f\|_{\rho+1/j}$, $j = 1, 2, \ldots$.

IMBEDDING DOMAINS INTO MANIFOLDS

László Lempert, Department of Analysis I
ELTE TKK, Múzeum-korùt 6-8, H-1088, Budapest, Hungary

Several notions of complex analysis are related to "canonical" (or "natural") imbeddings of arbitrary (say, bounded) domains $D \subset \mathbb{C}^n$ into "canonical" infinite-dimensional manifolds.

Example 1. Let H denote the Hilbert space of square integrable holomorphic functions on D. Then the Bergman kernel function defines a holomorphic imbedding $D \to H$.

Example 2. Let $\mathbb{P}H$ denote the projectivization of the above space H, $\pi: H \to \mathbb{P}H$ being the canonical projection. Composing the imbedding of Example 1 with this π, an imbedding $b: D \to \mathbb{P}H$ is obtained. This b can then be used to define the Bergman metric on D. Namely, the pullback of the Fubini-Study metric on $\mathbb{P}H$ along b will be the Bergman metric on D.

Example 3. Another intrinsic metric, the Carathéodory metric, can also be defined by a pullback. It is easy to construct a manifold M, which is a kind of infinite-dimensional polydisc, and a "canonical" imbedding $D \to M$ in such a way that the Carathéodory metric of D is the pullback along this imbedding of a certain metric ρ on M. This metric ρ can be defined explicitly; it looks like the Carathéodory metric of a (finite-dimensional) polydisc.

So we have seen examples of "canonical" imbeddings into simple infinite-dimensional manifolds: into a linear space, into a projective space and into a "polydisc". Another simple manifold is the ball, say, in a Hilbert space.

PROBLEM 1. Is it possible to construct "canonical" holomorphic imbeddings of bounded domains into the unit ball of a Hilbert space?

As to further problems, we mention that the imbeddings of Examples 1, and 3, are even proper, if the domain D is strictly pseudoconvex. Restricting our attention to such domains only, in Problem 1 one may drop the condition of being "canonical" but impose properness on the imbedding. The question thus would be if every strictly pseudoconvex domain $D \subset \mathbb{C}^n$ can be imbedded holomorphically and properly into the unit ball of, say, ℓ^2. This turns out indeed to be the case, as was shown in my talk. However, variations of this question are open.

PROBLEM 2. If $D \subset \mathbb{C}^n$ is strictly pseudoconvex and C^∞-bounded, does there exist a C^∞-imbedding of $\overline{D}$ into the closed unit ball of ℓ^2, which takes ∂D into the unit sphere, and which is holomorphic inside D?

PROBLEM 3. Is it possible to imbed holomorphically and properly any strictly pseudoconvex domain in $\mathbb{C}^n$ into a finite-dimensional ball (in some $\mathbb{C}^m$) ?

PROBLEM 4. Is it possible to imbed holomorphically and properly any strictly pseudoconvex domain in ℓ^2 into the unit ball of the same space ?

CONDITIONS (L_0) ET (L^*)

Nguyen Thanh Van, UER MIG, Université Paul Sabatier, 118, route de Narbonne F-31062 Toulouse Cedex, France

1) (avec Ahmed Zeriahi). Soit E un compact L-régulier de $\mathbb{C}^n$, soit μ une mesure de Radon positive sur E. Nous avons démontré dans l'article "Familles de polynômes presque partout bornées" (à paraître au Bull. Sci. Math. en 1983) le résultat suivant:

Inégalités de Bernstein-Markov. Si le couple (E,μ) vérifie la condition (L^*), alors

$$(B-M) \quad \begin{cases} \text{Pour tout } \lambda > 1 \text{ et tout } q > 0\text{, il existe une constante} \\ M > 0 \text{ et un voisinage } U \text{ de } E \text{ tels que} \\ \|P\|_U \leq M \cdot \lambda^{d^0 P} \cdot \left(\int |P|^q d\mu\right)^{1/q} . \end{cases}$$

PROBLÈME. Est-ce que la propriété $(B-M)$ entraîne que (E,μ) vérifie (L^*) ?

Remarque. Dans le travail "Familles de polynômes ponctuellement bornées" (Ann. Pol. Math. 1975), N.T.V. a affirmé par erreur qu'un raisonnement de Martineau permettait de répondre oui. Ce raisonnement ne marche que pour une fonction M (des variables λ et q) ayant un "bon" comportement par rapport à la variable q.

2. Pour un ensemble arbitraire E dans $\mathbb{C}^n$, est-ce que la Condition (L_0) en un point $a \in \mathbb{C}^n$ entraîne la L-régularité en ce point?

Remarque. Dans le travail "Lemme de Hartogs dans $\mathbb{C}^n$" (à paraître) N.T.V. a montré que la réponse est oui si on ajoute l'hypothèse que $E \setminus X$ vérifie (L_0) en a pour tout X pluripolaire. On y a montré aussi que l'inverse est vrai.

HULLS OF HOLOMORPHY

Peter Pflug, Fachbereich Mathematik der Universität Osnabrück - Abt. Vechta (Driverstraße 22), Postfach 1379, D-2848 Vechta, BRD

THEOREM (Pflug). Assume $G \subset G'$ to be domains in $\mathbb{C}^n$ with G' being a domain of holomorphy and suppose every holomorphic function f on G with

$$\sup_{z \in G} |f(z)| [\min(1, \operatorname{dist}(z, \partial G)) \cdot (1 + |z|^2)^{-1/2}]^{4n} < \infty$$

can be analytically extended to G'. Then it holds that $H(G) = G'$, where $H(G)$ denotes the hull of holomorphy of G.

QUESTION. Is it possible to prove the same result with an exponent n instead of $4n$?

HYPOELLIPTICITY OF THE $\bar\partial$-PROBLEM

Peter Pflug, address as above

THEOREM (J.J. Kohn). If G is a bounded smooth pseudoconvex domain and if for a boundary point $z^0 \in \partial G$ there exists a neighbourhood U such that $U \cap \partial G$ is real-analytic and does not contain any q-dimensional analytic subvarieties, then the $\bar\partial$-Neumann problem is hypoelliptic for (p,q)-forms; that means: for any $\alpha \in L^2_{p,q}(G)$, $\bar\partial\alpha = 0$, there exists $\beta \in L^2_{p,q-1}(G)$, $\bar\partial\beta = \alpha$, and sing supp $\beta \cap U =$ sing supp $\alpha \cap U$.

THEOREM (Diederich-Pflug). Assume G to be a pseudoconvex domain with C^2-boundary, $z^0 \in \partial G$ and suppose there exists a q-dimensional analytic subvariety $M \subset \partial G$, $z^0 \in M$. Then there exists $\alpha \in L^2_{pq}(G)$, $\bar\partial\alpha = 0$, such that for any solution $\beta \in L^2_{p,q-1}(G)$, $\bar\partial\beta = \alpha$, one has: sing supp $\beta \supsetneq$ sing supp α.

Example (Catlin). There is a domain in $\mathbb{C}^3$ without analytic sets in the boundary but for which the $\bar\partial$-Neumann problem for $(0,1)$-forms is not hypoelliptic, G has C^∞-boundary and is pseudoconvex.

QUESTION. Does hypoellipticity at z^0 for (p,q)-forms imply the non-existence of analytic sets of dimension q through z^0 of infinite order of tangency ?

BOUNDARY BEHAVIOUR OF THE BERGMAN KERNEL FUNCTION

Peter Pflug, address as above

THEOREM (Diederich). Assume G to be strictly pseudoconvex, then the Bergman kernel function $K(z,z)$ behaves like $1/\mathrm{dist}^{n+1}(z,\partial G)$ near the boundary.

THEOREM (Pflug). Let G be a pseudoconvex domain with C^2-boundary then, for any $\varepsilon > 0$, there is $U = U(\partial G)$ on which the following inequality holds: $K(z,z) \geq \dfrac{1}{\mathrm{dist}\,(z,\partial G)^{2-\varepsilon}}$.

Interpolation of both results gives for the exponent:

$\dim \mathbb{C}^n + 1 - \dim$ [zero space of the Levi form] .

THEOREM (Ohsawa). G as above. Let $z^0 \in \partial G$ and $k = \dim$ [zero space of the Levi form at z^0] then, for any $\varepsilon > 0$, it holds

$$\inf_{z \in G} K(z,z)|z - z^0|^{n+1-k-\varepsilon} > 0 .$$

QUESTIONS. a) Is it possible to cancel the ε ?
b) Is it possible to find sharper estimates ?

ABOUT THE NOTION OF COMPLETENESS W.R.T. THE CARATHÉODORY DISTANCE

Peter Pflug, address as above

Let c_G denote the Carathéodory distance of the domain $G \subset \mathbb{C}^n$. G is called "c_G-complete" iff any c_G-Cauchy sequence in G converges in G. G is called "c_G-finitely-compact" iff any c_G-ball with finite radius is relatively compact in G.

Remark. 1) "c_G-finitely-compact" implies "c_G-complete".
2) The inverse is true, if c_G is an inner-distance (Theorem of Hopf-Rinow).
3) In general: for c_G-finitely compact pseudoconvex domains the Carathéodory distance is not an inner distance (Barth, Vigué).

QUESTION. Give an example of a domain G in $\mathbb{C}^n$ such that G is c_G-complete but not c_G-finitely compact.

THEOREM (Pflug). For bounded Reinhardt domains $G \ni 0$ both notions are equivalent.

QUESTION. Find other classes for which both notions are equivalent.

LOGARITHMIC CAPACITY IN $\mathbb{C}^n$

Józef Siciak, Institute of Mathematics
Jagiellonian University, Reymonta 4, PL-30-059 Kraków, Poland

Put

$$c(E) = \liminf_{\|x\| \to \infty} \|x\| / \Phi_E(x) , \quad E \subset \mathbb{C}^n ,$$

where Φ_E is the extremal function associated with E (for the definition see [2]). We know [1] that $c(E) = 0$ if and only if E is pluripolar. If the dimension $n = 1$ then $c(E)$ is the logarithmic capacity of E. One can show that for all $n \geq 1$ the set function c satisfies the following two conditions:

I. $c(E) \leq c(F)$ if $E \subset F$,

II. $c(E_\nu) \uparrow c(E)$ if $E_\nu \subset E_{\nu+1}$, $E = \bigcup_{\nu=1}^{\infty} E_\nu$.

PROBLEM. Does it satisfy the condition

III. $c(K_\nu) \downarrow c(K)$, if $K_{\nu+1} \subset K_\nu$, K_ν is compact, $K = \bigcap_{\nu=1}^{\infty} K_\nu$?

In other words, we ask whether c is a Choquet capacity.

In order to solve this problem it would be enough to show that for every bounded pluripolar set $E \subset \mathbb{C}^N$ there exists a function W plurisubharmonic in $\mathbb{C}^N$ such that

(a) $W = -\infty$ on E,

(b) $W(z) \leq \text{const} + \log(1+|z|)$ for all z in $\mathbb{C}^N$,

(c) $\limsup_{\|z\|\to\infty} [W(z) - \log|z|] = 0$.

It is known [1] that there exists W satisfying (a) and (b).

REFERENCES

[1] J. Siciak, Extremal plurisubharmonic functions in $\mathbb{C}^n$. Ann.Polon.Math. 39 (1981), 175-211.

[2] --- , Extremal plurisubharmonic functions and capacities in $\mathbb{C}^N$. Lecture Notes of the Sophia Univ. Tokyo 1982, pp. 1-97.

ON A PROJECTIVE CAPACITY IN $\mathbb{C}^n$

Józef Siciak

address as above

Let σ denote the normalized surface measure on the unit sphere $S = \{z \in \mathbb{C}^n; |z_1|^2 + \ldots + |z_n|^2 = 1\}$. Let H denote the set of all functions $h \in PSH(\mathbb{C}^n)$ such that $h \not\equiv 0$, $h(\lambda z) = |\lambda| h(z)$ for all $\lambda \in \mathbb{C}$ and all $z \in \mathbb{C}^n$. Given a bounded subset E of $\mathbb{C}^n$ we define

$$\tau(E) = \inf\{\|h\|_E; h \in H, \int_S \log h \, d\sigma = 0\}.$$

If E is an unbounded subset of $\mathbb{C}^n$ we put

$$\tau(E) = \sup\{\tau(F); F \subset E, \ F \text{ is bounded}\}.$$

It is known [2] that τ is a Choquet capacity such that $\tau(E) = 0$ iff $E_c = \{e^{it}z; t \in \mathbb{R}, z \in E\}$ is pluripolar.

Given a compact set $K \subset \mathbb{C}^n$ we define

$$\gamma_\nu(K) = \inf\{\|Q\|_K; Q(z) = \sum_{|\alpha|=\nu} c_\alpha z^\alpha, \int \log|Q| d\sigma = 0\}, \ \nu = 1, 2, \ldots .$$

It was shown by H. Alexander [1] that the sequence $\{\sqrt[\nu]{\gamma_\nu(K)}\}$ is convergent and its limit $\gamma(K)$ has the property that $\gamma(K) = 0$ iff K_c is pluripolar.

It is also known [2] that

1°. $e^{\kappa_n}\gamma(K) \leq \tau(K) \leq \gamma(K)$ with $\kappa_n = \int \log|z_n| d\sigma$ for all compact subsets K of $\mathbb{C}^n$.

2°. If $n = 2$, then $\tau(K) = \gamma(K)$ for all compact subsets K of S. In particular γ can be extended to a Choquet capacity on S.

PROBLEM. Is it true that $\tau(K) = \gamma(K)$ for all compact subsets K of the unit sphere S in $\mathbb{C}^n$ with $n \geq 3$?

REFERENCES

[1] H. Alexander, On a projective capacity. Preprint 1980.

[2] J. Siciak, Extremal plurisubharmonic functions and capacities in $\mathbb{C}^n$. Lecture Notes of the Sophia Univ. Tokyo 1982, pp. 1-97.

SOME PROPERTIES OF THE SPACE $L^2H(D)$

M. Skwarczyński, Institute of Mathematics and Physics, Engineering High School (Malczewskiego 29), P.O. Box 245, PL-26-600 Radom, Poland

PROBLEM 1. Does there exist a domain $D \subset \mathbb{C}^n$ and a point $p \in \partial D$ such that in the space $L^2H(D)$ of all square integrable holomorphic functions the subset

$$\{ f \in L^2H(D);\ \limsup_{z\to p} |f(z)| < \infty \}$$

is not dense?

PROBLEM 2. Let D be the union of two discs $B_j = \{ z \in \mathbb{C};\ |z-j| < 1\}$, $j = 1, 2$. For $f \in L^2H(D)$ define

$$(\pi_j f)(z) = \begin{cases} f(z) & \text{if } z \notin B_j \\ \int_{B_j} f(u)\,\overline{K_{B_j}(u,z)}\,d\omega(u) & \text{if } z \in B_j \end{cases}$$

. The sequence $f_1 = \dfrac{\chi_{B_1}}{\text{vol } B_1}$, $f_2 = \pi_2 f_1$, $f_3 = \pi_1 f_2$, $f_4 = \pi_2 f_3$, ... is known to converge in $L^2_{loc}(D)$ to $K_D(z,1)$ (the Bergman function of D). Does it converge in $L^2(D)$?

PROBLEM 3. Does there exist a domain D in $\mathbb{C}^2$, $n \geq 2$, and two different points $p,q \in D$ such that the evaluation functionals

$$L^2H(D) \ni f \mapsto f(p) \in \mathbb{C} \quad \text{and} \quad L^2H(D) \ni f \mapsto f(q) \in \mathbb{C}$$

are proportional although both are non-zero?

See my paper Biholomorphic invariants related to the Bergman functions in Dissertationes Math. 173 (1980).

CAPACITY IN BANACH SPACES

Zbigniew Słodkowski, Institute of Mathematics of the Polish Academy of Sciences (Śniadeckich 8), P.O. Box 137, PL-00-950 Warszawa, Poland

Is it possible to extend Siciak's theory of ρ-capacity to the setting of complex Banach spaces ?

One can define $\rho(K) = \sup\{r;\ \overline{B(0,r)} \subset \hat{K}_c\}$, where $K_c = \{zy; |z| = 1, y \in K\}$ for bounded closed subsets K of a Banach space Y, and

$$\rho(\Omega) = \sup\{\rho(K);\ K \subset \Omega,\ K \text{ closed and bounded}\},$$

for open set Ω. Then for an arbitrary set $E \subset Y$ the functional $\rho(E)$ may be defined in the standard way. Is ρ a Choquet capacity ?

Another problem is whether a circled Borel set is polar in Y if and only if $\rho(E) = 0$.

POLAR SETS AND ZERO SETS IN BANACH SPACES

Zbigniew Słodkowski, address as above

In finite-dimensional spaces every polar set has Lebesgue measure zero. Although there is no simple integration theory in Banach spaces, there are some elementary and natural generalizations of the notion of a set of Lebesgue measure zero.

J.P.R. Christensen calls a Borel subset E of a Fréchet space Y a zero set if there is a Borel probability measure μ on Y such that for every $y \in Y$, $\mu(E + y) = 0$.

Are polar sets zero sets in this sense ?

A similar but different concept of a zero set in a Fréchet space was proposed by P. Mankiewicz (in a paper on differentiation of Lipschitz functions in Studia Mathematica, early seventies).

Are polar sets zero sets in Mankiewicz' sense ?

VALUE DISTRIBUTION THEORY AND INTERSECTION PROBLEMS

Chen-Han Sung

Dept. of Mathematics, University of California, Santa Barbara, CA 93106, USA

Let $f;\ \mathbb{C} \to \mathbb{CP}^2$ be a holomorphic curve and $\{D_\alpha\}_\alpha$ a family of algebraic curves of degree ν with simple normal crossings in $\mathbb{CP}^2$. The curve f is said to be nondegenerate to the order ν if $f(\mathbb{C})$ is not contained in any algebraic curve of order ν or less. It is known from algebraic geometry that the number of points (counting multiplicities) in the set of intersection between two algebraic curves in $\mathbb{CP}^2$ is the product of the degree of those two curves.

The problem here is about the non-compact case for the intersection between curves.

PROBLEM I. What is the sharp upper bound for the defect relation $\Sigma_\alpha \delta(D_\alpha, f) <$? It was conjectured by many people that the sharp bound should be $3/\nu$, when f is non-degenerate to the order ν.

PROBLEM II. Where will the curve $f(\mathbb{C})$ be situated, if it misses some number (e.g. the integer $[3/\nu]$) of the algebraic curves $\{D_\alpha\}_\alpha$ with normal crossings ?

Concerning Problem I we only know the answer for $\nu = 1$. In that case $\Sigma_\alpha \delta(D_\alpha, f) \leq 3$ for f being non-degenerate (to the order 1). This was proved by Henri Cartan in 1933.

It is also known from H. Cartan's work in 1933 that if $f(\mathbb{C})$ misses four $\mathbb{CP}^1$-lines in general position in $\mathbb{CP}^2$, then $f(\mathbb{C})$ must be a $\mathbb{CP}^1$-line and lie on one of the three "diagonal" lines (the dotted lines).

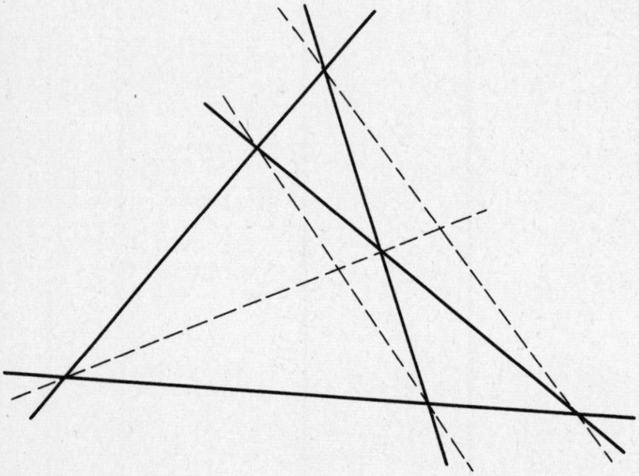

These four lines can be viewed as two degenerate conics (with simple normal crossings).

THE COMPLEX MONGE-AMPÈRE OPERATOR

Jerzy Kalina, Institute of Mathematics of the Polish Academy of Sciences
Łódź Branch, Kilińskiego 86, PL-90-012 Łódź, Poland

Let C be the closed convex cone of all nonnegative $n \times n$ Hermitian matrices $c = (c_{ij})$. Consider the function

$$\psi(c) := (\det c)^{1/n}, \quad c \in C.$$

Now, let μ be a vector-valued Borel measure on $\Omega \subset \mathbb{C}^n$ with values in the cone C, i.e., $\mu = (\mu_{ij})$ and for a Borel set $E \subset \Omega$ we have $(\mu_{ij}(E)) \in C$.

DEFINITION. $\psi(\mu)(E) := \inf\{\sum_{j=1}^{\infty} \psi(\mu(E_j)) : E = \bigcup_{j=1}^{\infty} E_j,\ E_j$ disjoint Borel sets of $\Omega\}$.

The function $\psi(\mu)$ is a nonnegative Borel measure (C. Goffman and J. Serrin: Sublinear functions of measures and variational integrals. Duke Math. J. 31, 159-178 (1964)).

If u is plurisubharmonic on Ω, then the matrix of Borel measures $\left(\frac{\partial^2 u}{\partial z_i \partial \bar{z}_k}\right)$ takes values in the cone C. Put

$$\psi(u) := \psi((\partial^2 u/\partial z_i \partial \bar{z}_k)).$$

This operator plays an important role in the construction of subsolutions of the nonlinear Dirichlet problem

$$\begin{cases} (dd^c u)^n = f dV & \text{on } \Omega \\ u \text{ is plurisubharmonic on } \Omega \\ u = \varphi & \text{on } \partial\Omega \end{cases}$$

where Ω is a strictly pseudoconvex domain in $\mathbb{C}^n$. (E. Bedford, B.A. Taylor: The Dirichlet problem for a complex Monge-Ampère equation, Inv. Math. 37, 1-44 (1976).)

PROBLEM 1. Let $u \in P(\Omega)$ (the set of plurisubharmonic functions) and $u_\varepsilon = u * \chi_\varepsilon$, where $\chi_\varepsilon(z) = \varepsilon^{-2n}\chi(z/\varepsilon) \geq 0$ is any smoothing kernel. Is it possible to select a sequence $\varepsilon_j \to 0$ such that $\psi(u_{\varepsilon_j}) \to \psi(u)$ in the weak sense?

PROBLEM 2. Is the measure $\psi(u)$ absolutely continuous with respect to Lebesgue measure for every $u \in P(\Omega)$?

PROBLEMS IN ANALYSIS ON COMPLEX MANIFOLDS

collected by Pierre Dolbeault (Paris) and prepared by Paweł Walczak (Łódź)

Here, some open problems in analysis on complex manifolds are collected. They were presented by twelve speakers at the problem session on August 26 during the 8th Conference on Analytic Functions at Błażejewko. Some of them are original, i.e. posed by the authors, while the others, although more or less known to the experts, are also included because of the interest they may cause. Some of these problems concern rather the theory of real differentiable manifolds. They are presented here because of their significance for or relationship with the theory of complex manifolds. In some cases, references are not complete for the texts were prepared during the conference without an access to the library. Anybody who needs more information about a particular problem is invited to contact the authors.

PRINCIPAL BUNDLES OVER STEIN MANIFOLDS

Bo Berndtsson
CTH, Dept. of Mathematics, S-412 96 Göteborg, Sweden

Let B be a principal fiber bundle with structure group $\mathbb{C}$, the additive group of complex numbers. Suppose B is a Stein manifold. Does it follow that the base is Stein, and hence that B is trivial ?

FONCTIONS J-PRESQUE HOLOMORPHES

Stančo Dimiev
Institute of Mathematics, Bulgarian Academy of Sciences, BG-1090 Sofia, P.O. Box 373, Bulgaria

Soit (M,J) une variété presque-complexe analytique, i.e. M est une variété analytique réelle equipée de la structure presque-complexe J à coefficients analytiques. On suppose que (M,J) admet des fonctions J-presque-holomorphes. Rappelons que le principe du maximum et le principe d'unité de la prolongation sont verifiés [1] et [2].

Etant données deux fonctions J-presque holomorphes f_1 et f_2 définies respectivement sur les domaines G_1 et G_2, $G_1 \subset M$, $G_2 \subset M$, on dit que chaque de ces deux fonctions est un prolongement J-analytique de l'autre, si l'on a un ouvert U contenu dans $G_1 \cap G_2$ sur lequel les restrictions de f_1 et f_2 coincident.

Pour un domaine D de M on dit que c'est un domaine de presque-holomorphie de type J s'il existe une fonction J-presque-holomorphe sur D, designée f, telle que pour tout sous-ensemble ouvert U de M, $U \cap D = \emptyset$, la restriction de f sur $U \cap D$ n'admet pas de prolongement J-analytique sur un domaine non-contenu dans D. Soit K un sous-ensemble compact du domaine Ω de M. Par $\hat{K}^J_\Omega$ est désigné l'ensemble suivant:

$$\hat{K}^J_\Omega = \{x \in \Omega : |f(x)| \leq \sup_{y \in \Omega} |f(y)|, \ \forall f \in AH_J(\Omega)\},$$

où $AH_J(\Omega)$ est l'ensemble des fonctions J-presque-holomorphes sur Ω. L'ensemble $\hat{K}^J_\Omega$ est un sous-ensemble fermé de Ω.

On dit que Ω vérifie la propriété de la J-pseudo-convexité, si pour tout sous-ensemble compact K de Ω, on a que K^J_Ω est compact.

THÉORÈME DE TYPE CARTAN-THULLEN. Tout domaine J-pseudo-convexe est un domaine de presque-holomorphie de type J (voir [1, 2]).

PROBLÈME 1. Est-ce que le théorème inverse est vrai pour toute structure presque-complexe analytique ?

PROBLÈME 2. Est-ce que l'absence de points J-singuliers, i.e.

$$\Gamma = \{a \in M : k(a) < k_U, \ a \in U\} = \emptyset \quad \text{(voir [3])}$$

pour tout $a \in M$ et pour tout voisinage ouvert U de a, est une condition suffisante pour avoir des fonctions J-presque-holomorphes globalement definies sur M ? Dans la classe des variétés presque-complexes analytiques trouver la notion analogue de la notion classique de la variété de Stein.

REFERENCES

[1] Dimiev, S., Fonctions presque-holomorphes, Semester on Complex Analysis 1979, Banach Center Publications, à paraitre.

[2] ——, Introduction dans l'analyse presque-complexe, Istituto Matematico "Guido Castelnuovo", Roma 1982.

[3] ——, Propriétés locales des fonctions presque-holomorphes, Conference on Analytic Functions, Błażejewko 1982, Proceedings, pp. 102-117.

BOUNDARIES OF HOLOMORPHIC CHAINS

Pierre Dolbeault
Mathématiques, L.A. 213 du C.N.R.S., Université Pierre et Marie Curie, 4, Place Jussieu, F-75230 Paris Cedex 05, France

The problem of constructing a holomorphic chain with a given boundary M in a complex manifold X is solved under some hypothesis in [1] and [2]. Solutions are found for manifolds with very peculiar topology and complex structure and for certain values of the dimension of the given boundaries. The method of the proof proposed by Harvey and Lawson is based on giving a global construction. An analogous problem has been solved in a real hyperplane of $\mathbb{C}^n$ [3]. In [2] intersections of chains are used.

PROBLEM 1. Compare the solutions of [1] and [2] when possible. In particular, compare the hypotheses.

PROBLEM 2. Find obstructions (topological and analytical), for instance as cohomology classes of M in X.

PROBLEM 3. Is it possible to make local constructions and then to patch them together, as we have tried to do in $\mathbb{P}^n\mathbb{C}$?

PROBLEM 4. To solve the problem on CR-manifolds. (Some results in this direction have been obtained by Benlarabi in Paris.)

PROBLEM 5. To build up an intersection theory for chains (linear combinations of either C^∞ or C^ω subvarieties with negligible singularities) as it can be done for instance for subanalytic chains.

REFERENCES

[1] Harvey, R., Holomorphic chains and their boundaries, Proc. Symp. Pure Math. (AMS) 30 (1979), Part 1, 309-382.
[2] Dolbeault, P., On holomorphic chains with a given boundary in $\mathbb{P}^n$C.
[3] Besnault, J. and P. Dolbeault, Symp. Math. 24 (1981), 205-213.

REGULARIZATION ON A MANIFOLD

Pierre Dolbeault, address as above

[1] gives an existence and properties of kernels regularizing distributions on a manifold.

PROBLEM. What can be done for the regularization of a plurisubharmonic function on a manifold ?

REFERENCES

[1] Laurent-Thiebaut, C., Regularisation sur une variété, C.R.A.S. Paris 292 1981 , 833-836.

HENKIN-LEITERER KERNEL ON A STEIN MANIFOLD

Pierre Dolbeault, address as above

In [1] the kernel is constructed for the forms of type (0,q). Construct a Henkin-Leiterer kernel for the forms of type (p,q).

REFERENCES

[1] Henkin, G.M. and J. Leiterer, Global integral formulae for solving the $\bar{\partial}$-equation on Stein manifolds, Ann. Polon. Math. 39 (1981), 93-116.

FINITELY SHEETED ENVELOPES OF HOLOMORPHY

Marek Jarnicki
Institute of Mathematics, Jagiellonian University, Reymonta 4, PL-30-059 Kraków, Poland

Let (X,p) be a connected Riemann domain spread over $\mathbb{C}^n$ and let $(\hat{X},\hat{p})$ denote its envelope of holomorphy.

PROBLEM. When $(\hat{X},\hat{p})$ is univalent or only finitely sheeted ?

RELATIVE CURRENTS

Wiesław Królikowski
Institute of Mathematics of the Polish Academy of Sciences, Łódź Branch, Kilińskiego 86, PL-90-012 Łódź, Poland

Let M be a complex manifold of complex dimension n. We denote by $D_i^c(M)$ the complex vector space of homogeneous currents (in the sense of de Rham) of degree i with compact support.

Let M_o be a subset of M such that $\mathrm{Int}\, M_o \neq \emptyset$. In $D_i^c(M)$, we introduce the following relation $\sim$:

$$(T_1 \sim T_2) \Longleftrightarrow (\mathrm{supp}(T_1 - T_2) \subset M_o) \quad \text{for} \quad T_1, T_2 \in D_i^c(M).$$

It is clear that this relation is an equivalence. Any equivalence class of currents $T \in D_i^c M$ will be called a relative current mod M_0 and be denoted by $[T]_s$.

It is easy to show that if $T_1 \sim T_2$ then $bT_1 \sim bT_2$, where b is the usual boundary operator for currents. According to the above property, we can define the boundary operator b for the relative currents mod M_0 by the formula

$$b[T]_s = [bT]_s.$$

Introducing, in a standard way, the notion of relative cycle mod M_0 and relative boundary mod M_0, we can define the relative homology group mod M_0, $H_i^c(M,M_0)$ as the quotient group $Z_i^c(M,M_0)/B_i^c(M,M_0)$, where $Z_i^c(M,M_0)$ is the group of all relative cycles mod M_0 and $B_i^c(M,M_0)$ is the group of all relative boundaries mod M_0.

We denote by $WH_i^c(M,M_0)$ the subgroup of $H_i^c(M,M_0)$ consisting of all elements Γ which are generated by relative currents $[T]_s$, where T is a cycle.

PROBLEM. When the identity

$$H_i^c(M,M_0) \equiv WH_i^c(M,M_0)$$

holds ?

CONNECTIONS CORRESPONDING TO HERMITIAN STRUCTURES

Julian Ławrynowicz
Institute of Mathematics of the Polish Academy of Sciences, Łódź Branch, Kilińskiego 86, PL-90-012 Łódź, Poland

The geometry of the Yang-Mills field is, at its present stage [1, 4], restricted in fact to the Minkowski space-time only. If we pass to complexification of some more general pseudo-riemannian manifolds, the difficulty lies in the fact that although we have the implication (cf. [2], case (b)):

(*) the connection corresponding to the Yang-Mills field $\Leftarrow$ the Yang-Mills field itself,

we cannot guarantee the inverse implication, which seems to be quite a serious problem. Of course, the connections in question have to be compatible with a given holomorphic structure, and the Yang-Mills field has to be expressed in terms of a suitable hermitian metric. The success in guaranteeing the implication (*) is due to the following proposition (see e.g. [1], p. 45 and [5], p. 78):

Given a holomorphic vector bundle $\mathbb{E} = (E,\pi,M)$ with E and M being complex manifolds, let h denote an hermitian structure on $\mathbb{E}$, i.e. $h : X \times X \longrightarrow C^{\infty}(M)$, where X is the modulus of cross-sections of M. Given $SU(n)$, $n = \dim \mathbb{E}$, consider also the unique $SU(n)$-structure determined by h as the reductions of the $SU(n)$-bundle of the bases for $\mathbb{E}$ to the bundle of unitary bases for $\mathbb{E}$. Suppose next that U is an open set in M and $\mathcal{O}(U)$ denotes the modulus of holomorphic cross-sections over U. Then there exists exactly one connection ∇ compatible with both structures of $\mathbb{E}$: the $SU(n)$-structure and the complex structure, that is

(i) $dh(s,t) = h(\nabla s,t) + h(s,\nabla t)$ for every pair of cross-sections $s,t \in X$, and

(ii) $\nabla'' s = 0$ for any cross-section $s \in \mathcal{O}(U)$,

respectively, where $\nabla = \nabla' + \nabla''$, $\nabla' : \mathcal{E}(M,E) \longrightarrow \mathcal{E}^{1,0}(M,E)$, $\nabla'' : \mathcal{E}(M,E) \longrightarrow \mathcal{E}^{0,1}(M,E)$, and $\mathcal{E}(M,E)$ denotes the modulus of all smooth cross-sections of $\mathbb{E}$. Moreover, the curvature form Ω^2 of ∇ is of type $(1,1)$.

Thus the problem of guaranteeing the converse to (*) is in fact the <u>PROBLEM</u> of finding the conditions under which, for a given connection satisfying the assumption (ii), there exists an hermitian metric h compatible with ∇, i.e. satisfying the assumption (i). The problem is not answered by Theorem (1.2) of [1], p. 46, which assures the existence of a unique holomorphic structure to a given complex vector bundle, using the Newlander-Nirenberg integrability theorem for complex structures [3].

The answer to the problem, apart from the uniqueness , can be formulated in terms of the holonomy groups of ∇ if we have no subsidiary conditions: all these groups have to be contained in $SU(n)$. This condition, however, seems to be unsatisfactory for the converse to (*) and, on the other hand, it seems natural to consider the <u>MODIFIED PROBLEM</u>: under the subsidiary condition that, in the metric h to be found, the Yang-Mills field lifted to $\mathbb{E}$ has to be source-free, i.e. the divergence of Ω^2 has to vanish.

<u>REFERENCES</u>

[1] Atiyah, M.F., Geometry of Yang-Mills fields, Lezioni Fermiane, Academia Nazionale dei Lincei - Scuola Normale Superiore, Pisa 1979.

[2] Gaveau, B., J. Ławrynowicz, and L. Wojtczak, On certain transformations of motion equations and of the corresponding manifolds, Conf. Analytic Functions Abstracts, Błażejewko 1982, pp. 12-15.

[3] Newlander, A. and L. Nirenberg, Complex analytic coordinates in almost complex manifolds, Ann. of Math. 65 (1957), 391-404.

[4] Гиндикин, С.Г. и Г.М. Хенкин, Преобразование Пенроуза и комплексная дифференциальная геометрия, in: Итоги науки и техники, Серия: Современные проблемы математики 17, Москва 1981, pp. 57-111.

[5] Wells, R.O., Jr., Differential analysis on complex manifolds, 2nd. ed. (Graduate Texts in Mathematics 65), Springer-Verlag, New York-Heidelberg-Berlin 1980.

SHORTEST CONNECTING CURVES

Peter Pflug
Universität Osnabrück-Abt. Vechta, Mathematisches Institut, Postfach 1249 (Driverstrasse 22), D-2848 Vechta, BRD

Let D be a domain in $\mathbb{C}^n$, $z_o \in \partial D$ and $U = U(\partial D)$. Let $ds^2(\ , \ ;D)$ denote the Bergman differential metric on D and $ds^2(\ , \ ;D\cap U)$ the Bergman differential metric on $D\cap U$. Then, using Hörmander's $\bar\partial$-methods, the following estimate can be proved:

$$(*) \qquad 0 < A \leq \frac{ds(z,X;D)}{ds(z,X;D\cap U)} \leq B$$

for all $z \in D\cap U$, near z_o and for all $X \in \mathbb{C}^n$.

PROBLEM. Does this inequality imply (by simple arguments of differential geometry) the following: if there is no Cauchy sequence $z_k \in U\cap D$, $z_k \rightarrow z_o$ with respect to the distance induced by $ds(\ , \ ; D\cap U)$, then there is no Cauchy-sequence $z_k \in D$, $z_k \rightarrow z_o$ with respect to the distance induced by $ds(\ , \ ;D)$?

Remark. By other argument, not using (*), the question can be answered in the affirmative way.

POINCARÉ LEMMA

Maciej Skwarczyński
Institute of Mathematics and Physics, Engineering High School, Malczewskiego 29, PL-26-600 Radom, Poland

Let ω be a smooth differential form on a manifold M such that for every open relatively compact subset $U \subset M$ there exists a smooth form α_U on U with the property

$$d\alpha_U = \omega.$$

PROBLEM. Does it follow that there exists a smooth form on M such that $d\alpha = \omega$? (M is not assumed to be paracompact!)

HOLOMORPHIC MAPPINGS INTO ALGEBRAIC SURFACES

Chen-Han Sung
Department of Mathematics, University of California, Santa Barbara, CA 93106, U.S.A.

Let M be an algebraic surface with positive canonical bundle.

PROBLEM. Does the image of a holomorphic mapping $f : \mathbb{C} \to M$ lie in an algebraic curve in M ?

Observe that for any $\mathbb{CP}^1$-line in $\mathbb{CP}^3$, there is a smooth surface of any given high degree which will contain this $\mathbb{CP}^1$.

GAUSS MAP OF A MINIMAL SURFACE

Chen-Han Sung, address as above

Let M be a non-flat complete minimal surface in $\mathbb{R}^{m+1}$. Its Gauss map can be viewed as a map from M into $\mathbb{CP}^{m-1}$.

PROBLEM. How many hyperplanes in general position in $\mathbb{CP}^{m-1}$ can the above Gauss map possibly miss ?

It is known that when $m = 2$, the complement of the image of the Gauss map of M contains at most six points of S^2 [Xavier, 1981]. The number of points can be sharpened when M is a parabolic surface [Sung, 1982].

COHOMOLOGY VANISHING THEOREMS (for positive semi-definite types)

Osamu Suzuki
Department of Mathematics, College of Humanities and Sciences, Nihon University, Setagaya-ku, Sakurajosui 3-25-40, J-156 Tokyo, Japan

In the following, X is assumed to be a weakly 1-complete manifold, i.e. X is a complex manifold with a complete plurisubharmonic function. K denotes the canonical line bundle over X.

PROBLEM. Can we prove that if a bundle $E \otimes K^{-1}$ is positive semi-definite on X and strictly positive definite except for some analytic set A (we mean positivity in the sense of Nakano), then

$$H^q(X,E) = 0, \quad q \geq 1 \ ?$$

Remark. If the set A is empty, we arrive at Nakano's cohomology vanishing theorem. If X is compact, the answer is affirmative.

PROBLEM. Can we answer the above question using some kind of the Gauss - Bonnet formula ?

In the case of a compact Riemann surface X, Kodaira's vanishing theorem can be shown as follows: We have

$$H^1(X,F) = H^0(X,F^{-1} \otimes K) \quad \text{(Serre duality theorem)}.$$

Therefore, if there could exist a non-trivial section $\varphi \in H^0(X,F^{-1} \otimes K)$, then the equality

$$\sum_j \deg \varphi|_{p_j} = (1/2\pi) \int_X C_1(F^{-1} \otimes K)$$

would hold, which proves the vanishing of φ. Can we find such a formula for sections of Hodge spaces, $\mathbb{H}^{0,q}(X,E)$? Can we apply it to our problem ?

MINIMAL FOLIATIONS

Paweł Walczak
Institute of Mathematics of the Polish Academy of Sciences, Łódź Branch, Kilińskiego 86, PL-90-012 Łódź, Poland

It can be proved [1] that if M is a non-negatively curved Riemannian manifold and F is a codimension-one foliation of M with all leaves compact and minimal, then all the leaves of F are totally geodesic. The analogous result in codimension two has not been obtained yet. Moreover, there are no examples of codimension-two foliations of non-negatively curved manifolds with all leaves compact and minimal, but not totally geodesic.

Holomorphic foliations of Kähler manifolds provide good examples of foliations with all leaves minimal. Therefore, it could be interesting to find an example of a holomorphic foliation F of a Kähler manifold M such that

(i) $\operatorname{codim}_C F = 1$,

(ii) the curvature (sectional, holomorphic or biholomorphic) of M is non-negative,

(iii) all the leaves of F are compact but some of them are not totally geodesic.

REFERENCES

[1] Walczak, P., On foliations with all leaves satisfying some geometrical conditions, preprint.

GROWTH OF HOLOMORPHIC MAPPINGS

Tadeusz Winiarski
Institute of Mathematics, Jagiellonian University, Reymonta 4, PL-30-059 Kraków, Poland

Let $f : \mathbb{C}^n \to \mathbb{C}^k$ be a holomorphic mapping of the finite order, i.e. there exist positive constants ϱ, A, B such that

$$\|f(x)\| \leq A \exp(B\|x\|^{\varrho}) \quad \text{for every} \quad z \in \mathbb{C}^n.$$

Let $B(r)$ be a ball around 0 of radius $r > 0$. Let

$$\Psi(r) = \mathrm{Vol}_{2n}(B(r) \cap \text{graph } f).$$

PROBLEM. Does there exist a connection between the growth of f and of the function Ψ ?

One can additionally assume that $n = k$ and f is an automorphism.

INTERSECTION MULTIPLICITY OF ANALYTIC SETS

Tadeusz Winiarski, address as above

Let us suppose that

1. Ω, Ω_1 are domains in $\mathbb{C}^n$, $\bar{\Omega}_1$ is compact, $\bar{\Omega}_1 \subset \Omega$,
2. D, D_1 are domains in $\mathbb{C}^k$, $\bar{D}_1$ is compact, $\bar{D}_1 \subset D$,
3. X, Y are analytic subsets of $\Omega \times D$ of pure dimension n, k, respectively, and such that $X \subset \Omega \times D_1$ and $Y \subset \Omega_1 \times D$.

Then $\pi_1 | X : X \ni (x,y) \mapsto x \in \Omega$ is an s_1-sheeted branched covering and and $\pi_2 | Y : Y \ni (x,y) \mapsto y \in D$ is an s_2-sheeted branched covering.

PROBLEM. Does the equality

$$\Sigma_{a \in X \cap Y}\, i(X \cdot Y; a) = s_1 \cdot s_2,$$

where $i(X \cdot Y;a)$ denotes the intersection multiplicity of the sets X, Y at the point a, hold ?

The answer to this question is positive in the case when Ω and D are balls with respect to arbitrary fixed real norms in $\mathbb{C}^n$, $\mathbb{C}^k$, respectively.

EXTENDING FUNCTIONS OF THE CLASS M(D)

Tadeusz Winiarski, address as above

Let D be a domain in $\mathbb{C}^n$. Let $M(D)$ denote the class of all functions of the form

$$|f_1|^{\alpha_1} \cdot |f_2|^{\alpha_2} \cdot \ldots \cdot |f_k|^{\alpha_k},$$

where $f_1, f_2, \ldots, f_k$ are holomorphic in D and $\alpha_1, \ldots, \alpha_k$ are positive real numbers.

Let V be an analytic subset of D. Let $\varphi \in M(D - V)$ be locally bounded in D.

PROBLEM. Does there exist a function $\varphi \in M(D)$ such that

$$\tilde{\varphi}|_{D-V} = \varphi \,?$$

Vol. 900: P. Deligne, J. S. Milne, A. Ogus, and K.-Y. Shih, Hodge Cycles, Motives, and Shimura Varieties. V, 414 pages. 1982.

Vol. 901: Séminaire Bourbaki vol. 1980/81 Exposés 561–578. III, 299 pages. 1981.

Vol. 902: F. Dumortier, P.R. Rodrigues, and R. Roussarie, Germs of Diffeomorphisms in the Plane. IV, 197 pages. 1981.

Vol. 903: Representations of Algebras. Proceedings, 1980. Edited by M. Auslander and E. Lluis. XV, 371 pages. 1981.

Vol. 904: K. Donner, Extension of Positive Operators and Korovkin Theorems. XII, 182 pages. 1982.

Vol. 905: Differential Geometric Methods in Mathematical Physics. Proceedings, 1980. Edited by H.-D. Doebner, S.J. Andersson, and H.R. Petry. VI, 309 pages. 1982.

Vol. 906: Séminaire de Théorie du Potentiel, Paris, No. 6. Proceedings. Edité par F. Hirsch et G. Mokobodzki. IV, 328 pages. 1982.

Vol. 907: P. Schenzel, Dualisierende Komplexe in der lokalen Algebra und Buchsbaum-Ringe. VII, 161 Seiten. 1982.

Vol. 908: Harmonic Analysis. Proceedings, 1981. Edited by F. Ricci and G. Weiss. V, 325 pages. 1982.

Vol. 909: Numerical Analysis. Proceedings, 1981. Edited by J.P. Hennart. VII, 247 pages. 1982.

Vol. 910: S.S. Abhyankar, Weighted Expansions for Canonical Desingularization. VII, 236 pages. 1982.

Vol. 911: O.G. Jørsboe, L. Mejlbro, The Carleson-Hunt Theorem on Fourier Series. IV, 123 pages. 1982.

Vol. 912: Numerical Analysis. Proceedings, 1981. Edited by G. A. Watson. XIII, 245 pages. 1982.

Vol. 913: O. Tammi, Extremum Problems for Bounded Univalent Functions II. VI, 168 pages. 1982.

Vol. 914: M. L. Warshauer, The Witt Group of Degree k Maps and Asymmetric Inner Product Spaces. IV, 269 pages. 1982.

Vol. 915: Categorical Aspects of Topology and Analysis. Proceedings, 1981. Edited by B. Banaschewski. XI, 385 pages. 1982.

Vol. 916: K.-U. Grusa, Zweidimensionale, interpolierende Lg-Splines und ihre Anwendungen. VIII, 238 Seiten. 1982.

Vol. 917: Brauer Groups in Ring Theory and Algebraic Geometry. Proceedings, 1981. Edited by F. van Oystaeyen and A. Verschoren. VIII, 300 pages. 1982.

Vol. 918: Z. Semadeni, Schauder Bases in Banach Spaces of Continuous Functions. V, 136 pages. 1982.

Vol. 919: Séminaire Pierre Lelong – Henri Skoda (Analyse) Années 1980/81 et Colloque de Wimereux, Mai 1981. Proceedings. Edité par P. Lelong et H. Skoda. VII, 383 pages. 1982.

Vol. 920: Séminaire de Probabilités XVI, 1980/81. Proceedings. Edité par J. Azéma et M. Yor. V, 622 pages. 1982.

Vol. 921: Séminaire de Probabilités XVI, 1980/81. Supplément: Géométrie Différentielle Stochastique. Proceedings. Edité par J. Azéma et M. Yor. III, 285 pages. 1982.

Vol. 922: B. Dacorogna, Weak Continuity and Weak Lower Semicontinuity of Non-Linear Functionals. V, 120 pages. 1982.

Vol. 923: Functional Analysis in Markov Processes. Proceedings, 1981. Edited by M. Fukushima. V, 307 pages. 1982.

Vol. 924: Séminaire d'Algèbre Paul Dubreil et Marie-Paule Malliavin. Proceedings, 1981. Edité par M.-P. Malliavin. V, 461 pages. 1982.

Vol. 925: The Riemann Problem, Complete Integrability and Arithmetic Applications. Proceedings, 1979-1980. Edited by D. Chudnovsky and G. Chudnovsky. VI, 373 pages. 1982.

Vol. 926: Geometric Techniques in Gauge Theories. Proceedings, 1981. Edited by R. Martini and E.M.de Jager. IX, 219 pages. 1982.

Vol. 927: Y. Z. Flicker, The Trace Formula and Base Change for GL (3). XII, 204 pages. 1982.

Vol. 928: Probability Measures on Groups. Proceedings 1981. Edited by H. Heyer. X, 477 pages. 1982.

Vol. 929: Ecole d'Eté de Probabilités de Saint-Flour X – 1980. Proceedings, 1980. Edited by P.L. Hennequin. X, 313 pages. 1982.

Vol. 930: P. Berthelot, L. Breen, et W. Messing, Théorie de Dieudonné Cristalline II. XI, 261 pages. 1982.

Vol. 931: D.M. Arnold, Finite Rank Torsion Free Abelian Groups and Rings. VII, 191 pages. 1982.

Vol. 932: Analytic Theory of Continued Fractions. Proceedings, 1981. Edited by W.B. Jones, W.J. Thron, and H. Waadeland. VI, 240 pages. 1982.

Vol. 933: Lie Algebras and Related Topics. Proceedings, 1981. Edited by D. Winter. VI, 236 pages. 1982.

Vol. 934: M. Sakai, Quadrature Domains. IV, 133 pages. 1982.

Vol. 935: R. Sot, Simple Morphisms in Algebraic Geometry. IV, 146 pages. 1982.

Vol. 936: S.M. Khaleelulla, Counterexamples in Topological Vector Spaces. XXI, 179 pages. 1982.

Vol. 937: E. Combet, Intégrales Exponentielles. VIII, 114 pages. 1982.

Vol. 938: Number Theory. Proceedings, 1981. Edited by K. Alladi. IX, 177 pages. 1982.

Vol. 939: Martingale Theory in Harmonic Analysis and Banach Spaces. Proceedings, 1981. Edited by J.-A. Chao and W.A. Woyczyński. VIII, 225 pages. 1982.

Vol. 940: S. Shelah, Proper Forcing. XXIX, 496 pages. 1982.

Vol. 941: A. Legrand, Homotopie des Espaces de Sections. VII, 132 pages. 1982.

Vol. 942: Theory and Applications of Singular Perturbations. Proceedings, 1981. Edited by W. Eckhaus and E.M. de Jager. V, 363 pages. 1982.

Vol. 943: V. Ancona, G. Tomassini, Modifications Analytiques. IV, 120 pages. 1982.

Vol. 944: Representations of Algebras. Workshop Proceedings, 1980. Edited by M. Auslander and E. Lluis. V, 258 pages. 1982.

Vol. 945: Measure Theory. Oberwolfach 1981, Proceedings. Edited by D. Kölzow and D. Maharam-Stone. XV, 431 pages. 1982.

Vol. 946: N. Spaltenstein, Classes Unipotentes et Sous-groupes de Borel. IX, 259 pages. 1982.

Vol. 947: Algebraic Threefolds. Proceedings, 1981. Edited by A. Conte. VII, 315 pages. 1982.

Vol. 948: Functional Analysis. Proceedings, 1981. Edited by D. Butković, H. Kraljević, and S. Kurepa. X, 239 pages. 1982.

Vol. 949: Harmonic Maps. Proceedings, 1980. Edited by R.J. Knill, M. Kalka and H.C.J. Sealey. V, 158 pages. 1982.

Vol. 950: Complex Analysis. Proceedings, 1980. Edited by J. Eells. IV, 428 pages. 1982.

Vol. 951: Advances in Non-Commutative Ring Theory. Proceedings, 1981. Edited by P.J. Fleury. V, 142 pages. 1982.

Vol. 952: Combinatorial Mathematics IX. Proceedings, 1981. Edited by E. Billington, S. Oates-Williams, and A.P. Street. XI, 443 pages. 1982.

Vol. 953: Iterative Solution of Nonlinear Systems of Equations. Proceedings, 1982. Edited by R. Ansorge, Th. Meis, and W. Törnig. VII, 202 pages. 1982.

Vol. 954: S.G. Pandit, S.G. Deo, Differential Systems Involving Impulses. VII, 102 pages. 1982.

Vol. 955: G. Gierz, Bundles of Topological Vector Spaces and Their Duality. IV, 296 pages. 1982.

Vol. 956: Group Actions and Vector Fields. Proceedings, 1981. Edited by J.B. Carrell. V, 144 pages. 1982.

Vol. 957: Differential Equations. Proceedings, 1981. Edited by D.G. de Figueiredo. VIII, 301 pages. 1982.

Vol. 958: F.R. Beyl, J. Tappe, Group Extensions, Representations, and the Schur Multiplicator. IV, 278 pages. 1982.

Vol. 959: Géométrie Algébrique Réelle et Formes Quadratiques, Proceedings, 1981. Edité par J.-L. Colliot-Thélène, M. Coste, L. Mahé, et M.-F. Roy. X, 458 pages. 1982.

Vol. 960: Multigrid Methods. Proceedings, 1981. Edited by W. Hackbusch and U. Trottenberg. VII, 652 pages. 1982.

Vol. 961: Algebraic Geometry. Proceedings, 1981. Edited by J.M. Aroca, R. Buchweitz, M. Giusti, and M. Merle. X, 500 pages. 1982.

Vol. 962: Category Theory. Proceedings, 1981. Edited by K.H. Kamps, D. Pumplün, and W. Tholen, XV, 322 pages. 1982.

Vol. 963: R. Nottrot, Optimal Processes on Manifolds. VI, 124 pages. 1982.

Vol. 964: Ordinary and Partial Differential Equations. Proceedings, 1982. Edited by W.N. Everitt and B.D. Sleeman. XVIII, 726 pages. 1982.

Vol. 965: Topics in Numerical Analysis. Proceedings, 1981. Edited by P.R. Turner. IX, 202 pages. 1982.

Vol. 966: Algebraic K-Theory. Proceedings, 1980, Part I. Edited by R.K. Dennis. VIII, 407 pages. 1982.

Vol. 967: Algebraic K-Theory. Proceedings, 1980. Part II. VIII, 409 pages. 1982.

Vol. 968: Numerical Integration of Differential Equations and Large Linear Systems. Proceedings, 1980. Edited by J. Hinze. VI, 412 pages. 1982.

Vol. 969: Combinatorial Theory. Proceedings, 1982. Edited by D. Jungnickel and K. Vedder. V, 326 pages. 1982.

Vol. 970: Twistor Geometry and Non-Linear Systems. Proceedings, 1980. Edited by H.-D. Doebner and T.D. Palev. V, 216 pages. 1982.

Vol. 971: Kleinian Groups and Related Topics. Proceedings, 1981. Edited by D.M. Gallo and R.M. Porter. V, 117 pages. 1983.

Vol. 972: Nonlinear Filtering and Stochastic Control. Proceedings, 1981. Edited by S.K. Mitter and A. Moro. VIII, 297 pages. 1983.

Vol. 973: Matrix Pencils. Proceedings, 1982. Edited by B. Kågström and A. Ruhe. XI, 293 pages. 1983.

Vol. 974: A. Draux, Polynômes Orthogonaux Formels – Applications. VI, 625 pages. 1983.

Vol. 975: Radical Banach Algebras and Automatic Continuity. Proceedings, 1981. Edited by J.M. Bachar, W.G. Bade, P.C. Curtis Jr., H.G. Dales and M.P. Thomas. VIII, 470 pages. 1983.

Vol. 976: X. Fernique, P.W. Millar, D.W. Stroock, M. Weber, Ecole d'Eté de Probabilités de Saint-Flour XI – 1981. Edited by P.L. Hennequin. XI, 465 pages. 1983.

Vol. 977: T. Parthasarathy, On Global Univalence Theorems. VIII, 106 pages. 1983.

Vol. 978: J. Ławrynowicz, J. Krzyż, Quasiconformal Mappings in the Plane. VI, 177 pages. 1983.

Vol. 979: Mathematical Theories of Optimization. Proceedings, 1981. Edited by J.P. Cecconi and T. Zolezzi. V, 268 pages. 1983.

Vol. 980: L. Breen. Fonctions thêta et théorème du cube. XIII, 115 pages. 1983.

Vol. 981: Value Distribution Theory. Proceedings, 1981. Edited by I. Laine and S. Rickman. VIII, 245 pages. 1983.

Vol. 982: Stability Problems for Stochastic Models. Proceedings, 1982. Edited by V.V. Kalashnikov and V.M. Zolotarev. XVII, 295 pages. 1983.

Vol. 983: Nonstandard Analysis-Recent Developments. Edited A.E. Hurd. V, 213 pages. 1983.

Vol. 984: A. Bove, J.E. Lewis, C. Parenti, Propagation of Singulariti for Fuchsian Operators. IV, 161 pages. 1983.

Vol. 985: Asymptotic Analysis II. Edited by F. Verhulst. VI, 497 pag 1983.

Vol. 986: Séminaire de Probabilités XVII 1981/82. Proceedin Edited by J. Azéma and M. Yor. V, 512 pages. 1983.

Vol. 987: C.J. Bushnell, A. Fröhlich, Gauss Sums and p-adic Divisi Algebras. XI, 187 pages. 1983.

Vol. 988: J. Schwermer, Kohomologie arithmetisch definierter Gru pen und Eisensteinreihen. III, 170 pages. 1983.

Vol. 989: A.B. Mingarelli, Volterra-Stieltjes Integral Equations a Generalized Ordinary Differential Expressions. XIV, 318 pages. 198

Vol. 990: Probability in Banach Spaces IV. Proceedings, 198 Edited by A. Beck and K. Jacobs. V, 234 pages. 1983.

Vol. 991: Banach Space Theory and its Applications. Proceeding 1981. Edited by A. Pietsch, N. Popa and I. Singer. X, 302 page 1983.

Vol. 992: Harmonic Analysis, Proceedings, 1982. Edited by G. Ma ceri, F. Ricci and G. Weiss. X, 449 pages. 1983.

Vol. 993: R.D. Bourgin, Geometric Aspects of Convex Sets with th Radon-Nikodým Property. XII, 474 pages. 1983.

Vol. 994: J.-L. Journé, Calderón-Zygmund Operators, Pseudo-Di ferential Operators and the Cauchy Integral of Calderón. VI, 12 pages. 1983.

Vol. 995: Banach Spaces, Harmonic Analysis, and Probability Theor Proceedings, 1980–1981. Edited by R.C. Blei and S.J. Sidne V, 173 pages. 1983.

Vol. 996: Invariant Theory. Proceedings, 1982. Edited by F. Gherar delli. V, 159 pages. 1983.

Vol. 997: Algebraic Geometry – Open Problems. Edited by C. Cil berto, F. Ghione and F. Orecchia. VIII, 411 pages. 1983.

Vol. 998: Recent Developments in the Algebraic, Analytical, an Topological Theory of Semigroups. Proceedings, 1981. Edited b K.H. Hofmann, H. Jürgensen and H. J. Weinert. VI, 486 pages. 1983

Vol. 999: C. Preston, Iterates of Maps on an Interval. VII, 205 pages 1983.

Vol. 1000: H. Hopf, Differential Geometry in the Large, VII, 184 pages 1983.

Vol. 1001: D.A. Hejhal, The Selberg Trace Formula for PSL(2, ℝ) Volume 2. VIII, 806 pages. 1983.

Vol. 1002: A. Edrei, E.B. Saff, R.S. Varga, Zeros of Sections o Power Series. VIII, 115 pages. 1983.

Vol. 1003: J. Schmets, Spaces of Vector-Valued Continuous Functions. VI, 117 pages. 1983.

Vol. 1004: Universal Algebra and Lattice Theory. Proceedings, 1982. Edited by R.S. Freese and O.C. Garcia. VI, 308 pages. 1983.

Vol. 1005: Numerical Methods. Proceedings, 1982. Edited by V. Pereyra and A. Reinoza. V, 296 pages. 1983.

Vol. 1006: Abelian Group Theory. Proceedings, 1982/83. Edited by R. Göbel, L. Lady and A. Mader. XVI, 771 pages. 1983.

Vol. 1007: Geometric Dynamics. Proceedings, 1981. Edited by J. Palis Jr. IX, 827 pages. 1983.

Vol. 1008: Algebraic Geometry. Proceedings, 1981. Edited by J. Dolgachev. V, 138 pages. 1983.

Vol. 1009: T.A. Chapman, Controlled Simple Homotopy Theory and Applications. III, 94 pages. 1983.

Vol. 1010: J.-E. Dies, Chaînes de Markov sur les permutations. IX, 226 pages. 1983.